Semi-Insulating III-V Materials

Kah-nee-ta 1984

ORGANIZING COMMITTEE

W. Ford, Conference Chairman (Harris Microwave Semiconductor)
C. Stolte, Assistant Chairman (Hewlett-Packard)
S. J. T. Owen, Local Arrangements (Oregon State University)
J. S. Blakemore, Programme Chairman (Oregon Graduate Center)
R. Y. Koyama, Treasurer (Tektronix)
R. Zucca, Public Relations (Rockwell International)
D. C. Look, Publications (Wright State University)
S. G. Bishop, Sponsorship (Naval Research Laboratory)

PROGRAMME COMMITTEE

The Programme Committee includes the members of the Organizing Committee, *ex officio*, together with the following:

P. K. Bhattacharya (University of Michigan)
R. D. Fairman (TRW Electronic Systems)
J. S. Harris (Stanford University)
H. V. Winston (Hughes Research Laboratories)

INTERNATIONAL ADVISORY COMMITTEE

R. Burmeister (Hewlett-Packard, USA)
M. Cardwell (Plessey Research, UK)
L. Cooper (Office of Naval Research, USA)
L. Eaves (University of Nottingham, UK)
H. Gatos (MIT, USA)
A. Hennel (Warsaw University, Poland)
T. Ikoma (University of Tokyo, Japan)
P. Jay (Thomson CSF, France)
G. Martin (LEP, France)
E. Omel'yanovski (State Scientific R & D Institute, USSR)
L. Samuelson (University of Lund, Sweden)
J. Schneider (Fraunhofer IAF, W. Germany)
E. Swiggard (Naval Research Laboratories, USA)
G. Witt (Air Force Office of Scientific Research, USA)
H. Wittmann (Army Research Office, USA)

Semi-Insulating III-V Materials

Kah-nee-ta 1984

Edited by

D C Look
Wright State University

J S Blakemore
Oregon Graduate Center

SHIVA
Shiva Publishing Limited

SHIVA PUBLISHING LIMITED
64 Welsh Row, Nantwich, Cheshire CW5 5ES, England

British Library Cataloguing in Publication Data

Look, D. C.
Semi-insulating III–V materials: Oregon 1984.
1. Gallium arsenide semiconductors
2. Chromium
I. Title II. Blakemore, J. S.
537.6'22 QC611.8.G3

ISBN 1–85014–031–6

Typeset and produced by J. W. Arrowsmith Ltd., Bristol

Contents

Acknowledgements

The organizing committee is pleased to acknowledge the support of the following organizations:

—Electron Devices Society/IEEE
—Air Force Office of Scientific Research (AFOSR)
—Army Research Office (ARO)
—Cominco Electronic Materials Inc.
—Ferrofluidics Corp.
—Hughes Aircraft Co.
—Morgan Semiconductor Inc.
—Office of Naval Research (ONR)
—Oregon Graduate Center
—Oregon State University
—Rockwell International Corp.
—Sumitomo Electric Industries Ltd.
—Tektronix Foundation

The AFOSR sponsorship was under Contract 84-0081, and the ARO and ONR sponsorships were under Contract N00014-84-G-0038. The views, opinions, and/or findings contained in this report are those of the author(s) and should not be construed as an official United States Government position, policy or decision, unless so designated by other documentation.

The spectacular setting of the Semi-Insulating III–V Materials Conference, Kah-nee-ta, 24–26 April 1984.

Preface

The third conference in this series was organized in response to a continued high interest in the progress of III–V semi-insulating materials. Following the unique and beautiful locations of the first two conferences, Nottingham, UK in 1980, and Evian, France in 1982, it was decided to hold the present gathering at a similarly unique American location, namely the Kah-nee-ta Resort on the Warm Springs Indian Reservation in Oregon. The excellence of the accommodations as well as a fine programme of contributed and invited papers attracted a new record number of delegates, 225, from 10 countries. Our Kah-nee-ta hosts did an outstanding job, not only of reducing technical problems to a minimum, but also of introducing us to several examples of authentic American Indian culture.

Much progress has been made in semi-insulating III–V technology since the first conference in 1980. For example, the electrical activity of Cr in GaAs is now quite well understood. Further progress has been made since the second conference in 1982, although the mystery centre at that time, EL2, has not succumbed to interpretation as readily as the 1982 delegates had hoped. Besides EL2, another topic of high interest at the present conference has been the role of dislocations in affecting GaAs device properties, and how to grow dislocation-free material. There was no firm resolution of either of these problems, and the attack must continue. Several new characterization techniques and refinements of old ones were offered. There will certainly be a need for a fourth conference in this series, and one is already being planned for Spring 1986, in Japan.

The editors wish to thank the authors of this volume who, as a whole, have cooperated fully in revising manuscripts as suggested by the reviewers or by us. In some cases we have made additional changes in grammar without consulting the respective authors in order that our publication timetable might be met. Also, in deference to our British publisher, we have chosen the English spelling alternative where applicable.

We are grateful to Stephen Bishop, from Shiva Publishing Limited, and Phil Won Yu who stayed with us at the conference grounds for two extra days to complete as much as possible of the editing process. We are also indebted to Bill Ford, the Conference Chairman, and all the members of the Organizing and Program Committees for their willing assistance, which has made our task much easier. Finally, we wish to thank our own institutions and research sponsors (DCL: Wright State University and the US Air Force Avionics Laboratory; JSB: Oregon Graduate Center and the National Science Foundation) for their indirect support in the preparation of this volume.

David C. Look, Dayton, Ohio
John S. Blakemore, Beaverton, Oregon

July 1984

Section 1:
Diffusion, gettering and annealing

Diffusion and Type Conversion

B. Tuck

Department of Electrical and Electronic Engineering, University of Nottingham, Nottingham NG7 2RD, UK

Abstract

After a brief description of basic diffusion theory, the published experimental work on the diffusion of Cr, Fe and Mn in GaAs and InP is reviewed. Both indiffusion and outdiffusion are considered and it is demonstrated that in general the profiles do not correspond to those predicted by elementary theory. Diffusion of transition elements out of substrates into growing epitaxial layers is described. The redistribution of atoms during the annealing cycle which follows ion implantation is then considered. This is a much more difficult problem than simple indiffusion and a short description is given of some of the mechanisms involved. Finally an account of the phenomenon of type conversion is presented, indicating its association with diffusion processes. Special consideration is paid to the role of Mn.

Introduction

The study of diffusion in the III–V semiconductors has given rise to an enormous volume of literature so, to keep this article to manageable proportions, discussion will be limited to GaAs and InP. The diffusants to be considered will be those which give rise to deep levels or are thought to be involved in the phenomenon of type conversion. This latter effect can be defined as the process whereby some semiconductor samples demonstrate significant changes in electrical conductivity after being subjected to heat treatments. It will be shown to be intimately related to diffusion.

Since so much of the work to be discussed involves either the diffusion of an element into a semiconductor, or diffusion out of it, it is perhaps worth starting by considering the elementary solutions to the diffusion equation for these two situations. The diffusion equation in one dimension can be written:

$$\frac{\partial C}{\partial t}=\frac{\partial}{\partial x}\left(D\frac{\partial C}{\partial x}\right) \tag{1}$$

where C is the concentration of diffusant and D is the diffusion coefficient. If D is a constant at constant temperature, then equation (1) becomes:

$$\frac{\partial C}{\partial t}=D\frac{\partial^2 C}{\partial x^2} \tag{2}$$

Now suppose that an impurity is diffused into a semi-infinite solid by maintaining a concentration C_0 at the surface from time $t=0$. This could be done, for instance, by bringing the solid into contact with a vapour containing the impurity, at some appropriately high temperature. The boundary conditions become:

$$C = C_0 \qquad t>0,\ x=0$$

$$C = 0 \qquad t=0,\ x>0$$

and the solution to equation (2) is:

$$C = C_0 \operatorname{erfc} \frac{x}{2\sqrt{Dt}} \tag{3}$$

The case of outdiffusion from an originally homogeneous sample is similar. Suppose the sample is doped to a level C_1 and it is raised to a high temperature at $t=0$. The dopant diffuses out and, if it is pumped away so that no solute vapour collects near the specimen, the boundary conditions become:

$$C = 0 \qquad t>0,\ x=0$$

$$C = C_1 \qquad t=0,\ x>0$$

and the solution is:

$$C = C_1 \operatorname{erf} \frac{x}{2\sqrt{Dt}} \tag{4}$$

The third situation of special interest here is diffusion from a homogeneously doped substrate into a growing epitaxial layer. If no dopant is lost from the surface of the layer, the solution becomes [1]:

$$C = \frac{C_1}{2}\left(1 + \operatorname{erf} \frac{x - vt}{2\sqrt{Dt}}\right) \tag{5}$$

where C_1 is the homogeneous level in the substrate and the origin of coordinates is taken as the growing surface, so that the layer-substrate interface recedes from the origin with velocity v.

Equations (3) and (4) are plotted in Figure 1. In the work to be described there are a great many profiles plotted of dopants diffusing in or diffusing out of semiconductor slices. It is important to note that almost none of them look like either of the profiles of Figure 1. What this indicates is that diffusion processes in semiconductors are rarely simple and that the temptation to describe a process at some given temperature with a single number called 'diffusion coefficient' should, more often than not, be resisted.

The first indications that diffusion processes might be important in SI (semi-insulating) material came from experiments in which epitaxial layers were grown on SI substrates [2–5]. Work of this type will be reviewed under the heading Diffusion during epitaxial growth, below. In the following section we examine the slightly more straight-forward experiments of indiffusion and outdiffusion.

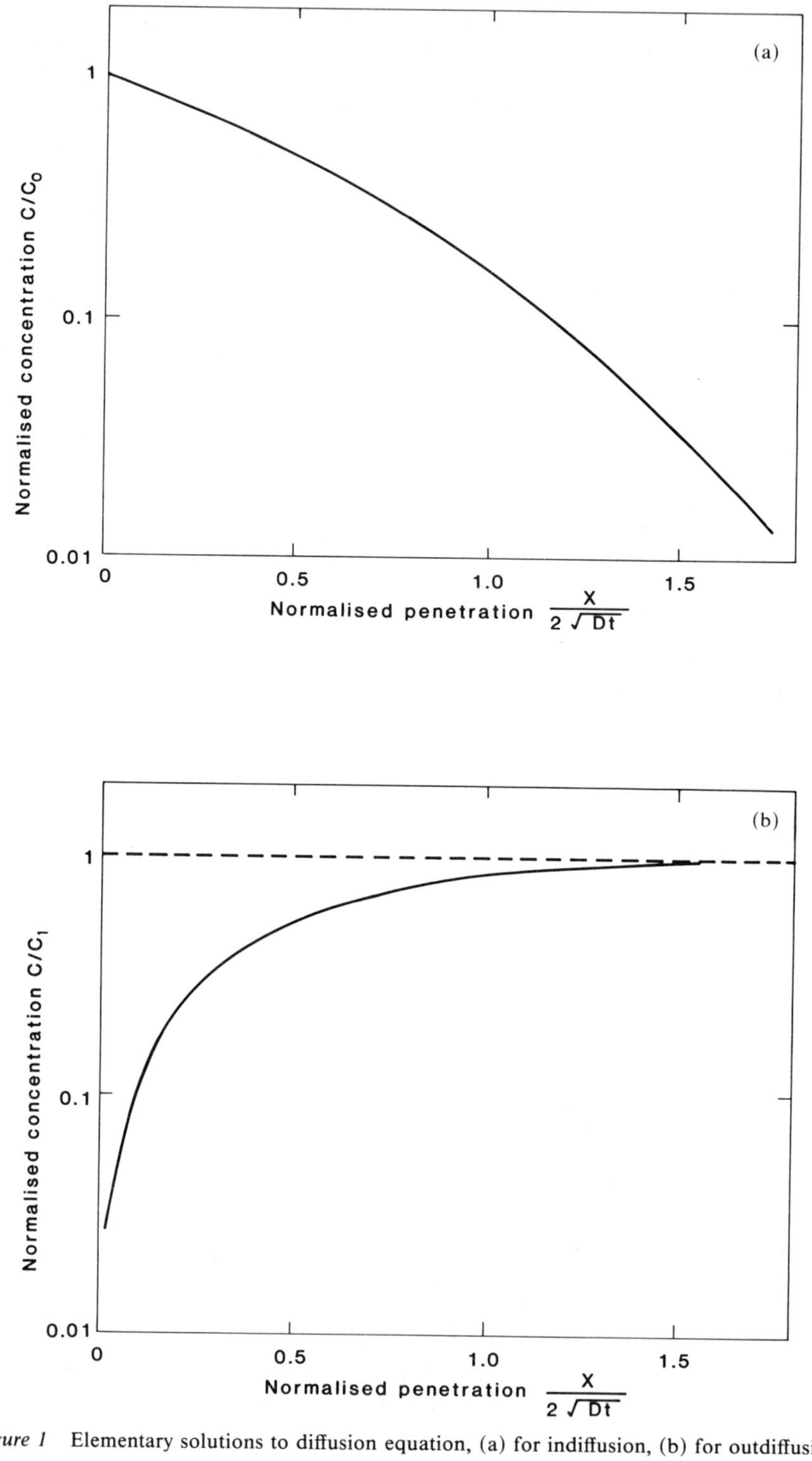

Figure 1 Elementary solutions to diffusion equation, (a) for indiffusion, (b) for outdiffusion.

Diffusion of Cr, Fe and Mn in GaAs and InP

There have been surprisingly few investigations in which indiffusion profiles have been plotted for the dopants of interest here. Work of this type has been reported for the Cr/GaAs [6, 7], Fe/GaAs [7–9], Cr/InP [7, 10], Fe/InP [7, 10, 11] and Mn/GaAs [12] systems. Of these the first is the best understood. Tuck and Adegboyega [6] carried out an extensive series of experiments in which radioactive Cr was diffused into GaAs. Profiles were plotted for the temperature range 800°C–1100°C and for times between 15 min and 10 h. A number of profiles for 4 h diffusions are shown in Figure 2. Each profile shows a high surface concentration with the surface peak dropping within about 20 μm to an almost constant 'bulk' concentration. Note that even for the lowest temperature recorded the Cr penetrated the whole of the slice. Results of this general form, showing extremely rapid diffusion by the bulk

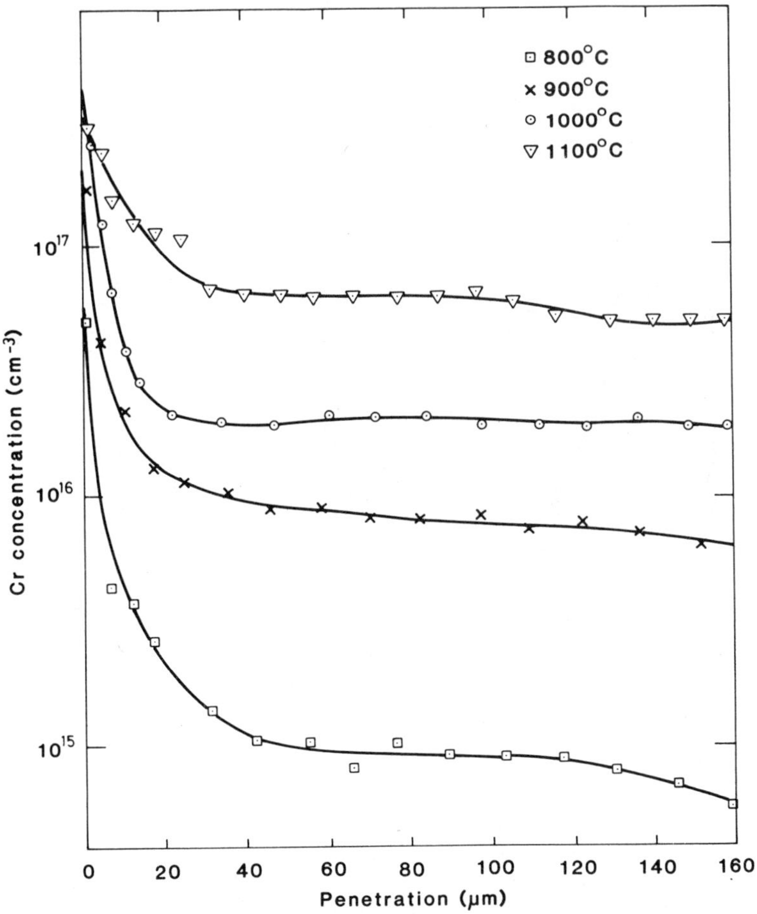

Figure 2 Diffusion profiles for Cr in GaAs for diffusion time of 4 h [6].

part of the profile, were also found for Fe in GaAs and both Cr and Fe in InP. A profile showing the diffusion of Fe into InP for 15 min at 900°C is given in Figure 3. For this shorter diffusion time the bulk section of the profile is sloping rather than horizontal, but the Fe has diffused through the specimen. The results for Mn are quite different. Diffusion of this element in GaAs gives curves which are often similar to the error function complement (erfc) curves of Figure 1a; some of Seltzer's profiles can actually be fitted quite accurately to erfc expressions [12], while others require one to describe the section of profile near the surface and another for the bulk part.

The very rapid diffusion of transition elements shown by these experiments obviously has implications for device work. These elements might be in the semiconductor by accident or by design but, either way, we would much prefer them to stay where they are during processing. In order to assess the significance of the phenomenon, it is necessary to establish whether the atoms

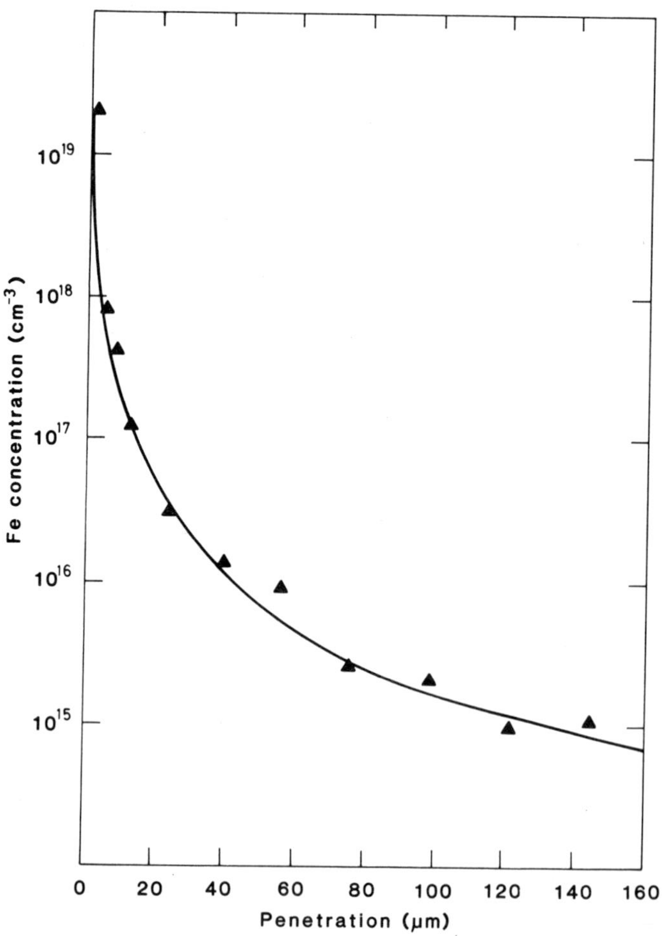

Figure 3 Diffusion profile of Fe in InP for diffusion time of 15 min at 900°C [13].

which arrive in a region by diffusion give rise to the same electrical effects as those incorporated during crystal growth. Tuck and Adegboyega [6] plotted carrier profiles in n-type GaAs into which Cr had been diffused at 1100°C. Their results showed that the sample became highly resistive in the region in which the Cr concentration exceeded the original donor doping. This suggested that the diffused Cr was acting in the 'normal' way as a deep acceptor. Later work by Skolnick *et al.* [14] confirmed this result for high diffusion temperatures but found that the acceptor action was absent for diffusion below 800°C. In addition, they found the characteristic 0.84 eV PL (photoluminescence) line, generally assumed to be due to the presence of Cr on a lattice site, only in samples diffused at 800°C and above. They suggested that for low-temperature diffusions a large proportion of the Cr atoms exist on electrically inactive interstitial sites. In a similar investigation Ke *et al.* [15] studied the DLTS (deep level transient spectroscopy) signal from GaAs diffused with Cr at 800°C and 1000°C. Interpretation of their results was not easy, but they came to the conclusion that they were observing signals from the same Cr species that normally occurs in GaAs doped with Cr during growth.

Experiments on Fe-diffused GaAs were performed by Brozel *et al.* [8]. They used electrical measurements to determine carrier concentrations and EPR (electron paramagnetic resonance) to measure the concentration profile of the Fe^{3+} ion and came to the conclusion that in the bulk of the specimen most of the Fe atoms were on Ga sites. Only about half of these appeared to give rise to deep acceptors, however. Near the surface of the diffused slice the situation was less clear; they suggested that the Fe was incorporated in the form of some complex.

The equivalent results for InP are most interesting. Experiments in which radioactive Fe was diffused into the semiconductor revealed no measurable acceptor activity of the Fe [10, 13]. Similar experiments using Cr gave the same result [10]. Further measurements were made on diffused material by Skolnick *et al.* [16]. The characteristic emission lines for substitutional Fe and Cr in InP were not detected. These results may relate to work published by Messham *et al.* [17]; they grew LPE (liquid phase epitaxial) layers of InP from Fe-doped melts and were also unable to find evidence of electrically active Fe. In another set of experiments, Debney and Jay [18] carried out mobility measurements on Fe-doped SI InP and estimated the concentration of electrically-active Fe atoms. They also performed chemical analyses of the specimens and came to the conclusion that the concentration of Fe atoms exceeded that of electrically active atoms by more than an order of magnitude. Thus it seems that when Fe and Cr are incorporated into InP, it is possible for a proportion to be situated on sites other than the In sublattice, and that this is especially so when the incorporation is by diffusion.

A good deal of work has been reported on the outdiffusion of these elements from GaAs and InP. A number of authors have examined the Cr/GaAs system by SIMS (secondary ion mass spectrometry) investigations on samples homogeneously doped with Cr during growth. This doping has been deliberate in some cases and accidental in others. The main points of interest have been the rapid diffusion of Cr and the unusual form of the outdiffusion profiles, which bears little resemblance to the curve of Figure 1b. After heat-treatments in the range 700–900°C most workers have found profiles with surface peaks of concentration at least an order of magnitude greater than the original homogeneous doping [19–22]. These peaks usually occupy a region within a micron or less of the surface. A number of the published profiles also show a dip just behind the peak; a good example is shown in Figure 4. The anneals were carried out both with and without capping of the

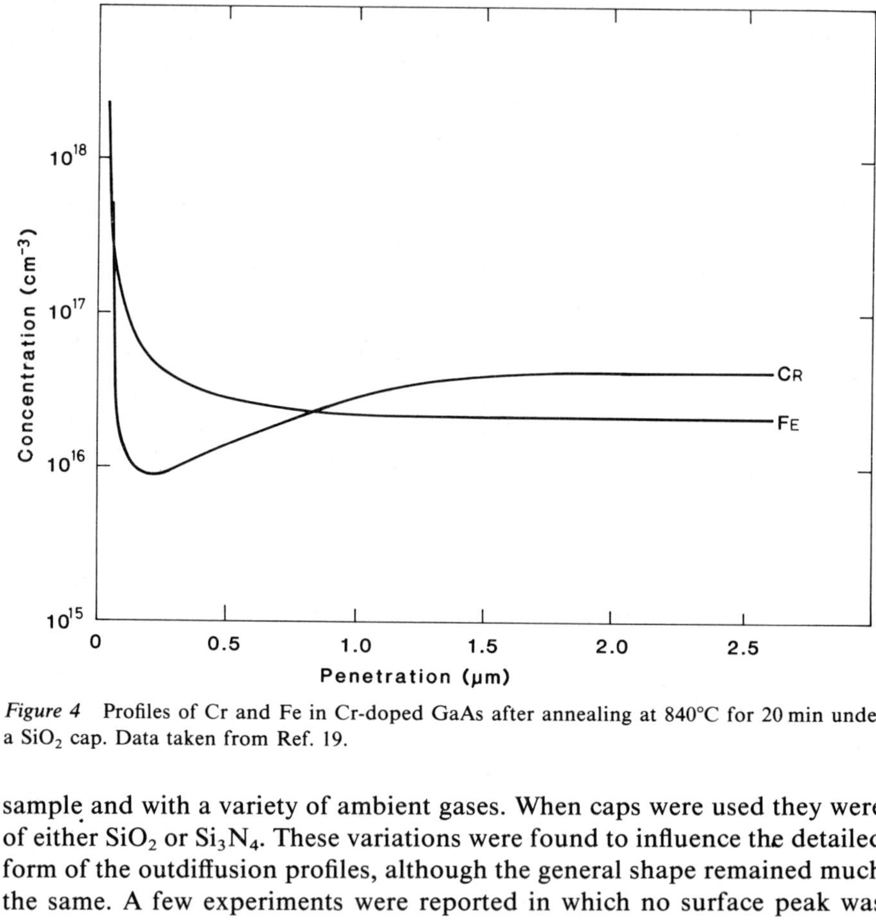

Figure 4 Profiles of Cr and Fe in Cr-doped GaAs after annealing at 840°C for 20 min under a SiO_2 cap. Data taken from Ref. 19.

sample and with a variety of ambient gases. When caps were used they were of either SiO_2 or Si_3N_4. These variations were found to influence the detailed form of the outdiffusion profiles, although the general shape remained much the same. A few experiments were reported in which no surface peak was found [23, 24]. Kasahara and Watanabe [24] suggested that the surface peak was due to defects at the cap-sample interface. This seems unlikely, however, in view of the fact that Vasudev *et al.* [22] performed similar annealing experiments both with and without capping and obtained similar results, with peaks, in the two cases.

Tuck *et al.* [5] reported an outdiffusion experiment in which radio-tracer Cr was employed. The Cr was first diffused into a GaAs sample, giving rise to a profile similar to those shown in Figure 2. About 50 μm was then etched from each surface, leaving a sample which was homogeneously doped with radioactive Cr to a level of about 5×10^{16} cm^{-3}. Outdiffusion anneals were carried out at 750°C, either in a closed ampoule or a VPE (vapour phase epitaxial) reactor. The result was qualitatively the same as the SIMS results quoted above; after the anneal a surface peak developed, with the surface concentration in excess of the original doping. The width of the surface peak was very much greater, however, penetrating about 20 μm into the sample, compared to <1 μm in the SIMS experiments. This implies a greater rate of diffusion in the radio-tracer experiment, underlining the fact that when Cr is introduced into GaAs by diffusion, the atoms do not necessarily enter the lattice in exactly the same way as they do when introduced by crystal growth.

An interesting variation on this experiment was reported by Magee *et al.* [25]. They performed annealing tests on Cr-doped GaAs after mechanically polishing one of the faces of the sample. They found, using SIMS, that the Cr moved preferentially towards the damaged face during the anneal, forming a large peak at the polished surface. Subsequently they used TEM (transmission electron microscopy) to study the damaged region and were able to show that the high Cr concentration corresponded to a region of high dislocation density [26].

Experiments in which the redistribution of Fe in GaAs was observed after annealing showed very similar results to the Cr/GaAs system [19, 27, 28] as did work on the Cr/InP and Fe/InP systems [29–31]. Mn also showed a surface build-up after annealing [20, 32, 33], although doubt was expressed by Clegg *et al.* [28] as to whether, in their case, the origin of the Mn at the surface was indeed the sample itself. They estimated that there was more Mn in their surface peak than had originally been in the whole sample. They concluded that the Mn they observed at the surface had diffused in from some outside source.

Diffusion during epitaxial growth

The first signs that transition elements are able to diffuse rapidly in the III–V semiconductors came from tests on GaAs layers grown on substrates containing Cr and Fe. Lang and Logan [2] found DLTS signals from Cu, Fe and Cr in LPE layers grown at 850°C. They attributed this to surface contamination of the substrate before the commencement of crystal growth, rather than outdiffusion from the substrate, however. White *et al.* [3] used a photocapacitance technique to examine VPE layers grown on Cr-doped substrates and reported a signal due to Cr in the layers. They believed that Cr had diffused out of the substrates during crystal growth. Cox and DiLorenzo [4] reported an interesting effect when growing undoped ‘buffer’ layers on SI substrates. The first few microns of undoped layer had a much higher resistivity than expected and they attributed this to outdiffusion of some acceptor from the substrate. In a later publication they indicated that this acceptor was probably Cr [34].

The proposition was tested in a very direct way by Tuck *et al.* [5]. Using diffusion techniques they produced substrates homogeneously doped with radioactive Cr at levels of 10^{16}–10^{17} cm^{-3}. Epitaxial layers were grown on these substrates and the layers were examined for the radio-tracer. In all cases rapid outdiffusion of Cr was found, from the substrate into the layer. The use of the radioactive element rendered the result quite unambiguous. The Cr found in the epitaxial layer could only have come from the substrate; it could not, for instance, have originated in the windings of the furnace or from any other external source.

SIMS has proved to be an invaluable tool for studying diffusion from substrates into growing layers. Virtually all investigations have reported that outdiffusion occurs. Results have been published showing the movement of Cr and Fe into epitaxial layers of InP [31, 35] and GaAs [36–39]. A peak in concentration is often seen at the layer-substrate interface; this seems to be influenced by the preparation of the substrate before growth. A good example of this is shown in Figure 5 which is for an InP layer grown by VPE onto a SI Fe-doped substrate. DLTS studies have also demonstrated in a very convincing manner the outdiffusion of Fe and Cr [40, 41].

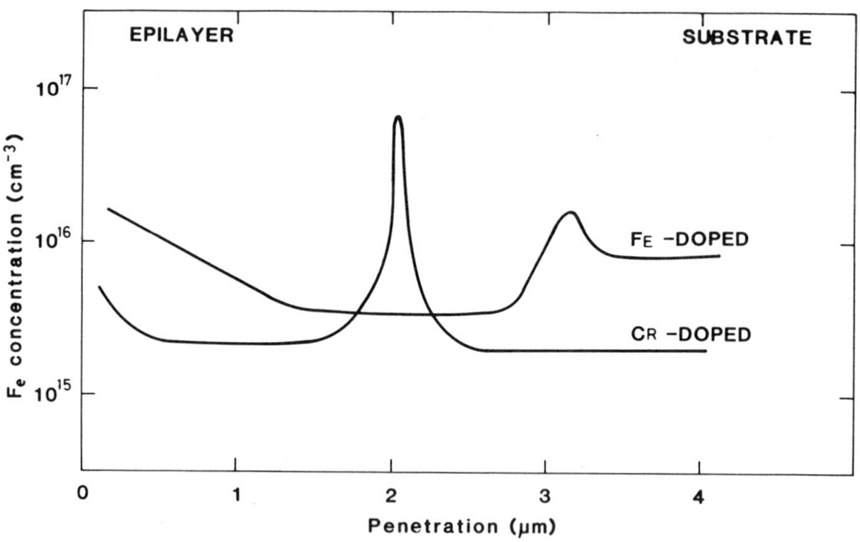

Figure 5 Profiles of Fe in VPE InP layers grown on Fe-doped and Cr-doped substrates. Note each one shows a peak at the substrate-layer interface. Data taken from Ref. 35.

Diffusion after ion implantation

Direct implantation into SI material provides an attractive technology for device manufacture. Unfortunately it is necessary to follow implantation with an annealing cycle at fairly high temperature in order to activate the implanted atoms. If the SI substrate is Cr-doped GaAs or Fe-doped InP, there is clearly a good chance of the Cr and Fe atoms redistributing themselves during this anneal. A good deal of work has been carried out studying the effect, almost all of it employing SIMS. The experiments fall into two main categories: those in which various elements are implanted into Cr-doped (or Fe-doped) material and those in which Cr (or Fe) is itself implanted into a substrate. In either case the Cr (and/or Fe) distribution is determined before and after annealing.

Most workers agree that when ions are implanted into Cr-doped GaAs substantial redistribution of the Cr takes place during anneal [42–50]. An example of Cr redistribution after S implantation and annealing at 840°C for 20 min is given in Figure 6. Generally speaking, Cr atoms move to the region

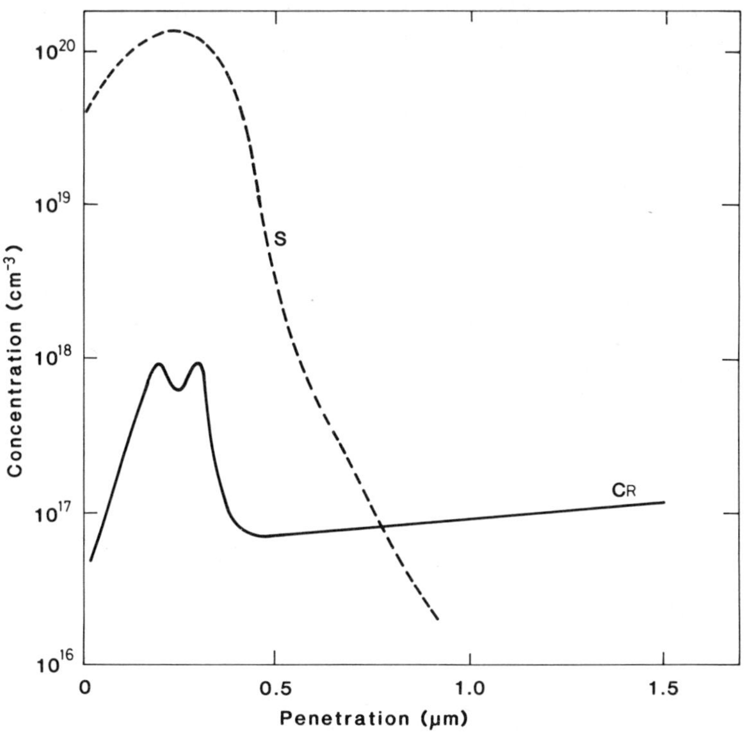

Figure 6 Cr redistribution after S implantation at 300 keV, 4×10^{15} cm^{-2} and annealing at 840°C for 20 min. Original S profile shown dotted. Data taken from Ref. 45.

of implantation damage and a concentration peak is observed in this region. A larger implantation fluence usually gives rise to a more pronounced peak. An exception to this rule is provided by Be implantation [43]; in this case the Cr redistributes in such a way as to reduce its concentration in the implanted region. This is presumably due to some electronic interaction between the Cr and Be atoms. A surface peak is often observed in the profile, similar to the one seen in simple outdiffusion experiments. Most of the anneals were carried out using a SiO_2 or Si_3N_4 cap and the surface peak may be due to gettering at the cap-semiconductor surface in these cases.

The relationship between the gettering of Cr and the damage caused by implantation was underlined in a paper of Sadena *et al.* [44]. They performed SIMS measurements to establish the Cr concentration and also did TEM work to locate regions of high damage. Peaks in the Cr concentration were found to correspond to the damaged regions. An attempt to explain this type of result was made by Magee *et al.* [49] using calculations published by Christel and Gibbons [51]. These workers made calculations of distributions of vacancies and interstitials in the implanted region. Magee *et al.* proposed that the gettering of Cr atoms should correspond to regions where the concentrations of interstitials are large.

Experiments in which Cr was implanted into GaAs showed very interesting effects on annealing [38, 52–57]. SIMS measurements immediately after implantation showed the usual LSS (Linhard, Scharff and Schiøtt) form. After annealing, however, the profiles were not what would be expected by a simple application of diffusion theory. A fairly typical result is shown in Figure 7. For short annealing times a peak develops at a point rather deeper

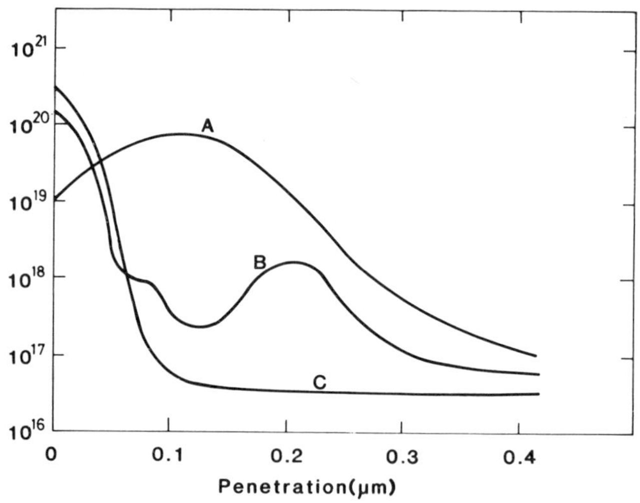

Figure 7 SIMS profiles after Cr implantation into GaAs at 190 keV, 1×10^{15} cm^{-2}. A—as implanted; B—after 15 min anneal at 870°C; C—after 90 min anneal at 870°C. Data taken from Ref. 57.

inside the specimen than the original as-implanted peak. Cr also builds up at the cap-GaAs interface. For longer anneal times the bulk peak disappears and only the interface peak remains. At this point the Cr distribution looks rather like a normal indiffusion profile (e.g. the 800°C result in Figure 2).

A good example of the high mobility of Cr after implantation was given by Simondet *et al.* [57]. Before carrying out the anneal, a GaAs sample was capped with Si_3N_4. The appropriate thickness of Si_3N_4 was deposited in

90 sec at 780°C. Even after this very short heat treatment, it was found that substantial redistribution of the Cr had taken place.

A few metallurgical studies have been made of Cr-implanted material [55, 56]. Pronko *et al.* [56] implanted at 120 keV with dose levels in the range 1×10^{14}–5×10^{16} cm^{-2}. Annealing was performed at 900°C using Si_3N_4 caps. They gave evidence of precipitation of Cr, the extent depending on the dose. For the highest dose, coarse non-uniform second phase precipitates were seen, of size 50–100 nm. At the lowest values of dose, they were much smaller, in the region of 6 nm. Perhaps the most surprising result was their estimate that at a dose of 5×10^{14} cm^{-2} as much as 40% of the Cr could be involved in the precipitates.

Work in which implantations were performed on Fe-doped GaAs [42, 50] and in which Fe was implanted into GaAs [58] gave results which were broadly similar to the Cr/GaAs case. There is very little work of this type published on InP. Gauneau *et al.* [59] implanted Fe and Cr into InP at 800 keV with a dose of either 10^{13} cm^{-2} or 10^{14} cm^{-2} and annealed at 750°C. Anneals were carried out under Si_3N_4 caps and also capless, with a phosphorus ambient. They found a surface peak and two internal peaks, one each side of the original internal peak. They attempted to explain their results using the theory of Christel and Gibbons [51].

Type conversion

When GaAs is annealed at temperatures in excess of about 700°C, changes in conductivity type are likely to occur. The phenomenon comes into that irritating category of effects which are not quite reproducible. GaAs samples from different sources show different conversion characteristics as, indeed, do samples obtained from the same source at different times, and sometimes samples can be raised to quite high temperatures without major changes in conductivity. Most early reports of type conversion concerned changes of n-type material to p-type after heat treatment [60–62]. At that time it was commonly suggested that Cu was the culprit, diffusing into the semiconductor from some source within the annealing furnace. More recently, interest has shifted to Cr-doped SI material. The most frequent report has been of conversion to p-type conductivity, but conversion to n-type has also been found on occasion [6, 64, 81] and, in one case, it was found that an initially SI slice could have a thin n-type skin at the surface with p-type conversion below [81].

Conversion of SI material has been studied by a variety of techniques, including Hall effect [32, 63–67], DLTS [68], SIMS [28, 32, 33, 63, 67], PL [32, 33, 63–67, 69] and SSMS (spark source mass spectrometry) [63]. In most of these studies it was found that the conversion was a surface phenomenon and that if a thickness of some microns was removed, the original high-resistivity material was retrieved. A number of authors have presented evidence to show that the extent of the conversion depends on whether the sample is capped and on the composition of ambient gases. Hallais *et al.* [65], for instance, found that conversion took place for uncapped specimens heated at 750°C in vacuum or in flowing hydrogen but that no change could be detected for specimens treated in argon. This result was confirmed by Klein *et al.* [32]. In later work, however, Mircea-Roussel *et al.* found conversion for SI material heated in argon; at temperatures below 870°C the samples became n-type and above this temperature they became p-type.

An interesting discussion has taken place in the literature concerning the role of Mn. Hallais *et al.* [65] found that after a 24 h anneal in vacuum the thickness of the converted (p-type) layer was about 1 μm. Both PL and Hall effect measurements were made on these layers. Conversion was attributed to a shallow acceptor level thought to be due to Ga vacancies. Their experiments in hydrogen showed a rather thicker p-type layer, about 12 μm, and electrical measurements indicated an acceptor at 110 meV above the valence band. This they attributed to Mn although they could not account for the presence of the element in their samples. In a later paper [64], the same workers presented the PL spectrum from a sample converted in hydrogen and showed, very convincingly, that it was virtually identical to the PL obtained from Mn-doped GaAs. These conclusions were confirmed by Zucca [66] who also used an ambient of flowing hydrogen. He found a converted p-type layer which he associated with an acceptor of activation energy 98 meV and a 'Mn' PL spectrum at 1.4 eV. When the layer was removed by etching both the acceptor level and the characteristic PL disappeared. The same result was found by other workers, but not necessarily given the same interpretation. Koschel *et al.* [69] attributed the 1.4 eV PL line to a C acceptor–As vacancy complex. They acknowledged, however, that the line was very similar to that due to Mn, but could see no reason for Mn contamination to occur in their specimens.

More recently, SIMS studies have shown that, whatever the source, much SI GaAs contains Mn both before and after anneal. Before heat treatment the level is typically 10^{15} cm^{-3}. After treatment, a surface peak is observed which presumably gives rise to the 'Mn' PL and acceptor effects noted above. Unfortunately this shape of profile can correspond to both indiffusion and outdiffusion, so it is not obvious whether or not the source of Mn in the peak is the bulk of the semiconductor. As noted above, Clegg *et al.* [28] were of the opinion that the Mn surface peak which they observed after anneal contained more Mn than had previously been in the entire specimen. They therefore concluded that indiffusion was being observed. Strong evidence that a Mn surface peak could develop by transport from the interior to the surface was given by Nordquist *et al.*, however [33]. An initial heat treatment at 740°C gave rise to a high concentration peak of Mn at the surface (shown by SIMS) and the 1.4 eV PL line. If the specimen was etched to remove the peak, the PL line disappeared. Subsequent annealing caused a renewed SIMS peak which was smaller than before, and revived the PL line, but at a weaker intensity than before. This suggests that the surface was gettering Mn from the bulk during anneal.

Just as things were seeming to become a little more clear, Look *et al.* [63, 67] published work in which they used electrical measurements to determine the concentration of the 100 meV acceptor level and SIMS to find the concentration of Mn. They found that the acceptor concentrations were significantly greater than the Mn concentrations, in one sample by a factor of 20. It does not seem possible, therefore, that the 100 meV level is caused solely by Mn atoms. They did not discount the role of Mn in type conversion, however, but suggested that a number of native defects may give rise to similar energy levels.

Very little has been published on the effect of heat-treatments on InP, but it appears that InP provides fewer problems than GaAs in this respect. Rumsby *et al.* [70] described the growth of SI Fe-doped InP by the LEC method and noted that the material retained its high resistivity after annealing at 650°C for $1\frac{1}{2}$ h in flowing hydrogen. This was confirmed by Morioka *et al.* [71] who reported that their InP remained stable after heating for 30 min under a SI_3N_4 cap.

Discussion

The high diffusion rates of Cr and Fe in GaAs and InP are now well established and there is general agreement that such rates are only attainable by a diffusion mechanism which is primarily interstitial. The accumulation of the diffusing element at the surface during outdiffusion also seems to be accepted as a genuine result. Attempts have been made to develop a theory which satisfactorily describes these effects [6, 72, 73]. Account has to be taken of the diffusion of the element in both its substitutional and its interstitial forms, of the interaction between the two forms and of interaction between the element and electrons in the semiconductor. The mathematics becomes rather complicated and it is usually necessary to make some fairly gross approximations if analytical solutions are to be obtained. Nevertheless, theoretical profiles have been published which reproduce the experimental work in a qualitative manner. The redistribution which takes place during the annealing phase after implantation presents an even greater problem of description. Here the atoms are diffusing in severely damaged material and the mechanisms are much more complex. A start has been made along these lines [49, 74] using the theory of Christel and Gibbons [51] to estimate the distributions of defects after implantation. There seems to be a need for further work in this area.

The situation with regard to type conversion is a little clearer than it was a few years ago, although not much. The evidence linking Mn to p-type surface layers is strong, but there is also good evidence to suggest that it is not the sole cause of the effect. This is to be expected, since all atoms and defects giving rise to energy levels in the semiconductor are likely to redistribute during an annealing process. The result observed at the surface is the sum of all these movements. The net effect is normally such as to create an excess of acceptors over donors at the surface, but the reverse sometimes happens, turning the surface n-type. The development of 'undoped' SI GaAs seems to represent a real step forward. This material is rendered SI by the presence of a deep donor, EL2, which compensates a small concentration of residual acceptors. It is generally agreed that EL2 is associated with one or more native defects [75]. Undoped SI GaAs seems to be a good deal more resistant to type conversion than Cr-doped, providing it is grown from an As-rich melt [76, 77].

It is interesting that work has recently been published indicating that EL2 has at least one thing in common with Cr; it diffuses in GaAs. Makram-Ebeid *et al.* [78] measured the concentration profile of the defect after annealing and presented profiles looking rather like the outdiffusion profile of Figure 1b. Even this type of result may not be as simple as it seems, however, since work has also been published suggesting that EL2 can be 'passivated' by hydrogen diffusing in from the surface [68, 75, 79]. A reduction in EL2 concentration near the surface after anneal could, therefore, be due to hydrogen diffusing in rather than a defect diffusing out. Indiffusion of hydrogen might also explain the anomaly in type-conversion experiments outlined in the last section, whereby heat treatments in hydrogen gave results which were different to those found for treatments in other ambients.

The information on InP, which is relatively slight, indicates that the transition elements diffuse just as fast in this semiconductor as they do in GaAs. The apparent low electrical conductivity of Fe and Cr introduced by diffusion into InP is puzzling and should reward further study. No InP equivalent of the undoped GaAs material has been reported and it is still necessary to dope fairly heavily with Fe to obtain SI crystals. On the other hand, for unknown reasons, thermal conversion seems almost absent from InP; this must be one of the very few respects in which InP is less troublesome than GaAs.

In the first paper in the first volume of this series, Dr White referred to the fast diffusion of Cr and type conversion as the two most unpopular phenomena in GaAs work [80]. Nothing has happened in the intervening four years to make them any more popular. Indeed, as the size of devices becomes smaller, one can only envisage diffusion and diffusion-related phenomena becoming even more significant.

References

1. Tuck B. (1974). *Introduction to Diffusion in Semiconductors*; Peter Peregrinus, Stevenage
2. Lang D. V. and Logan R. A. (1976). *J. Appl. Phys.*, **47,** 1533
3. White A. M., Porteous P. and Dean P. J. (1976). *J. Electron. Mat.*, **5,** 91
4. Cox H. M. and DiLorenzo J. V. (1977). *Inst. Phys. Conf. Ser.*, **33b**, 11
5. Tuck B., Adegboyega G. A., Jay P. R. and Cardwell M. J. (1979). *Inst. Phys. Conf. Ser.*, **45,** 114
6. Tuck B. and Adegboyega G. A. (1979). *J. Phys. D: Appl. Phys.*, **12,** 1895
7. Brozel M. R., Tuck B and Foulkes E. J. (1981). *Electron. Lett.*, **17,** 532

8. Brozel M. R., Foulkes E. J., Tuck B., Goswami N. K. and Whitehouse J. E. (1983). *J. Phys. D: Appl. Phys.*, **16,** 1085
9. Boltaks B. I., Kulikov G. S., Nikulitsa I. N. and Shishiyanu F. S. (1975). *Inorg. Mat.*, **11,** 292
10. Brozel M. R., Foulkes E. J. and Tuck B. (1982). *Phys. Stat. Solidi,* **(a)72,** K159
11. Shishiyanu F. S., Gheorghiv V. Gh., and Palazov S. K. (1977). *Phys. Stat. Solidi,* **(a)40,** 29
12. Seltzer M. S. (1965). *J. Phys. Chem. Sol.*, **26,** 243
13. Foulkes E. J. (1983). *PhD Thesis, Nottingham University*
14. Skolnick M. S., Brozel M. R. and Tuck B. (1982). *Solid-State Commun.*, **43,** 379
15. Ke W. K., Hamilton B., Peaker A. R., Brozel M. R., Tuck B. and Wight D. R. *Solid St. Electron,* in press
16. Skolnick M. S., Foulkes E. J. and Tuck B., *J. Appl. Phys.*, in press
17. Messham R. L., Majerfeld A. and Bachmann K. J. (1982). In *Semi-Insulating III–V Materials, Evian,* p. 75 (S. Makram-Ebeid and B. Tuck, eds.); Shiva Publishing, Nantwich, UK
18. Debney B. T. and Jay P. R. (1980). In *Semi-Insulating III–V Materials, Nottingham,* p. 305 (G. J. Rees, ed.); Shiva Publishing, Nantwich, UK
19. Wilson R. G. (1982). *Inst. Phys. Conf. Ser.*, **63,** 1
20. Watanabe K., Matsuoka Y., Imamura Y. and Ito T. (1982). *Inst. Phys. Conf. Ser.*, **63,** 383
21. Huber A. M., Morillot G., Linh N. T., Favennec P. N., Deveaud B. and Toulouse B. (1979). *Appl. Phys. Lett.*, **34,** 858
22. Vasudev P. K., Wilson R. G. and Evans C. A. (1980). *Appl. Phys. Lett.*, **37,** 837
23. Fairman R. D. and Oliver J. R. (1980). In *Semi-Insulating III–V Materials, Nottingham,* p. 83 (G. J. Rees, ed.); Shiva Publishing, Nantwich, UK
24. Kasahara J. and Watanabe N. (1980). *Jap. J. Appl. Phys.*, **19,** L151
25. Magee T. J., Peng J., Hong J. D., Evans C. A., Deline V. R. and Malbon R. M. (1979). *Appl. Phys. Lett.*, **35,** 277
26. Magee T. J., Hung J., Deline V. R. and Evans C. A. (1980). *Appl. Phys. Lett.*, **37,** 53
27. Yu P. W. (1981). *J. Appl. Phys.*, **52,** 5786
28. Clegg J. B., Scott G. B., Hallais J. and Mircea-Roussel A. (1981). *J. Appl. Phys.*, **52,** 1110
29. Shanabrook B. V., Klein P. B., Siebenmann P. G., Dietrich H. B., Henry R. L. and Bishop S. G. (1982). In *Semi-Insulating III–V Materials, Evian,* p. 310 (S. Makram-Ebeid and B. Tuck, eds.); Shiva Publishing, Nantwich, UK
30. Oberstar J. D., Streetman B. G., Baker J. E. and Williams P. (1981). *J. Electrochem. Soc.*, **128,** 1814
31. Holmes D. E., Wilson R. G. and Yu P. W. (1981). *J. Appl. Phys.*, **52,** 3396
32. Klein P. B., Nordquist P. E. R. and Siebenmann P. G. (1980). *J. Appl. Phys.*, **51,** 4861
33. Nordquist P. E. R., Klein P. B., Bishop S. G. and Siebenmann P. G. (1981). *Inst. Phys. Conf. Ser.*, **56,** 569
34. Cox H. M. and DiLorenzo J. V. (1980). In *Semi-Insulating III–V Materials, Nottingham,* p. 41 (G. J. Rees, ed.); Shiva Publishing, Nantwich, UK
35. Chevrier J., Armand M., Huber A. M. and Linh N. T. (1980). *J. Electron. Mat.*, **9,** 745
36. Palmateer S. C., Schaff W. J., Galuska A., Berry J. D. and Eastman L. F. (1983). *Appl. Phys. Lett.*, **42,** 183
37. Huber A. M., Morillot G., Bonnet M., Merenda P. and Bessonneau G. (1982). *Appl. Phys. Lett.*, **41,** 638
38. Deveaud B. and Favennec P. N. (1977). *Solid-State Commun.*, **24,** 473
39. Wilson R. G., Vasudev P. K., Jamba D. M., Evans C. A. and Deline V. R. (1980). *Appl. Phys. Lett.*, **36,** 215
40. Bremond G., Nouailhat A. and Guillot G. (1982). *Inst. Phys. Conf. Ser.*, **63,** 239
41. Tegude F. J., Baston J., Arnold N. and Heime K. (1982). In *Semi-Insulating III–V Materials, Evian,* p. 291 (S. Makram-Ebeid and B. Tuck, eds.); Shiva Publishing, Nantwich, UK
42. Feng M., Eu V., Kanber H. and Henderson W. B. (1981). *J. Electron. Mat.*, **10,** 973
43. Kanber H., Feng M. and Whelan J. M. (1983). *Inst. Phys. Conf. Ser.*, **65,** 605
44. Sadena D. K., Washburn J. and Zee T. (1982). *Inst. Phys. Conf. Ser.*, **63,** 359
45. Evans C. A., Hopkins C. G., Norberg J. C., Deline V. R., Blattner R. J., Wilson R. G., Jamba D. M. and Park Y. S. (1980). In *Semi-Insulating III–V Materials, Nottingham,* p. 138 (G. J. Rees, ed.); Shiva Publishing, Nantwich, UK

46. Evans C. A., Deline V. R., Sigmon T. W. and Lidow A. (1979). *Appl. Phys. Lett.*, **35,** 291
47. Magee T. J., Lee K. S., Ormond R., Blattner R. J. and Evans C. A. (1980). *Appl. Phys. Lett.* **37,** 447
48. Magee T. J., Lee K. S., Ormond R., Evans C. A., Blattner R. J. and Hopkins C. (1980). *Appl. Phys. Lett.*, **37,** 635
49. Magee T. J., Kawayoshi H., Ormond R. D., Christel L. A., Gibbons J. F., Hopkins C., Evans C. A. and Day D. S. (1981). *Appl. Phys. Lett.*, **39,** 906
50. Favennec P. N., Gauneau M., L'Haridon H., Deveaud B., Evans C. A. and Blattner R. J. (1981). *Appl. Phys. Lett.*, **38,** 271
51. Christel L. A. and Gibbons J. F. (1980). *J. Appl. Phys.*, **52,** 5050
52. Deveaud B., Favennec P. N., L'Haridon H., Armand C. and Martinez A. (1982). In *Semi-Insulating III–V Materials, Evian*, p. 269 (S. Makram-Ebeid and B. Tuck, eds.); Shiva Publishing, Nantwich, UK
53. Simondet F., Venger C., Martin G. M. and Chaumont J. (1980). In *Semi-Insulating III–V Materials, Nottingham*, p. 100 (G. J. Rees, ed.); Shiva Publishing, Nantwich, UK
54. Vasudev P. K., Wilson R. G. and Evans C. A. (1980). *Appl. Phys. Lett.*, **37,** 308
55. Rai A. K., Bhattacharya R. S. and Pronko P. P. (1982). *Appl. Phys. Lett.*, **41,** 1086
56. Pronko P. P., Rai A. K., Holland O. W., Appleton B. R. and Narayan J. (1982). *J. Appl. Phys.*, **53,** 5621
57. Simondet F., Venger C. and Martin G. M. (1980). *J. Appl. Phys.*, **23,** 21
58. Wilson R. G. and Evans C. A. (1981). *Inst. Phys. Conf. Ser.*, **56,** 563
59. Gauneau M., L'Haridon H., Rupert A. and Salvi M. (1982). *J. Appl. Phys.*, **53,** 6823
60. Edmond J. T. (1960). *J. Appl. Phys.*, **31,** 1428
61. Hwang C. J. (1968). *J. Appl. Phys.*, **39,** 5347
62. Toyama M. (1969). *Jap. J. Appl. Phys.*, **8,** 1000
63. Look D. C., Yu P. W., Ehret J. E., Yeo Y. K. and Kwor R. (1982). In *Semi-Insulating III–V Materials, Evian*, p. 372 (S. Makram-Ebeid and B. Tuck, eds.); Shiva Publishing, Nantwich, UK
64. Mircea-Roussel A., Jacob G. and Hallais J. P. (1980). In *Semi-Insulating III–V Materials, Nottingham*, p. 133 (G. J. Rees, ed.); Shiva Publishing, Nantwich, UK
65. Hallais J., Mircea-Roussel A., Farges J. P. and Poiblaud G. (1977). *Inst. Phys. Conf. Ser.*, **33b,** 220
66. Zucca R. (1977). *Inst. Phys. Conf. Ser.*, **33b,** 228
67. Look D. C. and Pomrenke G. S. (1983). *J. Appl. Phys.*, **54,** 3249
68. Hughes B. and Li C. (1983). *Inst. Phys. Conf. Ser.*, **65,** 57
69. Koschel W. H., Bishop S. G., McCombe B. D., Lum W. Y. and Wieder H. H. (1977). *Inst. Phys. Conf. Ser.*, **33a,** 98
70. Rumsby D., Ware R. M. and Whittaker M. (1980). In *Semi-Insulating III–V Materials, Nottingham*, p. 59 (G. J. Rees, ed.); Shiva Publishing, Nantwich, UK
71. Morioka M., Kikuchi K., Kohe K. and Akai S. (1982). *Inst. Phys. Conf. Ser.*, **63,** 37
72. Jordan A. S. (1982). In *Semi-Insulating III–V Materials, Evian*, p. 253 (S. Makram-Ebeid and B. Tuck, eds.); Shiva Publishing, Nantwich, UK
73. Zahari M. D. and Tuck B. (1983). *J. Phys. D: Appl. Phys.*, **16,** 635
74. Kanber H., Feng M. and Whelan J. M. (1984). *J. Appl. Phys.*, **55,** 347
75. Walukiewicz W., Lagowski J. and Gatos H. C. (1983). *Appl. Phys. Lett.*, **43,** 112
76. Ta L. B., Thomas R. N., Eldridge G. W. and Hobgood H. M. (1983). *Inst. Phys. Conf. Ser.*, **65,** 31
77. Ta L. B., Hobgood H. M., Rohatgi A. and Thomas R. N. (1982). *J. Appl. Phys.*, **53,** 5771
78. Makram-Ebeid S., Gautard D., Devillard P. and Martin G. M. (1982). *Appl. Phys. Lett.*, **40,** 161
79. Lagowski J., Kamiska M., Parsey J. M., Gatos H. C. and Lichtensteiger M. (1982). *Appl. Phys. Lett.*, **41,** 1078
80. White A. M. (1980). In *Semi-Insulating III–V Materials, Nottingham*, p. 3 (G. J. Rees, ed.); Shiva Publishing, Nantwich, UK

Different Mn-Diffusion Mechanisms in Semi-Insulating GaAs Studied by Photoluminescence

T. Hariu, T. Sato, Y. Sasaki, K. Matsushita and Y. Shibata

Department of Electronic Engineering, Tohoku University, Sendai 980, Japan

Abstract

A method of quantitative analysis of Mn impurity in GaAs by PL (photoluminescence) intensity ratio between shallow donor-shallow acceptor and shallow donor-Mn acceptor pair emissions was proposed and experimentally verified by the diffusion of Mn into LEC (liquid encapsulated Czochralski) GaAs. The concentration of Mn piled-up near the surface of LEC GaAs by heat treatment in hydrogen without intentional doping was estimated to be $1\text{–}2 \times 10^{16}\ \mathrm{cm}^{-3}$ at 850°C and $2\text{–}3 \times 10^{15}\ \mathrm{cm}^{-3}$ at 750°C. Different Mn diffusion mechanisms are discussed in relation to the site-conversion of Si in HB (horizontal Bridgman) GaAs and the different behaviour of deep level luminescence bands.

Introduction

Manganese is a typical residual impurity often encountered in surface layers of SI GaAs which are subjected to heat treatment [1] as well as in ion-implanted layers [2] and also in MBE (molecular beam epitaxy) grown layers [3]. Manganese diffusion or redistribution has been intensively studied in relation to thermal conversion to p-type [1, 4–6], but the variety of results seems to indicate that its mechanism is quite complicated, probably owing to the different nature of the various GaAs wafers and heat treatments studied by different investigators. In this paper, we first propose a method of quantitative analysis of Mn impurity in GaAs by using PL measurement, and then discuss the interaction of Mn with other impurities or defects by consulting the simultaneous PL change in the regions of edge emission and deep level luminescence.

Quantitative analysis of Mn

Photoluminescence has been extensively employed to detect impurities in semiconductor crystals because of its non-destructive nature and high sensitivity. However, its use for quantitative analysis has been limited to very pure material [7] because several coexisting competitive recombination processes, radiative or non-radiative, make it almost impossible to analyse the density of a particular impurity quantitatively in typical semiconductor crystals. We here propose a method in which the measurement of PL is applied for the quantitative analysis of Mn in GaAs by taking the ratio of shallow donor-Mn acceptor pair emission (I_{Mn}) to shallow donor-shallow acceptor pair emission (I_{DA}). The present proposal is based upon the consideration that each intensity of the above two emission bands is proportional to the density of the respective acceptor, multiplied by the spatial extension of the wave function of the respective acceptor state, with the same proportionality constant which depends upon the density of shallow donors and other recombination processes including non-radiative recombination. Then, because the spatial extension of each acceptor state as a point defect is independent of the density, the ratio of I_{Mn} to I_{DA} gives the density of Mn as a function of the density of the shallow acceptor, independent of other recombination processes as well as of donor density, as long as the PL is measured at low temperatures and low excitation intensities where D–A pair emission is dominant over band-to-acceptor or donor-to-band emission. The present method has another advantage: it gives a result which does not depend on measurement systems, because the relative PL intensity, rather than absolute intensity, is required for the analysis.

Experimental

Two types of SI GaAs, HB and LEC, including wafers from different suppliers, were subjected to heat treatment or the diffusion of Mn which was vacuum-deposited on their surfaces, at different temperatures in a hydrogen atmosphere. The PL was measured at 4.2 K, with the sample immersed in liquid helium and excited by a CW argon laser at a wavelength of 514.5 nm. It was confirmed, from the emission spectrum dependence on excitation intensity and temperature, that both I_{DA} and I_{Mn} were due to donor-acceptor pair emissions. The van der Pauw method was employed for the determination of the hole density at room temperature. As described later, this corresponds directly to the density of Mn in Mn-diffused LEC GaAs. In-depth profiles of PL and hole density were obtained by successive step etching.

Results and discussion

A very different behaviour of Mn diffusion was observed in the present HB and LEC GaAs. From the PL spectra of the original wafers without heat treatment, it was deduced that shallow acceptors in HB GaAs are carbon and silicon. When the HB wafers were subjected to heat treatment for Mn diffusion, I_{DA} due to Si was enhanced and became dominant. This enhancement was much more marked in Mn-diffused samples than in a control sample which was subjected to heat treatment without Mn diffusion. It is thus suggested that the replacement of Si from Ga sites to As sites is closely related to the mechanism of Mn diffusion in HB GaAs. In contrast to the case in HB GaAs, the shallow acceptor in LEC GaAs detected by PL is carbon. No appearance of I_{DA} due to Si was observed upon Mn diffusion.

In Figure 1, in order to eliminate the effect of the introduction of non-radiative centres near the surface region, I_{Mn}/I_{DA} in LEC GaAs is plotted against the normalized depth $u = x/2(Dt)^{1/2}$, where x is the depth from the surface, t the diffusion time and D the diffusion constant of Mn at each diffusion temperature given in Sze's text [8]. The in-depth profiles of Mn concentration were also estimated from the hole concentration obtained by the differential Hall effect measurement and are shown in Figure 1 by solid lines. In this calculation, the contributions of residual acceptors and donors to conductivity and Hall effect were neglected, compared with Mn acceptors in Mn-diffused samples. As shown in Figure 1, the agreement between the

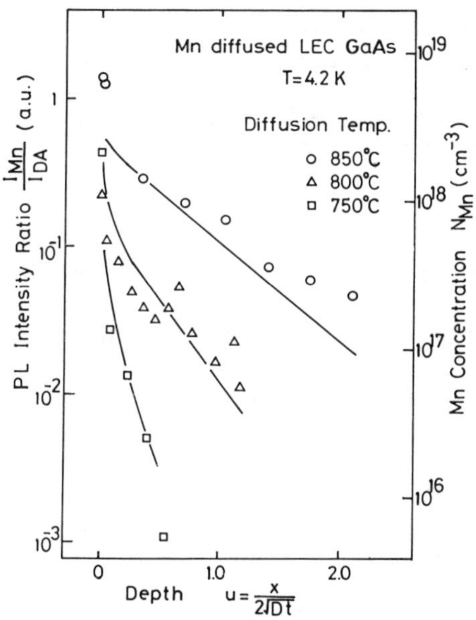

Figure 1 In-depth profiles of I_{Mn}/I_{DA} in Mn-diffused LEC GaAs. The density profiles of Mn estimated from hole concentrations are also shown by solid lines, for which the ordinate scale on the right hand side should be read.

two measurements is excellent. It was also confirmed that the spatial depth of the excited region, which is determined by the absorption coefficient for excitation light at a wavelength of 514.5 nm in GaAs, changes the above profiles only very little. It is also suggested that the diffusion length of excited excitons or free carriers is short enough to be neglected in obtaining the above profiles.

The PL intensity ratio I_{Mn}/I_{DA} versus Mn concentration estimated from Hall measurements shown in Figure 2 comprises data for different diffusion temperatures and different depths. This curve shows that I_{Mn}/I_{DA} is proportional to the Mn concentration and can be used to obtain the Mn concentration from a PL measurement. The Mn concentration piled-up near the surface of a control sample upon heat treatment in hydrogen for 15 min. This mechanism

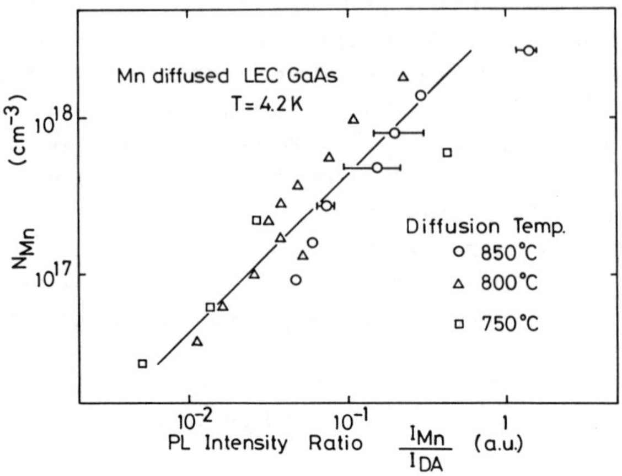

Figure 2 I_{Mn}/I_{DA} versus the concentration of Mn estimated from hole concentration, comprising data for different diffusion temperatures and for different depths.

has been considered to be responsible for the p-conversion of SI GaAs [1, 6]. The Mn surface concentration can be estimated to be $1\text{–}2 \times 10^{16}$ cm^{-3} at 850°C and $2\text{–}3 \times 10^{15}$ cm^{-3} at 750°C.

The present method of quantitative analysis is not valid when new dominant shallow acceptors are introduced by Mn diffusion. One such example might be the site-conversion of Si in the present HB GaAs, as stated above. Figure 3 shows the in-depth profiles of I_{Mn} in HB wafers subjected to Mn diffusion.

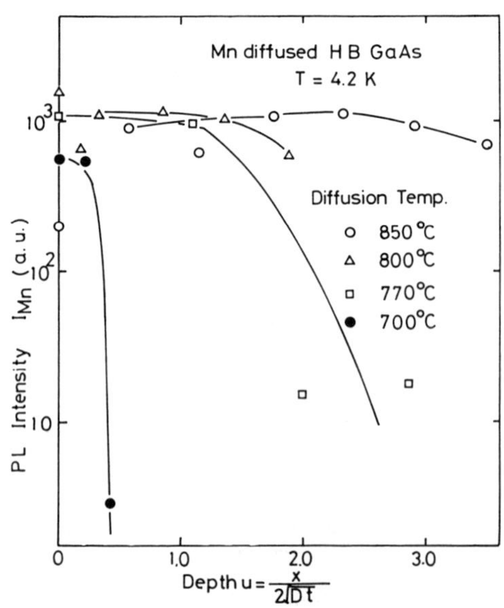

Figure 3 In-depth profiles of I_{Mn} in HB GaAs in which Mn was diffused at different temperatures.

The Mn diffusion extends much deeper in HB GaAs than in the LEC described above, and I_{DA} due to Si becomes dominant over I_{DA} due to C up to the depth at which I_{Mn} is quite high. Beyond this region, both I_{DA} due to Si and C are again observed as in the original sample. It is also to be noted that the saturation of I_{Mn} is attained near the surface even at 770°C, contrary to the case for LEC GaAs. This mechanism of enhanced diffusion of Mn in HB GaAs is considered to be responsible for the thicker converted surface layer observed in the present HB control samples as compared with LEC samples.

Figure 4 shows the in-depth profiles of I_{Mn} and the 0.65 eV band luminescence intensity when two kinds of LEC GaAs from different suppliers were

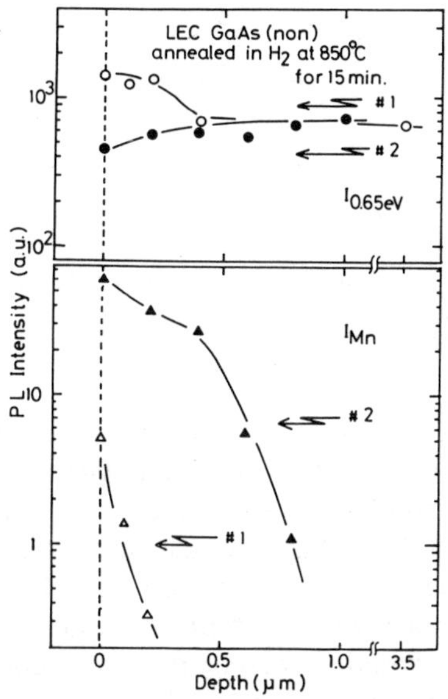

Figure 4 In-depth profiles of I_{Mn} and 0.65 eV band in two kinds of LEC GaAs from different suppliers which were subjected to heat treatment in hydrogen at 850°C for 15 min without intentional doping ('non'). Sample 1 is the same as used to obtain Figures 1 and 2.

subjected to the same heat treatment at 850°C for 15 min. The 0.65 eV band is likely due to EL2 [9]. In sample 1 which was used to obtain Figures 1 and 2, the pile-up of Mn is less than for sample 2. Note that the 0.65 eV band increases toward the surface for sample 1, but decreases for sample 2. A possible explanation will be given later in connection with the mechanism of Mn diffusion.

When SI GaAs wafers are heat-treated in hydrogen at temperatures above 400°C, we can observe that a new emission band with a peak at about 1.0 eV begins to appear, but then decreases for still higher temperatures. This band, we believe, is related to V_{Ga}, since the same band is observed in our plasma-assisted epitaxially grown layers [10] when the supply ratio As/Ga is large, but disappears when the supply ratio is sufficiently reduced. Figure 5 shows the change of the PL depth-profile in this band as well as in the 0.65 eV band accompanied by Mn diffusion into LEC sample 1 at 800°C and 850°C.

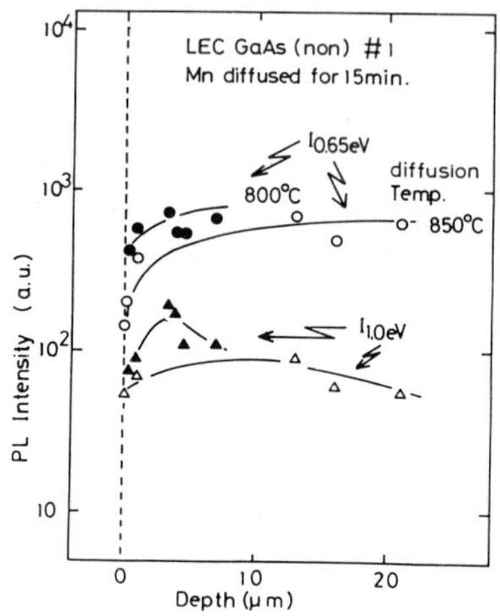

Figure 5 In-depth profiles of 1.0 eV and 0.65 eV bands in LEC GaAs sample 1 into which Mn was diffused at 800°C and 850°C.

The mechanism of Mn diffusion or redistribution is now discussed in connection with the above observations. It is generally accepted that Mn substitutes at a Ga site and behaves as an acceptor at about 0.1 eV above the top of valence band. It is then plausible that the Mn-diffusion is affected by the density of the residual Ga vacancies, and how they are supplied during the Mn diffusion. Si is an amphoteric impurity in GaAs and there is no large difference of free energy, whether it substitutes at a Ga site or an As site, as also suggested by the p-n conversion in LPE growth. Therefore, when the density of residual Si (in Ga sites or as precipitates at dislocations) is high as in the present HB GaAs, it can enhance the Mn diffusion even when As vacancies are created.

When GaAs is heat-treated in hydrogen, a Ga vacancy is more likely to be created [11, 12], probably by formation of a gallium hydride. This is considered to be the origin of the 1.0 eV band related to V_{Ga}. As the temperature of heat treatment is raised further, facilitating Mn or As diffusion, these species can occupy the V_{Ga}, and the Mn band or 0.65 eV band will be increased. However, if the density of Mn is sufficiently high (as in the case of intentional Mn diffusion) it can take over the Ga site of As_{Ga} or EL2 and then decrease the 0.65 eV band. The behaviours of the 1.0 eV and 0.65 eV bands in Figure 5 can then be well explained. The different behaviours of the 0.65 eV band in different LEC samples shown in Figure 4 can be explained by the different density of residual Mn, as indicated by the larger pile-up of Mn in sample 2.

Conclusions

A method of quantitative analysis of Mn impurity in GaAs by the PL intensity ratio between shallow donor-Mn acceptor pair emission and shallow donor-shallow acceptor pair emission was proposed and experimentally verified by the diffusion of Mn into LEC GaAs. A calibration curve obtained by collecting the data for different temperatures and different depths indicated the density of Mn piled-up near the surface of the LEC GaAs, after heat treatment in hydrogen for 15 min, to be $1\text{–}2 \times 10^{16}\ cm^{-3}$ at 850°C and $2\text{–}3 \times 10^{15}\ cm^{-3}$ at 750°C. This method is valid unless new dominant acceptors are induced by Mn doping, as was in fact observed by Mn diffusion into HB GaAs.

A very different behaviour of Mn diffusion was observed in HB GaAs. The I_{DA} due to Si acceptor became dominant after Mn diffusion, in addition to those due to C and Si in the original HB wafers. This enhancement is much more remarkable in Mn-diffused samples than in control samples which were subjected to heat treatment without Mn diffusion. It is then concluded that the replacement of residual Si from Ga sites to As sites enhances the diffusion of Mn. This mechanism is considered to be responsible for a converted layer of larger depth in HB control samples as compared with LEC samples.

Other mechanisms of Mn-diffusion were also discussed concerning how Ga vacancies are supplied for Mn substitution, based on the simultaneous observations of the 0.65 eV and 1.0 eV bands.

ACKNOWLEDGEMENTS

The authors wish to thank Dr K. Kaminishi of Oki Electric Co. for his encouragements and for partial support.

References

1. Klein P. B., Nordquist P. E. R. and Siebenman P. G. (1980). *J. Appl. Phys.*, **58,** 4861
2. Kanber H., Feng M. and Whelan J. M. (1982). *Appl. Phys. Lett.*, **40,** 960
3. Ilegems M., Dingle R. and Rupp Jr L. W. (1975). *J. Appl. Phys.*, **46,** 3059
4. Zucca R. (1976). *Inst. Phys. Conf. Ser.*, **33b,** 228
5. Lum W. Y., Wieder H. H., Koschel W. H., Bishop S. G. and McCombe D. (1977). *Appl. Phys. Lett.*, **30,** 1
6. Lum W. Y. and Wieder H. H. (1977). *Appl. Phys. Lett.*, **31,** 213
7. Tajima M. (1978). *Appl. Phys. Lett.*, **35,** 719
8. Sze S. M. (1981). In *Physics of Semiconductor Devices* (2nd edition), p. 68; John Wiley & Sons, New York
9. Yu P. W. (1982). *Solid-State Commun.*, **43,** 953
10. Hariu T., Matsushita K., Komatsu Y., Shibuya S., Igarashi S. and Shibata Y. (1982). *Inst. Phys. Conf. Ser.*, **65,** 141
11. Watanabe K., Matsuoka Y., Imamura Y. and Ito T. (1981). *Inst. Phys. Conf. Ser.*, **63,** 383
12. Bringans R. D. and Bachrach R. Z. (1983). *Solid-State Commun.*, **45,** 83

Migration of Iron in Iron-Doped InP after Implantation of Acceptor, Donor and Neutral Impurities

P. N. Favennec and M. Salvi

CNET (LAB/ICM), 22301 Lannion, France

Abstract

Iron-doped substrates are implanted with high doses of dopant impurities and we show the influence of the implanted impurity on the iron redistribution. The Fe redistribution depends strongly on the nature of the implanted impurity and not only on the surface and the damage induced by the implantation. Three groups of impurities in which the Fe redistribution are similar have been determined: (a) donor impurities; (b) acceptor impurities; and (c) O and Fe impurities. A model with three flows of Fe is used to explain the complex redistribution of Fe after implantation and annealing.

Introduction

Ion implantation into SI III–V materials is an attractive means for making various devices. It has been shown that chromium redistribution during post-implantation annealing in bulk Cr-doped GaAs can lead to drastic variations in the transport of the implanted layers and can have serious consequences on the performance of the devices [1–4]. In InP, Fe which produces a deep acceptor, is added to the melt to compensate the background donor impurities and to obtain SI substrates. Like Cr in GaAs, Fe in InP can rapidly migrate when substrates undergo high temperature annealing after implantation [5–7], thus affecting the electrical activity of the implanted impurities and also the device performances. This paper describes Fe redistribution in implanted InP substrate. The Fe-doped substrates are implanted with high doses of ions and we wish to show the influence of the implanted impurity on the Fe redistribution.

Experimental

The substrates used in our experiments were (100) wafers of Fe-doped InP grown by the LEC (liquid encapsulated Czochralski) technique at CNET laboratories. These substrates were SI and the Fe concentration was about $5 \times 10^{16}\ cm^{-3}$. During implantation, substrates were tilted 7° from the (100) axis to avoid channelling effects. Implantation was carried out at room temperature. The implanted wafers were annealed capless in a flowing high purity 0.4% PH_3 in H_2 stream containing <1 ppm H_2O at atmospheric pressure. These conditions ensured that the surface conditions were preserved. All anneals were carried out at 750°C for 15 min. Concentration depth profiles were measured by SIMS (secondary ion mass spectrometry) using an oxygen beam or a cesium beam according to the impurities to be analysed. The CAMECA IMS 3f ion microanalyser was used in this study. Calibration of atomic concentrations was established using the unannealed depth profiles as a reference. The sputter rate was determined by a Talystep which measured crater depths to within 6%.

Results

INTRODUCTION OF ACCEPTOR IMPURITIES INTO IRON-DOPED InP

We introduced acceptor impurities into Fe-doped InP substrates and measured the acceptor impurity and the induced Fe depth profiles. In Figure 1a, we show the profiles after a Zn diffusion from a Zn-doped silica film deposited on the substrate surface. The thermal diffusion was at 450°C for 120 min. In Figure 1b, we show the profiles after a Zn diffusion, from Zn vapour at 580°C, for 20 min into an implanted substrate. The diffusion process was an open-ampoule diffusion and it is described elsewhere [8]. The implantation characteristics were: O ions, 700 keV and a dose of 10^{14} ions cm^{-2}. The implanted sample was not annealed before the thermal diffusion of zinc. In Figures 1c and 1d, we show the induced profiles after implantation and annealing at 750°C during 15 min for the Zn and Be implanted samples. Zinc ions were implanted at 700 keV with a dose of 10^{15} ions cm^{-2}. Beryllium ions were implanted at 1300 keV with a dose of 10^{15} ions cm^{-2}. For these four processed samples, the common point is the diffusion of acceptor impurities, which are coming either from the surface or from the implanted layer.

In the depth profiles of iron, we note four different regions:

1. Region I is the bulk, Zn is below the detection limit and the Fe concentration is constant at about 5×10^{16} cm^{-3}.
2. Region II, observed in all samples, diffused or implanted, is a region with a very low concentration of Fe (below the detection limit) as well as a region with a Zn concentration higher than 10^{16} cm^{-3}. The Fe concentration is seen to return quickly to the bulk level in a region that coincides with an abrupt drop from some 10^{16} cm^{-3} down to the detection limit. The low concentration of Fe in Region II seems to be related to the presence of the acceptor impurities Zn or Be.
3. Region III is observed only when there is an implanted layer. Iron is accumulated in the implanted layer. This accumulation seems to go hand in hand with implantation.

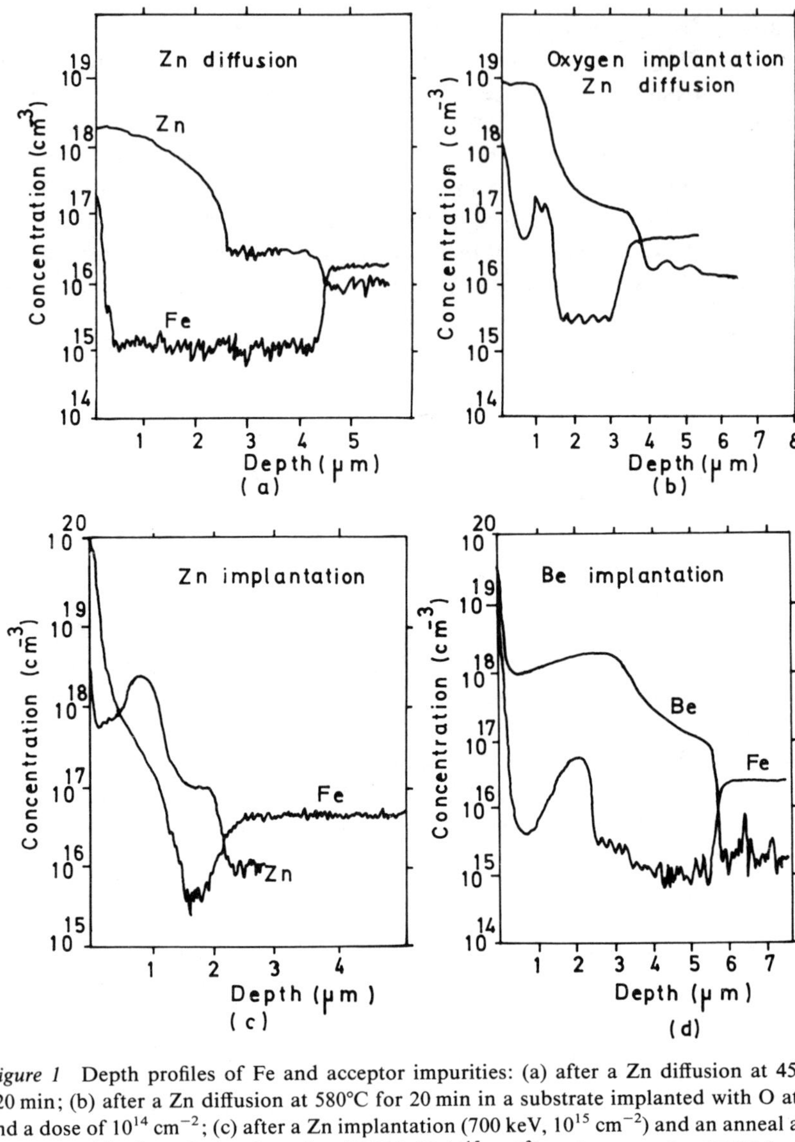

Figure 1 Depth profiles of Fe and acceptor impurities: (a) after a Zn diffusion at 450°C for 120 min; (b) after a Zn diffusion at 580°C for 20 min in a substrate implanted with O at 1 MeV and a dose of 10^{14} cm^{-2}; (c) after a Zn implantation (700 keV, 10^{15} cm^{-2}) and an anneal at 750°C for 15 min; (d) after a Be implantation (1300 keV, 10^{15} cm^{-2}) and an anneal at 750°C for 15 min.

4. Region IV is the surface layer. In all cases, we have an accumulation of Fe at the surface.

The first two figures corresponding to a Zn diffusion from the surface have been discussed previously [7]. In the case of an implantation of acceptor impurities, the implanted layer has two roles:

1. It is a diffusion source; impurities diffuse from the implanted layer towards the surface and the bulk.

2. It is a source of defects; some Fe impurities are trapped by these defects, the maximum concentration of these defects being located at about $0.8\bar{R}_p$, where $\bar{R}_p$ is the projected range of the implanted ions.

IMPLANTATION OF NEUTRAL IMPURITIES

To show the effects of defects induced by implantation on the Fe redistribution, and in order to understand if the accumulation of Fe in the layers implanted with Zn or Be is related to the defects, we have implanted Ne with a dose of 10^{15} cm^{-2}, and a high energy of 1 MeV. After annealing, the depth profile of Fe (Figure 2) shows a slight accumulation of Fe beyond the implanted layer between the depths of 1.6 and 1.9 μm ($\bar{R}_p = 1.37$ μm).

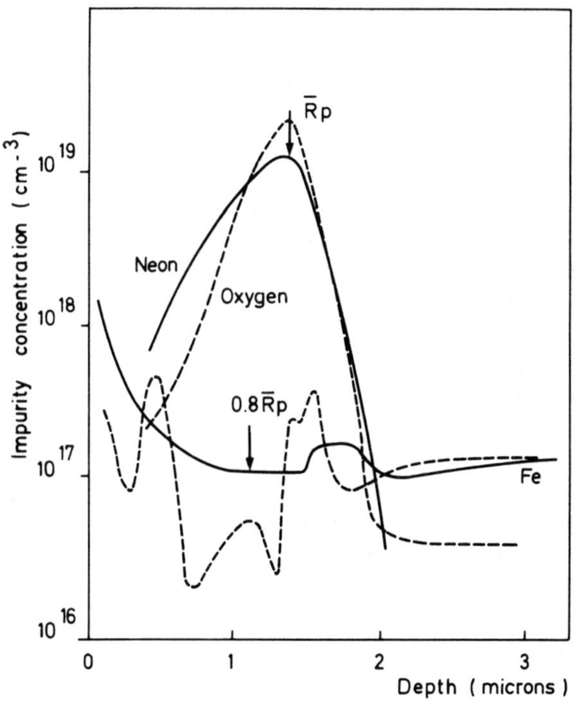

Figure 2 Depth profiles after implantation at 1 MeV with 10^{15} cm^{-2} and an anneal at 750°C for 15 min. Solid line—Ne and Fe profiles after a Ne implantation; dashed line—O and Fe profiles after an O implantation.

OXYGEN IMPLANTS

Our previous results on O-implanted InP [9] have shown that the implanted O impurities do not have an electrical behaviour in the undoped InP substrates. It was thus useful to compare the influence of an O implantation to that of the Ne implantation since both of these impurities should be neutral in the InP crystals. Figure 2 shows the O and Fe depth profiles in O-implanted and annealed material for a dose of 10^{15} cm^{-2} and an energy of 1 MeV. The O profile exhibits a nearly Gaussian profile, so at 750°C neither the implanted O nor the Ne diffuse in InP.

Despite the O profile not changing, in accordance with furnace annealing, a complicated structure appears in the Fe profile as shown in the Figure 2. The strong perturbation of Fe distribution is seen as a dip surrounded by two peaks. It can be thought of as a depletion of Fe, in the centre, and an accumulation on either side. In the Fe-depleted layer, there is no single valley, but a single peak, centred at $0.8\bar{R}_p$ ($\bar{R}_p$ being 1.37 μm), with two valleys on each side. The O profile seems to be like the Fe profiles in Fe-implanted and annealed material, as published by Gauneau *et al.* [10]. These stratified structures of Fe were observed only after implantation of O or Fe.

IMPLANTATION OF DONORS

Silicon and Se are donor impurities in InP. We have implanted these impurities in Fe-doped InP and, after annealing, we have measured profiles of implanted impurities and Fe. Figure 3 shows the depth profiles obtained after a Se implantation. The implantation was 400 keV with a dose of 5×10^{14} cm^{-2}. On the Fe profile we see an accumulation of Fe at the surface and an Fe peak located just beyond the implanted layer. The electron carrier profile as measured by differential Hall measurements [11] is also shown in Figure 3. We note that, beyond $\bar{R}_p$, the electrical and Se profiles are identical. Increasing the Se always induced an Fe accumulation beyond the implanted layer and, whatever the Se dose, the peak of the Fe seems to be located just at the interface between the n-doped layer and the SI substrate. After Si implantation, we have determined the same shape of the Fe profiles. Also, it seems to us that the accumulation of Fe near the n/SI interface is probably due to the fact that these two impurities, Si and Se, are donor impurities. Lastly, after annealing, Se takes the place of a P atom (Se_P), and Si takes the place of In atoms (Si_{In}). The Fe redistribution is probably not due to the location of impurity in the lattice (Si_{In}, Se_P).

Discussion

We have clearly demonstrated that Fe, a major impurity in SI InP, after annealing at high temperature, redistributes itself in the implanted layer. This Fe redistribution is strongly influenced by the nature of the implanted impurity and not only by the surface and the damage induced by the implantation. We can show three groups of impurities in which the Fe redistributions are similar after Fe implantation at high dose: (a) Si, Se and Ne; (b) Be and Zn; and (c) O and Fe.

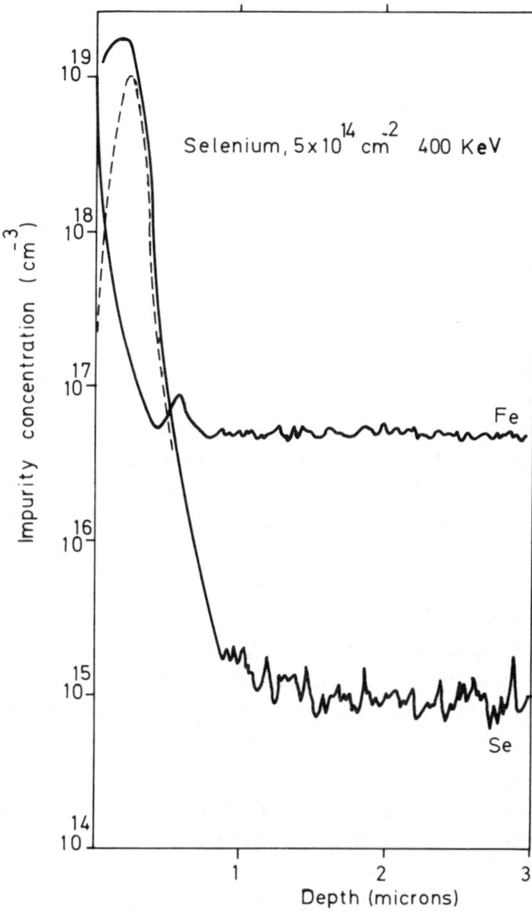

Figure 3 Depth profiles of Se and Fe after a Se implantation (400 keV, 5×10^{14} cm^{-2}) and an anneal at 750°C for 15 min. The dashed curve shows the electron carrier profile.

The Fe diffusion process in InP can be described by the interstitial-substitutional diffusion mechanism, as for Zn and Mn in GaAs [12, 13]. The primary stable Fe species are ionized Fe_i^+ interstitial donor and Fe_s^- substitutional acceptors on In lattice sites. The interstitial diffusion constant is much greater than the substitutional one. The overall diffusion constant of Fe, D_{Fe} is the concentration weighted average of the interstitial and substitutional diffusion constants, $D_{Fe_i^+}$ and $D_{Fe_s^-}$:

$$D_{Fe} = \frac{[Fe_i^+]D_{Fe_i^+} + [Fe_s^-]D_{Fe_s^-}}{[Fe_i^+] + [Fe_s^-]}$$

Under 'normal conditions', the Fe substitutional acceptors are in equilibrium with the In vacancies and Fe interstitial donors. The equilibrium reaction can be expressed by:

$$Fe_s^- + 2h^+ \rightleftarrows Fe_i^+ + V_{In}$$

and then the concentration ratio of interstitial donors to substitutional acceptors $[Fe_i^+]/[Fe_s^-]$ would vary as the square of the hole concentration p^2. Normally, this ratio is much less than one, and the diffusion coefficient can be approximated as:

$$D_{Fe} \simeq \frac{[Fe_i^+]}{[Fe_s^-]} \times (D_{Fe_i^+} + D_{Fe_s^-})$$

It should then be true that

$$D_{Fe} \text{ in } p^+\text{-type} > D_{Fe} \text{ in p-type} > D_{Fe} \text{ in SI type} > D_{Fe} \text{ in n-type}$$

In an implanted layer, it is possible to have 'abnormal conditions'. The layer is perturbed (maximum damage density is at about $0.8\bar{R}_p$) and most Fe impurities became interstitials ($[Fe_i^+] \gg [Fe_s^-]$) and these interstitials can be complexed with some defects (D) inducing a modification of the normal diffusion constant of Fe. As a consequence of these possible variations of the Fe diffusion constant, we have tried to explain the Fe redistribution occurring in an implanted layer in an Fe-doped substrate.

For Si and Se implantations, the Fe redistribution is identical, so this indicates that Fe redistribution is independent of the implanted impurity's position in the lattice. The implanted layer is doped n-type. During annealing, the Fe_i^+ interstitials contained in the bulk diffuse quickly towards the surface and the implanted layer. As the diffusion constant is lower in the n-type layer than in the SI substrate, the Fe interstitials are slowed down at the edge of the n-type layer and as a consequence there is an Fe accumulation at the n/SI interface.

For Ne implantations, the same mechanism can explain the Fe redistribution, but in this case we must assume that the implanted layer is less resistive (n^--type) than the substrate. This assumption is difficult to verify as we have no way to affirm that the Ne implanted layer has strictly the same resistivity as the high resistivity substrate.

For an O-implanted layer, there is a mechanism which could explain the shape of the Fe profile. This could happen if, in the region of maximum defect density, the substitutional Fe is changed into another faster diffusing species. During bombardment, at $0.8\bar{R}_p$, the damage density is high and most Fe impurities became interstitials either directly or indirectly through collision. We propose that these interstitials or complexes of interstitials with defects ($Fe \times D$) flow out from $\bar{R}_p$ towards the surface and the bulk faster than substitutional Fe could flow in, leaving a net deficit of impurities. Qualitatively, this explanation is obviously too crude, but it can explain the shape of the Fe profile. Baruch *et al.* [14, 15] have previously used this explanation to explain the complex redistribution of B in B-doped Si when the substrate is bombarded with protons at high temperature. The shapes of B profiles are similar to those shown in Figures 2 and 4: a dip surrounded

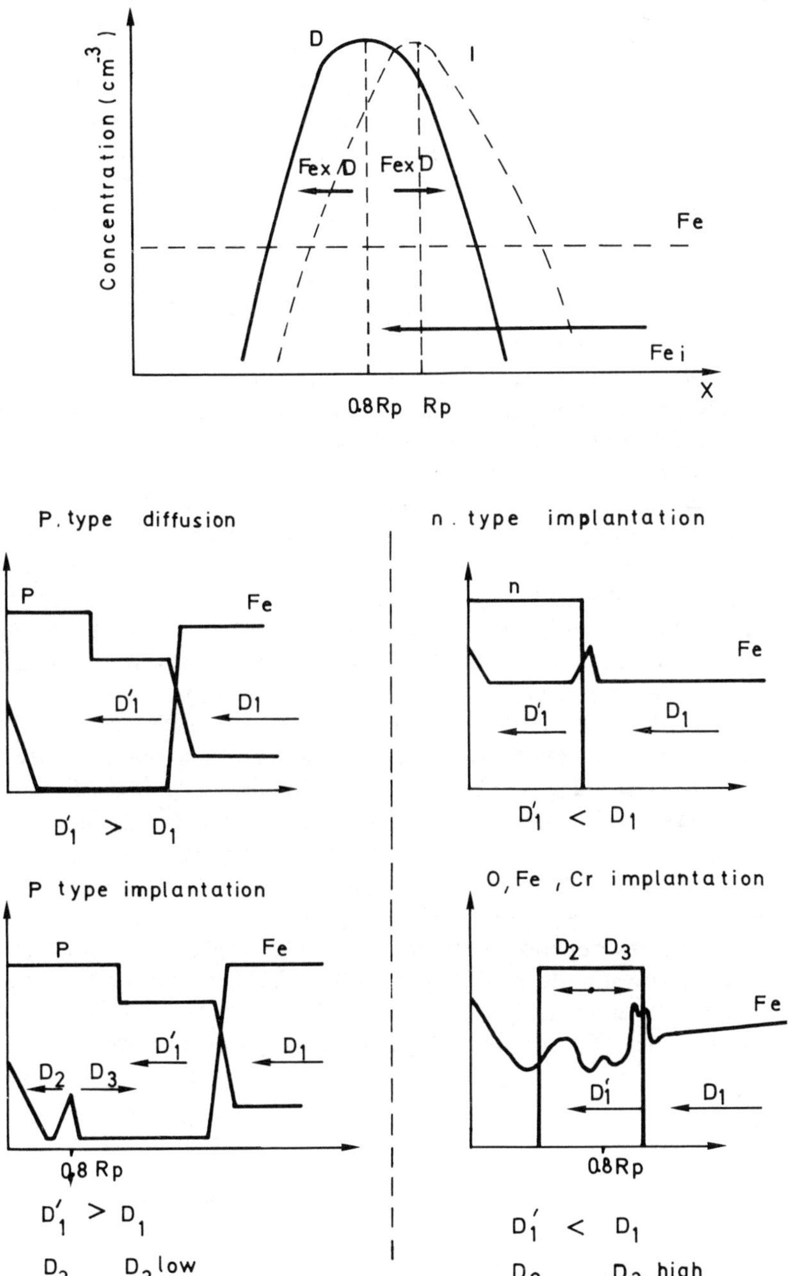

Figure 4 Three flow model: Fe_i flows from the bulk towards the surface (D_1 is the diffusion constant in the bulk, D'_1 is the diffusion constant in the doped layer); $Fe \times D$ flow from the damaged layer towards the surface, D_2 is the diffusion constant of $Fe \times D$; $Fe \times D$ flow from the damaged layer towards the bulk, D_3 is the diffusion constant of $Fe \times D$.

by two shoulders. In the depleted layer there is not a single valley but a single peak centred at $0.8\bar{R}_p$, with two valleys on each side. In the case of an implantation in InP, it is presumed that Fe flows out from the damaged region as Fe_i (or $Fe_i \times D$) resulting in a net depletion of total Fe. In the undamaged part of the crystal, Fe_i returns in an equilibrium substitutional configuration. In addition to these two interstitial flows coming from the implanted layer, there must be an interstitial flow from the bulk towards the implanted layer and this should be responsible for the accumulation of Fe at a greater depth (see, for example, Figure 2). This mechanism with three flows should be similar to that predicted by Armand and Martinez [16] for Cr in GaAs when the substrate is implanted with O.

When we have a p-type layer in a SI substrate, the Fe_i^+ interstitials contained in the bulk diffuse, towards the surface for annealing, and, in addition, during the diffusion of the acceptor impurities (implanted or not), as explained in [7], we would have an exchange between Zn_{In} or Be_{In} and Fe_{In}, Fe_{In} becoming Fe_i, and so $[Fe_i]/[Fe_{In}] \gg 1$, and the Fe diffusion constant can be approximated to $D_{Fe_i^+}$. Iron diffuses quickly towards the surface. As a result, there is an Fe depletion in the acceptor impurity-rich layer and an accumulation at the surface. This mechanism can explain the Fe profiles in the case of a Zn diffusion into Fe-doped InP (Figure 1a). If the impurities are first implanted and then diffused, the same mechanism is observed, but in addition there should be a formation of Fe-defect ($Fe \times D$) complexes, with a low diffusion constant, centered at $0.8R_p$.

We have tried to explain our complex results of redistribution of Fe after implantation. A model with three flows of Fe is summarized in Figure 4. The first Fe flow comes from the bulk towards the surface, and the Fe_i diffusion constant depends strongly on the doping level. The second flow of Fe_i or $Fe \times D$ comes from the damaged layer and goes towards the surface. The third flow of Fe_i or $Fe \times D$ comes from the damaged layer and goes towards the bulk. We can therefore conclude by proposing that the nature of the $Fe \times D$ complexes depends on the nature of the implanted impurities. Thus, Zn and Be create complexes with a low diffusivity; O and Fe create complexes (or interstitials) with a high diffusivity; and Se, Si and Ne do not create complexes of $Fe \times D$, or if they do, only in low concentrations.

Conclusions

We have shown that after implantation and annealing, the Fe redistribution in Fe-doped InP depends strongly on the nature of the implanted impurity and not only on the surface and the damage induced by the implantation. Three groups of impurities in which the iron redistributions are similar have been determined: (a) Be and Zn; (b) Si, Se and Ne; and (c) O and Fe. A unifying model for explaining the Fe migration has been tried. The Fe could migrate with three flows, these flows depending on the doping level of the crossed region and the nature of the implanted layer. The Fe redistribution during post-implantation annealing in bulk Fe-doped InP can lead to drastic variations in the electrical properties of the implanted layers and this explains the difficulty in interpreting electrical measurements after implantation in Fe-doped substrates.

References

1. Huber A. M., Morillot G., Linh N. T., Favennec P. N., Deveaud B. and Toulouse B. (1979). *Appl. Phys. Lett.*, **34,** 858
2. Evans C A., Deline V. R., Sigmon T. W. and Lidow A. (1979). *Appl. Phys. Lett.*, **35,** 291
3. Favennec P. N. and L'Haridon H. (1979). *Appl. Phys. Lett.*, **35,** 699
4. Wilson R. G., Vasudev P. K., Samba D. M., Evans C. A. and Deline V. R. (1980). *Appl. Phys. Lett.*, **36,** 215
5. Oberstar J. D., Streetman B. G., Baker J. E. and Williams P. (1981). *J. Electrochem. Soc.*, **128,** 1814
6. Obserstar J. D., Streetman B. G., Baker J. E. and Williams P. (1982). *J. Electrochem. Soc.*, **129,** 1312
7. Favennec P. N., Henry L., Salvi M., Huber A. M. and Morillot G. (1983). *Solid St. Electron.*, **26,** 771
8. Favennec P. N., Henry L., Gauneau M., L'Haridon H. and Pelous G. (1980). *Electron. Lett.*, **16,** 832
9. Favennec P. N., L'Haridon H., Salvi M. and Gauneau M. (1984). *Rev. Phys. Appl.*, **19,** 33
10. Gauneau M., L'Haridon H., Rupert A. and Salvi M. (1982). *J. Appl. Phys.*, **53,** 6823
11. Bellec M., Le Cleach X., Favennec P. N. and L'Haridon H. (1983). *Nuclear Inst. Meth.*, **209,** 629
12. Van Gurp G. J., Van Ommen A. M., Boudewijn P. R., Oosthoek D. P. and Willemsen M. F. C. (1984). *J. Appl. Phys.*, **55,** 338
13. Kanber H., Feng M. and Whelan J. M. (1984). *J. Appl. Phys.*, **55,** 347
14. Baruch P., Monnier J., Blanchard B. and Castaing C. (1975). *Inst. Phys. Conf. Ser.*, **23,** 453
15. Baruch P. (1977). *Inst. Phys. Conf. Ser.*, **31,** 126
16. Armand C. and Martinez A. Private communication

The Heat Treatment of Fe-Doped InP Substrates for the Growth of Higher Purity $Ga_{0.47}In_{0.53}As$ by MBE

A. S. Brown, S. C. Palmateer, G. W. Wicks, L. F. Eastman, A. R. Calawa*
and C. Hitzman†

School of Electrical Engineering, Cornell University, Ithaca, NY 14853, USA

Abstract

Impurity outdiffusion, in particular Fe, Mn and Cr, from Fe-doped InP into GaInAs epitaxial layers during MBE (molecular beam epitaxy) growth can account for as much as two-thirds of the carrier concentration and can reduce electron mobilities by as much as 40%. By heating the InP before growth and then polishing to remove the impurities which have been gettered at the surface, epitaxial layers with unintentional carrier concentrations less sensitive to substrate temperature and higher mobilities can be obtained.

Fe-doped InP has been annealed under pure H_2, a 92% Ar–8% H_2 mixture, and a PH_3 ambient to relate the efficiency of impurity gettering to the ambient heating. Hall measurements show that the carrier mobilities, in epitaxial layers grown on InP annealed in a pure H_2 ambient, increase more than in those layers grown on InP heated in an Ar–H_2 mixture. This suggests that the formation of phosphorus vacancies aids the surface gettering of outdiffusing impurities.

Introduction

It has been shown that Fe, Cr, Mn and possibly Si and S outdiffuse from Fe-doped InP into $Ga_{0.47}In_{0.53}As$ layers grown on the InP and severely degrade GaInAs electron mobilities [1]. The growth of high purity GaInAs by MBE requires limiting the outdiffusion of impurities from the InP substrates on which the epitaxial layers are grown. A substrate treatment process which involves gettering impurities at the InP surfaces by heating, and then removing the surfaces by polishing, has been found to inhibit impurity outdiffusion during MBE growth [1, 2].

It is important to gain an understanding of the mechanisms which cause impurities to accumulate at the InP surfaces during heating for a number of reasons. First, in order to exploit fully the heat treatment process and prepare the purest substrate possible, the conditions which cause impurity accumulation must be known so that they can, within reason, be maximized. Second, knowledge of the driving forces for the movement of impurities in the substrate may, by analogy, offer insight into ways to control impurity movement and incorporation during epitaxial growth.

* Lincoln Laboratory, MIT, Lexington, Massachusetts 02173, USA.
† Charles Evans Associates, San Mateo, California 94402, USA.

Preliminary investigations of the growth of GaInAs on heat-treated InP:Fe indicated that electron mobilities show greater improvement in GaInAs grown on InP in a H_2 ambient, than in epitaxial layers grown on InP heated in an Ar(92%)–H_2 ambient [1]. The predominance of Ar limits the P_2 loss, as evidenced by less decomposition of the InP during annealing. This seemed to indicate that the heat treatment process is more efficient, i.e. more impurities are accumulated at the surfaces, when a greater loss of P_2 from the surface occurs. To determine the magnitude of this effect, InP:Fe samples were heated under H_2, Ar–H_2, and PH_3 ambients and examined by SIMS (secondary ion mass spectrometry) and PL (photoluminescence).

Experimental

InP:Fe samples were first chemi-mechanically polished in a bromine 0.5% methanol solution to a thickness of approximately 500 μm. Immediately before heating they were degreased with acetone, methanol and de-ionized water, and then loaded proximately into a slotted graphite boat. The samples were heated at either 600°C for six hours or 550°C for 24 hours in a H_2, Ar(92%)–H_2(8%), or PH_3 ambient.

Photoluminescence

The samples were optically excited at 4 K with the 6328 Å line of a He–Ne laser with a power density of 500 mW cm^{-2}. The PL was detected through a 0.5 m grating spectrometer by a cooled S-1 photomultiplier. Figure 1 shows the PL spectra for four samples described below:

1. Annealed, H_2 ambient, 600°C, 6 hours (a).
2. Annealed, Ar–H_2 ambient, 550°C, 24 hours (b).
3. Annealed, PH_3 ambient, 600°C, 6 hours (c).
4. Unheated (d).

The samples are listed with reference to the unannealed sample in decreasing order of degradation of surface morphology as expected with a deviation from stoichiometry due to P_2 loss [3].

From Figure 1 it can be seen that the luminescence intensity of the exciton line at 1.414 eV decreases progressively with increasing P_2 loss during annealing. A factor of five decrease in the emission intensity from the unheated to the hydrogen heated sample is observed. All of the samples exhibit a peak at ~1.375 eV which is due to either C or Zn [4]. The most interesting characteristic is the presence of a peak at ~1.391 eV in all the heated samples, but which is absent in the unheated sample. This peak has been observed in vapour-grown, undoped InP and was seen to increase in intensity when the

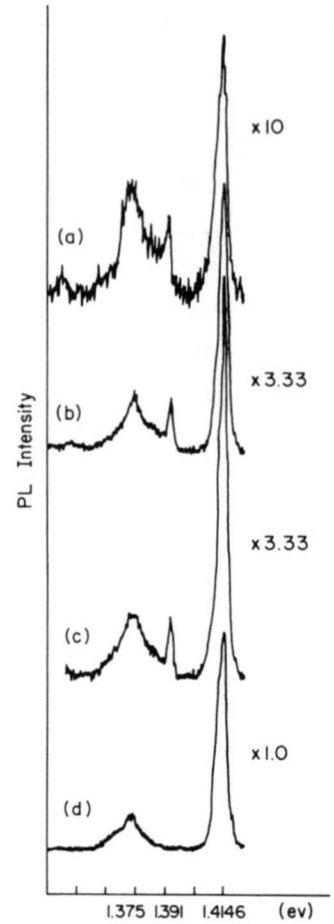

Figure 1 Photoluminescence intensity versus energy for annealed and unannealed InP samples.

sample was heated in hydrogen [5]. A similar transition observed at 1.394 eV has been attributed to an exciton bound to a deep acceptor [6].

Secondary ion mass spectrometry (SIMS)

SIMS measurements were used to determine the distribution of impurities in the heated samples. Figure 2 shows the concentration of Fe and Cr in the samples in the same order as Figure 1. It can be seen that the surface accumulation of these impurities also increases with increasing P_2 loss during annealing. While heating under a PH_3 ambient causes no observable redistribution of these impurities, heating under H_2 causes concentration increases by over two orders of magnitude at the surface, decreasing monotonically over a depth of 2 μm or more.

Figure 3 shows scans of the same samples for Si and S. Here again there is extreme surface accumulation during the H_2 heating while no redistribution is observed for heating under PH_3. Also, the accumulation of these donors is almost twice as deep as that of the acceptors, indicating faster movement of these impurities. Absolute concentration values cannot be inferred from these scans. Changes from run-to-run do occur and may be an artifact of the SIMS measurement.

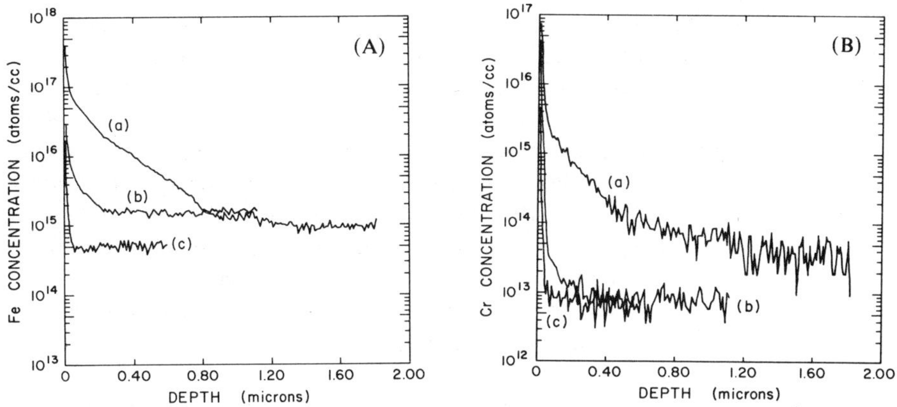

Figure 2 Concentration of (A) Fe and (B) Cr for InP: Fe anneals under conditions as described in the text.

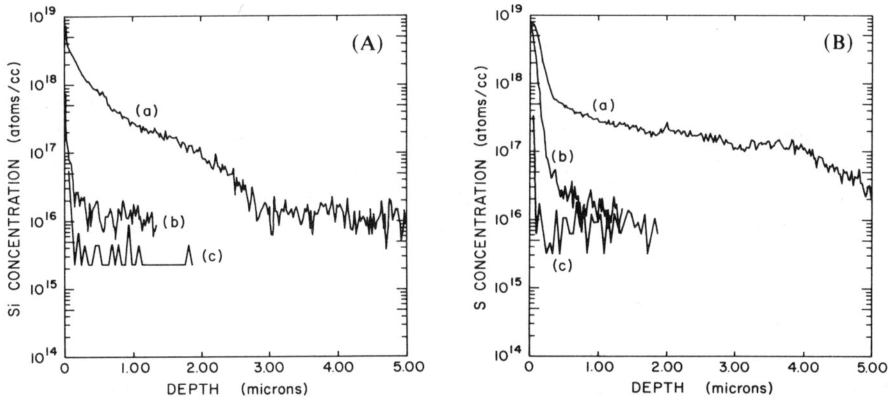

Figure 3 Concentration of (A) Si and (B) S for InP: Fe anneals under conditions as described in the text.

Discussion of results

The results of these experiments are that both acceptors and donors can be made to accumulate at the surface of InP. The amount of impurity accumulation is dependent on a deficiency of phosphorus at the InP surface. If there is no appreciable phosphorus deficiency, such as is the case for the InP annealed in a PH_3 ambient, then impurity accumulation is not observed, at least within the limits of our SIMS measurements. The presence of the 1.391 eV luminescence line in the PH_3 annealed sample indicates that some change is also occurring in that sample. A further observation is that the donor accumulation is two to three times as deep as that seen for the acceptors. It is tempting to suggest phosphorus vacancies are formed which are occupied by the impurities.

If the concentration profiles of the impurities, however, were fully phosphorus-vacancy concentration dependent then it is difficult to understand why the donor accumulation would be so much deeper into the crystal. One could postulate that the InP surface which remains within its solidus field also contains indium vacancies with a different diffusion rate than that of phosphorus vacancies, similar to the case suggested for GaAs [7]. The donor and acceptor movement could then be associated with the different vacancies.

Another possible mechanism for the surface accumulation is a thermodynamic equilibrium driving force rather than a kinetic driving force. The drastic loss of phosphorus during the hydrogen heat treatment may cause changes in the solubilities of the impurities such that the free energy is lowered by impurity movement to the InP surface.

Further investigation into these phenomena under more controlled conditions, such as anneals under a constant, known P_2 overpressure, should yield important information as to the mechanism controlling this process.

Conclusions

The accumulation of Fe, Cr, Mn, Si and S at the surfaces of annealed Fe-doped InP is observed to be dependent on the annealing ambient. Deep accumulation is produced by annealing under H_2, while no accumulation is observed for the PH_3 anneal. Photoluminescence measurements reveal a peak at 1.391 eV in all of the annealed samples, which is absent in the bulk InP.

ACKNOWLEDGEMENTS

We would like to thank J. D. Berry for invaluable technical assistance, E. J. Weaver for typing the manuscript, and J. Jorgensen for the artwork. This work was supported by the Army Research Office under contract no. DAAG29-82-K-011 and the Office of Naval Research under contract no. N00014-75-C0739.

References

1. Brown A. S., Palmateer S. C., Wicks G. W., Eastman L. F., Calawa A. R. and Hitzman C. Unpublished
2. Palmateer S. C., Schaff W. J., Galuska A., Berry J. D. and Eastman L. F. (1983). *Appl. Phys. Lett.*, **42 (2)**, 183
3. Samples were also annealed in an Ar–H_2 ambient at 600°C for six hours, however the surface showed less degradation than case (b)
4. Skromme B. J., Stillman G. E., Oberstar J. D. and Chin S. S. (1984). *Appl. Phys. Lett.*, **44 (3)**, 319
5. Williams E. W., Elder W., Astles M. G., Webb M., Mullin J. B., Straughan B. and Tufton P. J. (1973). *J. Electrochem. Soc.*, **120 (12)**, 1741
6. Fischbach J. U., Benz G., Stath N. and Pilkuhn M. H. (1972). *Solid-State Commun.*, **11**, 725
7. Rohdin H., Muller M. W. and Wolfe C. M. (1982). Technical Report ONR-82-5. Unpublished

Acceptor Redistribution and Optimization of Annealing Parameters in Si-Implanted Semi-Insulating GaAs

F. Vidimari

Telettra Spa, Vimercate (Milano), Italy

Abstract

A shallow Mg pile-up was found to be responsible for surface compensation in some samples of Si-implanted GaAs. It was believed to come from the GaAs substrate and to redistribute upon annealing. The process limiting step was probably due to surface state induced electric fields. However it was possible to minimize this effect by decreasing the annealing time to 1 minute. At proper annealing temperatures, implanted layers suitable for FET (field effect transistor) channels were obtained almost independently of the surface conditions.

Ion implantation of silicon into undoped LEC (liquid encapsulated Czochralski) semi-insulating GaAs can produce shallow FET channels with very abrupt interfaces which are the key to very low noise performance at microwave frequencies [1]. However, the potential reproducibility inherent to ion implantation was routinely obtained only at implant energies relatively higher than those required to achieve the above mentioned purposes. Typically, 100 keV Si implants, face to face annealed at 850°C, showed unsatisfactory doping uniformity from ingot to ingot and sometimes even from wafer to wafer. In this work some wafers coming from two GaAs ingots (undoped LEC semi-insulating) which gave comparable results only at energies above 150 keV, were implanted with a Si dose of 3×10^{12} ions cm^{-2} at 100 keV. Ingot A showed an apparent activation comparable to that obtained at higher energy ($\eta > 90\%$) and always higher than ingot B. The wafers, all adjacent, were heat treated under the following annealing conditions:

annealing type: face to face in $H_2 + AsH_3$
annealing temperature: 850°C or 900°C
AsH_3 pressure: from 1 to 24 torr
annealing time: from 1 to 30 minutes

Each wafer was cut into four parts and each part was then annealed simultaneously with another one belonging to the other ingot. Carrier concentration and Hall mobility profiles were taken from each sample by means of suitable gated structures. Chemical profiles of ^{52}Cr, ^{55}Mn, ^{24}Mg and ^{28}Si were obtained by SIMS (secondary ion mass spectrometry) [2]. The background concentrations were respectively: Cr $\sim 1\times10^{15}$ cm^{-3}; Mn $\sim$ 1–2 $\times10^{15}$ cm^{-3}; Mg $\sim$ 1–2 $\times10^{14}$ cm^{-3}; and Si $\sim 1\times10^{15}$ cm^{-3} for both ingots.

In spite of the lower background level, Mg showed a larger pile-up towards the GaAs surface than did the other acceptors in sample B. In sample A very little redistribution occurred. By using a simple model based on the charge neutrality equation, on the experimental data, and on the fact that η, the doping efficiency, which takes into account the fact that Si can occupy either Ga or As sites, depends on AsH_3 pressure [3], it was possible to demonstrate that Mg pile-up can determine the surface compensation and activation of low dose 100 keV Si implants into GaAs. The equation used was:

$$n(x) = \eta Si(x) - Mg(x) + p(x) - k(x)$$

where n(x) and p(x) are respectively the electron and hole concentrations, Si(x) and Mg(x) the Si and Mg concentrations, respectively, and k(x) the contribution due to other donors and acceptors. Since the samples were always n-type and k(x) was more than one order of magnitude lower than maximum Si and Mg concentrations, the last two terms of the equation could be neglected.

Figure 1 is a synthesis of the most notable effects detected after annealing experiments. Magnesium accumulated at the surface to concentrations as high as 10^{17} cm^{-3}. The extent of this accumulation was more pronounced for annealing temperatures of 850°C and for lower AsH_3 pressures. Furthermore short annealing times were effective in reducing the surface concentration. The data reported in Figure 1 must be taken on qualitative terms, since identical annealing conditions on samples of ingot B could give rise to different surface compensation. For comparison also, a Mg profile measured on a sample from ingot A is shown in Figure 1. The concentration spread of the other acceptors is represented by a shaded area.

Although Mg accumulations towards the surface have previously been detected both in bulk [4] and ion implanted [5] LEC semi-insulating GaAs, it is not yet clear where this element comes from. In this work, a contaminated environment could not supply the Mg because samples showing Mg pile-up were annealed simultaneously with samples that did not show it, even though the background levels were comparable; therefore in this case Mg must have come from the substrate. Should Mg be driven toward the surface according to kinetic limitations as proposed by Jordan [6] for Mn, one should expect to observe an increase of the pile-up by increasing the annealing temperature as well as the AsH_3 pressure, as a consequence of an increased generation of V_{Ga} near the surface. The experimental behaviour indicated an opposite trend. Moreover the characteristic vacancy diffusion lengths are well in excess of the depths observed for Mg accumulation; therefore the short range and the reverse temperature dependence of the Mg accumulation suggest that surface state electrostatic fields could be the limiting factor of the phenomenon. A similar hypothesis was made by C. M. L. Yee *et al.* [7] for Cr redistribution in GaAs.

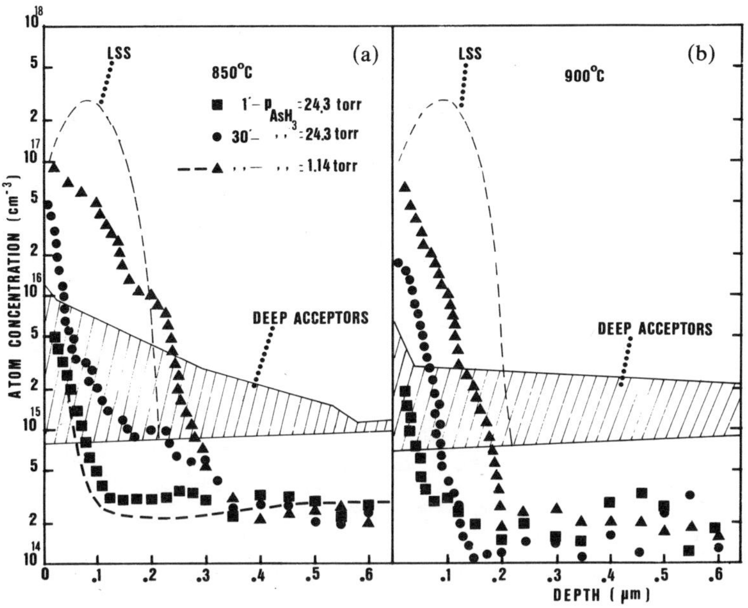

Figure 1 Magnesium profiles in Si ion implanted LEC undoped semi-insulating GaAs annealed at different AsH_3 pressures (a) at 850°C and (b) at 900°C. Ingot A (——); ingot B (■ ● ▲).

As to the different behaviour of the materials of ingots A and B with respect to Mg redistribution, it is impossible, on the basis of the available data, to make any sound hypothesis, even though the concentration of some structural defects could differentiate the surface properties of the two ingots. However, it was possible to reduce significantly Mg accumulation and its related effects, by annealing the GaAs wafers either at 850°C or 900°C for 60 seconds (Figure 1).

Carrier concentration and Hall mobility profiles were measured on samples annealed at 850°C and 900°C for 60 seconds. Typical results are shown in Figure 2. Though the doping profiles did not differ significantly in the two cases, the samples annealed at the lower temperature showed a degraded mobility towards the interface, probably associated with implantation damage not completely annealed out; however, 900°C heat treatments gave good mobilities ($\mu > 4000\ cm^2\ V^{-1}\ s^{-1}$) throughout the whole layer. Furthermore, unlike the behaviour observed for longer annealing times, only minor modifications could be introduced on the layer properties by different AsH_3 pressures.

Short annealing times and high temperatures (60 s, 900°C) seem therefore to be able to guarantee reproducible results on low dose and shallow Si implants into LEC undoped semi-insulating GaAs, thus avoiding undesirable impurity redistributions and giving quality layers for low noise microwave devices.

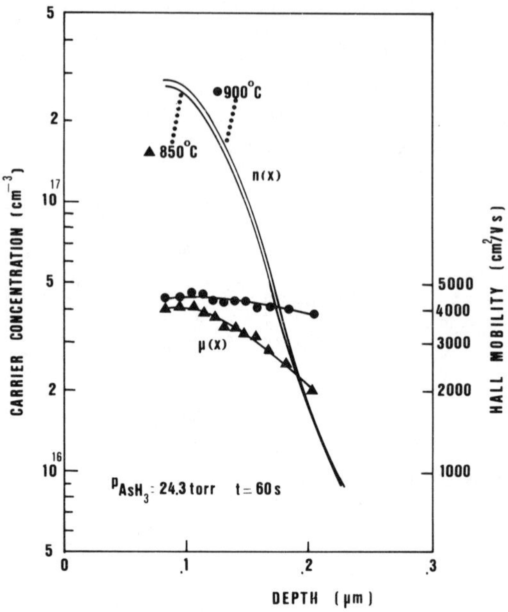

Figure 2 Carrier concentration and mobility profiles of 100 keV ($N_d = 3 \times 10^{12}$ cm^{-2}) Si implanted GaAs annealed for 60 s at 850°C and 900°C (ingot B).

ACKNOWLEDGEMENTS

The author wishes to thank CISE Spa for allowing him to complete this work at CISE facilities and especially F. Crociani and C. Angione for having made this work possible by their valuable technical support.

References

1. Feng M., Eu V. K., Kamber H., Watkins E., Schellenberg J. M. and Yamasaki H. (1982). *Appl. Phys. Lett.*, **40,** 802
2. Charles Evans and Associates, San Mateo, California
3. Vidimari F. (1982). In *Semi-insulating III–V Materials, Evian,* p. 300 (S. Makram-Ebeid and B. Tuck, eds.); Shiva Publishing, Nantwich, UK
4. Huber A. M., Morillot G., Merenda P. and Perrocheau J. (1980). *Inst. Phys. Conf. Ser.*, **56,** 579
5. Huber A. M. Private communication
6. Jordan A. S. (1982). In *Semi-insulating III–V Materials, Evian,* p. 253 (S. Makram-Ebeid and B. Tuck, eds.); Shiva Publishing, Nantwich, UK
7. Yee C. M. L., Fedders P. A. and Wolfe C. M. (1983). *Appl. Phys. Lett.*, **42,** 377

SiO_xN_y Capped Annealing for Si-Implanted GaAs

M. Kuzuhara, H. Kohzu, T. Tsuji and A. Higashisaka

Microelectronics Research Laboratories, NEC Corporation, 4-1-1, Miyazaki, Miyamae-ku, Kawasaki 213, Japan

Abstract

Enhanced electrical activation of Si implants in GaAs is demonstrated by using silicon oxynitride (SiO_xN_y) encapsulant. Annealing with SiO_xN_y encapsulant having ~1.75 refractive index results in higher electrical activation than that obtained with either SiO_2 or Si_3N_4 encapsulant. A maximum electron concentration of 3×10^{18} cm^{-3} is obtained after SiO_xN_y capped 850°C, 15 min annealing for a 1.4×10^{14} cm^{-2} dose. Auger depth profiling studies suggest that the enhanced activation can be ascribed to the generation of an optimized concentration of Ga vacancies caused by Ga outdiffusion during annealing.

Ga outdiffusion from encapsulated GaAs was first identified by Gyulai *et al.* [1], and has been considered to be detrimental in effectively activating the implanted dopants in GaAs. Therefore, an oxygen-free Si_3N_4 encapsulant has been proposed to be preferable in suppressing Ga outdiffusion during annealing [2–4]. It is obvious that an excessive amount of Ga vacancy formation might produce a highly compensated or resistive layer near the GaAs surface. However, an improvement in electrical activation can be anticipated when the generated Ga vacancy concentration is optimized to enhance the substitution of implanted dopants into Ga lattice sites. In this case, however, the implanted species are limited to Ga site substitutional dopants, such as Si, Sn, Mg and Zn.

In this study, a ternary SiO_xN_y dielectric film has been introduced as an annealing encapsulant to investigate the effect of Ga vacancy formation on the annealing behaviour of Si implants in GaAs.

SiO_xN_y films were deposited by the pyrolitic reaction of the SiH_4–NH_3–O_2 system at a substrate temperature of 700°C. The film compositions were controlled by varying the O_2 flow rate, while keeping the SiH_4 and NH_3 flow rates constant. The deposited SiO_xN_y films were characterized by ellipsometry and Auger electron spectroscopy for determining thickness, refractive index and O/N atomic ratio in the film. Typical results are listed in Table 1.

Si-implanted GaAs samples (5×10^{12} cm^{-2}, 100 keV), encapsulated with SiO_xN_y films with various oxygen contents, were annealed at 850°C for 15 min in a flowing N_2 atmosphere. The electrical activation reaches 87% for an SiO_xN_y encapsulant with ~1.75 refractive index, while the electrical activations for SiO_2 ($n = 1.46$) or Si_3N_4 ($n = 2.03$) encapsulants are 58% and 31%,

Table 1 *SiO_xN_y film properties and their AES results*

Ref. index	O/N ratio	AES Ga signal
1.46	∞	A, B
1.49	4	A, B
1.53	3	B only
1.59	2	n.d.
1.66	1.2	n.d.
1.75	0.6	n.d.
1.83	0.35	n.d.
2.03	~0	n.d.

A—detected at the film surface; B—detected in the film layer; n.d.—not detected.

respectively (Figure 1). Similar results were also observed for the case of high dose Si implants (7×10^{13} cm^{-2}, 150 keV). A maximum electron concentration of 2.5×10^{18} cm^{-3} with 70% activation is obtained after SiO_xN_y (n = 1.75) capped annealing, as shown in Figure 2, where the result for the Si_3N_4 encapsulant is also depicted for comparison. Furthermore, a peak concentration of 3×10^{18} cm^{-3} has been achieved for a 1.4×10^{14} cm^{-2} dose.

In order to examine the Ga outdiffusion extent, Auger depth profiling studies were performed for annealed SiO_xN_y/GaAs structures with various oxygen contents. As listed in Table 1, the Ga signal can be detected at the film surface and/or within the film layer for SiO_xN_y films with a refractive index below 1.52. However, no Ga signal can be detected within the Auger

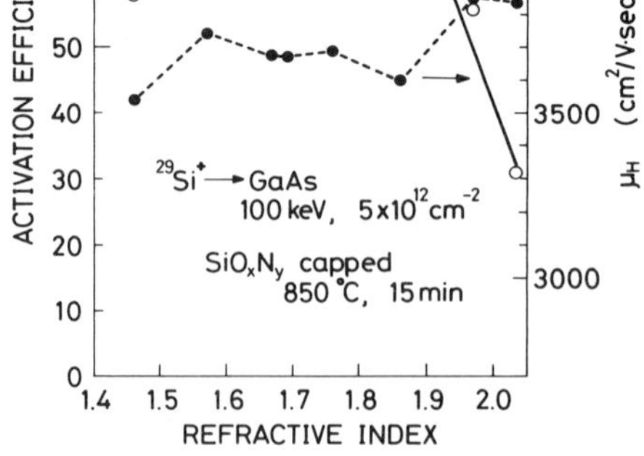

Figure 1 Electrical activation dependence on SiO_xN_y encapsulant refractive index.

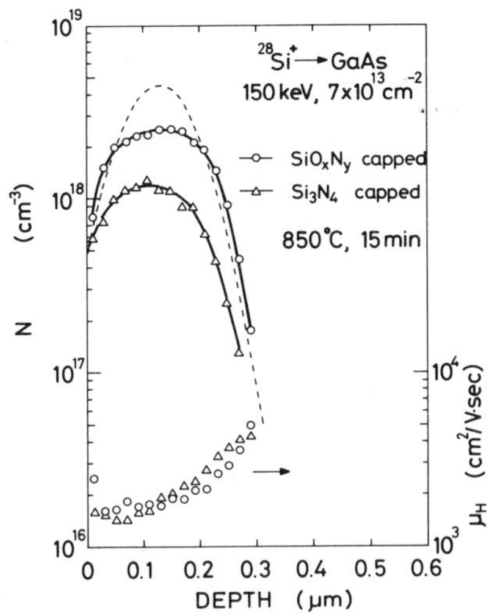

Figure 2 Carrier concentration profiles for Si implants in GaAs (7×10^{13} cm^{-2}, 150 keV) annealed with SiO_xN_y encapsulant (n = 1.75) and with Si_3N_4 encapsulant (n = 2.03).

detection limit (~0.5 at. %) for films with refractive index over 1.59. These results suggest that the amount of Ga outdiffusion can be gradually depressed by decreasing the O/N ratio in the SiO_xN_y encapsulant. Although no Ga signal has been observed in the SiO_xN_y (n = 1.75)/GaAs structure due to the Auger detection limit, some concentration of Ga vacancies must be produced in the GaAs substrate caused by a controlled, but not excessive, amount of Ga outdiffusion during annealing.

The SiO_xN_y encapsulations were also examined for other dopants, such as Sn, Mg and S. As expected, SiO_xN_y (n = 1.75) capped annealing led to enhancement in activation for Sn and Mg implants, while no enhanced activation was observed for S implants, because S is an As site substitutional dopant in GaAs. These results further support the concept that the enhanced activation of Si implants in annealed SiO_xN_y capped GaAs can be ascribed to the generation of an optimized amount of Ga vacancies caused by Ga outdiffusion during annealing, and not to a thermal conversion of the substrate nor a Si indiffusion from the encapsulant.

In conclusion, favourable annealing conditions for Si-implanted GaAs, where the Si substitution into Ga lattice sites is enhanced, has been established by adjusting the ternary SiO_xN_y film composition so as to optimize the concentration of Ga vacancies generated in the GaAs substrate during annealing.

ACKNOWLEDGEMENTS

The authors would like to thank Dr Y. Takayama for his continuous advice and encouragement during the course of this work.

References

1. Gyulai J., Mayer J. W., Mitchell I. V. and Rodriguez V. (1970). *Appl. Phys. Lett.*, **17,** 332
2. Donnelly J. P., Lindley W. T. and Hurwitz C. E. (1975). *Appl. Phys. Lett.*, **27,** 41
3. Inada T., Miwa H., Kato S., Kobayashi E., Hara T. and Mihara M. (1978). *J. Appl. Phys.*, **49,** 4571
4. Vaidyanathan K. V., Helix M. J., Wolford D. J., Streetman B. G., Blattner R. J. and Evans Jr C. A. (1977). *J. Electrochem. Soc.*, **124,** 1781

Annealing Behaviour of Magnesium-Implanted InP

P. J. Topham

Plessey Research (Caswell) Limited, Allen Clark Research Centre, Caswell, Towcester, Northants NN12 8EQ, UK

Abstract

Iron doped, SI InP has been implanted at 400 keV with a dose of 10^{14} Mg^+ cm^{-2} either at room temperature or at 200°C. After annealing in phosphine the maximum sheet carrier concentration was found to be 4×10^{13} holes cm^{-2} for either implant. The room temperature implant required annealing at 670°C or greater to produce conductivity whilst the hot implant was activated at 600°C. The room temperature implant gave a peak carrier concentration of 8×10^{17} holes cm^{-3} which was slightly larger than that for a hot implant. The magnesium diffuses rapidly for annealing temperatures above 700°C with the hot implant diffusing more rapidly than the room temperature implant. A common peak in the diffusion profile is found to correspond to the damage profile observed by RBS (Rutherford back scattering).

Ion implantation offers a convenient and controllable method of doping InP with acceptors. The principal species used to date has been beryllium [1] and good activity was produced. Due to the toxicity of beryllium the alternative acceptor, magnesium, may be preferred. Initial work has shown that p-type layers can be produced by implanting magnesium [2, 3] although SIMS (secondary ion mass spectrometry) profiling has shown that diffusion is significant [4].

The InP substrates were obtained as (100) polished slices doped with iron ($\rho > 10^7$ Ω cm), also some zinc-doped ($p \sim 6 \times 10^{16}$ holes cm^{-3}) slices were used. A dose of 10^{14} Mg^+ cm^{-2} was implanted at 400 keV in a non-channelling direction into substrates at room temperature or heated to 200°C. Annealing was carried out in a PH_3 ambient for a time of ten minutes. Electrical contacts were made using sintered AuZn allowing Hall measurements to be made on the Fe-doped samples whilst the p-type substrates were used for electrochemical profiling. RBS and SIMS were carried out on various samples using 1.5 meV He^+ ions and 15 keV O^+ ions respectively.

The maximum sheet carrier concentration after annealing produced by implanting 10^{14} Mg^+ cm^{-2} was about 4×10^{13} holes cm^{-2} (Figure 1), whether the substrates were heated or not. However, the temperature dependence of annealing differed in that the samples implanted at room temperature required annealing at 670°C to produce any activity, with samples annealed at lower temperatures remaining high resistance, whilst samples implanted hot produced 6×10^{12} holes cm^{-2} when annealed at 600°C . The maximum activity from the room temperature implant was produced by annealing at 730°C

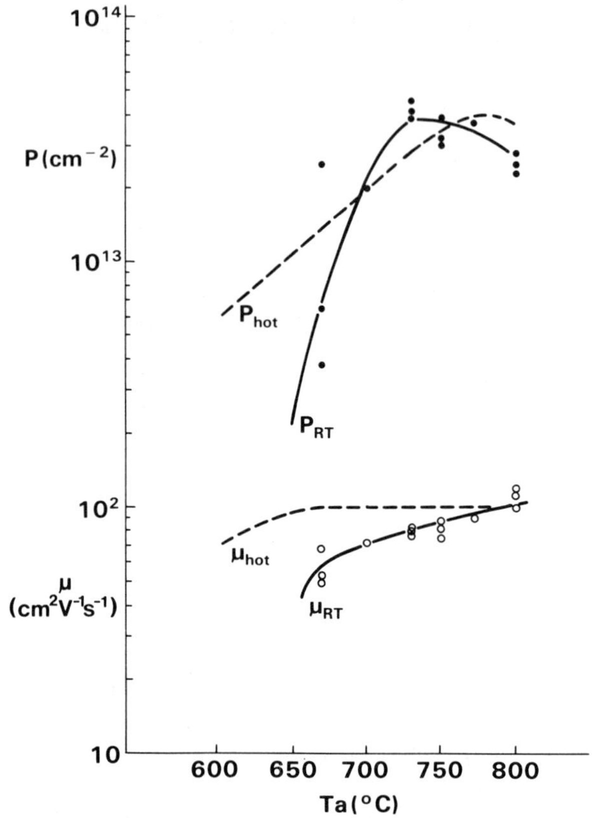

Figure 1 Sheet carrier concentration and mobility versus annealing temperature for InP implanted with 10^{14} Mg^+ cm^{-2} at 400 keV into substrates at room temperature (RT) or 200°C (hot).

whilst the hot implant required annealing at 780°C to obtain the best activation. The hot implant gives a mobility of about 100 $cm^2\,V^{-1}\,s^{-1}$ for most annealing temperatures but this mobility is only produced by the room temperature implant when annealed at 800°C.

The magnesium implanted at room temperature diffuses slightly when annealed at 700°C and the diffusion becomes severe when annealed at 800°C (Figure 2) with a tail at about 3×10^{16} Mg cm^{-3} extending to 3 μm deep. The iron was depleted from this region and accumulated at a depth of about 0.7 μm in a manner similar to that seen by Oberstar *et al.* [4]. The carrier concentration profile has the same shape as the impurity profile although the depth scales do not agree exactly. The peak carrier concentration from this anneal was 8×10^{17} holes cm^{-3} and similar values were obtained for 750°C and 700°C anneals. The correspondence between impurity and carrier profiles was noted at the lower annealing temperatures although the percentage of electrically active magnesium at any point on the profile falls with decreasing annealing temperature.

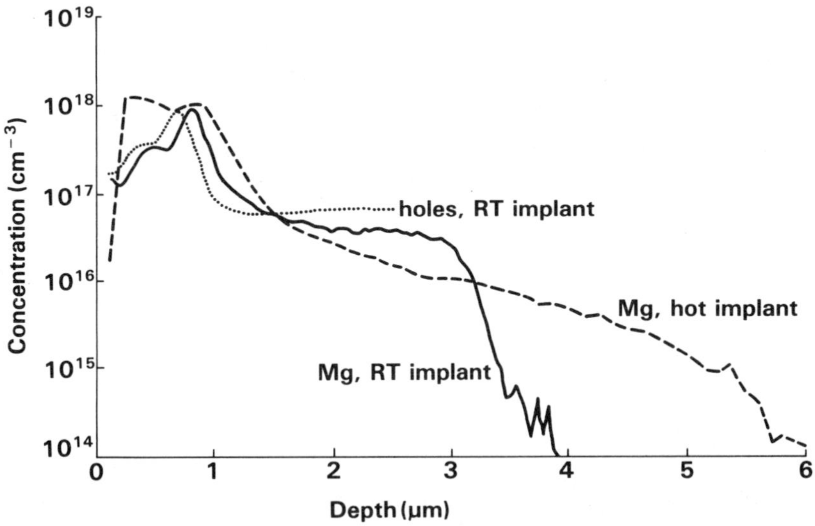

Figure 2 Impurity profiles in samples annealed at 800°C for ten minutes, with a corresponding carrier concentration profile for an implant into the zinc-doped substrate.

The magnesium implanted at 200°C has a similar impurity profile to the room temperature implant, with no diffusion occurring during implantation. However, the hot implanted magnesium begins to diffuse when annealed at 600°C for ten minutes and a significant diffusion tail is produced when annealed at 700°C. When annealed at 800°C the diffusion tail extends to about 5.5 μm in the zinc-doped samples although the diffusion front is not as sharply defined as for the room temperature implant. The highest carrier concentration from a hot implant was 6×10^{17} holes cm^{-3} obtained by annealing at 800°C, falling with decreasing annealing temperatures to a value of 1.5×10^{17} holes cm^{-3} for a 600°C anneal. Again the similarity in shape between impurity and carrier profiles was marked at all annealing temperatures.

The damage introduced by the room temperature implant does not amorphise the surface layer, as observed by RBS, but produces a broad disorder peak centred at a depth of about 0.3–0.4 μm. The damage profile is thus somewhat shallower than the projected range of about 0.55 μm. After annealing at 600°C for ten minutes the damage parameter X_{min} is returned to the single crystal value of 4.5% but significant dechannelling occurs between depths of 0.2 to 0.5 μm showing that disorder is still present in the region corresponding to the damage profile. This dechannelling is still discernable in samples annealed at 670°C, although the surface layer is indistinguishable from the virgin crystal. This damage peak at about 0.3–0.4 μm may explain the shallower of the two peaks visible in both the magnesium and carrier profiles (Figure 2) if magnesium is gettered on to the damage profile. The deeper peak is more difficult to explain and a tentative suggestion is that phosphorus is recoil implanted more deeply into the InP, creating a net deficiency of indium, allowing magnesium to occupy indium sublattice sites.

The influence of damage upon diffusion of the magnesium would be expected to be less for the hot implant and the impurity profile (Figure 2) does have less pronounced peaks than the room temperature implant.

In conclusion, it has been found that carrier concentrations approaching 10^{18} holes cm^{-3} can be produced by annealing implanted magnesium. Heating the substrates during implantation produces better electrical activity at annealing temperatures below 700°C but diffusion is more pronounced for the hot implant. However, for hot or room temperature implants the magnesium diffuses rapidly at temperatures above 700°C. A commonly occurring peak in the impurity profiles of annealed samples matches the damage profile observed by RBS.

ACKNOWLEDGEMENTS

The Electronic Engineering Department of Surrey University is thanked for assistance with the implants and N. Hoskins of Plessey Research (Caswell) Ltd. for the electrochemical profiles. This work has been carried out with the support of the Procurement Executive, Ministry of Defense, sponsored by DCVD.

References

1. Donnelly J. P. and Armiento C. A. (1979). *Appl. Phys. Lett.*, **34,** 96
2. Inada T., Taka S. and Yamamoto Y. (1981). *J. Appl. Phys.*, **11,** 6623
3. Donnelly J. P. and Hurwitz C. E. (1977). *Appl. Phys. Lett.*, **31,** 418
4. Oberstar J. D., Streetman B. G., Baker J. E. and Williams P. (1982). *J. Electrochem. Soc.*, **129,** 1320

The Characteristics of Carbon Implants into Semi-Insulating GaAs

W. M. Paulson and G. Tam

Semiconductor Research and Development Laboratories, Motorola Inc.,
5005 E. McDowell Road, Phoenix, Arizona 85008, USA

Abstract

Carbon implants into LEC (liquid encapsulated Czochralski) SI GaAs show p-type characteristics and activation efficiencies that depend upon dose. Implanted carbon is an efficient compensator of both Si and Se implants and reduces the Hall mobility of these n-type implants. Light dose carbon implants significantly reduce the pinchoff voltage of n-type implants because of compensation and also the p-n junction formed at the backside of the donor implant.

Carbon is one of the primary residual acceptor impurities in SI ingots. The effects of carbon are not obvious since carbon can be an amphoteric dopant. The properties of SI GaAs are dependent upon the concentration of both the deep levels (Cr, EL2) and shallow levels (C, Si). Carbon affects the properties of undoped SI GaAs [1] and directly affects the pinchoff voltage for n-type FET (field effect transistor) layers [2]. The carbon concentration is a function of the crystal growth process [3] and may vary from the seed to the tang of the crystal and, in addition, varies from ingot to ingot. Ion implanted carbon in SI GaAs consistently gives p-type layers [4–6] although the properties of the layers are dependent upon the dose, anneal temperature and the substrate material. Carbon implant layers were reported by Stringfellow *et al.* [7] to give p-type layers with about 50% activation. These references show the influence of carbon on the properties of SI GaAs and n-type implanted layers. The purpose of this paper is to characterize implanted carbon in undoped LEC SI GaAs. In addition, the influence of carbon on the properties of layers with implanted Si and Se will be reported.

The starting substrates were undoped LEC GaAs. Bulk SIMS (secondary ion mass spectrometry) analysis [8] gave a background carbon concentration of 7×10^{15} cm^{-3} for this ingot. Single energy C implants were done at 80 keV to doses of 3×10^{13}–5×10^{13} cm^{-2}. Multiple energy carbon implants were used to give approximately flat profiles to about 0.4 μm. The calculated LSS (Linhard, Scharff and Schiøtt) profiles are shown in Figure 1. To obtain the profile at 1×10^{16} cm^{-3} (profile a, Figure 1) we implanted ^{12}C at 50 keV and 1.14×10^{11} cm^{-2} and at 150 keV and 3.28×10^{11} cm^{-2}. The doses were scaled upwards to yield flat profiles at approximately 2×10^{16} and 5×10^{16} cm^{-3}, profiles b and c in Figure 1, respectively. The effects of the donor (Si or Se), the acceptor (C) and the interaction of C+Si (or C+Se) were obtained by physically masking parts of the wafer during the appropriate implantation step. Following the implant the wafers were encapsulated with sputtered Si_3N_4 and annealed at 850°C for 23 min. The wafers were evaluated using capacitance-voltage measurements to determine the pinchoff voltage and the carrier concentration profiles along with Hall measurements to obtain mobilities (μ), sheet resistances (R_s) and sheet carrier concentrations (N_s). SIMS analysis was used to measure the depth profiles for C, Si and Se.

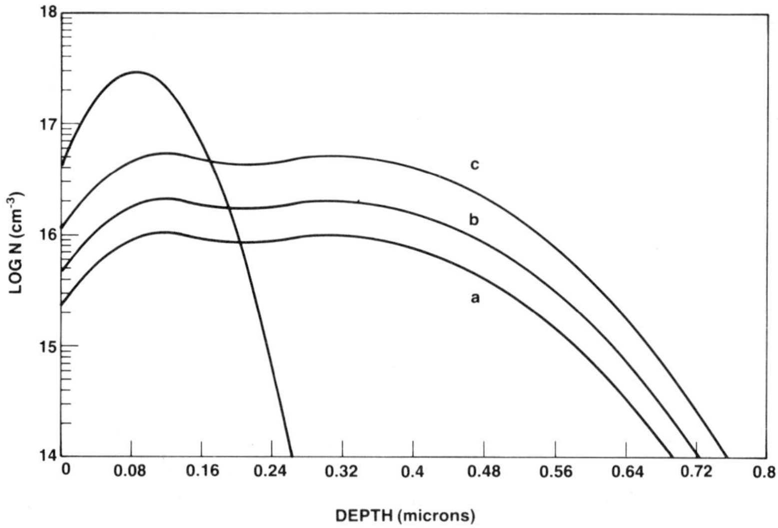

Figure 1 Calculated LSS profiles for the donor implant (Si or Se) and multi-energy carbon implants.

The Hall data for the carbon implants are summarized in Table 1. The implants at 80 keV are p-type. For the multi-energy implants the lightest two doses (a and b in Figure 1) showed high sheet resistances, which may be due to surface depletion effects and compensation by deep levels such as EL2. The profile at 5×10^{16} was p-type with a mobility of 335 $cm^2\,V^{-1}\,s^{-1}$. The activation efficiency is 49% if the surface depletion charge (assuming a surface potential of 0.48 for p-type GaAs) is added to the measured sheet carrier

Table 1 *Hall data for carbon implants into undoped LEC GaAs*

Energy (keV)	Dose (cm^{-2})	$R_s(\Omega/\square)$	μ ($cm^2\,V^{-1}\,s^{-1}$)	N_s (cm^{-2})
80	3.00×10^{13}	7652	305	2.67×10^{12}
80	5.00×10^{13}	8277	280	2.69×10^{12}
50	5.70×10^{11}	31 314	335	5.95×10^{11}
150	16.40×10^{11}			

concentration. Therefore, carbon implants in LEC SI GaAs exhibit p-type conduction but have activation efficiencies below 50%.

SIMS depth profiles of Si, Se and C were obtained for the samples that had the multi-energy carbon implants [8]. The Si and Se profiles were in good agreement with LSS calculations for wafers with or without the implanted carbon. Unfortunately, the background carbon concentration was between 5×10^{16} and 1×10^{17} cm^{-3}, which was larger than the implanted carbon concentration at the highest doses (profile c, Figure 1). Because of this background, the carbon peaks were difficult to resolve clearly. However, with the profile at 5×10^{16} cm^{-3} there was a distinct increase in carbon at a depth that agrees with the LSS calculations (see Figure 1). Therefore, we have assumed that carbon has not moved significant distances during the anneal step.

The effects of the carbon implants on the characteristics of the implanted Si or Se were measured on the same wafer. The pinchoff voltage for the Si implant was about 5.5 V, while for the Se implant was 4.5 V. The pinchoff voltage steadily decreased with increasing carbon concentration. The pinchoff voltages were 3.5, 2.0 and 0.8 V for the Si implant plus the carbon profiles at 1×10^{16}, 2×10^{16} and 5×10^{16} cm^{-3}, respectively. The compensation of the Se implants by carbon gave similar pinchoff voltages. The corresponding Hall data are shown in Figure 2 for the Si and Si+C samples. The symbols

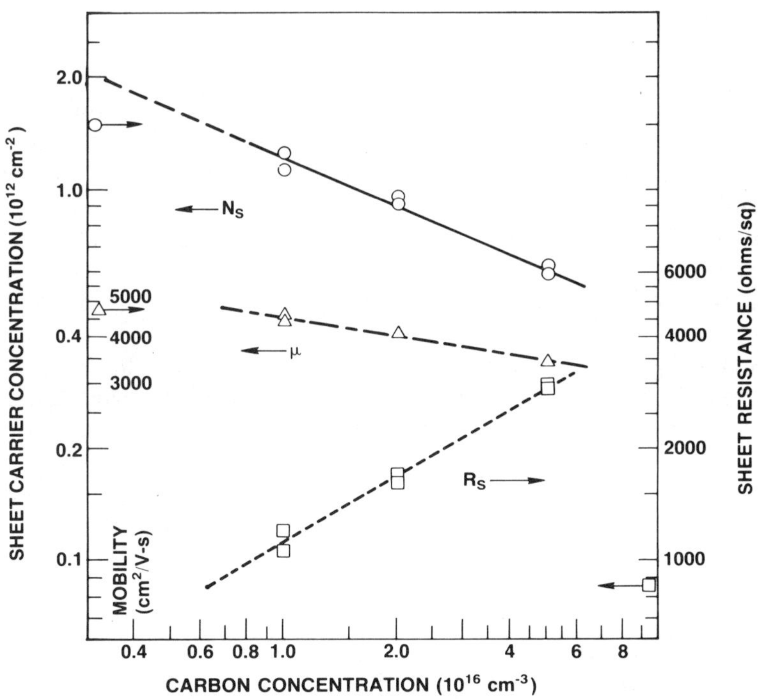

Figure 2 The sheet carrier concentration, sheet resistance and mobility for Si implants (on the axes) and Si+C implants.

with the arrows along the vertical axes are μ, N_s and R_s for the implants directly into the LEC substrates. R_s steadily increases while N_s and μ decrease with increasing carbon concentrations. The same trend in the Hall data was observed for the Si implanted samples.

The carbon implants into undoped LEC GaAs were clearly p-type but the activation varied from 50% for light dose implants ($<1\times10^{12}$ cm^{-2}) to less than 10% for higher dose implants ($>1\times10^{13}$ cm^{-2}). The R_s was higher at a dose of 5×10^{13} than at 3×10^{13} which may indicate solid solubility effects at this concentration. The light dose multiple energy carbon implants will also be influenced by the background values of the deep levels such as EL2, which are typically in the low 1×10^{16} cm^{-3} range [1]. The carbon concentrations need to exceed the deep donor levels to show p-type characteristics. The low activation efficiencies are typical of amphoteric dopants.

There are two effects that will influence the n-type implant layers. First, the carbon acceptors will compensate the donors, $(N_d - N_a)$. Second, the deep carbon profiles were p-type so there will be a p-n junction at the backside of the Si or Se implant profile. This junction depletion region will extend into the n-implant region as well as into the p-doped region. The effects of these carbon implants on the pinchoff voltage can be calculated as follows. The depletion layer width is calculated by equating the charge on both sides of the junction. In addition, the total potential drop across the depletion region is equal to the junction voltage which was assumed to be 1.30 V. The resulting pinchoff voltage is calculated by integrating the charge from the surface to the edge of the junction depletion region. Assuming the profiles in Figure 1, the calculated pinchoff voltages were 2.47, 2.14 and 1.52 for the profiles a, b and c, respectively. These calculated voltages show the same trend as the measured values. In these calculations of the pinchoff voltage we have assumed that all the carbon is activated. If 50% of the carbon dose were active, which is shown by the Hall data, then the junction depth and the amount of compensation would change and the calculated pinchoff voltage would be slightly larger. However, the activation of the carbon acceptors may be different in the presence of the implanted donors (Si or Se).

In summary, we have shown that carbon implants into high purity LEC SI GaAs are p-type with dose dependent activation efficiencies below 50%. Carbon is an efficient compensator of both Si and Se doped GaAs. A p-n junction forms between light dose carbon implants and n-type Si or Se implants. The decreasing pinchoff voltages with increasing concentrations of carbon are caused by both compensation and the junction depletion region.

ACKNOWLEDGEMENTS

We would like to thank B. Lorigan for the implants, S. Aynes for the processing and measurements and M. Scott for typing the paper.

References

1. Holmes D. E., Chen R. T., Elliott K. R., Kirkpatrick C. G. and Yu P. W. (1982). *IEEE Trans. El. Dev.*, **ED-29,** 1045
2. Chen R. T., Holmes D. E., Asbeck P. M. and Kirkpatrick C. G. (1983). *Electronic Materials Conference, Burlington, VT*
3. Hunter A. T., Kimura H., Baukus J. P., Winston H. V. and Marsh O. J. (1984). *Appl. Phys. Lett.*, **44,** 74
4. Shin B. K. (1976). *Appl. Phys. Lett.*, **29,** 438
5. Sansbury J. D. and Gibbons J. F. (1971). In *Ion Implantation*, p. 251 (F. H. Eisen and L. T. Chadderton, eds.); Gordon and Breach, New York
6. Shin B. K., Ehret J. E., Park Y. S. and Stefiniw M. (1978). *J. Appl. Phys.*, **49,** 2988
7. Stringfellow G. B., Koschel W., Briones F., Gladstone J. and Patterson G. (1981). *Appl. Phys. Lett.*, **39,** 581
8. C. A. Evans and Associates, San Mateo, California

Section 2:
Dislocations and extended defects

On the Dislocation Density in Melt-Grown GaAs

J. Lagowski, H. C. Gatos, T. Aoyama* and D. G. Lin

Massachusetts Institute of Technology, Cambridge, Massachusetts 02139, USA

Abstract

We have discovered a striking relationship between the dislocation density and the Fermi energy in melt-grown GaAs. Thus, a shift of the Fermi energy from 0.1 eV above to 0.2 eV below its high temperature (1100 K) intrinsic value increases the dislocation density by as much as 5 orders of magnitude. The Fermi energy shift was achieved by n and p-type doping at a level of about 10^{17} cm^{-3}, i.e. one to two orders of magnitude below that commonly employed for impurity hardening. The pronounced role of the Fermi energy on dislocation densities must be associated with the dependence of the point defect interactions on the charge state of these defects, i.e. on the occupancy of the associated electronic levels. It appears most likely that these interactions and thus the dislocation density are controlled in a critical manner by the concentration of gallium vacancies. Similar post-solidification defect interactions were recently proposed in order to explain the formation of the antisite defect As_{Ga} and the associated midgap level EL2. Our findings permit the identification of regimes of thermal stress, stoichiometry, Fermi energy and impurity hardening which are of fundamental importance in the formation of dislocations in the melt-grown GaAs.

Introduction

A reduction of the dislocation density in melt-grown single crystals is perhaps the most important materials problem in GaAs IC technology. Dislocation-free GaAs is commercially available, but only as heavily doped material, with a donor concentration ranging from 10^{18} to 10^{19} cm^{-3}. Semi-insulating material, currently used for ICs, has been recently obtained dislocation-free by utilizing doping with isoelectronic impurities (group III and V elements) at a level of 10^{19} to 10^{20} cm^{-3} [1].

It has been a common trend to treat dislocations in melt-grown crystals, and particularly in crystals grown by the LEC (liquid encapsulated Czochralski) technique, within a framework of critical stress models [2, 3]. Such models assume that dislocations are generated in the solidified material when thermal stress exceeds certain critical values, which is supposed to be a well-defined characteristic of the material.

Impurity effects on dislocation density [1, 4] have been attributed to various mechanisms of blocking the propagation of dislocations (e.g. strong localized bonding [5], differences in tetrahedral radii between the impurity and the host atoms [6], and impurity-defect blocking complexes [7]). They are

* On leave from Hitachi Research Laboratories, Ibaraki, Japan.

phenomenologically treated as impurity hardening effects which increase the value of the critical stress. Such treatment is an extension of that developed for elemental semiconductors (Ge and Si) and it does not take into account the unique stoichiometry aspects of GaAs related to a finite existence region. It is of interest to note that the effects of stoichiometry on the dislocation density in GaAs were originally reported in the mid-60s [8]. The systematic demonstration of the critical role of stoichiometry in the generation of dislocations has been provided in our recent study on GaAs grown under minimized thermal stress by the HB (horizontal Bridgman) method in which stoichiometry is controlled (and varied) by controlling (varying) the arsenic source temperature, T_{As}. We have found that the dislocation density exhibits a sharp and very pronounced minimum at an optimum arsenic temperature $T_{As} = 617 \pm 1°C$ [9]. We have further found that growth under 'off stoichiometry' conditions leads to high dislocation densities which exceed those found in crystals grown under pronounced thermal stress [9, 10]. It has been suggested that stoichiometry also plays an important role (in the form of stoichiometry dependent critical stress) in dislocation formation during LEC growth of GaAs [11].

In the present communication we present experimental results which shed a new light on the role of defects in the dislocation formation in GaAs grown under low thermal stress; we show that in moderately doped crystals the dislocation density is controlled by the Fermi energy.

Experimental

Single crystals of GaAs with a cross-sectional area up to 2 cm^2 were grown using a HB apparatus especially designed for the precise control of growth parameters [9]. The growth was carried out in a high purity fused quartz boat (especially treated to prevent the melt from wetting the boat) or in a PBN (pyrolitic boron nitride) boat.

A series of experiments was carried out to investigate the effects of stoichiometry on the properties of GaAs. In these experiments the arsenic source temperature, T_{As}, was decreased (increased) from 623°C to 610°C (from 610°C to 623°C) in discrete steps of 1–2°C at 1–2 hour intervals, corresponding to 1–2 cm of growth. The corresponding melt composition ranged from approximately 0.49 (610°C) to 0.51 (623°C) atom-fraction of arsenic.

Undoped, Si-doped and Zn-doped growth melts were employed in these experiments. The crystals with free carrier concentrations (both holes and electrons) below 5×10^{16} cm^{-3} were grown with addition of Ga_2O_3 to the melt in order to reduce the silicon contamination. The carrier concentration was determined from Hall effect and conductivity measurements. Dislocations were revealed using molten KOH.

Results and discussion

EFFECT OF STOICHIOMETRY

Our previous investigation of lightly doped n-type crystals has shown a striking correlation between deviations from stoichiometry and dislocation density. We have established an optimum arsenic source temperature, 617 ± 1°C, which leads to a minimum dislocation density [9, 10]. The present growth experiments confirm these results and further show that the same behaviour is also characteristic in lightly doped p-type GaAs crystals.

In Figure 1 we show the dependence of the dislocation density on the arsenic temperature T_{As} for p-type and n-type GaAs crystals with free carrier concentrations of about 10^{16} cm^{-3}. A minimum dislocation density of about

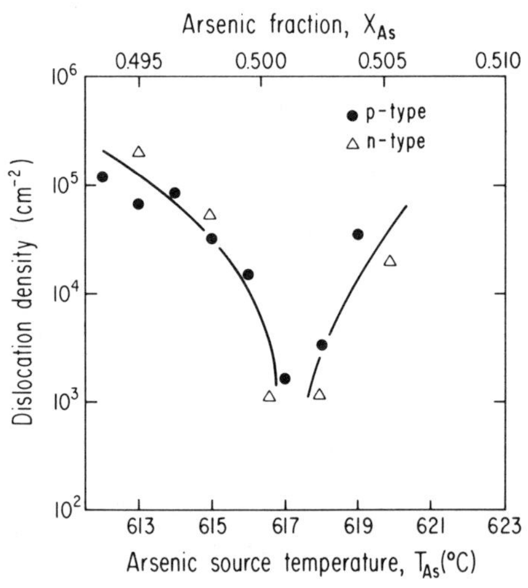

Figure 1 Dislocation density versus arsenic source temperature (stoichiometry) for lightly doped n and p-type GaAs.

10^3 cm^{-2} is obtained at an optimum temperature 617 ± 1°C. Deviation from this temperature leads to an increase of the dislocation density by as much as two orders of magnitude, i.e. to dislocation densities which are comparable to (or even higher than) those found in GaAs crystals grown under large thermal stresses [1–3].

In Figure 2 we show Nomarski contrast photomicrographs of etched surfaces of three segments of a p-type crystal. These photomicrographs show sharp changes in the dislocation density during growth in response to stoichiometry changes. The dislocation density rapidly decreases in the crystal as T_{As} becomes closer to the optimum value. On the other hand, dislocations are generated as T_{As} is shifted away from an optimum value. Exactly the same behaviour has been observed previously in lightly doped n-type GaAs.

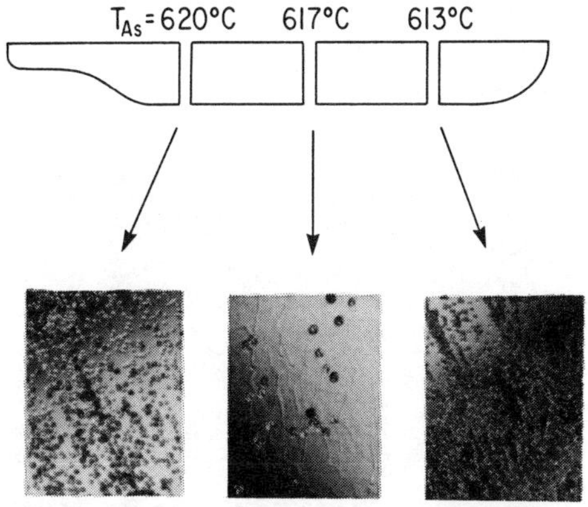

Figure 2 Photomicrographs of etched crystal surfaces from different segments of a GaAs crystal.

These findings indicate that stress (non-stoichiometry) induced dislocations do not readily propagate with progressing crystal growth; accordingly they must be in a form of dislocation loops rather than edge dislocations.

EFFECTS OF DOPING

We have reported previously that, in crystals grown under carefully minimized thermal stress, n-type doping constitutes a very effective means for the virtual elimination of non-stoichiometry induced dislocations [10]. This dramatic effect of doping is observed for a donor concentration of about $10^{17}\ cm^{-3}$, i.e. significantly smaller than those involved in impurity hardening phenomena believed to block the propagation of stress induced dislocation [1, 4–7]. This relatively low level of doping is quite sufficient to convert crystals with high dislocation densities, N_D, grown under arsenic deficient conditions ($T_{As} \simeq 613°C$, $N_D \sim 10^5$–$10^6\ cm^{-2}$) to dislocation-free material ($N_D < 100\ cm^{-2}$). The same is true for crystals grown under arsenic rich conditions.

The effect of p-type doping is very different from that of n-type doping. The differences are evidenced by the Nomarski contrast photomicrographs of Figure 3. It is seen that for crystals grown under optimum stoichiometry, an increase in the acceptor concentration increases the dislocation density, in contrast to the effect resulting from an increase of the donor concentration. It is also seen in Figure 3 that GaAs with a hole concentration of about 5×10^{17} exhibits a high dislocation density (5×10^4–$5 \times 10^5\ cm^{-2}$) which could not be minimized to any significant extent by the optimization of the arsenic source temperature. The effects of the conductivity type and carrier concentration on the dislocation density in GaAs grown under optimum T_{As} are quantitatively summarized in Figure 4. It is seen that the transition from n-type to p-type material enhances the dislocation density by as much as five orders of magnitude.

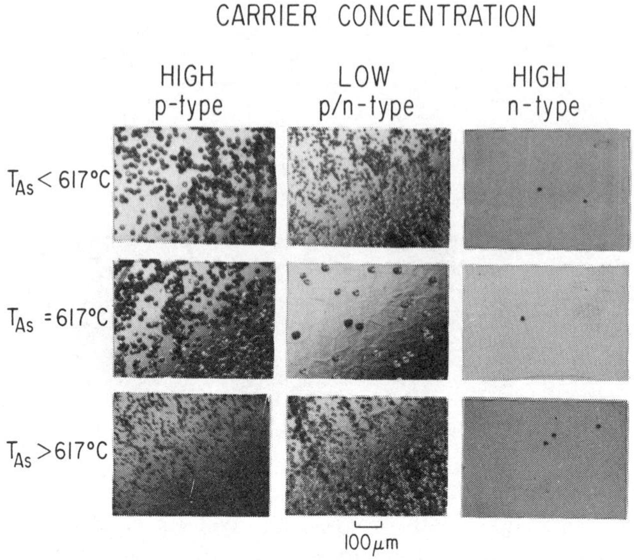

Figure 3 Photomicrographs of etched surfaces of GaAs crystals grown under different arsenic source temperatures and under different doping conditions.

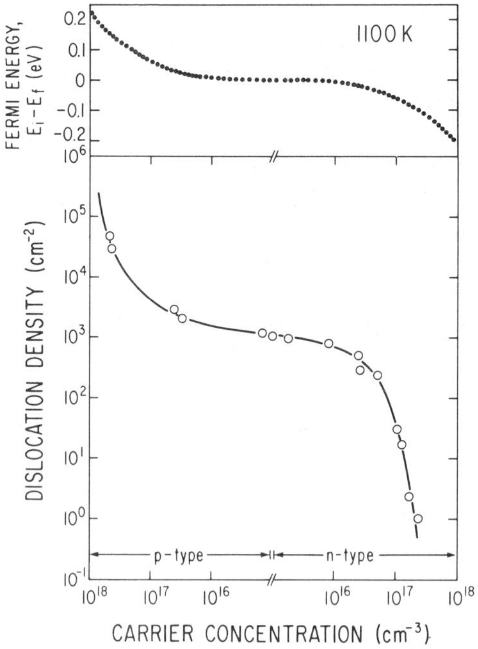

Figure 4 Dislocation density versus 300 K free carrier concentration. Upper portion shows the corresponding values of the Fermi energy at 1100 K.

THE ROLE OF FERMI ENERGY

The key to understanding the formation of dislocations in GaAs grown under low thermal stress lies in the finite existence region of GaAs. Accordingly, the type and concentration of point defects (especially of vacancies and their complexes) in a solidifying crystal is controlled by the melt composition (i.e. the arsenic source temperature). During the post-solidification cooling, defects in the crystal become supersaturated and interact, forming other defects such as defect complexes and dislocations.

We have suggested, in conjunction with the antisite defect model of EL2, that the relevant defect interactions are controlled by the migration of gallium vacancies [12]. The same model of gallium vacancy migration can explain the effects of the conductivity type and of the carrier concentration on the dislocation density in GaAs. The formation of dislocations in GaAs upon condensation of vacancies requires the participation of defects from both the Ga and As sublattices. However, the process can be initiated by only one type of defect, such as V_{Ga}, since the corresponding defect, V_{As}, can be created upon migration of V_{Ga}. The point defects in GaAs lead to localized electronic levels. The occupancy of the levels, i.e. the charge state of defects, is controlled by the Fermi energy. The charge state of the various defects determines their concentration, migration and ability to coalesce.

Transition from p to n-type GaAs caused by intentional doping shifts the Fermi energy upwards in the energy gap increasing the concentration of negatively charged acceptor-type defects and decreasing the concentration of ionized deep donor levels. According to Ref. 12 critical post-solidification defect interactions take place at a temperature of about 1100 K. Thus, in Figure 4 we have indicated the Fermi energy values of 1100 K corresponding to the carrier concentration at 300 K [13]. The correlation between the Fermi energy and dislocation density is evident. The plateau of dislocation density coincides with a constant Fermi energy. In the n-type region the dislocation density decreases when the added donor concentration becomes comparable to (or higher than) the intrinsic carrier concentration at 1100 K, n_i (1100 K), while on the p-type region the dislocation density increases when the acceptor concentration becomes comparable to n_i (1100 K). This relationship is consistent with the critical role of a gallium vacancy which in GaAs leads to an acceptor level close to the middle of the energy gap. The downward shift of Fermi energy caused by n to p-type transition increases the fraction of neutral gallium vacancies which promote condensation and migration of vacancies required for the formation of dislocations.

It is of interest to note that the importance of the gallium vacancy has previously been suggested in accounting for the stoichiometry effects [8] and also in conjunction with donor impurity hardening in dislocation blocking by complexes of donor atom-gallium vacancy [7]. The role of the Fermi energy has been discussed in the dislocation model as a factor altering the propagation of dislocation and thus also affecting the low temperature value of the critical stress [4]. To our knowledge the correlation between Fermi energy and dislocation formation in GaAs reported here is established for the first time.

DIFFERENT REGIMES OF IMPURITY: EFFECTS ON DISLOCATIONS

The present finding together with literature data made it possible to define regimes of fundamental importance in the formation of dislocations in melt-grown GaAs. Two mechanisms of dislocation formation should be distinguished: (a) stress-induced glide dislocations which are generated by an excessive thermal stress; and (b) nonstoichiometry-induced dislocations which are generated upon condensation of excessive vacancies into dislocation loops. As shown in Figure 5, dislocation densities corresponding to both

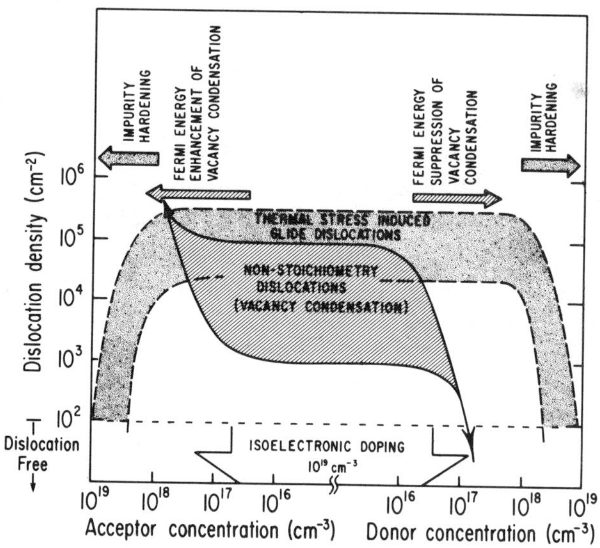

Figure 5 Schematic diagram of dislocation density in GaAs versus free carrier concentration and conduction type.

of these mechanisms overlap over a wide range of the dopant concentration. In strictly undoped material (without electrically active or isoelectronic impurities) the lowest dislocation density is controlled by the stoichiometry. The Fermi energy effects on the nonstoichiometry-induced dislocations are significant for dopant concentrations exceeding about 8×10^{16} cm^{-3}. A donor-type impurity at a concentration of 2×10^{17} cm^{-3} is sufficient to eliminate completely nonstoichiometry dislocations while an acceptor impurity increases the dislocation density.

The thermal stress-induced dislocations are generated at a temperature which is much closer to the melting point than 1100 K, corresponding to nonstoichiometry dislocations. Thus, the effects of Fermi energy on propagation and multiplication of stress-induced dislocations would require a high concentration of an electrically active dopant, exceeding 10^{18} cm^{-3}. As shown in Figure 5, this concentration range overlaps with that considered in other 'impurity hardening' phenomena. GaAs free of stress-induced dislocations is typically grown with electrically active dopant of concentrations well above 10^{18} cm^{-3} or with an even higher concentration of isoelectronic dopants. The effectiveness of impurity hardening in GaAs is a function of stoichiometry.

ACKNOWLEDGEMENTS

The authors are grateful to the National Aeronautics and Space Administration and the Office of Scientific Research for financial support.

References

1. Jacob G. (1982). In *Semi-Insulating III–V Materials, Evian*, p. 2 (S. Makram-Ebeid and B. Tuck, eds.); Shiva Publishing, Nantwich, UK
2. Jordan A. S., Caruso R. and Von Neida A. R. (1980). *Bell Syst. Tech. J.*, **59,** 593
3. Mil'vidskii M. G. and Bochkarev E. P. (1978). *J. Cryst. Growth*, **44,** 61
4. Mil'vidskii M. G., Osvensky V. B. and Shifrin S. S. (1981). *J. Cryst. Growth*, **52,** 396
5. Seki Y., Watanabe H. and Matsui J. (1978). *J. Appl. Phys.*, **49,** 822
6. Kirkby P. A. (1975). *IEEE J. Quantum Electron.*, **QE-11,** 562
7. Swaminathan V. and Copley S. M. (1967). *J. Appl. Phys.*, **47,** 4405
8. Brice J. C. and King G. D. (1966). *Nature*, **209,** 1346; Brice J. C. (1970). *J. Cryst. Growth*, **7,** 9
9. Parsey Jr J. M., Nanishi Y., Lagowski J. and Gatos H. C. (1981). *J. Electrochem. Soc.*, **128,** 936; **129,** 388
10. Parsey Jr J. M., Lagowski J. and Gatos H. C. (1984). *J. Electrochem. Soc.*, in press
11. Duseaux M. and Jacob G. (1982). *Appl. Phys. Lett.*, **40,** 789
12. Lagowski J., Gatos H. C., Parsey J. M., Wada K., Kaminska M. and Walukiewicz W. (1982). *Appl. Phys. Lett.*, **40,** 342
13. Fermi energy values were calculated using GaAs parameters compiled by Blakemore J. S. (1982). *J. Appl. Phys.*, **53,** R123

Structure of Lattice Defects in Semi-Insulating LEC GaAs

F. A. Ponce*, F-C. Wang† and R. Hiskes

Hewlett-Packard Laboratories, Palo Alto, California 94304, USA

Abstract

The lattice structure of defects in undoped, semi-insulating LEC (liquid encapsulated Czochralski) GaAs has been studied using conventional and high resolution TEM (transmission electron microscopy). The material exhibits a cellular structure along the ⟨100⟩ growth direction. The central portion of the cells are virtually free of extended defects. In the vicinity of the cell boundaries, large densities of various defects are present, among them dislocations and small precipitates. Precipitates ranging from less than 1 nm to about 20 nm in diameter and with a different structure than that of GaAs have been observed using lattice imaging techniques. In addition, small microdefects with typical diameters between 2 and 3 nm have been observed. The latter appear to be small loops lying along (111) planes, of both the vacancy and the interstitial types.

Introduction

In recent years there has been an increasing interest in SI GaAs. Its high electron mobility makes it suitable for high speed integrated circuit applications. Typical IC technologies, however, require high quality, homogeneous and uniform materials [1, 2]. Recent work [3, 4] has demonstrated that there is a close correlation between EPD (etch pit densities) and electrical and optical parameters such as resistivity, Hall mobility and photoluminescence efficiency. It is interesting to note that these parameters improve in regions of the crystal where the dislocation density (EPD) is highest, causing doubt as to whether minimization of dislocation densities would actually improve the electric performance of the material.

Lattice defects can be classified into three categories: (a) extended defects, such as dislocations and stacking faults, which are either linear or planar in nature; (b) microdefects, such as interstitial and vacancy loops and small inclusions or precipitates, with dimensions under 100 nm; and (c) point defects, such as vacancies, interstitials, and antisite defects. Various techniques are used to observe lattice defects in crystals. X-ray topography (XRT) is very useful in determining the macroscopic arrangement of extended defects, with a typical resolution of about one micron. Transmission electron

* Present address: Xerox Palo Alto Research Center, Palo Alto, California 94304, USA.
† Santa Rosa Technology Center, Hewlett-Packard Company, Santa Rosa, California 95401, USA.

microscopy allows the study of both extended defects and microdefects. In the high resolution mode, TEM can provide direct lattice images with point resolutions of about 0.2 nm, which is close to the interatomic separation in most semiconductors. It is believed that point defects can be observed by electrical or optical techniques. However, little is known today about the actual lattice structure of defects in SI GaAs materials.

We report here on a study of the electric and structural characteristics of undoped, SI LEC GaAs. Our emphasis is on the detailed lattice structure of defects and its variation across the length and diameter of the crystal, and its correlation with electrical and structural properties.

Experimental procedure

The unintentionally doped GaAs materials used in this investigation were grown in the ⟨100⟩ direction in PBN crucibles using a low pressure (2 atm) LEC method with *in situ* synthesis. Nitrogen gas was used as the ambient control atmosphere. The ingot was approximately 6 cm in diameter and 2 kg in weight and passed Hewlett-Packard standard acceptance tests for SI quality and thermal conversion.

⟨110⟩ slices were cut along the growth axis of the crystal. Transmission XRT was performed to determine variations of the defect structure along the growth direction (longitudinal) and perpendicular to it (radial). TEM specimens were prepared from selected regions, based on the XRT information. The specimens were thinned mechanically to about 20 μm and were then thinned to electron transparency using Ar^+ ions at 3–5 keV accelerating potential. Some samples were prepared by chemical etching techniques in order to assess the effect of ion bombardment which was found to be negligible. Transmission electron microscopy was performed in a JEOL 200CX instrument equipped with a high resolution top entry stage operating at 200 kV. The electron source was a LaB_6 filament. The images were recorded under axial illumination, with a beam divergence of 1.2 mrad. The instrumental point resolution is about 0.25 nm. The details of the image formation process and image characteristics are available elsewhere [5, 6].

⟨100⟩ slices, cut normal to the growth direction, were used for electrical characterization. A 4 mm wide strip was sliced from a test wafer for etching and van der Pauw measurements. The dislocation EPD was obtained by counting etch pits on photographs covering 1.1 mm square areas taken after molten KOH etching at 460°C for 10 min. Subsequent to KOH etching, wafers were solvent cleaned, etched and cut into 4 mm square samples suitable for van der Pauw measurements to determine the as-grown Hall mobility and free carrier concentration.

The rest of the test wafer was used to produce discrete MESFETs (metal-semiconductor field effect transistor). The FET geometry is typical for low-noise devices with a 1 × 350 μm square gate and 3.5 μm source-drain spacing. Every six devices (1.7 mm), a Schottky barrier test pattern covering a 70 × 70 μm square area enabled the measurement of capacitance. The active layer and the N^+ contact layer were obtained by implanting silicon at 200 and 60 keV with a $5 \times 10^{12}\ cm^{-2}$ dose for each energy. Ti/Pt/Au and Ni/Cr/Ge/Au metallizations were deposited for the gates and ohmic contacts respectively. Transconductance, threshold voltage and saturation currents were measured on the MESFETs. Capacitance was measured on the test patterns.

Electrical characterization

The results of all material and device measurements in adjacent areas were compared. A positive correlation from the experimental data between EPD, mobility and transconductance was observed and is illustrated in Figure 1.

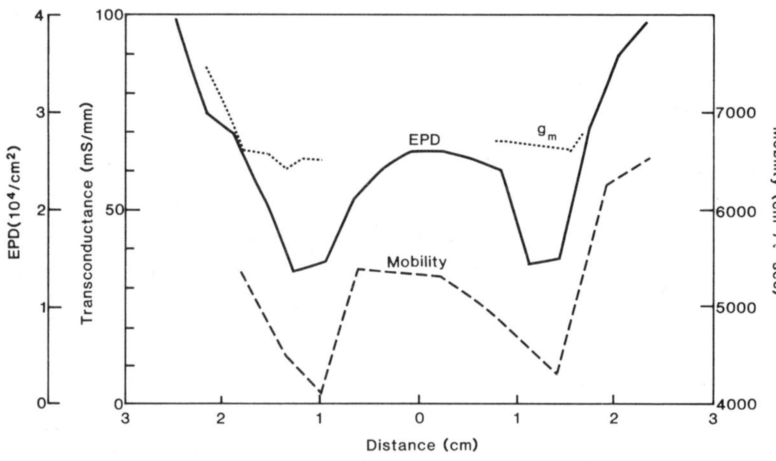

Figure 1 Radial position dependence of dislocation etch pit density, as-grown substrate mobility, and MESFET transconductance across the wafer.

All three parameters have high values at the edge and centre of the crystal and have minima at about 0.5 R (where R = radius of ingot). No other clear correlations were observed among other parameters. Similar phenomena had already been observed on GaAs grown in the same puller but in quartz crucibles [4]. The authors had speculated that microscopic imperfection interaction was responsible for the observed experimental correlation between EPD, mobility and transconductance. We believe that the same origins are responsible here. Namely, it is a combination of the intrinsic material properties determined during material formation and the extrinsic growth environments (higher thermal gradients) causing the microscopic interactions between imperfections. That, in turn, affects the electrical properties, specifically mobility and FET performance.

Structural characterization

Figure 2 shows a transmission X-ray topograph of a ⟨110⟩ slice cut along the crystal growth axis. The picture was obtained using a (220) GaAs reflection through the 0.3 mm thick slice. This figure clearly shows the complex defect distribution characteristic to undoped LEC GaAs crystals. Figure 3 is a similar X-ray topograph of a ⟨100⟩ slice cut normal to the growth direction, about 1.5 cm from the seed end of the crystal shown in Figure 2. The X-ray intensity follows the familiar W-shape reported in the literature [2–4, 7]. The topographs suggest a cellular structure, where cell dimensions range from less than 0.1 up to a few mm, the largest occurring at $r \sim 0.5$ R.

A transmission electron micrograph of the region close to a cell boundary is shown in Figure 4. Typically the cell boundaries appear to be flat and do not necessarily exhibit dislocations. Dislocations have been observed close to these boundaries and, on occasion, dislocation tangles are present. The cell boundaries do not always follow any particular crystallographic plane. Figure 5 shows the structure of a cell boundary intersecting a thin portion of the crystal. This figure is a high-resolution lattice image taken in the ⟨110⟩ projection of GaAs. Each spot represents a pair of Ga and As atomic columns

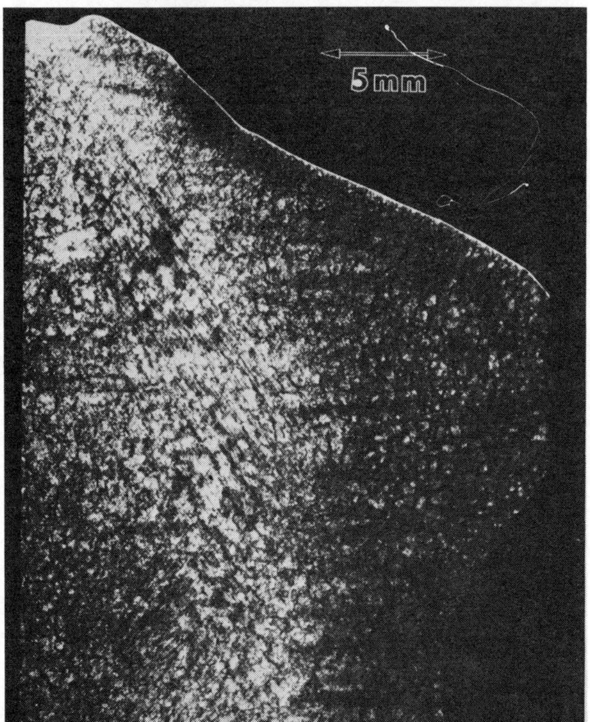

Figure 2 Transmission X-ray topograph of ⟨110⟩ longitudinal section of undoped, LEC GaAs crystal. The seed end is up.

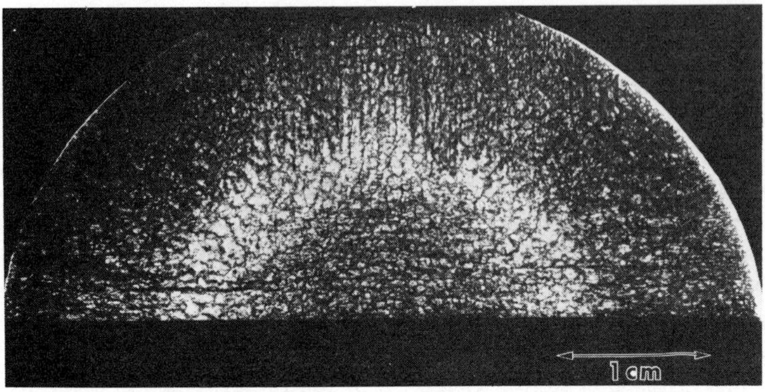

Figure 3 Transmission X-ray topograph of ⟨100⟩ section perpendicular to growth direction.

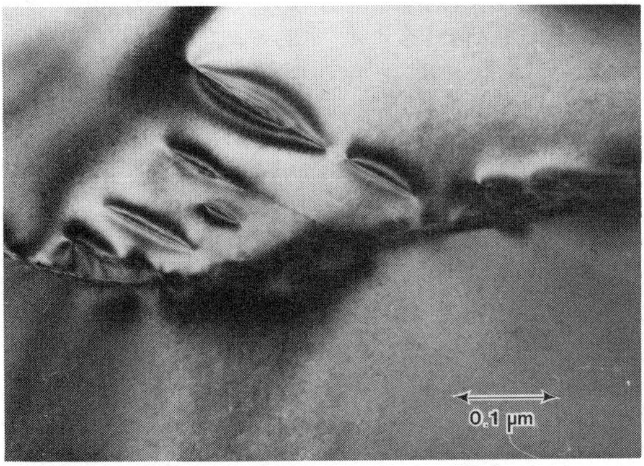

Figure 4 Transmission electron micrograph of ⟨110⟩ section at r ~ 0.5 R. The defect structure in the vicinity of a cell boundary is observed.

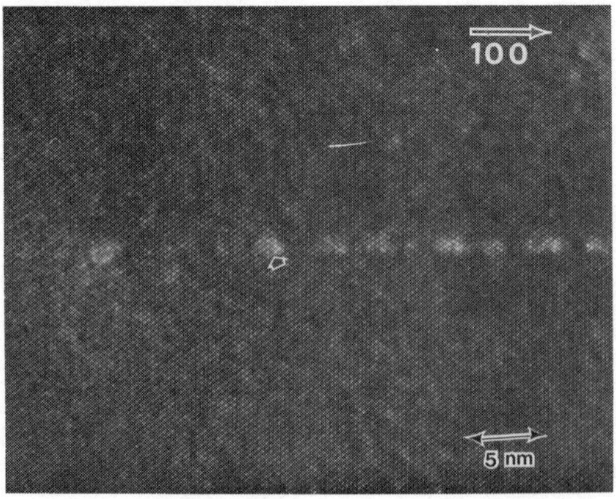

Figure 5 High resolution lattice image of region along cell boundary. Small coherent particles (~ 1–2 nm in diameter) are observed.

seen in projection. The cell boundary is composed of small particles whose structure appears amorphous. Typical dimensions are 1–10 nm. They are clearly seen when the cell boundary is oriented along the growth direction as in Figure 5. There they are aligned in the ⟨100⟩ direction. When the boundary deviates from such direction, larger precipitates are observed occasionally, as seen in Figure 6a.

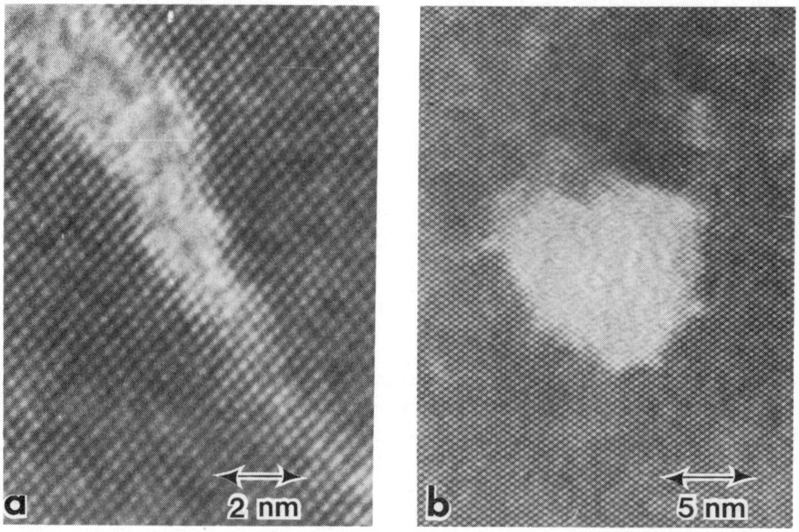

Figure 6 Amorphous precipitates in undoped SI GaAs: (a) along the cell boundary; (b) in the interior of a cell located close to periphery of crystal ($r \sim 0.8$ R).

Similar precipitates have also been observed in the interior of the cells, especially in the centre ($r \leq 0.25$ R) in the periphery of the crystal ($r \geq 0.75$ R). One such instance is shown in Figure 6b. This precipitate has a diameter of about 10 nm. These precipitates, or second phases, have a structure that appears amorphous in high resolution TEM, and we have been unable to obtain evidence of a crystalline phase. They appear to be coherent (no dislocations or extra planes are observed in their periphery), with very little strain field associated with them. Burger's circuits around them are complete ($b = 0$). Therefore, they cannot be associated with decorated dislocations. Chemical analysis has not been possible on these precipitates because of their small size and instability under beam irradiation. They could be As or Ga precipitates and be associated with deviations from stoichiometry.

In addition, small microdefects such as shown in Figure 7a are very common in the interior of the cells. These lie along (111) planes and can be identified from the image contrast as interstitial loops, typically 2–5 nm in diameter. Similar loops exhibiting a vacancy contrast have also been observed but more rarely. Because their dimensions are of similar magnitude to the specimen thickness needed for high resolution TEM ($t \leq 10$ nm), it is difficult to assess

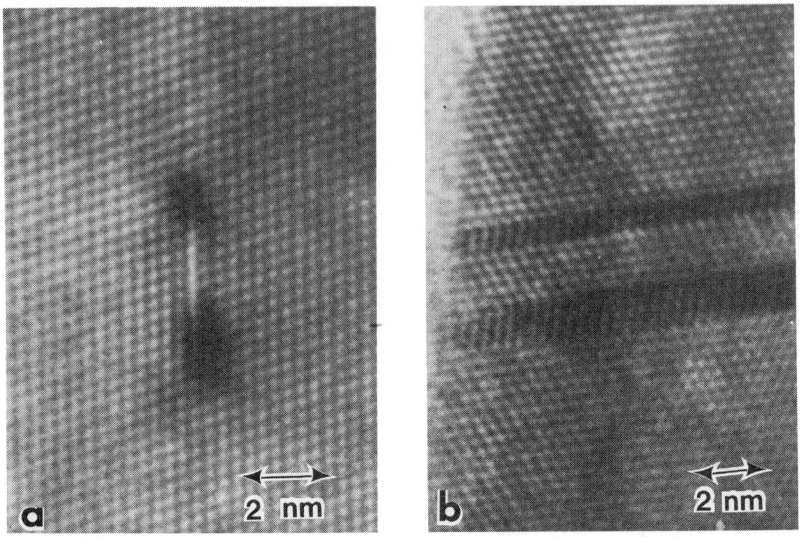

Figure 7 High resolution lattice images of defects along (111) planes: (a) microdefect with a diameter of ~3.3 nm which shows contrast similar to that of interstitial loops; (b) multiple stacking faults (microtwins) associated with crystallographic slip caused by the propagation of Shockley partial dislocations (r ~ 0.8 R).

their structure in detail. Because of their size and technical limitations, it is not possible to do direct chemical analysis using X-ray techniques in the TEM.

A commonly observed feature in the longitudinal X-ray topographs are the slip lines. These are most common in regions close to the ingot surface (r ~ R). Figure 7b shows the structure resulting from crystallographic slip, two thin microtwins which can be thought of as multiple intrinsic stacking falts, each having been produced by the passage of a Shockley partial dislocation [5, 8]. The defect structure described here is unique to undoped LEC material. GaAs doped with Cr, Si, Te and Se have also been investigated in a similar way, but the observed structures were quite different [9].

Discussion and conclusions

The determination of the mechanism of generation of the observed defect structure is quite difficult due to the complexity of the observed structures. Two mechanisms are suggested which could account for the cellular structure. The first is due solely to the interaction of dislocations which originate at the seed and at the ingot surface, and which propagate and multiply, driven primarily by the large thermal stresses present in the crystal. Calculations have been performed which match fairly well the defect density distributions [2, 7]. If this were the prevailing mechanism for defect generation, we would expect to observe large densities of dislocations arranged in polygonal distributions [8]. In our study, dislocations are but a small constituent of the large variety of defects observed. Also, most of the defects do not exhibit non-zero Burger's vectors which should be associated with dislocations.

The second mechanism which could explain the observed cellular structure is constitutional supercooling [10]. Deviations from stoichiometry in the melt could lead to the formation of a thin layer which is rich in one of the constituents. Small fluctuations could lead to the breakdown of the growth interface and to the formation of cellular structure. Thus the cell boundaries and the observed amorphous phases would be explained. Constitutional supercooling would be enhanced by a lowering of the thermal conductivity in the enriched layer in the liquid [10]. There is, however, little data at present regarding thermal conductivity for various compositions of the liquid.

In summary, we have observed a close correlation between the electrical properties and the structural characteristics of the material. The structure has been observed using high resolution TEM, and shows a cellular-like structure. Various types of precipitates and microdefects have been observed, as well as microtwins generated by crystallographic slip. These defects appear in numbers several orders of magnitude higher than the dislocation density. Two mechanisms could be responsible for this: (a) polygonization of dislocations generated by excessive thermal stresses; and (b) constitutional supercooling. The observed structure seems more consistent with the latter.

ACKNOWLEDGEMENTS

The authors are grateful to G. Anderson and R. Smith for technical assistance. Dr M. Bujatti provided valuable help in the processing and measurements of MESFETs. Fruitful discussions with M. Scott and R. Burmeister are also gratefully acknowledged.

References

1. Fairman R. D., Chen R. T., Oliver J. R. and Ch'en D. R. (1981). *IEEE Trans. El. Dev.*, **ED-28,** 135
2. Jacob G. (1982). In *Semi-Insulating III-V Materials, Evian*, p. 2 (S. Makram-Ebeid and B. Tuck, eds.); Shiva Publishing, Nantwich, UK
3. Tajima M. (1982). *Jap. J. Appl. Phys.*, **21,** L227
4. Wang F-C. and Bujatti M. (1984). *IEEE El. Dev. Lett.* In press
5. Ponce F. A., Yamashita T., Sinclair R. and Bube R. H. (1981). In *Defects in Semiconductors*, p. 503 (J. Narayan and T. Y. Tan, eds.); North-Holland, Amsterdam
6. Spence J. C. H. (1981). *Experimental High-Resolution Electron Microscopy*; Clarendon, Oxford
7. Jacob G., Farges J. P., Schemali C., Duseaux M., Hallais J., Bartels W. J. and Roksnoer P. J. (1982). *J. Cryst. Growth*, **57,** 493
8. Hirth J. P. and Loethe J. (1982). *Theory of Dislocations*, 2nd edition; John Wiley
9. Ponce F. A. and Yamashita T. (1984). Unpublished
10. Laudise R. A. (1970). *The Growth of Single Crystals*, pp. 104–9; Prentice-Hall

The Structural Characterization of Semi-Insulating GaAs by Double Crystal X-Ray Topography

G. T. Brown, M. S. Skolnick, G. R. Jones, B. K. Tanner* and S. J. Barnett*

Royal Signals and Radar Establishment, St. Andrews Road, Great Malvern, Worcestershire WR14 3PS, UK

Abstract

Double crystal reflection XRT (X-ray topography) has been used to study the distribution and nature of defects in SI GaAs. Small diameter samples have been examined using a conventional X-ray source, while two and three inch diameter wafers have been studied using synchotron radiation. The particular problems associated with the high strain sensitivity of this technique are discussed. Dislocations are observed in cellular (polygonized) configurations, slip bands and linear arrays (lineage). Comparison of the topographs with IR micrographs reveals that all of these dislocation configurations are correlated with strong 1 μm absorption indicating an enhancement in the local neutral EL2 concentration. The lineage features are associated with a small lattice tilt which has been determined to be of the order of 0.01°.

Introduction

Double crystal X-ray topography (DCXRT) is a powerful method for imaging variations in both long and short range strain, and thus is particularly important for the assessment of semiconductor substrate surfaces. The major advantage that this technique offers over alternative reflection XRT methods is the high strain sensitivity. Hart has shown that it is possible to image strains of one part in 10^7 [1] with DCXRT. However, long exposures are frequently required, causing problems with stability. These problems can be solved by sampling part of the second-diffracted image and holding the diffraction conditions with a microcomputer [2]. For large diameter wafers, where the source to sample distance has to be increased, the exposure time can be as large as 20 h. In these cases the use of a synchotron radiation source (SRS) can reduce exposure times by more than an order of magnitude with consequent improvement in stability.

The application of DCXRT to the assessment of SI GaAs is of particular importance in evaluating the potential of this material as a substrate for IC fabrication using ion-implantation methods. It has recently become apparent that there is an inhomogeneous distribution of a number of important properties in SI GaAs [3–8]. The distribution of many measured parameters varies either in direct correlation or anticorrelation with the dislocation (EPD)

* Department of Physics, University of Durham, South Road, Durham DH1 3LE, UK.

distribution. Perhaps the most striking correlation of this type reported to date is the direct correlation of FET (field effect transistor) threshold voltage with dislocation density [9, 10]. The DCXRT technique reveals high contrast images of dislocations lying close to the surface and such topographic images are not confused by overlapping of images, which is a serious limitation to transmission methods for the dislocation densities involved here (10^4–10^5 cm^{-2}). This technique should therefore be particularly useful as a means of establishing the behaviour of dislocations during each device fabrication stage, e.g. implantation, capping and annealing.

Experimental methods

The double crystal camera used for the laboratory studies is based on a design by Hart [11] and has been described fully elsewhere [2]. The double crystal camera used in conjunction with the SRS at Daresbury Laboratory was designed and built by Bowen *et al.* [12] and both pieces of equipment are fully microcomputer controlled. For all the experiments discussed in this study the dispersive (angular and wavelength) asymmetric (+ −) setting was employed. The possibility of defects in the reference crystal appearing in the final image was eliminated by using high uniformity dislocation-free NTD (phosphorus doped) FZ silicon (100 Ω cm) which was chemically polished. Thick sections were used (~5 mm) to ensure that no strains were introduced on mounting the reference crystal. The reflection combinations were chosen with due consideration to rocking curve width (which determines the strain sensitivity), the diffracted intensity from the sample and image foreshortening. $CuK\alpha_1$ was used for all topography on a laboratory scale and radiation at 1.5 Å was selected for SRS experiments.

The SI GaAs samples were supplied by Cambridge Instruments and were grown in a high pressure puller by the LEC (liquid encapsulated Czochralski) technique using PBN crucibles and boric oxide encapsulant. ⟨100⟩ wafers 3 mm thick were supplied, polished on both sides. Topographic images were recorded on L4 emulsion plates and exposures using conventional X-ray tube sources were 15 h; however, by using the Daresbury SRS at 2 GeV and 200 mA, exposures were reduced to 20 mins. Long range strain mapping was achieved by taking Polaroid topographs at 20 second steps through the total sample rocking curve.

Results and discussion

MAPPING OF LONG RANGE STRAINS

The DCXRT method is probably the only means of rapidly mapping iso-lattice parameter areas on a wafer. However there are two important practical considerations. The first problem that has to be overcome is the introduction of strain during sample mounting. This is of particular importance for larger diameter wafers. In these studies, samples (which were deliberately wafered to a thickness of 3 mm) were mounted with low temperature wax at three points. Since no threefold symmetry was apparent in the strain patterns observed in the ⟨100⟩ topographs it was inferred that no mounting strains were introduced. The second consideration is warpage, or bow. Clearly, for severely warped samples the wafer may be bent away from the Bragg condition at points where the warpage is greatest. Sample warpage is more serious for samples which are single-sided polished [13] and/or are less than 1 mm thick [14]. Thus all samples in this study were double-sided polished. However, some warpage was still present and Figure 1 shows a map of the warpage

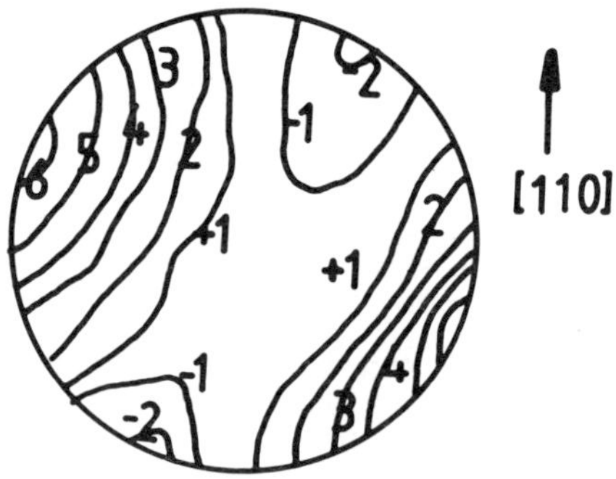

Figure 1 Map of the warpage present in the sample topographed in Figure 2.

determined by the scanning laser method using a Canon LSF500. The greatest peak-valley measurement recorded was 7.5 μm. The wafer had a slight saddle shape which is asymmetric. This type of warpage has been observed previously in much thinner polished wafers of LEC SI GaAs [15] but the symmetry was rotated by 45° to that observed here. The ideal solution to warped wafers is to hold down the sample with a vacuum chuck and efforts are in hand to achieve this. However the warpage did not prove a serious limitation since the pattern of the warpage (Figure 1) was not reflected in the topographic strain mapping (Figure 2). The bright areas in Figure 2a correspond to the diffracting areas on the wafer as the sample is tilted through the 422 rocking curve. Figure 2b is a strain contour map which is a combination of all the topographs in Figure 2a and which effectively plots areas of equal Bragg angle for the (211) planes contributing to the images in Figure 2a. This map bears some similarity to dislocation EPD maps [3] and also bears a striking resemblance to the distribution of the EL2 deep level as mapped by Holmes *et al.* [6].

DISLOCATION IMAGES (SHORT RANGE STRAIN FIELDS)

The detailed contrast features observed in the diffracting areas of topographs correspond to dislocations. These dislocations were observed in three different

(a)

(b)

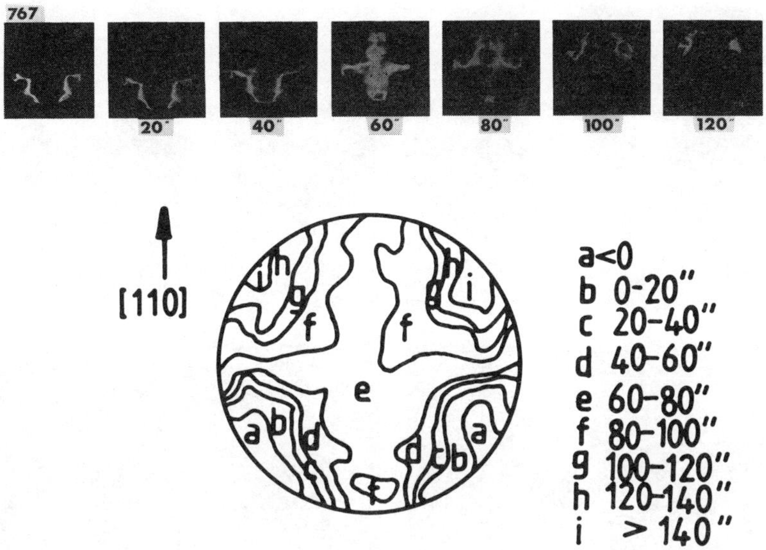

Figure 2 (a) SRS topographs of GaAs ⟨100⟩ wafer showing the areas diffracting from different angular settings. The diffraction geometry is 440 Si/422 GaAs. (b) Combined map including all of the diffracting areas in (a).

configurations as found previously in etch-pitting experiments on LEC GaAs [15]. These are cellular or polygonized structures, linear arrays (lineage), which are sometimes crystallographically aligned, and slip bands lying along ⟨110⟩ and which are more prevalent in an annular region halfway along a radius. Examples of the images of these three dislocation configurations are given in Figure 3. These are all areas from topographs taken with the SRS since the contrast is generally better than for topographs taken using a conventional X-ray tube source due to the smaller beam divergence and hence narrower rocking curve. The important feature of these topographs is that individual dislocation images can be discerned within all of these configurations. These characteristic DCXRT dislocation images with strong black/white contrast have been discussed by Bonse [16].

The lineage feature shown in Figure 3a is associated with a sharp contrast reversal across the feature. This is invariably observed for such linear arrays and is due to a misorientation between the areas of crystal either side of the feature. This misorientation was measured for the lineage features in the sample topographed using an X-ray tube source and shown in Figure 4. The measurement was made using synchotron radiation (1.5 Å), thus eliminating any confusion associated with the Kα_2 line. The incoming once-diffracted beam was collimated by a slit 85 μm wide and allowed to impinge on an area containing lineage. The rocking curve for the surface symmetric 400 reflection was found to be split into a doublet and the separation of the two peaks (which corresponds to diffraction from crystal either side of the lineage feature) was determined to be 0.01° (36 sec of arc).

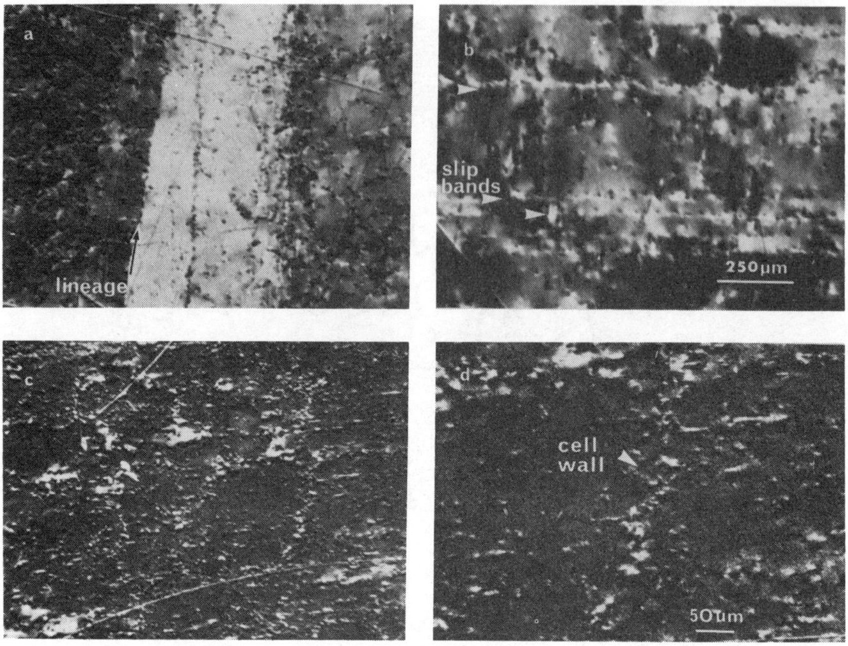

Figure 3 SRS topographs (diffraction geometry 440 Si/442 GaAs) showing different dislocation configurations: (a) lineage; (b) slip bands; (c) cell structure; and (d) cell structure at higher magnification showing individual dislocation images within cell walls.

A further point of interest is that when the diffraction vector is parallel to lineage features, the misorientation is not imaged, which suggests that there is a strong tilt component contributing to this misorientation. This is particularly apparent in Figure 4a which is a topograph imaged with the 422 diffraction vector. The lineage in this wafer is particularly prevalent along the four ⟨110⟩ directions and this can be observed in the equivalent IR transmission micrograph of this wafer in Figure 4b. On the topograph the

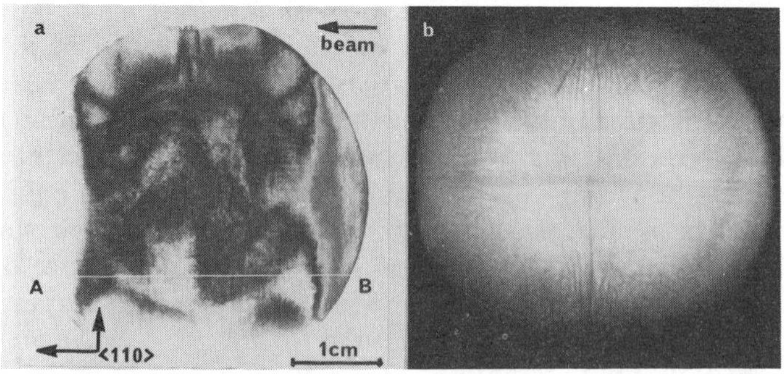

Figure 4 (a) Topograph taken using an X-ray tube source with 422 Si/422 GaAs diffraction geometry and (b) transmission IR micrograph of the same wafer.

misorientation associated with the lineage perpendicular to the diffraction vector is imaged but this is not the case for the lineage parallel to the diffraction vector. When this sample was rotated through 90° the set of lineage features which were previously exhibiting no misorientation contrast now showed strong contrast. When a 440 diffraction vector was used, which is 45° to both sets of lineage features, the misorientation contrast was poor. These points all suggest that the misorientation across lineage, at least when aligned along ⟨110⟩, contains a strong tilt component.

The micrograph in Figure 4b is the transmission IR image of the same wafer topographed in Figure 4a. This technique has been used to study SI LEC GaAs previously [7, 8] and has been shown to represent an image of the neutral EL2 distribution [17,18]. It has already been established that strong 1 μm absorption is observed where lineage and cellular dislocation configurations are observed [7, 8] but it is clear by comparison of Figures 4a and 4b that the slip bands observed on the topograph are not present on the IR micrograph. At first sight it would seem that there is thus no enhanced neutral EL2 concentration associated with dislocations in slip bands. However, if the sample is sectioned along a ⟨110⟩ direction normal to the growth axis through an area where slip bands are observed on the topograph (e.g. AB in Figure 4a), and a transmission IR micrograph is now taken, the slip bands are clearly observed to correspond to strong absorption (Figure 5). The reason that no contrast was observed for ⟨100⟩ sections is due to the

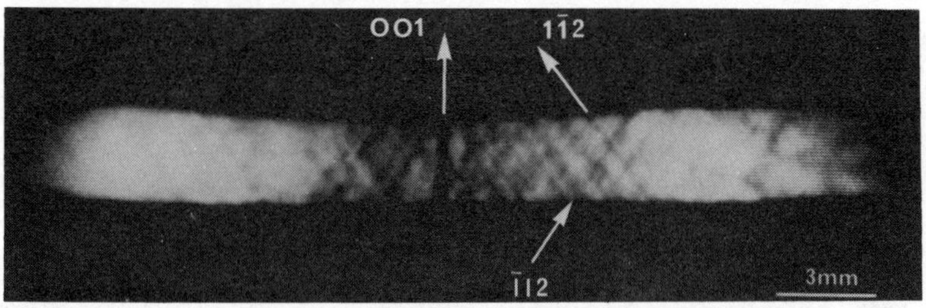

Figure 5 ⟨110⟩ cross-section cut normal to the growth axis taken through an area containing slip bands and imaged using transmitted IR.

smearing out of the contrast because the (111) slip planes are inclined to the surface. For a ⟨110⟩ section the (111) planes are parallel to the incident IR beam leading to enhanced localized absorption. Similar effects have been observed for slip bands in small diameter SI LEC GaAs crystals grown at near-atmospheric pressure [17].

Conclusions

Double crystal reflection XRT has been shown to be a useful method of mapping long range strain in SI LEC GaAs wafers. Dislocations are also clearly imaged and have been shown to be distributed in linear and cellular arrays and slip bands. The linear arrays have been shown to be associated with a matrix misorientation which contains a high tilt component and has been determined to be of the order of 0.01°. Comparisons of topographs with IR micrographs reveal that all dislocation configurations are associated with enhanced 1 μm absorption and hence enhanced neutral EL2 concentration. This is the first time that slip bands have been shown to correlate with $[EL2]^{o}$ in two inch wafers.

ACKNOWLEDGEMENTS

The authors would like to thank Cambridge Instruments for the provision of material and colleagues of the UK GaAs Consortium for helpful discussions. Financial support from SERC (SJB) and access to the SRS at Daresbury Laboratory is gratefully acknowledged and M. Hill, P. Moores and Mrs S. D. Baker are acknowledged for invaluable experimental assistance.

References

1. Hart M. (1968). *Science Progress*, **56,** 429
2. Jones G. R., Young I. M., Cockayne B. and Brown G. T. (1981). *Inst. Phys. Conf. Ser.*, **60,** 265
3. Blunt R. T., Clark S. and Stirland D. J. (1982). *IEEE Transactions on Microwave Theory and Techniques*, **MTT-30,** 943
4. Tajima M. (1982). *Jap. J. Appl. Phys.*, **21,** L227
5. Okada Y. (1983). *Jap. J. Appl. Phys.*, **22,** 413
6. Holmes D. E., Chen R. T., Elliott K. R. and Kirkpatrick C. G. (1983). *Appl. Phys. Lett.*, **43,** 305
7. Skolnick M. S., Brozel M. R., Reed L. J., Grant I., Stirland D. J. and Ware R. M. (1984). *J. Electron. Mat.*, **13,** 107
8. Brozel M. R., Grant I., Ware R. M. and Stirland D. J. (1983). *Appl. Phys. Lett.*, **42,** 610
9. Nanishi Y., Ishida S., Honda T., Yamazaki H. and Miyazawa S. (1982). *Jap. J. Appl. Phys.*, **21,** L335
10. Miyazawa S., Ishii Y., Ishida S. and Nanishi Y. (1983). *Appl. Phys. Lett.*, **43,** 853
11. Hart M. Private communication
12. Bowen D. K., Clark G. F., Davies S. T., Nicholson J. R. S., Roberts K. J., Sherwood J. N. and Tanner B. K. (1982). *Nucl. Inst. and Meth.*, **195,** 227
13. Takusa S., Otsuka H., Yoshihiro N. and Oku T. (1980). *Jap. J. Appl. Phys.*, **20,** Supplement 20–1, 25
14. Obakata T., Emori H. and Fukuda T. (1983). *Jap. J. Appl. Phys.*, **22,** L168
15. Chen R. T. and Holmes D. E. (1983). *J. Cryst. Growth*, **61,** 111
16. Bonse U. (1962). In *Direct Observations of Imperfections in Crystals*, p. 431 (J. B. Newkirk and J. H. Wernick eds.); New York
17. Skolnick M. S., Hope D. A. O. and Cockayne B. (1984). This volume, p. 446
18. Skolnick M. S., Reed L. J. and Pitt A. D. (1984). *Appl. Phys. Lett.* In press

A Low Melting Hydroxide Dislocation Etch for Single Crystal GaAs

H. Lessoff and R. Gorman

Naval Research Laboratory, Washington DC 20375, USA

Abstract

A low melting dislocation etch for GaAs single crystal wafers has been developed which consists of 50 mole per cent KOH and 50 mole per cent NaOH. This etch is a liquid at 170°C which is reproducible, controllable and reveals greater details than other dislocation etches currently in use.

Introduction

Since there is a correlation of dislocations in GaAs substrates with its electrical properties and the properties of devices made on the substrate, it is increasingly important to be able to determine reproducibly the dislocation distribution [1–3]. A number of etches have been developed for SI material with different crystallographic orientations. Richards and Crocker [4] and Schell [5] have developed etches suitable on the ⟨111⟩ face. Both of these etches are based on a fairly concentrated nitric acid solution. Other etches are reviewed by Stirland and Straughan [6]. The dislocation etching of SI ⟨100⟩ GaAs has been limited mainly to the etches developed by Abrahams and Buiocchi (A/B) [7] and Grabmaier and Watson (KOH) [8]. Each of the ⟨100⟩ dislocation etches has limitations. The A/B etch is difficult to interpret and has a dislocation 'etch memory' due to the preferential removal of perfect material during the etching [6]. The A/B etch reveals dislocations at the surface and also striae and microstructure but the interpretation is difficult [9].

The molten KOH etch has been reported to have been used over a wide range of temperatures from 300°C [8] to 450°C [1]. The range of temperatures reported would indicate that since the melting point of KOH is 360°C, the etch used at temperatures below 360°C must contain water. As the water is volatilized from the molten KOH/H_2O mixture, the resultant material solidifies and sample breakage can occur. Generally KOH etch is only used for a single etch in order to ensure reproducibility. According to T. Takenaka *et al.* an optimum KOH solution etching temperature of 350°C is required for the clearest etch pit definition [10]. Thus there are experimental difficulties reproducing etching results in KOH. The KOH etch is reported to be effective only in revealing dislocations at the surface and is particularly sensitive to strain. The A/B etch will also disclose impurity segregation, micro defects and striae [9].

The purpose of this paper is to describe a new etch which eliminates many of the reproducibility and experimental difficulties in the use of the KOH at or below its melting point. The etchant is a mixture of KOH and NaOH. A mixture of 50 mole per cent KOH and 50 mole per cent NaOH melts at 170°C although the separate constituents NaOH and KOH melt at 318°C and 360°C respectively. The melt is stable and can be used for successive etching. The etching rate is slow enough that the degree of etching can be readily controlled. The KOH/NaOH etch is effective on both ⟨100⟩ and ⟨111⟩ faces of conducting and SI GaAs.

Experimental results

Samples of ⟨100⟩ and ⟨111⟩ GaAs were prepared by chemical polishing in bromine methanol after cutting and lapping GaAs crystals. Crystals used were prepared by both the Bridgman and the LEC (liquid encapsulated Czochralski) methods and included chromium doped and undoped samples. The etchant was put into a platinum crucible which was held in a resistance heated furnace. The temperature was measured by a Pt/Pt/Rh thermocouple in the melt and temperature controlled to ±4°C. Upon stabilization of the etch temperature (300–375°C) the GaAs specimen was immersed in the molten etchant for 3–60 min, depending on crystal orientation. The specimen was then removed from the etchant and allowed to cool to room temperature, rinsed with de-ionized water, acetone, methanol and electronic grade isopropyl alcohol prior to warm air drying. The samples were observed using a microscope with Nomarski interference contrast. Correlation of etch patterns was made with specimens also etched with KOH and the Richards and Crocker etch.

After immersion in the etch for 45 min at 350°C, the etch pit patterns developed on ⟨100⟩ LEC and Bridgman SI GaAs are shown in Figures 1a and 1b. The KOH/NaOH developed etch pits are of the typical hexagonal configuration developed by molten KOH [8]. At 350°C etching times of 15–60 min are effective; increased etching time enlarges the etch pit structure. The non-etched areas of GaAs maintains a smooth and specular surface. Additional structure is apparent including shallow rectangular depressions having their lengths aligned on the $\langle\bar{1}10\rangle\langle 1\bar{1}0\rangle$ and the widths on the $\langle\bar{1}\bar{1}0\rangle$ ⟨110⟩ directions (Figure 1b). The rectangular depressions also form line segments. The KOH/NaOH etch is also effective on ⟨111⟩ GaAs, although the time of etch must be reduced, 0.5–3 min, to develop typical triangular pits. Prolonged etching on the ⟨111⟩ face causes extensive surface roughening and increases the difficulty in observing the etch pits.

The rate of etching of the KOH/NaOH solution was determined by measuring the loss of weight of a sample with etching time. Knowing the dimensions of the sample and the density of GaAs one can calculate the etch

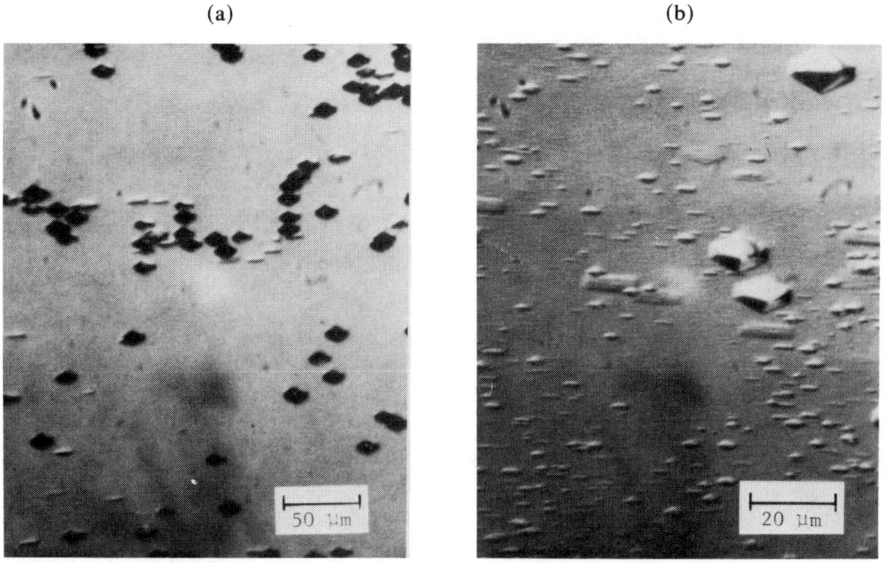

Figure 1 (a) ⟨100⟩ semi-insulating undoped LEC GaAs etched for 45 min at 350°C; (b) ⟨100⟩ semi-insulating Cr doped Bridgman GaAs etched for 45 min at 350°C.

rate. At 350°C, the etch rate for the NaOH/KOH etch was found to be 0.08 ± 0.01 μm/min for a ⟨100⟩ surface of GaAs. Ishii *et al.* have reported an etch rate of 0.98 μm/min at 350°C for KOH on ⟨100⟩ GaAs [11]. The mixed hydroxide etch is nearly an order of magnitude slower than KOH and thus allows a controllable etch of thin GaAs epitaxial films to reveal dislocations.

Conclusions

A 50 mole per cent KOH/50 mole per cent NaOH melt is an effective etchant for revealing dislocations in GaAs. The etchant is useful from 300–400°C with the clearest etch patterns being developed at 350°C. The etching time ranges from 15–60 min for ⟨100⟩ faces and 0.5–3 min for ⟨111⟩ faces. The etch leaves the dislocation-free areas of GaAs with minimum attack and allows the observation of additional defect structures.

References

1. Miyazawa S., Mizutami T. and Yamazaki H. (1983). *Jap. J. Appl. Phys.*, **21,** L542
2. Nanishi Y., Ishida S., Honda T., Yamazaki H. and Miyazawa S. (1983). *Jap. J. Appl. Phys.*, **21,** L335
3. Nanishi Y., Ishida S. and Miyazawa S. (1983). *Jap. J. Appl. Phys.*, **22,** L54
4. Richards J. L. and Crocker A. J. (1980). *J. Appl. Phys.*, **31,** 611
5. Schell H. A. (1957). *Z. Metallkd.*, **48,** 158

6. Stirland D. J. and Straughan B. W. (1976). *Thin Solid Films*, **31,** 139
7. Abrahams M. S. and Buiocchi C. J. (1965). *J. Appl. Phys.*, **36,** 2285
8. Grabmaier J. G. and Watson C. B. (1969). *Phys. Stat. Solidi*, **32,** K13
9. Wagner W. R., Greene K. I. and Koszi L. I. (1981). *J. Electrochem. Soc.*, **128,** 1091
10. Takenaka T., Hayashi H., Yamamoto S., Murata K. and Inoguchi T. (1978). *Jap. J. Appl. Phys.*, **17,** 447
11. Ishii M., Hirano R., Kan H. and Ito A. (1976). *Jap. J. Appl. Phys.*, **15,** 645

Dislocation Mapping in Semi-Insulating LEC GaAs Wafers Using Thermal-Wave Imaging

Y. E. Strausser and A. Rosencwaig

Therma-Wave Inc., Fremont, California 94539, USA

Abstract

We have applied a new imaging technique, thermal-wave microscopy, to the analysis of GaAs wafers. We have found substantial correlation between thermal-wave images and XRT (X-ray topography) images of these dislocations.

Thermal-wave microscopy is a recently developed technique which images subsurface thermal features in materials [1]. Thermal waves are generated from the absorption in the sample of the energy from a beam-blanked electron beam in a scanning electron microscope. This produces periodic heating in the sample with subsequent periodic heat flow in the form of thermal waves. These thermal waves will reflect and scatter from surface and subsurface thermal features. Thermal features are regions of the sample that exhibit local variations in volume specific heat and thermal conductivity. This includes subsurface cracks or voids, grains and grain boundaries; crystalline defects such as twins, stacking faults and dislocations; and aggregates of lattice point defects including vacancies, and interstitial or substitutional impurities, such as dopants.

The scattered and reflected thermal waves change the local temperature profiles in the sample. These variations in the local temperature profiles are detected through the effects on the thermoacoustic waves that are generated by the local stress-strain perturbations produced by the thermal waves. The thermoacoustic waves, in turn, are detected by means of an ultrasonic transducer in contact with the sample. Thus by recording the amplitude of the thermoacoustic waves as a function of electron beam position on the sample surface we can record the local variations in the sample temperature, and thus record a thermal-wave image.

We have applied thermal-wave microscopy to the analysis of dislocations in SI (semi-insulating), LEC (liquid encapsulated Czochralski) GaAs wafers. As a calibration of the technique, we have correlated the thermal-wave images with conventional techniques, such as XRT, KOH etching and cathodoluminescence.

X-ray topography is a technique which uses X-ray diffraction to image features which produce diffraction contrast (i.e. produce changes in diffracted beam angles) [2]. Thus it is applied primarily to single crystal or, at least, highly oriented polycrystalline samples. Contrast is produced at crystalline defects such as dislocations, twins or stacking faults which are regions of high strain and often the location of precipitate particles. In general, the XRT images local variations in crystal strain and is thus widely used to image dislocations in GaAs crystals.

The X-ray topograph is usually recorded by exposing a photographic emulsion, thus producing a unity magnification image of the sample. Further magnification is then obtained photographically by enlarging the image during printing. The magnification and resolution limits are imposed by the grain size in the emulsion and this typically limits the resolution to approximately 1–5 μm.

Figure 1 is a part of an X-ray topograph taken on an LEC-grown GaAs crystal. This is a Lang transmission X-ray topograph using the 220 diffracted

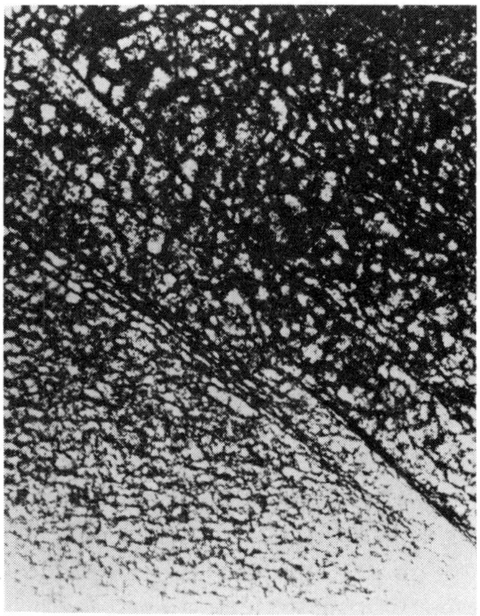

Figure 1 X-ray topograph of a section of an LEC-grown GaAs crystal at 20× magnification, showing a dense network of dislocations.

beam from Mo $\kappa\alpha_1$ radiation. A 30 hour exposure was required using Ilford L4 50 μm thick emulsion. The patterns seen in this topograph are primarily very dense dislocation networks. These dislocation networks seem to grow in the direction of the long axis of the crystal, since successive slices of the crystal produce almost identical dislocation topographs. Because of this, a reflection topograph from either surface of the sample produces an almost identical pattern.

Figure 2 is a thermal-wave micrograph of the same area of the crystal. The thermal-wave micrograph was obtained in two minutes (as compared with the 30 hours needed for the X-ray topograph). Clearly, the same features are

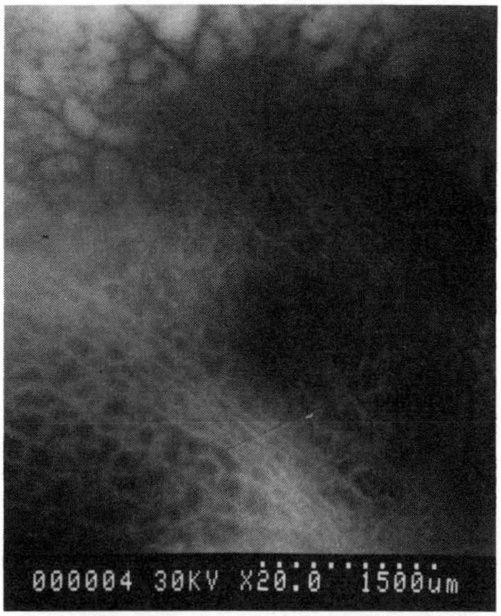

Figure 2 Thermal-wave image of the same region of the GaAs sample shown in Figure 1, showing the same dislocation networks.

visible in the two imaging techniques and, in fact the dislocation networks appear to be better resolved in the thermal-wave image. This is partly due to the topograph being a transmission image and the dislocation networks having, therefore, a thicker projection when integrated through the sample than when only being integrated through the thermal-wave sampling volume. We have also seen individual dislocations at higher magnification in the thermal-wave image, as shown in Figure 3, where the individual dislocations appear as small points.

We thus find substantial correlation between thermal-wave microscopy and XRT in the imaging of dislocations in GaAs crystals. The thermal-wave images show dislocation networking with varying sized subgrains distributed over the wafer in patterns such as are observed by XRT. Higher magnification thermal-wave imaging of the networks (Figure 3) shows individual dislocations as high contrast features connected together in the networks by a web of lesser contrast, all surrounding regions which contain few or no contrast-producing features. We believe the web to be material into which the dislocations have exerted enough of a strain field to have gettered impurity atoms and point defects. The contrast free regions surrounded by the web would then correspond to the most defect-free material.

Thermal-wave microscopy appears to be a very useful tool in the characterization of GaAs crystals, and has significant benefits over other characterization methods. These benefits include: no sample preparation; non-destructive analysis; and a much shorter time for data acquisition. Furthermore, these benefits are achieved with no loss of spatial resolution or sensitivity.

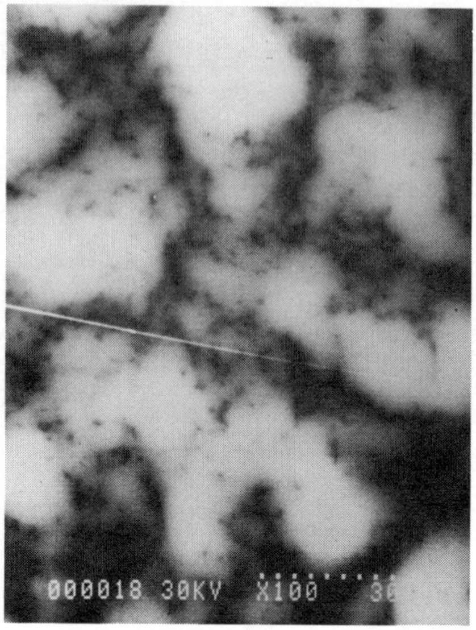

Figure 3 High magnification thermal-wave image of the LEC-grown crystal in the region of the crack seen in Figures 1 and 2, showing dislocation networks, features (individual dark points) related to individual dislocations and regions of high and low defect density.

ACKNOWLEDGEMENTS

We would like to thank Martin Scott of Hewlett Packard for supplying the GaAs samples and the X-ray topograph.

References

1. Rosencwaig A. (1982). *Science*, **218**, 223
2. Tanner B. K. and Bowen D. K. (1980). In *Characterization of Crystal Growth Defects by X-ray Methods*; Plenum Press, New York

Dislocation Configurations and Complex Decoration Effects in Undoped, Semi-Insulating, LEC GaAs

D. J. Stirland, P. D. Augustus, M. R. Brozel* and E. J. Foulkes*

Plessey Research (Caswell) Ltd., Allen Clark Research Centre, Caswell, Towcester, Northants NN12 8EQ, UK

Abstract

Blocks of LEC (liquid encapsulated Czochralski) GaAs with (001), (110) and (1$\bar{1}$0) faces, containing ⟨110⟩ lineage features, have been examined by optical and infrared microscopy. Concentrations of [EL2] are reduced in regions bordering 'sheet' lineage. High resolution optical and electron microscopic examinations have shown that the etched appearance of dislocations depends on local precipitation effects. A typical precipitate has been identified as a single crystal of hexagonal arsenic. There is evidence that some dislocations may have moved after precipitation.

Recent work [1–6] has indicated that dislocations in 2 or 3 inch diameter, undoped, SI LEC GaAs are found in cellular arrangements [1–6] and lineage structures [2–6], the latter classification being subdivided into 'sheets' and 'streamers' [6]. In addition, Okada [3] has postulated that two *different kinds* of dislocation exist in LEC (not boat-grown) GaAs, since an A/B etch [7] produces ridges at some dislocations and grooves at others. Here, we describe detailed observations of sheets and streamers, together with results which indicate that localized precipitation effects determine etching behaviour at dislocations.

A block specimen was cut from a 6 mm thick (001) slice of a 3 inch diameter ingot to exhibit a prominent lineage feature along ⟨$\bar{1}$10⟩. The detailed configuration of the lineage shown schematically in Figure 1a was deduced by optical examinations of all the etched faces, some parts of which are shown in Figure 1b. Distinction between clustered (sheet) and extended (streamer) dislocation arrays is exemplified by features ps and qr respectively. Transition point X on (001) face ABCD lies vertically above corresponding position X′ on (001) face EFHG within accuracy of measurement, thus demonstrating that the sheet/streamer structure extends down the ⟨001⟩ growth axis for at least 6 mm. The transition is less sharply defined when individual dislocations comprising the lineage structure are resolved (Figure 1d). However, 1 μm wavelength transmission infrared examination [5, 6] reveals another distinction. Features with low infrared absorption flank the sheets, but not the

* Department of Electrical and Electronic Engineering, Trent Polytechnic, Burton St., Nottingham NG1 4BU, UK.

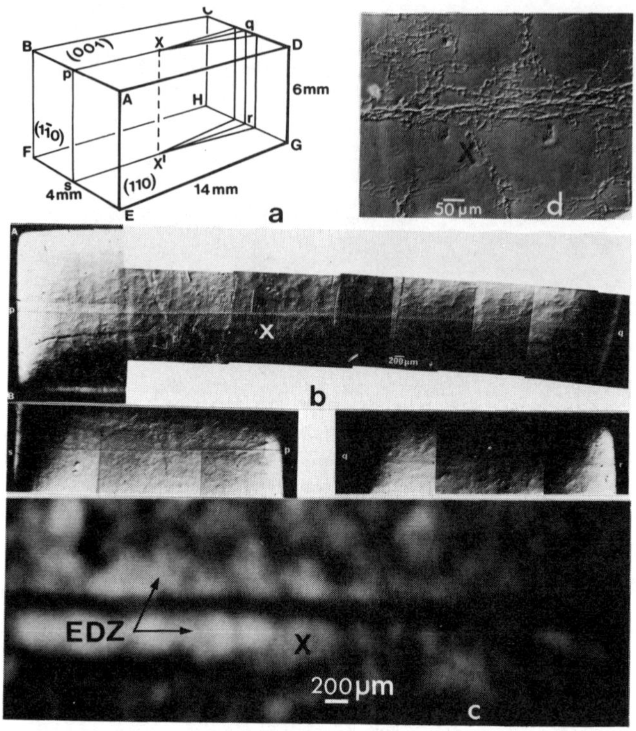

Figure 1 (a) A/B etched block specimen, (b) optical micrographs of indicated etched faces, (c) infrared transmission micrograph, (d) transition region.

streamers (Figure 1c). They are 200–300 μm wide zones of both low dislocation density and [EL2], which extend down the entire sheet length [8]. We have named them 'extended denuded zones' (EDZs). Regions containing cellular structure show higher absorption at cell interiors than at EDZs, although cell interiors themselves are essentially dislocation-free. This is because, for a 6 mm thick specimen, the transmitted infrared radiation passes normally through the walls of 20–30 cells, whose dimensions are ~200–300 μm. The EDZs are bounded on one side by sheet-forming dislocations and on the other by cellular wall dislocations [8].

Figure 2a shows an A/B etched area at higher magnification containing one dislocation (C_1C_2) which is part of a cellular network and another (D_1D_2) which is not. Interference (Nomarski) contrast is reversed between (C_1C_2) and (D_1D_2): the former is grooved, the latter is ridged. Features P_1 and P_2 on (C_1C_2) are etch pits. Figure 2b has a single dislocation along which contrast reversal occurs, and close examination shows that region (FG) contains many contiguous etch pits. An explanation has been given previously [9]. In general, the A/B etch produces ridges at dislocations. However, if dislocations are decorated with precipitates which etch preferentially then pits are formed at the top of ridges. If the precipitates are small and closely

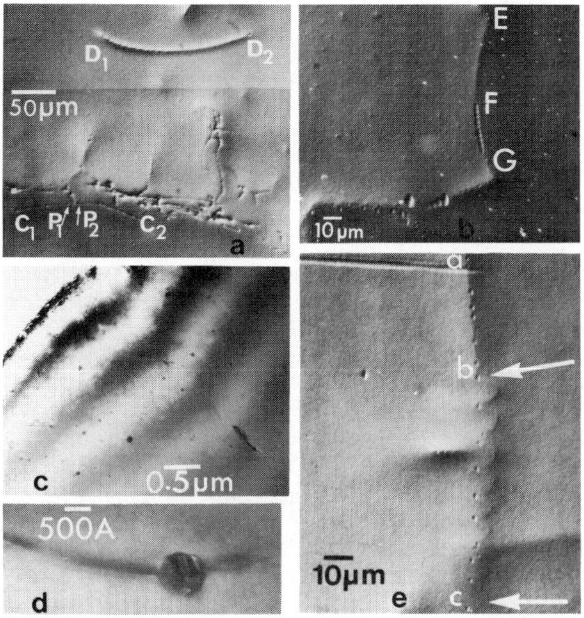

Figure 2 (a, b) A/B etched dislocations ⟨001⟩ surface, (c, d) TEM of dislocation region, (e) A/B etched dislocation ⟨110⟩ surface.

spaced ($\leqslant 1$ μm) a continuous *groove* is formed. Some of these explanations were confirmed by TEM (transmission electron microscopy) examinations of etched specimens. Thus, thickness fringes in Figure 2c delineate a groove, the effect of etching a decorated dislocation. The part of the dislocation embedded in the thin foil contains a precipitate, shown further magnified in Figure 2d. This precipitate was identified by electron diffraction to be a single crystal of elemental hexagonal arsenic, thus substantiating an earlier identification [10] of a precipitate on a dislocation in boat-grown, Cr-doped GaAs. Other micrographs show elliptical pits on grooves, presumably resulting from etched-out precipitates, as at P_1 and P_2 (Figure 2a).

Some A/B etched dislocations on (110) surfaces, and on (001) surfaces, have a configuration similar to that in Figure 2e. Conflicting explanations for the bowed segments of the dislocation linking the etch pits within the arrowed section are possible. Either they represent sections of a gliding dislocation pinned in motion by precipitates, or they represent projected parts of a helical climbing dislocation. In the latter case, the precipitates might be debris generated during climb. Because the set of etch pits from a–c lie on a continuous line there is a strong presumption that precipitation occurred *before* dislocation bowing. TEM observations on such dislocation configurations are now required to distinguish between these mechanisms, which clearly are of prime importance in elucidating the link between EL2 and dislocations.

ACKNOWLEDGEMENTS

This work was partially supported by Procurement Executive, Ministry of Defence, sponsored by DCVD.

References

1. Clark S. and Stirland D. J. (1981). *Inst. Phys. Conf. Ser.*, **60,** 339
2. Chen R. T. and Holmes D. E. (1983). *J. Cryst. Growth*, **51,** 111
3. Okada Y. (1983). *Jap. J. Appl. Phys.*, **22,** 413
4. Brozel M. R., Grant I., Ware R. M. and Stirland D. J. (1983). *Appl. Phys. Lett.*, **42,** 610
5. Stirland D. J., Grant I., Brozel M. R. and Ware R. M. (1983). *Inst. Phys. Conf. Ser.*, **67,** 285
6. Skolnick M. S., Brozel M. R., Reed L. J., Grant I., Stirland D. J. and Ware R. M. (1984). *J. Electron. Mat.*, **13,** 107
7. Abrahams M. S. and Buiocchi C. J. (1965). *J. Appl. Phys.*, **36,** 2855
8. Stirland D. J., Brozel M. R. and Foulkes E. J. Unpublished
9. Augustus P. D. and Stirland D. J. (1980). *J. Micros.*, **118,** 111
10. Cullis A. G., Augustus P. D. and Stirland D. J. (1980). *J. Appl. Phys.*, **51,** 2556

Section 3:
Crystal growth and related problems

Crystal Growth of Semi-Insulating Gallium Arsenide

R. K. Willardson

Willardson Consulting, Spokane, Washington 99216, USA

Abstract

Semi-insulating 76 mm diameter GaAs substrates can be produced by zone levelling, Bridgman, static and electrodynamic gradient freeze methods as well as by low and high pressure liquid-encapsulation techniques. Each growth method has unique salient features, which with the appropriate temperature gradient modification and control can give high quality crystals with few dislocations. Micro electrical homogeneity presents a new challenge. Acceptance of a particular method for growing material for use in integrated circuits will be based on cost and final device yield.

Introduction

Gallium arsenide is beginning to emerge as a realistic technological basis for the next family of very-high-speed integrated circuits for analog, digital and optoelectronic applications. In the past two years significant progress has been made in crystal growth techniques. However, this progress merely emphasizes the magnitude of the challenge to grow crystals that are not only dislocation free but also extremely uniform and controlled with respect to all of the electrical parameters required by the new GaAs foundaries as they begin regular production runs.

A number of techniques have been developed to grow 'dislocation free' crystals and to more than double the size of the crystals to over 10 kg. Large crystals of 100 and 125 mm diameter still have high densities of dislocations and there is a question as to whether such large diameter substrates will have sufficient strength to survive a real production line with reasonable yields. It is certain that some methods of crystal growth are not adaptable to such large diameters. But even more significant some of the growth techniques used to reduce dislocations also reduce the concentration of the EL2 level and modify other defect as well as impurity levels.

Homogeneity is a key factor. As temperature variations, gradients and other conditions are improved to reduce dislocations, the growing interface can change from concave to convex. Many papers have been published on InSb crystal defects, such as impurity striations and facets [1, 2], but convex interfaces in pulled GaAs crystals are a new development. The consequences of improved growth interfaces will unfold during the next two years.

A number of papers in this volume highlight some of the recent advances in GaAs crystal growth and in techniques for the determination of the homogeneity of characteristics of major concern such as concentrations of the EL2 level and FET (field effect transistor) threshold voltages. This article presents historical, current and future perspectives for these developments. Special attention will be given to new designs for equipment and the applicability of certain growth methods to the production of crystals with diameters greater than 100 mm and weights greater than 20 kg.

Dislocation density

For the past quarter of a century a controversy has ensued as to whether boat grown or pulled crystals were the best. The simple concensus was that boat grown crystals had the lowest dislocation density, while pulled crystals had the highest purity, particularly after the advent of PBN (pyrolitic boron nitride) crucibles and *in situ* compounding in high pressure systems. Simply stated this meant 10^3 versus 10^5 dislocations cm^{-2} and over 10^{16} versus less than 10^{15} silicon impurities cm^{-3}. While the concensus was almost correct for commercially available GaAs, in the mid-1960s researchers at IBM produced both Czochralski and Bridgman grown GaAs with silicon concentrations less than 10^{15} cm^{-2} and Steinemann and Zimmerli [3] pulled 'dislocation free GaAs'.

Steinemann and Zimmerli showed that dislocation-free crystals of doped and undoped GaAs can be grown by the Czochralski method from a carbon crucible with a surface coating of pyrolytic carbon. They pointed out that the required growth conditions are: (a) precise control of stoichiometric melt composition, i.e. constant partial pressure of As within the sealed system; (b) low temperature gradient in the solidified portion of the crystal near the solid-liquid interface; (c) minimization of radial heat flow in the crystal; and (d) a nearly horizontal solid-liquid interface.

These investigators produced a single crystal seed by reducing the diameter immediately after seeding to 1–2 mm and keeping this value for the next 10–20 mm of growth. The crystal was then tapered out to a diameter of 15 mm, and 100 g of crystal was grown from a 125 g melt. The key to temperature gradient control was a radiation shield or lid on top of the crucible. This lid had a 25 mm hole in its centre through which the crystal was pulled. Crystals were grown in the ⟨111⟩, ⟨100⟩ and ⟨110⟩ directions.

Although the crystals produced by Steinemann and Zimmerli were relatively small compared to the 10 kg ingots grown today, their quality and size were as good or better than today's industry standard for laser diodes—the Sumitomo class DF Bridgman grown substrates which are guaranteed to have a 20×20 mm area with less than 100 dislocations cm^{-2}.

More recently Kirkpatrick and co-workers [4] reduced the radial gradients near the crystal-melt interface in a Melbourn puller by doubling the thickness of the B_2O_3 layer, i.e. putting a better lid over the melt. By increasing the layer thickness from 9 to 17 mm, the etch pit density at the top of the crystal in the central section of a 75 mm diameter crystal was decreased from about

1×10^5 to 1.7×10^4. Growth was in the ⟨100⟩ direction. Similar results have been reported by Peigen *et al.* [5] for crystals of a smaller diameter. These crystals grown in a LP (low pressure) LEC (liquid encapsulated Czochralski) system showed a decrease in the density (cm^{-2}) of dislocations from 10^5 to 10^4 for 60 mm diameter crystals as the B_2O_3 layer was increased from 12 to 25 mm and from 30 000 to 2000 cm^{-2} for 30 mm diameter crystals.

In 1982 Hiskes *et al.* (as noted in Refs. 6 and 7) reported the growth of 65 mm LP LEC GaAs with dislocation densities under 200 cm^{-2} over 80% of a substrate cut from the tail of a Si-doped crystal. Typical dislocation densities for semi-insulating material were in the low 10^4 range with regions on the axis of the ingot near 1×10^3 cm^{-2}. Unfortunately details of this work have not been published.

Another approach to the reduction of dislocation densities is the use of impurity hardening effects [7–10]. Hobgood *et al.* [9] have used iso-electronic dopants (Sb, B, In) to harden the lattice and prevent thermally induced glide dislocations. Indium is very effective in reducing dislocations. Crystals virtually free of dislocations (100–400 cm^{-2}) were grown. At the periphery dislocation densities are found to increase and they are revealed as linear arrays aligned along the two orthogonal 110-glide directions lying in the (100) surface. It is important to note that these crystals were grown in the ⟨100⟩ direction in a HP LEC puller and had a diameter of 43 mm. With the exception of dislocation density, the In-doped substrates are reported to be indistinguishable from undoped GaAs. Boron additions only reduced the dislocation density by a factor of 6 compared to undoped GaAs.

Kirkpatrick *et al.* report a reduction by a factor of two in the dislocation density at all three points of the W configuration at the top of a 75 mm diameter crystal as the inert gas pressure is reduced from 300 to 50 psi [4]. Later more detailed studies have been made by Emori *et al.* [11].

It is well known from Bridgman GaAs crystal growth that the arsenic pressure can strongly affect the dislocation density and that the optimum pressure corresponds to a temperature of 617°C for the arsenic reservoir. Stoichiometry is expected to become a major factor in low dislocation LEC GaAs.

The application of a magnetic field to the melt during growth results in remarkable improvements in the quality of silicon crystals. This technique has been extended to GaAs crystal growth by Terashima *et al.* [12] with striking results. A drastic reduction occurs in the temperature fluctuations in the melt. Temperature fluctuations of 18°C at zero magnetic field were reduced continuously down to 0.1°C as the magnetic field was increased to 1250 Oe. The EL2 level was reduced from about 1.2×10^{16} to 1.3×10^{15} cm^{-3}. Also significant was the reduction in dislocation density from about 1×10^5 to less than 3000 cm^{-2}.

Nishida [13] has recently reported that by temperature-control improvements within the crystal growth furnace it is possible to produce high-quality 65 mm diameter LEC GaAs with dislocations ranging from 0 to 200 cm^{-2}. These improvements mainly consist of using the LEF process [5], a high pressure Melbourn puller, and a lid over the crucible as in the work of Steinemann and Zimmerli [3]. This lid has the shape of a top hat and provides an afterheater similar to those used at Bell & Howell to control axial gradients for LP LEC growth in 1970. The thick (100 mm) B_2O_3 layer prevents loss of arsenic from the exposed surface of the crystal at high temperatures. The hole in the lid only has to accommodate the pull rod and the seed, since the crystal remains in the afterheater until the pull is completed.

Other reports [14] credit Sumitomo Electric with the growth of GaAs crystals 127 mm in diameter, 15 cm in length and weighing 9.5 kg. These crystals are grown by the LEC method using a computer controlled temperature environment, but have high dislocation densities (60 000 cm^{-2}). Their commercial production 76 mm LEC crystals have dislocation densities of 20 000 to 30 000 cm^{-2}.

Dislocation densities can be reduced in LEC growth by control of stoichiometry, thermal gradients and fluctuations using minimum argon pressure, thick boric oxide layers, magnetic fields, precise automated temperature and diameter control as well as impurity hardening additions. The present status regarding the independent effects of some of these variables is summarized in Table 1.

Crystal growth equipment

Gallium arsenide crystals can be grown from the melt in two basic ways. In one, which includes the Czochralski (CZ) and floating zone, the growing interface and the resulting crystal do not touch the container (boat or crucible). These growth processes are discussed in the next two sections. In the other methods, the melt and the crystal are contained on at least two sides and twinning can be nucleated at the point of contact of the boat with the growth interface. Yields are reduced 40 to 90% because of twinning. Techniques such as sandblasting the inside of quartz boats roughen the surface and the surface tension of the melt permits it to ride on the peaks. This procedure reduces twinning.

Boat growth methods include Bridgman, where the furnace moves; Stockbarger where the crucible moves; and gradient freeze, where the temperature gradient moves. Zone levelling can utilize any one of these methods to move a molten zone along an ingot to convert it to a single crystal with uniform impurity concentrations. These methods can be used horizontally, vertically or at any angle in between.

The boat growth methods have the advantage of having a support for the weight of large crystals, but the disadvantage of twinning and contamination from both the boat and the hot walls of the enclosure tube. Silicon contamination from hot quartz walls usually ranges from 0.5 to 5 ppma in GaAs, but can be reduced an order of magnitude by use of As_2O_3 or related approaches [15].

Table 1 *Crystal growth of GaAs*

Growth technique	Crystal diameter (mm)	Crystal weight (g)	Crystal length (cm)	Growth direction	Dislocation[c] density (cm^{-2})	Dislocation[d] distribution	Reference
Gremmelmeir							
Graphite crucible	15	100	11	111	10	—	3
with lid				110			
				100			
Horizontal Bridgman							
3T-Technique	20 × 20[b]			111	100	—	DF[a]
Si doped for	30 × 30[b]			111	500	—	AAA[a]
Laser Diodes	40 × 40[b]			111	2 000	—	AA[a]
3T-technique	40			111	10 000	—	R-40[a]
with As_2O_3	50		100	111	20 000	—	R-50[a]
for SI	76			111	30 000	—	R-76[a]
	76	8800	30	111	5 000	—	15
Low Pressure LEC							
(?) B_2O_3 Si	65	1800	10		200[e]		6
(?) B_2O_3 SI	65	1800	10		1 000		6

Growth technique	Crystal diameter (mm)	Crystal weight (g)	Crystal length (cm)	Growth direction	Dislocation[c] density (cm^{-2})	Dislocation[d] distribution	Reference
Low Pressure LEF							
12 mm B_2O_3	30			110	30 000	U	5
25 mm B_2O_3	30			110	2 000	U	5
15 mm B_2O_3	60			110	100 000	U	5
25 mm B_2O_3	60			110	10 000	U	5
High Pressure LEC							
9 mm B_2O_3	76			100	100 000	W	4
17 mm B_2O_3	76			100	17 000	U	4
(?) B_2O_3	65				100		13
(?) B_2O_3	127	9500			60 000		14
0.1% In addition	50	2000	20	100	200	U	9
High Pressure MLEC							
0 Oe	50	800	8	100	100 000		12
1300 Oe	50	800	8	100	3 000		12

a–Sumitomo standard products designations; b–guaranteed area on a 53 × 43 mm D-shaped wafer; c–dislocation density taken from centre of slice near top of crystal; d–distribution uniform (—), W-shaped (W) or U-shaped (U); e–measured at bottom of crystal.

Horizontal Bridgman (HB) growth is the most commonly used method for boat growth of GaAs. This technique has produced low dislocation substrates for optoelectronics for the past two decades. New developments in Bridgman equipment include systems using large 90 mm wide, 600 mm long boats, which have produced 8.8 kg single crystals and systems using 53 or 65 mm wide boats that are one metre in length. These systems have three computer-controlled temperature zones [15].

Unfortunately, Bridgman crystals of large cross-section have relatively high dislocation densities (5000 cm^{-2}). In addition if 75 mm or greater diameter wafers become standard, most of the current production equipment will be too small to be of commercial value. Also, while crystals can be grown in the ⟨100⟩ direction, yields are so low that ⟨111⟩ or ⟨110⟩ growth directions are usually used.

Vertical Bridgman growth units with tapered crucibles have been used to obtain large round ingots, but have not proven commercially successful.

Another new development is a commercial electrodynamic gradient (EDG) system manufactured by the Mellen Company of Webster, New Hampshire. In this EDG system, the temperature adjacent to each furnace winding in the growth zone is measured at four thermocouple ports radially located 90°

Figure 1 Electrodynamic gradient crystal grower in horizontal position. Unit can be rotated 90° to platform on the right for vertical operation. Courtesy of the Mellen Company.

from each other. The temperature data are fed into a computer which controls the current to each non-inductive winding to produce a linear, electronically movable precision temperature profile.

In a typical EDG operation, the gradient could be 1°C cm^{-1} moving at 1 or 5 mm h^{-1} with the melt kept at 1240 ± 1°C, the solid at 1230 ± 1°C and the arsenic reservoir at 617 ± 0.1°C. After the growth is completed, a cooling cycle of 10°C h^{-1} (or any other) can be used. Precision profiles are easily programmed into the computer and can be readily changed.

Planar heat flow, i.e. heat flow parallel to the bore, is established by control of both heating and watercooling circuits. Crystals of GaAs oriented ⟨100⟩ have been grown in this sytem, which can be operated either horizontally as shown in Figure 1 or vertically.

The heat exchanger method (HEM) was developed by Crystal Systems for the growth of single crystal sapphire and has more recently been used to produce silicon crystals for photovoltaic applications. This technique is now being applied for the growth of large crystals of GaAs by Ghemini Technologies—a joint venture between Crystal Systems and Cabot Crystals, Inc. In the HEM process the solid–liquid interface expands from a seed at the bottom of a crucible. As growth proceeds, impurities are rejected to the outside surfaces, which are removed when the crystal is processed. Temperature fluctuations are minimized by this motionless crystal growth process, promoting excellent crystalline perfection. Large crystals and square substrates can be produced economically, but it would appear that they would be more applicable to solar cells than integrated circuits.

High pressure LEC systems

High pressure (HP) LEC systems were initially developed by Arthur D. Little and the Royal Radar and Signals Establishment for GaP and InP. Cambridge Instruments is the major commercial supplier. In the late 1970s, Swiggard, Aucoin and others [7, 16] introduced the concept of HP *in situ* compound synthesis from the elemental Ga and As components. The synthesis occurs rapidly and exothermally at about 820°C under 60 atmospheres of argon pressure which in conjunction with the B_2O_3 layer is sufficient to minimize the escape of arsenic.

High purity PBN crucibles are a key to the preparation of high quality semi-insulating GaAs. One should be cautious in purchasing and handling PBN crucibles so as to obtain and maintain the highest purity. These crucibles are expensive but can be reused 20 or more times. The purity of the B_2O_3 is equally important.

The Malvern, with a charge capacity of 1 kg, was introduced by Metals Research over a dozen years ago. It was followed by the Melbourn, which has a resistance heated 150 mm diameter crucible capable of containing charges up to 10 kg. Since Metals Research was acquired by Cambridge Instruments, the Malvern has become the CI-351. The Melbourn has been modified to make it less expensive, but more difficult to clean. It comes in two versions, the CI-354 and CI-358. The outside chambers and other equipment are identical, but the internal heater assembly of the CI-354 is designed for a 4 kg capacity and 50 mm diameter crystals, while the capacity of the CI-358 is 8 kg of GaAs, and it is designed for 75 mm diameter crystals.

The work of Thomas *et al.* [16] and Kirkpatrick *et al.* [4] as well as the indium doping studies for low dislocation densities [9] were done with a Melbourn puller, while the magnetic field studies [12] utilized an inhouse modified MSR-6RA. Figure 2 shows a schematic drawing of the newly

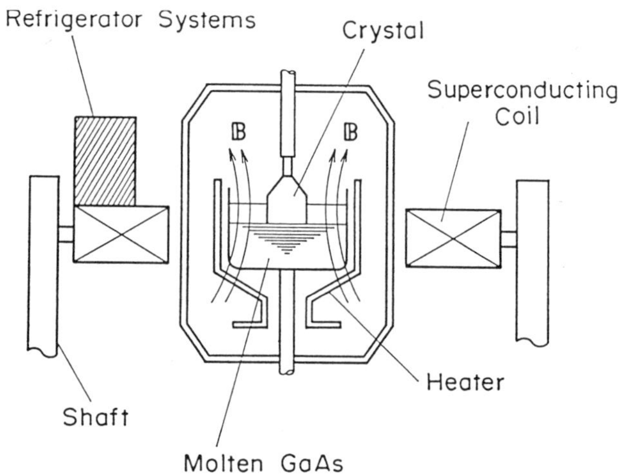

Figure 2 Schematic vertical magnetic field applied to LEC crystal puller. After Terashima *et al.* [17].

developed vertical MLEC apparatus. The magnet can supply up to 5000 Oe to the crucible centre. When a magnetic field of 1000 Oe is applied, temperature fluctuations both in a 1 kg melt in a 100 mm diameter crucible and in a 3 kg melt in a 150 mm crucible are reduced from 16°C to 0.3°C. A reduction to 1°C is made by only 500 Oe. However, with a 5 rpm crucible rotation it requires 900 Oe for the fluctuation reduction to 1°C [17].

An automated growth process with computerized diameter control has been patented by MITI. The process is in use at the Optoelectronic Joint Research Laboratory at Kawasaki. This system permits crystal diameter control to within ± 1 mm.

The major advantage of the HP LEC system is the ability to grow crystals in the ⟨100⟩ direction with high yield. A major problem is heat transfer from the resistance furnace to the part of the crystal above the B_2O_3 by the high pressure of argon. The result is the loss of arsenic from the surface of the crystal. In extreme cases, enough arsenic is lost to cause droplets of gallium to form. These gallium droplets can move down the side of the crystal into the melt or actually penetrate the crystal and form gallium inclusions. Penetration of gallium inclusions apparently occurs only when the diameter of the crystal increases markedly [4]. In general, dislocation clusters are formed around surface gallium inclusions. In addition to generating strains throughout the crystal, dislocations and other defects on the periphery of the ingot, gallium droplets returning to the melt can modify the stoichiometry of the melt, especially near the solid-liquid interface.

Other disadvantages of the HP LEC system are the lack of access to the growing chamber while the GaAs is molten. Crystals cannot be removed or additions made to the melt. Stoichiometry modification or control is very limited.

Low pressure LEC units

For the past 15 years the LP LEC technique has been used to produce GaAs, mainly for LEDs (light emitting diodes). Initially, this process required that the compound be formed in a separate system, but recently the use of an arsenic cell injection system [6] has been used to compound the GaAs *in situ.* The arsenic cell can be inserted using the seed rod and a vacuum lock. The cell is used to bubble arsenic into the hot gallium to form the compound and is subsequently removed from the system. In another version the arsenic cell is mounted within the pulling chamber and can be used to inject arsenic into the melt continuously or periodically. A preferred arrangement is to have the arsenic cell external to the system. In the latter case it is more convenient to obtain precise control of the temperature and hence the arsenic pressure.

By proper design of the injection nozzle contamination of the melt can be minimized and stoichiometry of the GaAs can be maintained or modified. Precise stoichiometry control requires an injection tube to penetrate the B_2O_3 layers at all times. This tube, of course, stirs the melt and requires a larger diameter crucible. Thick layers of B_2O_3 present additional difficulties.

Most LP LEC pullers are converted silicon pullers or of a proprietary laboratory design. During the past year Hamco has announced a commercial GG80 LP GaAs puller, complete with arsenic injection cell. (Of course this unit is a modified version of their old reliable GG800 8 kg silicon crystal grower.) Hamco is developing a GG850 unit which has a chamber designed for pressures up to 5 atm. They are reported to be developing an automatic diameter control for these units. The new Wacker GaAs facility is using special LP LEC units made by Leybold-Hereaus. Harris Microwave has a custom designed, 30 kg capacity LP LEC puller from which W. M. Ford has grown a 14 kg GaAs single crystal [18].

The LP systems have the advantage over HP systems of greater flexibility because of the greater space available within the chamber as well as the ability to use vacuum locks and arsenic-injection feed-through tubes. Another important advantage is the opportunity to utilize large silicon puller designs, heater assemblies and even stock items such as quartz crucibles. With dislocation densities reduced by thick B_2O_3 layers, wet B_2O_3 can be used to minimize silicon contamination. The danger of crucible breakage is increased with thick B_2O_3 layers, and relatively inexpensive quartz crucibles become even more attractive.

The major disadvantage of LP pullers is the low yield obtained when growing crystals more than 100 mm long in the ⟨100⟩ direction. However, it is expected that this problem will be solved and that 20 kg dislocation-free semi-insulating GaAs crystals grown in the ⟨100⟩ direction will become standard by 1990.

Homogeneity

Another approach to the reduction of convection currents is to stir the melt more vigorously to decrease the thickness of the boundary layer at the crystal-melt interface [19]. In heavily tin doped InP, the short range longitudinal variations in dopant concentrations or stria have been reduced to a few per cent by increasing the rates of seed or crucible rotation, ω_s and ω_c, respectively. The magnitude of these variations is determined by scanning the length of a longitudinal section of the crystal with a focussed CO_2 laser. The intensity of the transmitted radiation can be used to determine the absorption coefficient and from it the carrier concentration (i.e. tin concentration). The results of a typical rotation experiment are shown in Figure 3. Notice that for a seed rotation of 80 rpm also for a crucible rotation of 40 rpm the magnitude of the carrier concentration variations is greatly reduced.

The distribution of the EL2 level in 75 mm diameter crystals of semi-insulating GaAs grown by the HP LEC method has been studied by whole slice imaging, which was performed using a silicon vidicon TV camera [20]. Different kinds of fine structure, including cell structure and bands of infrared ('sheets' and 'streamers') lying in (110) planes running down the ⟨001⟩ growth directions, were observed. Infrared line scans taken at a 1.0 μm wavelength show a familiar W pattern (seen for dislocations in HP LEC GaAs) for EL2 concentration. This technique is a powerful tool for studying the effects of changes in growth conditions on the distribution of the EL2 level. Details of the techniques and the results obtained can be found in references [20, 21].

Native defect EL2 concentrations have been found to be a function of growth conditions and melt stoichiometry. Even in the presence of a magnetic field the EL2 concentrations can be controlled from less than 1×10^{15} to 5×10^{17} cm^{-3} [22]. Infrared imaging and line scans in low dislocation GaAs will aid greatly in optimizing growth processes.

The uniformity of HP LEC semi-insulating GaAs is greatly improved by annealing the crystals after growth. In as-grown crystals there is a correlation between the concentration of the EL2 level and the dislocation density. On annealing at temperatures of 750°C or above this correlation is reduced, and the variation in resistivity and mobility decreases [23].

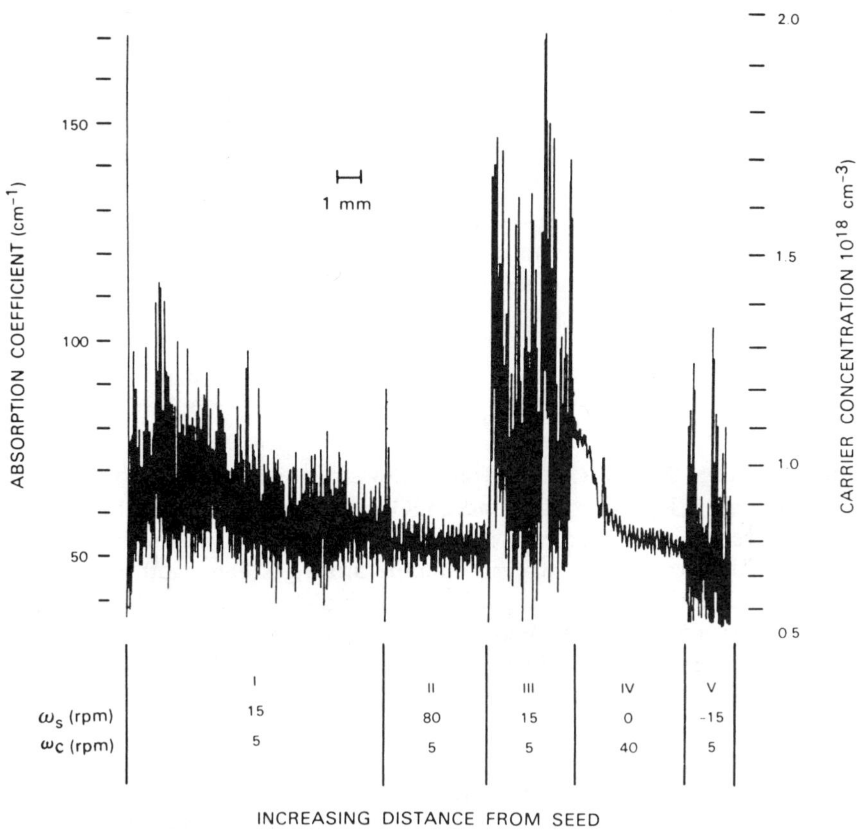

Figure 3 Absorption coefficient at 10.5 μm and carrier concentration versus distance from seed end of tin doped LEC InP crystal. After Iseler *et al.* [19].

As the diameter of the crystal increases, so do the thermal stresses which occur during cool down. Some crystals are so strained that they shatter when first contacted by an inside diameter (ID) diamond saw. If the slicing operation begins at the tail end of the crystal, the yield is increased. In other cases the crystal can only be cut after high temperature annealing.

The processing of larger wafers also results in increased strain. Since the surface fracture strength of GaAs is only one third that of silicon, breakage during processing can be significant in determining overall yield. Residual strain from a particular growth process could be the ultimate factor in determining its merit. Impurity hardening processes could strengthen the substrates sufficiently to improve processing yields.

In addition to gallium inclusions mentioned earlier, carbon microinclusions have been observed in samples of HB and LEC crystals as well as VPE (vapour phase epitaxial) and LPE (liquid phase epitaxial) films. Electron diffraction analysis shows that the microinclusion is primarily cubic with a lattice constant of 5.54 Å [24]. Since impurities such as oxygen and silicon were found adjacent to the microinclusions, it appears the carbon is a nucleation centre for other precipitates.

Comparisons of the homogeneity of GaAs crystals grown by various techniques have been made. However, because of improvements being made in growth processes, most have little validity. In addition, such things as single crystal length for LP LEC are avoided when it is compared to HP LEC growth [25]. Also opposing conclusions are obtained in comparisons of Bridgman and LEC crystal uniformity of threshold voltage [4, 26]. Again one is led to the conclusion that it is the particular process and not the crystal growing method that is significant.

Summary

Semi-insulating, three inch diameter GaAs substrates can be produced by zone levelling, Bridgman, static and EDG freeze methods as well as by low and high pressure LEC techniques. Each growth method has unique salient features, which with the appropriate temperature gradient modification and control can give high quality crystals with few dislocations. Dislocation-free crystals up to 65 mm in diameter have been produced by boat and LEC growth techniques. Only HP LEC provides good yield of $\langle 100 \rangle$ grown crystals over 100 mm in length. Crystals 127 mm in diameter and weighing 14 kg have been grown by LP LEC. Automatic computerized diameter control can provide 78 ± 2 mm diameter LEC crystals.

As each major growth perturbation, which affects the properties of a crystal, such as impurity content or dislocation density, is reduced other factors assume a dominant role. The effect of stoichiometry on the dislocation density in crystals prepared by non-Czochralski methods is well known. It could now be a major factor in the variable nature of the low dislocation densities produced by using a magnetic field to dampen thermal oscillations in HP LEC growth or the thick boric oxide layers used to reduce thermal gradients in LP LEC pullers. The use of indium or antimony additions to reduce dislocation densities can change the stoichiometry at the growing interface.

With thermal conversion during annealing better understood, attention now turns to residual strain in the crystal and to local variations in the uniformity of impurity and defect concentrations. The effects of random convection currents and the resultant impurity stria can be reduced by stirring the melt more vigorously with rapid seed or crucible rotation to decrease the thickness of the boundary layer at the melt-crystal interface.

Significant micro non-uniformities in the EL2 distributions in GaAs have been found. These distributions fall into three distinct classes, termed cellular structure, streamers (lineages) and sheets. Micro electrical homogeneity presents yet another challenge for the crystal grower. Also, the mechanical strains introduced by thermal treatments during and immediately following growth, and during anneals are significant factors in determining device yield. A method for quantitatively determining the magnitude of strain in a substrate is needed. This measurement should be quick, inexpensive and non-destructive.

Since a number of methods for crystal growth can produce large, high-purity crystals of undoped semi-insulating GaAs, the major factors relating to success in acceptance of a particular method for growing material for use in ICs are: material cost (single crystal yield); cost of processing (electrical and mechanical yield); and final IC cost (homogeneity and performance yield). Clearly, the GaAs substrates available today need to be improved; but rapid progress is being made. It is expected that by 1990 homogeneous 20 kg dislocation-free GaAs crystals grown in the ⟨100⟩ direction will be standard for IC production.

ACKNOWLEDGEMENTS

The author gratefully acknowledges valuable discussions with, and the contribution of both published and unpublished information from, S. Akai, W. M. Ford, G. W. Iseler, R. L. Lane, R. Mellen, E. M. Swiggard, K. Terashima, R. N. Thomas, G. Woodhouse and Z. Yuanxi.

References

1. Mullin J. B. (1962). In *Compound Semiconductors,* p. 365 (R. K. Willardson and H. L. Goering, eds.); Reinhold Publishing
2. Terashima K. (1982). *J. Cryst. Growth,* **60,** 363
3. Steinemann A. and Zimmerli U. (1967). In *Crystal Growth,* p. 81 (H. S. Peiser, ed.); Pergamon Press
4. Kirkpatrick C. G., Chen R. T., Holmes D. E., Asbeck P. M., Elliott K. R., Fairman R. D. and Oliver J. R. (1984). In *Semiconductors and Semimetals Vol. 20,* p. 159 (R. K. Willardson and A. C. Beer, eds.); Academic Press, New York
5. Peigen M., Ju W., Yuanxi Z., Jinhua Y. and Souchun L. (1984). This volume, p. 134
6. Stolte C. A. (1984). In *Semiconductors and Semimetals Vol. 20,* p. 89 (R. K. Willardson and A. C. Beer, eds.); Academic Press, New York
7. Swiggard E. M. (1983). *Digest GaAs IC Symposium,* p. 26; IEEE
8. Jacob G. (1982). In *Semi-Insulating III–V Materials, Evian,* p. 2 (S. Makram-Ebeid and B. Tuck, eds.); Shiva Publishing, Nantwich, UK
9. Hobgood H. M., Thomas R. N., Barrett D. L., Eldridge G. W., Sopira M. M. and Driver M. C. (1984). This volume, p. 149
10. Duseaux M. and Martin S. (1984). This volume, p. 118
11. Emori H., Terashima K., Orito F., Kikuta T. and Fukuda T. (1984). This volume, p. 111
12. Terashima K., Katsumata K., Orito F., Kikuta T. and Fukuda T. (1983). *Jap. J. Appl. Phys.,* **22,** 401
13. Nishida Y. (1984). *Microwave Systems News,* **14,** 38
14. Sumitomo Press Release, Tokyo, March 30, 1984
15. Nishine S., Kito N., Fujita K., Sekinobu M., Shikatani O. and Akai S. (1982). In *Proc. IEEE Gallium Arsenide Integrated Circuit Symposium,* p. 58; IEEE
16. Thomas R., Hobgood H. M., Eldridge G. W., Barrett D. L., Braggins T. T., Ta L. B. and Wang S. K. (1984). In Ref. 4, p. 1
17. Terashima K., Katsumata T., Orito F. and Fukuda T. (1984). *Jap. J. Appl. Phys. Lett.,* in press

18. Ford W. M. (1984). Private communication
19. Iseler G. W., Taylor M. S. and Delaney E. J. (1983). *Solid State Research, Lincoln Laboratory Second Quarterly Technical Summary Report*, 10 August 1983, p. 15
20. Skolnick M. S., Brozel M. R., Reed L. J., Grant I., Stirland D. J. and Ware R. M. (1983). *J. Electron. Mat.*, **13,** 107
21. Skolnick M. S. (1984). This volume, p. 446
22. Terashima K. (1984). Private communication
23. Rumsby D., Ware R. M., Smith B., Tylberg M., Brozel M. R. and Foulkes E. J. (1983). Ref. 7, p. 34
24. Qun W., Yuzhen L. and Guiru S. (1983). *1983 National Symposium on Gallium Arsenide and Related Compounds, China*
25. Duncan W. M., Westphal G. H. and Sherer J. B. (1983). *IEEE El. Dev. Lett.*, **EDL-4,** 199
26. Yamazaki H., Nanishi Y., Miyazawa S. and Ishii Y. (1983). Ref. 7, p. 30

Effect of Ambient Gas on LEC Undoped Semi-Insulating GaAs

Haruo Emori, Kazutaka Terashima, Fumio Orito, Toshio Kikuta and Tsuguo Fukuda

Optoelectronics Joint Research Laboratory, 1333 Kamikodanaka Nakahara-ku, Kawasaki 211, Japan

Abstract

Undoped GaAs single crystals were grown by the LEC technique from PBN crucibles in N_2, Ar or Kr ambient gases. The influence of the ambient gas on the properties of the crystals was examined by Hall effect and resistivity measurements, SIMS (secondary ion mass spectrometry) analysis, LVM (localized vibrational mode) IR absorption measurement, low temperature PL (photoluminescence) measurement and EPD (etch pit density) count. All crystals showed SI properties, but the reduced resistivity was observed in the case of N_2 ambient. These results can be interpreted in terms of the variation of impurity concentration and the type of deep level. Consequently, SI GaAs crystal with the highest resistivity and the lowest dislocation density was obtained in the case of Kr ambient.

Introduction

Reliable large-area SI GaAs substrates are required to establish the device fabrication technology by direct ion implantation processing. It is of prime importance to establish a reproducible growth technique for homogeneous undoped SI GaAs crystals.

GaAs single crystals are normally grown using N_2 or Ar as an ambient gas. It is possible to grow the single crystals even in Kr or He ambient. However, no clear study has been reported on the relationship between the kind of ambient gas and the electrical properties of undoped LEC GaAs crystals, while it has been shown that the properties of undoped crystals depend critically upon melt stoichiometry [1, 2], the water content of B_2O_3 encapsulant [3, 4] and impurity concentration in Ga and As [5]. This study has been made to clarify the effect of ambient gas on the properties of undoped LEC GaAs single crystals. It was revealed that the ambient gas affected electrical resistivity, impurity concentration, the type of deep level and dislocation density. This investigation gives much information useful in selecting the kind of ambient gas and growth condition for the LEC GaAs.

Experimental procedures

GaAs single crystals used in this study were grown in 5–9s N_2, 5–9s Ar or 4–9s Kr ambient from PBN crucibles by the LEC technique with *in situ* synthesis and unintentional doping. *In situ* synthesis was carried out from 6–9s Ga (500 g) and 6–9s As (555 g) using super dry (<100 wt. ppm H_2O) B_2O_3 as an encapsulant. After completion of the compounding reaction, two inch diameter ⟨100⟩ GaAs single crystals were pulled under an ambient gas pressure of 20 kg cm^{-2} utilizing the advanced automatic diameter control technique.

Undoped GaAs crystals were characterized by the Hall effect and resistivity measurements in the temperature range 295–430 K, SIMS impurity analysis, LVM IR absorption measurements, low temperature PL measurements and EPD count.

Results and discussion

ELECTRICAL PROPERTIES

Figure 1 shows the resistivity variation along the ⟨100⟩ growth direction of the crystals grown in Ar, Kr or N_2 ambient from PBN crucibles. The crystals are all n-type SI, and their resistivities determined at 295 K are in the range of 3×10^8–8×10^7 (Kr), 1.3×10^8–2.5×10^7 (Ar) and 4×10^7–8×10^6 (N_2) Ω cm. The As atomic fraction of the initial melt for these crystals was 0.499 (Kr),

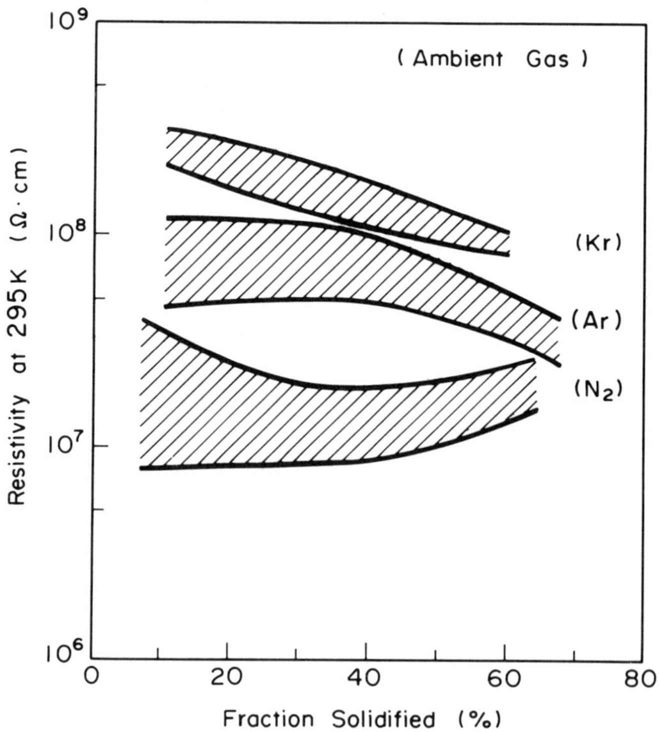

Figure 1 Resistivity variation in SI LEC GaAs single crystals against fraction solidified.

0.499 (Ar) and 0.497 (N_2). The initial melt composition was determined on the assumption that weight loss after the growth process was all As loss during *in situ* synthesis [1]. As will be described later in this paper, a large number of Ga atoms seem to escape from the melt during the N_2 ambient growth, compared with the rare gases growth. Therefore, the actual melt should be As-rich in the case of N_2 ambient growth. It was noticed that the crystal grown in Kr ambient was uniform SI with the highest resistivity from seed end to tail end, and that the crystal grown in N_2 ambient exhibited about one order of magnitude lower resistivity.

Hall effect measurements as a function of temperature were used to determine the activation energy of the SI GaAs crystals. Samples were cut from the wafer centre at the front parts of the SI crystals. The experimental values of the slope of the Arrhenius plot as $n \times T^{-3/2}$ against 1000/T were 0.77 (Kr), 0.77 (Ar) and 0.64 (N_2) eV. The measured activation energy of the crystals grown in Kr and Ar ambient is close to 0.76 eV which is the reported value [6, 7] for the EL2 level, whereas the lower energy is obtained in the crystal grown in N_2 ambient.

IMPURITY CONCENTRATIONS

Concentrations of carbon on the As sublattice [8] in the SI crystals were obtained by LVM measurements, as shown in Figure 2. Carbon concentrations in the crystal grown in N_2 ambient are less than the detection limit (4×10^{14} cm^{-3}) from seed end to tail end, against 5×10^{15} cm^{-3} in the front parts and 1×10^{15} cm^{-3} in the tail parts of the crystal grown in Kr ambient. The intermediate values were observed in the crystal grown in Ar ambient.

Boron concentrations in the crystals were obtained through SIMS analysis. Low boron contamination, less than 2×10^{15} cm^{-3}, was observed in the crystal grown in N_2 ambient, whereas high boron concentrations from 1×10^{17} cm^{-3} to 2×10^{17} cm^{-3} were observed in the crystals grown in rare gases. No significant differences in other impurities in the SI crystals can be detected by SIMS analysis. That fact suggests that the decrease in carbon concentration causes the reduced resistivity of the SI crystal grown in N_2 ambient. It should be noticed that, if boron acts as a double acceptor on the As sublattice, the reduced resistivity of the crystal grown in N_2 ambient can be ascribed to the decrease in the boron concentration.

When quartz crucibles are used, lower residual silicon concentration has been observed [9] in the SI crystal grown in N_2 ambient, compared with the conductive crystals grown in rare gases. It has been reported [3, 4, 10, 11] that the incorporation of boron and carbon into the crystals grown from PBN crucibles, or silicon from quartz crucibles, can be suppressed by the formation of Ga_2O, especially in wet B_2O_3 encapsulant, during growth. It is explained [12] on the basis of the free energy of formation that Ga_2O suppresses dissociation of B_2O_3 encapsulant and quartz crucibles and also removes silicon and carbon by oxidation.

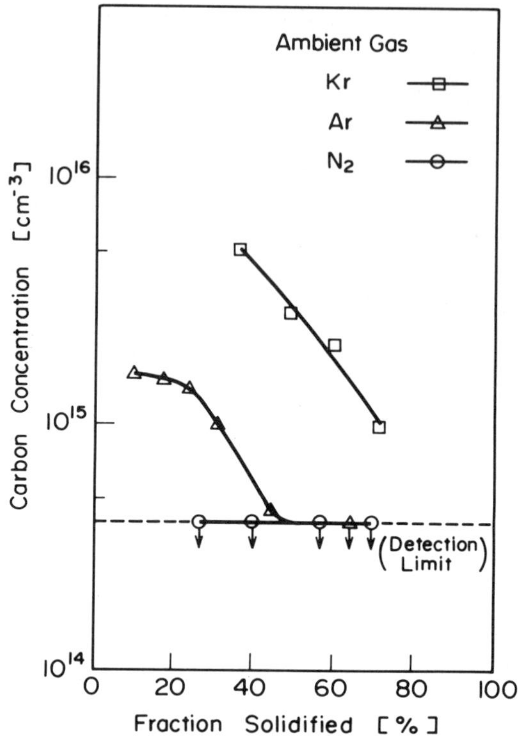

Figure 2 Carbon concentration profiles measured by LVM IR absorption along growth direction of SI GaAs crystals. The detection limit is 4×10^{15} cm^{-3}.

The crystals grown in N_2 ambient using dry B_2O_3 showed an extremely low concentration of carbon and boron, suggesting a large amount of Ga_2O formation in the B_2O_3 encapsulant as compared with the crystals grown in Ar and Kr ambient.

OPTICAL ASSESSMENT

PL excitation spectra were obtained at 4.2 K from the centre part of wafers cut from GaAs crystals pulled in Kr or N_2 ambient, as shown in Figure 3. In addition to the 0.8 eV deep emission band observed in both SI crystals, the 0.65 eV deep emission band was dominantly present in the crystal grown in N_2 ambient. The 0.8 eV band has been dominantly observed [13] in SI crystals grown from PBN crucibles in Ar ambient using dry B_2O_3 as well as the crystal grown in Kr ambient. Although the origin of the 0.65 eV band has not been clarified, its presence was reported [4] in crystals grown using wet B_2O_3. Therefore, it is suggested that the deep levels in the crystal grown in N_2 ambient are similar to those in the crystals grown using wet B_2O_3.

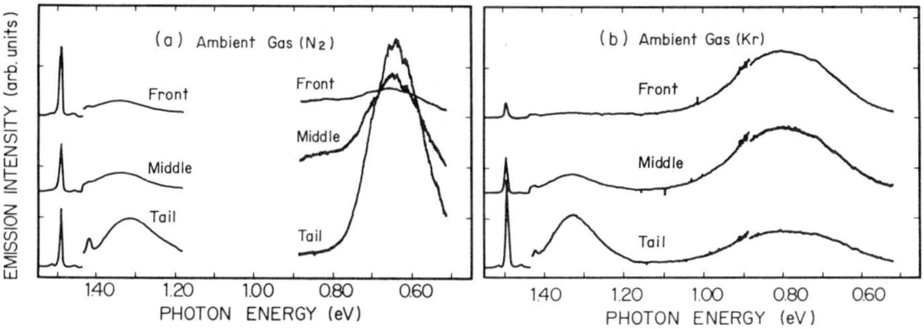

Figure 3 Two typical PL spectra from SI GaAs crystals. (a) 0.65 eV band dominant in crystal grown in N_2 ambient. (b) 0.8 eV band dominant in crystal grown in Kr ambient.

MELT STOICHIOMETRY

The resistivity of the crystal grown in N_2 ambient decreases in the tail part at a solidified fraction of 0.8. The slope of the Arrhenius plots of $n \times T^{-3/4}$ and Hall mobility are shown in Figure 4. A low activation energy of 0.455 eV

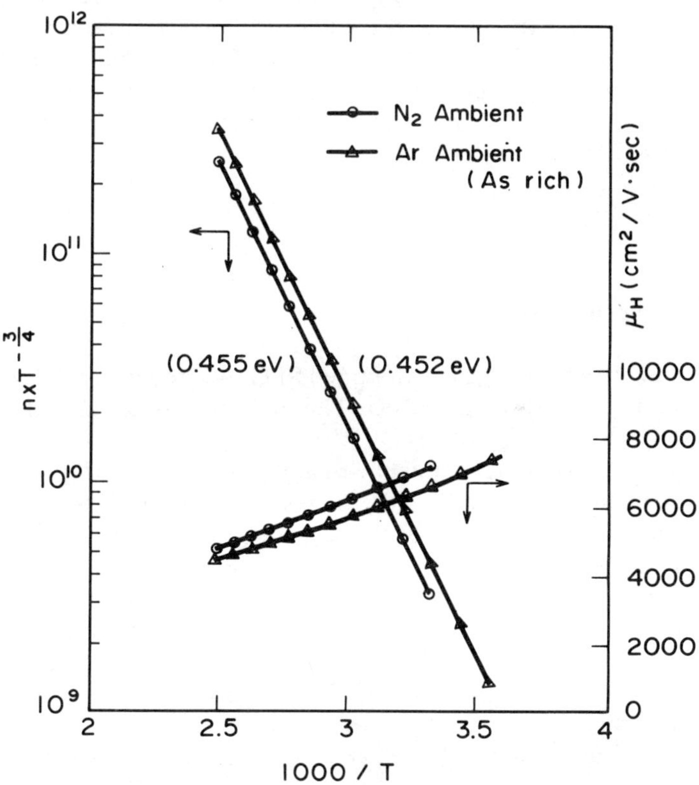

Figure 4 Plot of Hall mobility and carrier concentration times $T^{-3/4}$ against reciprocal temperature for low-resistivity GaAs crystals.

and a high Hall mobility of 7000 $cm^2\,V^{-1}\,sec^{-1}$ at 300 K were observed, and these results agreed very closely with the values for the low resistivity tail parts of the crystals which were pulled in Ar ambient from the heavily As rich condition with initial melt of 0.509 As atomic fraction. Since the deep level responsible for the apparent activation energy of 0.42 eV was suggested [6, 14] to be a point defect or point defect/impurity complex in the materials grown using the methods favouring As-rich growth, this level might be associated with our measured level with an activation energy of 0.455 eV. Therefore, it is very likely that the Ga atoms in the GaAs melt escaped into the N_2 ambient through the B_2O_3 encapsulant, and that the growth in N_2 ambient caused an As-rich melt in this experiment.

DISLOCATION DENSITIES

The effect of the ambient gases on the generation of dislocations was examined. The EPD was observed on the (100) surface of 2 inch diameter cross-sectional wafers cut from the crystals that were pulled in Kr, Ar or N_2 ambient at a pressure of 20 $kg\,cm^{-2}$. The distribution of the dislocation densities in the crystals grown in Kr, Ar or N_2 ambient increased in this order. We have shown [9] that the EPD in the crystals grown from quartz crucibles also increases in the order $Kr < Ar < N_2$, and that this order can be explained by a high temperature gradient caused by the high heat transfer coefficient. As a result, the dislocation generation is suppressed by using Kr as an ambient gas.

Summary and conclusions

The effect of the ambient gas on undoped SI GaAs grown from PBN crucibles was examined using N_2, Ar and Kr. The main results are as follows:

1. The crystal grown in Kr ambient was uniform SI with the highest resistivity from seed end to tail end, while the resistivity of the crystal grown in N_2 ambient was less than that of the crystals grown in rare gases.
2. The crystal grown in N_2 ambient showed a relatively low level of contamination by both boron and carbon. The low activation energy (<0.64 eV) was observed only when N_2 was used as an ambient gas. Moreover, PL measurements revealed the presence of a dominant 0.65 eV line in the crystal grown in N_2 ambient, and a dominant 0.8 eV line in the crystals grown in rare gases.
3. The ability to suppress the generation of dislocations was in the order $Kr > Ar > N_2$, as expected from their heat transfer efficiencies.

ACKNOWLEDGEMENTS

The authors are indebted to I. Hayashi and T. Iizuka for useful discussions. They would like to thank M. Tajima for helpful discussions concerning measurements of LVM IR absorption. They also thank T. Shimada for his useful advice related to crystal growth technique.

The present research effort is part of a major research and development project on optical measurement and control systems, conducted under a program set up by the Ministry of International Trade and Industry's Agency of Industrial Science and Technology.

References

1. Holmes D. E., Chen R. T., Elliot K. R., Kirkpatrick C. G. and Yu P. W. (1982). *IEEE Trans. El. Dev.*, **ED-29,** 1045
2. Ta L. B., Hobgood H. M., Rohatgi A. and Thomas R. N. (1982). *J. Appl. Phys.*, **53,** 5771
3. Rumsby D. H. and Ware R. M. (1981). *Inst. Phys. Conf. Ser.*, **63,** 573
4. Oliver J. R., Fairman R. D., Chen. R. T. and Yu P. W. (1981). *Electron. Lett.*, **17,** 839
5. Farges J. P., Jacob G., Schemali C., Martin G. M., Mircea-Roussel A. and Hallais J. (1982). In *Semi-Insulating III–V Materials, Evian*, p 45 (S. Makram-Ebeid and B. Tuck, eds.); Shiva Publishing, Nantwich, UK
6. Martin G. M., Farges J. P., Jacob G., Hallais J. P. and Poiblaud G. (1980). *J. Appl. Phys.*, **51,** 2840
7. Hughes B. and Li C. (1982). *Inst. Phys. Conf. Ser.*, **65,** 57
8. Theis W. M., Bajai K. K., Litton C. W. and Spitzer W. G. (1982). *Appl. Phys. Lett.*, **41,** 70
9. Emori H., Matsumura T., Kikuta T. and Fukuda T. (1983). *Jap. J. Appl. Phys.*, **22,** 1652
10. Weiner M. E., Lassota D. T. and Schwartz B. (1971). *J. Electrochem. Soc.*, **118,** 301
11. Hunter A. T., Kimura H., Baukus J. P., Winston H. V. and Marsh O. J. (1984). *Appl. Phys. Lett.*, **44,** 74
12. Bass S. J. and Oliver P. E. (1968). *J. Cryst. Growth*, **3, 4,** 286
13. Kikuta T., Terashima K. and Ishida K. (1983). *Jap. J. Appl. Phys.*, **22,** L409
14. Martin G. M., Mitonneau A. and Mircea A. (1977). *Electron. Lett.*, **13,** 191

Growth and Characterization of Large Dislocation-Free GaAs Crystals for Integrated Circuits Applications

M. Duseaux and S. Martin

Laboratoires d'Electronique et de Physique Appliquée, 3 Avenue Descartes, 94450 Limeil-Brévannes, France

Abstract

We have shown that it is possible to grow small (d = 20 mm) dislocation-free GaAs ingots with the LEC (liquid encapsulated Czochralski) technique by doping with the isovalent element indium. In this paper, we report recent results concerning the growth of two inch diameter In doped crystals, free of dislocations ($\sim 5 \times 10^2$ dislocations cm^{-2}). X-ray topography analysis and TEM (transmission electron microscopy) studies do not reveal any microprecipitates in the matrix. Some of these single crystals exhibit semi-insulating properties without adding deep acceptor impurities. The homogeneity of the electrical properties is studied by Hall effect measurements from top to tail and across the wafers. The carbon concentration is deduced from LVM (localized vibrational mode) absorption and mapping of the deep donor EL2 concentration is obtained by near infrared absorption. It has been shown that the concentration of these species is quite low (in the 10^{15}–5×10^{15} cm^{-3} range).

Introduction

The detrimental effect of the presence of dislocations in GaAs substrates on the life time of GaAlAs/GaAs lasers is well known. More recently, a correlation between the dislocation distribution in a semi-insulating GaAs wafer and FET (field effect transistor) performance has been investigated [1, 2]. These studies have shown that one of the main objectives in the bulk GaAs field is to obtain a uniform distribution of dislocations across the wafer.

The best way to succeed with the LEC technique is to make dislocation-free material [3], since many studies have shown that the inhomogeneity of the dislocation distribution (W shape) is related to the thermal stresses occurring during the growth [4, 5, 6].

In order to obtain dislocation-free GaAs, a necking procedure has been tried [7, 8], but it has been shown that the dislocations were nucleated as soon as the crystal diameter was in the 4–15 mm range [9].

Another way to decrease the dislocation density is to dope the crystal with isoelectronic impurities. The two main advantages of the isoelectronic elements are:

1. Their large solubility in GaAs.
2. They do not bring any modification in the electrical properties of the material.

Several isoelectronic species were tested: N [10], B, P, In, Sb [3, 8]. Small ingots (d = 15–20 mm) were pulled and the best candidate allowing growth of single crystals free of dislocations seemed to be indium. In order to examine the crystallographic quality of such crystals, TEM observations were made. It was shown that the matrix was free of microdefects [11].

We will now present our results on the growth and characterization of 2 inch diameter indium doped GaAs crystals.

Crystal growth

It was necessary to follow the decrease of dislocation density as a function of indium concentration. Therefore the indium concentration and distribution were measured with different techniques. Firstly, by using the fact that the energy gap depends on the indium concentration, OA (optical absorption) showed that the indium concentration varies from top to tail in the 10^{19}–2×10^{20} at. cm^{-3} range. Measurements from one edge to the other gave an M shaped variation [12]. However, experiments have recently been made with a better resolution by EMA (electron microprobe analysis) which have not confirmed this M shape distribution of indium [13]. As a result of this analysis, we can now state that:

1. The minimum indium concentration for dislocation-free GaAs ($<500\ cm^{-2}$) is 4–5×10^{19} at. cm^{-3} and 7–8×10^{19} at. cm^{-3} for 20 mm and 50 mm crystal diameters, respectively.
2. The indium profile across a slice is flat.

Moreover, the distribution coefficient of indium has been calculated as $K_{In} \sim 0.1$.

Due to the large amount of indium which is required in the GaAs matrix, and to the relatively small value of the distribution coefficient, some problems may occur during the growth itself since supercooling occurs. In particular, when the crystal diameter increases, it becomes more and more difficult to find good experimental conditions allowing growth of single crystals. As a matter of fact, when the indium concentration increases along the crystal and reaches a given value (larger than 2×10^{20} at. cm^{-3} in our case), the crystal becomes polycrystalline.

Each ingot was pulled in a MSR 6 puller under the following conditions:

quartz crucible;
1.2 cm thick wet B_2O_3 layer;
pulling rate: $0.9\ cm\ h^{-1}$;
growth axis: $\langle 001 \rangle$;
starting load: $m_{Ga} + m_{As} \sim 1.05$ kg; $30 < m_{InAs} < 70$ g.

With these conditions, it has been possible to pull eight 500 g single crystals over nine runs, one of them being twinned. In each case, the maximum value of the crystal diameter was in the 50–55 mm range.

Transmission X-ray topographs are presented in Figure 1. Figure 1a represents a part of a d = 50 mm (100) slice. The observed growth striations correspond to fluctuations of the indium concentration. Moreover, even if most of the wafer is free of dislocations, slip bands are still present on the edges, in the ⟨110⟩ directions. They are probably due to thermal stresses occurring during cooling. Figure 1b represents a longitudinal (110) section, parallel to the growth axis. Besides striations and dislocations, it is possible to see the location where the crystal becomes poly, at the tail side of the ingot.

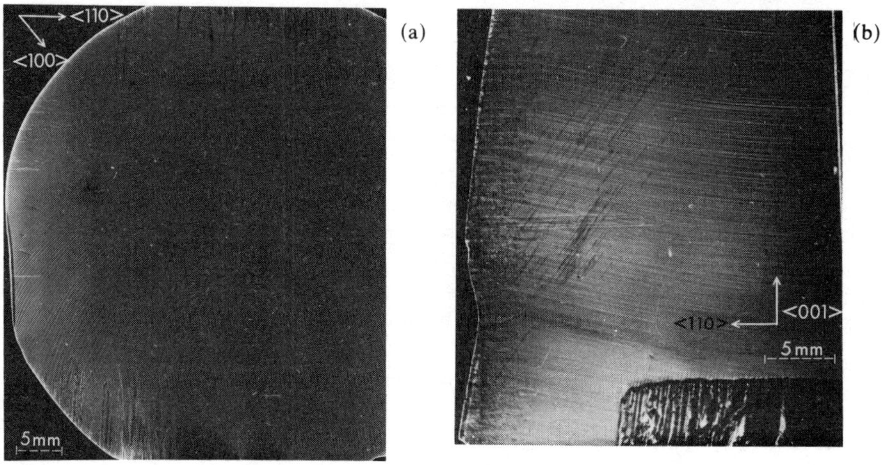

Figure 1 X-ray topographs of GaAs:In. (a) d = 50 mm (100) slice; (b) (110) longitudinal cross-section.

Characterization

Two ingots have been extensively studied. Maps of different physical species will be presented in this chapter. These two crystals (CZ 1, CZ 2) have not been intentionally doped with impurities other than In.

The starting conditions were almost the same, and the reproducibility will be discussed:

$m_{Ga} + m_{As} \sim 1050$ g
$m_{InAs} \sim 70$ g
CZ 1: As = 49.80% (mole/mole)
CZ 2: As = 50.00% (mole/mole)

In each case, the characterizations were made on two (three) different slices coming from the middle and bottom (top, middle, bottom) of the crystals.

Each position in the ingot corresponds to a given value of the (solidified/starting load) ratio:

$$x_i = m_{iGaAs}/(m_{Ga} + m_{As} + m_{InAs})$$

The x_i values are 0.05, 0.2, and 0.4.

NEAR INFRARED ABSORPTION

Since the semi-insulating properties of the undoped LEC GaAs are obtained by the compensation of the shallow acceptors by the deep donor level EL2, the distribution of this level is of highest importance. Some reports have dealt with the non-uniformity of the EL2 concentration and a correlation with the dislocation distribution across such slices [14, 15, 16, 17]. Moreover, EL2 was detected in a dislocation-free n-type ($n = 5 \times 10^{15}$ cm^{-3}) material not doped with In [15]. Thus, it seemed particularly useful to study the EL2 variation in our dislocation-free In doped ingots.

This distribution was determined by optical absorption at 1.1 μm using the cross-section reported by Martin *et al.* [15]. The slices used were 4 mm thick and cut perpendicular to the growth direction. Contour plots of the EL2 concentration have been determined from absorption measurements made every 5 mm^2 on each sample. The positions of the Fermi level given by Hall effect measurements indicated that in all cases EL2 is virtually 100% occupied (see Electrical Measurements, below).

The contour plots are presented for CZ 1 ($x_1 = 0.2$, $x_2 = 0.4$) and CZ 2 ($x_1 = 0.2$, $x_2 = 0.4$) in Figures 2a and 2b, respectively. Comparisons with results

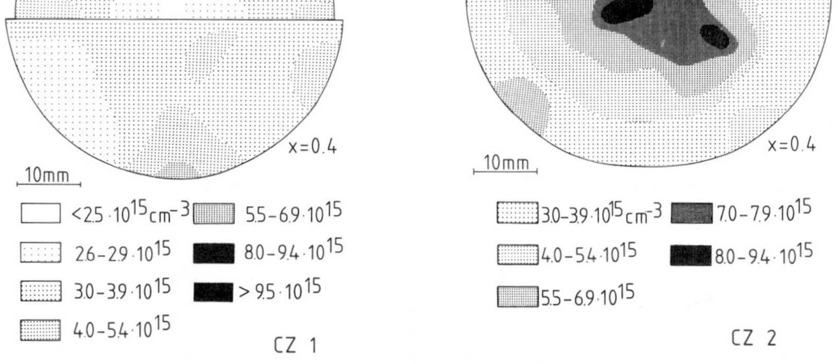

Figure 2 EL2 distribution in (a) CZ 1 and (b) CZ 2.

already observed on dislocated LEC GaAs crystals encourage us to draw the following conclusions:

1. The EL2 concentration is on an average much lower and more homogeneous.
2. The increase of the EL2 concentration near the ⟨100⟩ edges is in agreement with previously reported observations [17].
3. The W shape already reported [15] for EL2 distribution across a slice, does not appear here. This affirmation does not take into account the results obtained on the tail of CZ 2 which became poly in the centre (since the reflections increase the absorption and the EL2 concentrations deduced are overestimated).
4. This sheds some light on the mechanisms of EL2 formation. Since EL2 is still present in these dislocation-free crystals, it is not only connected with the dislocations. However, in these ingots, the EL2 concentration is lower than in dislocated ones. So, we can assume that this low concentration corresponds to a background of EL2 due to a thermodynamic equilibrium. Furthermore, the presence or the motion of dislocations create an additional EL2 concentration which enhances this background.

FAR INFRARED ABSORPTION (LVM)

The main acceptor in undoped LEC GaAs crystals is carbon which plays a significant role in the compensation of the deep level EL2. In order to determine its concentration in our material, we measured LVM absorption at 582 cm^{-1}. We used the calibration reported by Brozel *et al.* [18] in order to convert absorption into concentration. Our measurements were made at room temperature whereas Brozel's were made at 77 K. Nevertheless, we applied Brozel's calibration directly without any correction. This may lead to some underestimation of concentration.

All carbon concentrations determined were below 5×10^{15} cm^{-3}. These low values, observed on crystals which were pulled under wet B_2O_3, are in agreement with a recent report of Kimura *et al.* [19] where it is shown that an increase of the water content of the B_2O_3 encapsulant causes a decrease of the carbon concentration in the crystal.

Studies of radial concentrations indicate that carbon uniformity is achieved in each slice. Such a result has already been reported [20]. On the other hand, a decrease of carbon concentration from seed to tail was noticed. This result seems to confirm the hypothesis of a distribution coefficient for carbon higher than one. Nevertheless, since the carbon concentrations observed were near the detection limit, this may not be a definite result.

ELECTRICAL MEASUREMENTS

In order to get a good idea of the homogeneity of the electrical properties, five to ten samples were measured on each slice.

The samples were taken along the ⟨100⟩ and ⟨110⟩ radius of each slice. The results concerning the resistivity at room temperature are presented in Figures 3a and 3b for CZ 1 and CZ 2 respectively. The resistivity varies gradually from the centre to the edge of each slice: the resistivity is high in the centre (10^8 Ω cm), where the material is semi-insulating, and decreases to reach 10^4 Ω cm at the edge. No difference appears in its behaviour along the ⟨100⟩ and ⟨110⟩ directions: the values of the resistivity seem to follow a circular symmetry around the ⟨001⟩ axis. Without taking into account the seed slices which are too small, we can state that the low resistivity region in a slice decreases from the seed to the tail.

Hall effect measurements versus temperature show that in the high resistivity regions (dark ones in Figure 3), the Fermi level is pinned close to the deep donor EL2, whereas it is pinned close to EL6 in the low resistivity

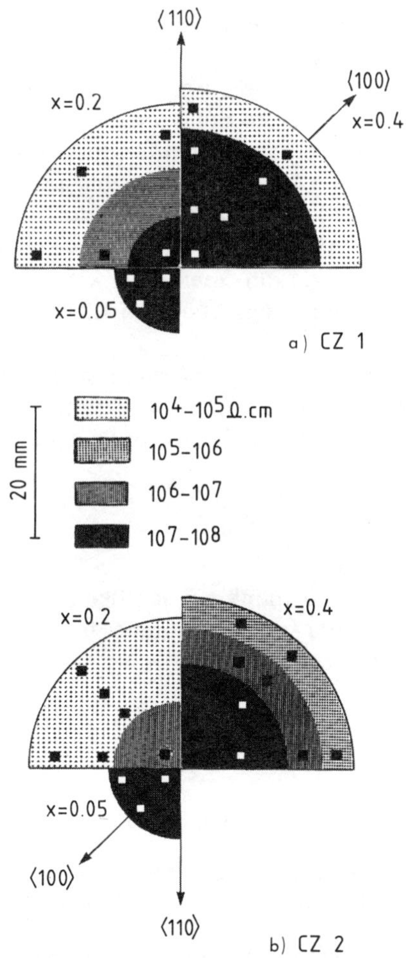

Figure 3 Resistivity distribution in (a) CZ 1 and (b) CZ 2.

regions. If the variations of the electrical properties from top to tail can be related to differences of the impurity distribution coefficients, then they are not related to cross-section variations.

In order to explain the decrease of the resistivity from centre to edges, if N_A is an acceptor concentration and N_{DS} the concentration of a donor shallower than EL2, we can assume that the difference $(\Sigma N_{DS}^+ - \Sigma N_A^-)$ increases from centre to edge. Moreover, since no cross-section variation of the main acceptor C is observed, we must have an increase of ΣN_{DS}^+ from centre to edge. Especially, if EL6 is a donor, its radial variations may have a great importance for the homogeneity of the electrical properties of undoped GaAs.

Conclusion

This work shows that indium doping in GaAs allows us to pull 2 inch diameter crystals almost free of dislocations ($<500\ cm^{-2}$) and free of microprecipitates. The intrinsic properties of GaAs are not distinctly affected by the indium atoms. Note that it is possible to dope the crystal with other species in order to obtain any electrical type. Our attention has been focused on the characterization of 'undoped' GaAs: In. These ingots are not homogeneously semi-insulating, and we have proposed that variations of donors shallower than EL2 are responsible for that. Moreover, we have shown that the deep donor EL2 is still present in dislocation-free GaAs, with a mean concentration lower than the usual one. This fact suggests that EL2 is a defect due to a thermodynamic equilibrium.

Future investigations will be concerned with the properties of devices fabricated on this kind of GaAs: In substrate. Particularly, we hope to demonstrate the exact effect of dislocations on such devices.

ACKNOWLEDGEMENTS

The authors would like to thank C. Schiller for the X-ray topographs, C. Schemali and J. Maluenda for their comments on the manuscript and G. M. Martin for helpful discussions.

References

1. Nanishi Y., Ishida S., Honda T., Yamazaki H. and Miyazawa S. (1982). *Jap. J. Appl. Phys.*, **21(6),** 335
2. Nanishi Y., Ishida S. and Miyazawa S. (1983). *Jap. J. Appl. Phys.*, **22(1),** 54
3. Jacob G., Duseaux M., Farges J. P., Van Den Boom M. B. and Roksnoer P. J. (1983). *J. Cryst. Growth.*, **61,** 417
4. Jordan A. S., Caruso R. and Von Neida A. R. (1980). *Bell Syst. Tech. J.*, **59(4),** 593
5. Duseaux M. (1982). *Thesis, Paris VI*

6. Duseaux M. (1983). *J. Cryst. Growth*, **61,** 576
7. Jacob G., Farges J. P., Schemali C., Duseaux M., Hallais J. P., Bartels W. J. and Roksnoer P. J. (1982). *J. Cryst. Growth*, **57,** 245
8. Jacob G. (1982). In *Semi-Insulating III-V Materials, Evian*, p. 2 (S. Makram-Ebeid and B. Tuck, eds.); Shiva Publishing, Nantwich, UK
9. Duseaux M. and Jacob G. (1982). *Appl. Phys. Lett.*, **40(9),** 790
10. Jacob G. (1982). *J. Cryst. Growth*, **59,** 669
11. Duseaux M., Schiller C., Cornier J. P., Chevalier J. P. and Hallais J. P. (1983). *J. Physique, Colloque C4*, 397
12. Martin G. M., Duseaux M. and Farges J. P. (1983). *Proceedings of the IOOC Conference, Tokyo*
13. Schiller C. and Duseaux M. (1983). *Proceedings of the 10th ICXCOM, Toulouse*
14. Holmes D. E., Elliot K. R., Chen R. T. and Kirkpatrick C. G. (1982). In *Semi-Insulating III-V Materials, Evian*, p. 19 (S. Makram-Ebeid and B. Tuck, eds.); Shiva Publishing, Nantwich, UK
15. Martin G. M., Jacob G., Goltzene A. and Schwab C. (1980). In *Inst. Phys. Conf. Ser.*, **59,** 28
16. Brozel M. R., Grant I., Ware R. M. and Stirland D. J. (1983). *Appl. Phys. Lett.*, **42,** 610
17. Holmes D E., Chen R. T., Elliot K. R. and Kirkpatrick C. G. (1983). *Appl. Phys. Lett.*, **43(3),** 305
18. Brozel M. R., Clegg J. B. and Newman R. C. (1978). *J. Phys.*, D**11,** 1331
19. Hunter A. T., Kimura H., Baukus J. P., Winston H. V. and Marsh O. S. (1984). *Appl. Phys. Lett.*, **44(1),** 74
20. Chen R. T., Holmes D. and Kirkpatrick C. G. (1983). *VII International Conference on Crystal Growth, Stuttgart*

Homogeneous GaAs Grown by Vertical Magnetic Field Applied LEC Method

J. Osaka and K. Hoshikawa

Electrical Communication Laboratory, Nippon Telegraph and Telephone Public Corporation, 1839, Ono, Atsugi-shi, Kanagawa, 243-1, Japan

Abstract

The effects of a vertically applied, axially symmetric magnetic field on melt stability and crystal properties of GaAs were investigated by using a newly developed, vertical magnetic field applied LEC (liquid encapsulated Czochralski) apparatus. The presence of residual circulating steady thermal convection in a magnetic field applied melt was found from the melt temperature measurements. Striation-free undoped, SI crystals were obtained by reducing the influence of this thermal convection. Semi-insulating behaviour in the vertical magnetic-field applied crystal was assumed to result from the compensation of the shallow acceptor carbon by the deep donor EL2.

Introduction

In recent years there has been a growing interest in the MF (magnetic field) applied crystal growth technique as a promising technique for improving semiconductor crystal quality [1–6]. In the case of silicon growth by pulling from the melt, it was reported [1–4] that this technique has practical advantages in connection with homogenization of a microscopic impurity distribution and with control of oxygen contamination. It is well known that MF remarkably reduces the temperature fluctuation in an electrically conductive melt. As a result, irregular growth striation, which is composed of concentration variation of impurities and/or native defects, can be suppressed. It is anticipated that this technique also has special advantages in GaAs crystal growth because residual impurities and point defects in high purity GaAs grown by the LEC (liquid encapsulated Czochralski) technique play an important role in its SI mechanism. However, there has been little information about the effect of the MF applied to GaAs on crystal quality and the SI mechanism [5].

We have developed a vertical MF applied LEC (VMLEC) apparatus and investigated these problems. In this paper we report the effect of the MF on melt stability and crystal homogeneity of GaAs. We also discuss the SI mechanism in undoped SI VMLEC crystals.

Growth apparatus and experimental procedure

Figure 1 schematically shows the growth apparatus developed in our laboratory, which consists of a solenoid coil and a high pressure puller. Up to 2000 Oersted (Oe) vertical MF was applied perpendicular to the growth interface using a coil surrounding the chamber as shown in Figure 1. The MF is then axially symmetric. The resistance-heating puller was designed so as to be able to synthesize GaAs up to 12 kg in weight and to grow crystals up to 5 inches in diameter using a crucible of 8 inches in diameter. About 80 mm diameter GaAs crystals were grown in the ⟨100⟩ direction with a 150 mm diameter PBN crucible under 8 atm argon. The growth rate was 10 mm h^{-1}. A melt of 4 kg was prepared by *in situ* synthesis under 70 atm argon. Dry (100–200 ppm wt. H_2O) 700 gram B_2O_3 encapsulant was used. On the basis of the experimental results on As evaporation losses during the compounding cycle (about 9 g/2 kg As) and growth process (about 3 mg h^{-1} cm^{-2}) [7], the melt composition (Ga/GaAs) was controlled to be 0.5 ± 0.005 by choosing the proportions of starting Ga and As charges.

In several crystals, elemental Se of 2×10^{16} at. cm^{-3} at the seed end of the crystal was doped to facilitate the detection of striations. Striation patterns of samples cut parallel to the growth direction were revealed by chemical etching under illumination using an AB [8] etchant in order to investigate microscopic homogeneity. The electrical and optical properties of crystals were evaluated by Hall effect measurements, near infrared absorption measurements and photoluminescence measurements at 4.2 K. SIMS (secondary ion mass spectrometry) analysis was carried out for determining impurity concentrations in the grown crystals.

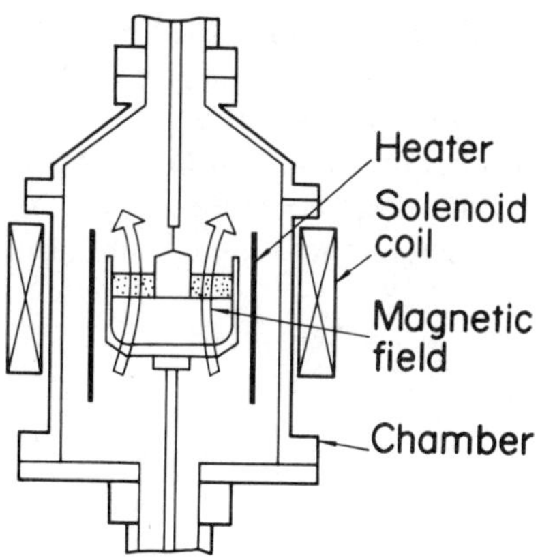

Figure 1 Schematic diagram of vertical MF LEC growth apparatus.

Effect of magnetic field on melt stability and crystal homogeneity

Striation patterns observed in Se doped crystals grown with and without MF are shown in Figure 2. Both crystals were grown with rotation of the crystal

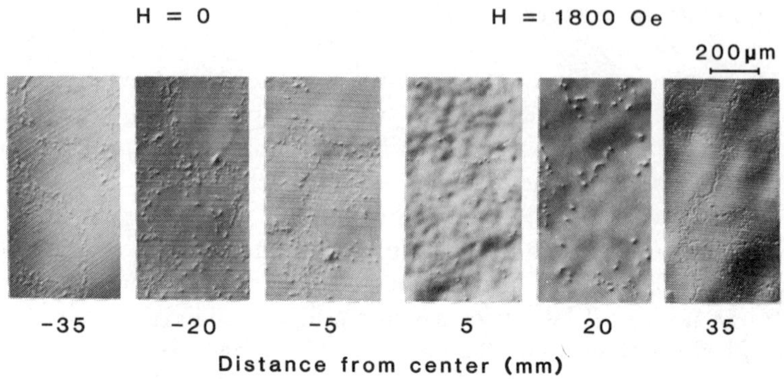

Figure 2 Photographs of striation patterns observed in Se doped crystals grown with and without MF.

and crucible at 11 and 1 rpm, respectively, in the same direction. Striations in the conventional LEC crystals exhibited irregular intervals and strengths throughout the crystal, while the VMLEC crystal has a striation-free area in the centre but not in the peripheral region. At the periphery, rotational striations with nearly identical interval and strength were clearly observed. The presence of rotational striations have been reported in axially asymmetric horizontal MF applied CZ crystals [6]. It was considered that the rotational striations would be caused by an axial asymmetric temperature distribution in a melt [6]. The present result suggests that the axial asymmetric temperature distribution exists even in the axially symmetric MF applied melt. To clarify this hypothesis, temperature measurements were carried out at 25 mm under the surface of a 4 kg melt.

Figure 3 shows the melt temperature variation as a function of MF strength. It can be noticed that the OTC (oscillating thermal convection), which is detected as the temperature fluctuation by a thermocouple, was remarkably reduced when an MF of more than 800 Oe was applied. Furthermore, one notes that the temperature itself decreased as the MF strength increased. It can be readily supposed that temperature distribution changes as MF strength

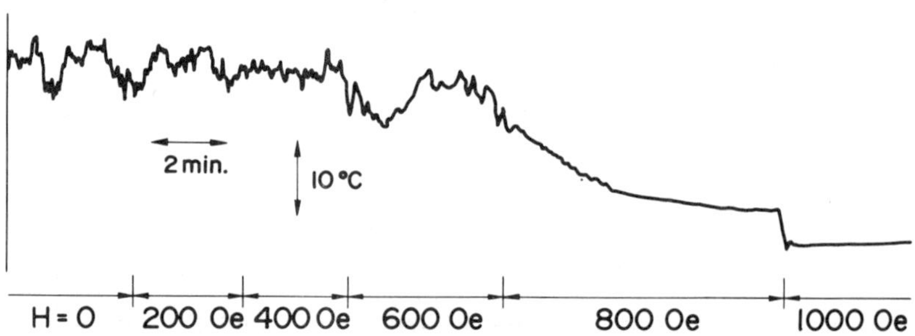

Figure 3 Melt temperature variation as a function of MF strength. The strength was changed within a few seconds. The temperature measurement was carried out at 25 mm under the surface of a 4 kg melt.

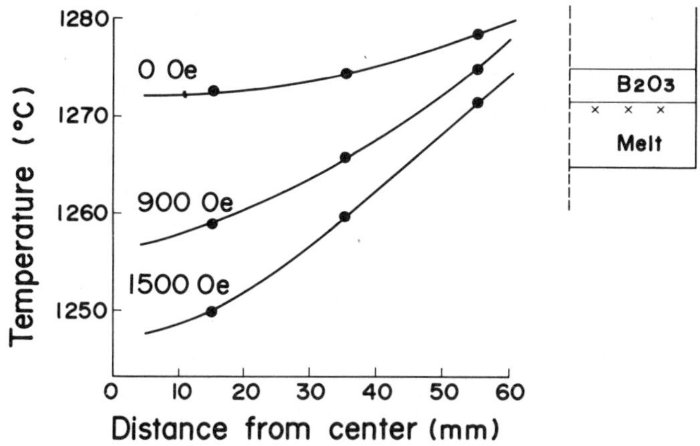

Figure 4 Temperature distribution in the melt at 5 mm below the surface and its dependence on the MF strength. Cross points in the inserted figure correspond to measured points.

is changed. The temperature distribution in the melt at 5 mm below the surface and its dependence on the MF strength were also measured as shown in Figure 4. Figure 4 suggests that circulating steady (laminar) thermal convection (CSTC), that reduces the temperature gradient in the melt, cannot be suppressed so easily by the MF, but it gradually decreases as the MF is increased. Therefore, it is considered that the temperature distribution in the melt gradually changes toward a final state which is determined only by a thermal conduction in the melt as the MF is increased, as shown in Figure 4. From these results, we have concluded that the CSTC still exists in the strongly MF applied melt up to 1500 Oe in the present results. Here, we propose a thermal convection model as schematically shown in Figure 5. (a) The OTC is superimposed on the CSTC when MF was not applied. The CSTC generally has axial asymmetry caused by axial asymmetric heating. (b and c) This axial asymmetric CSTC remains in the MF applied melt, but the increase in MF suppresses the OTC.

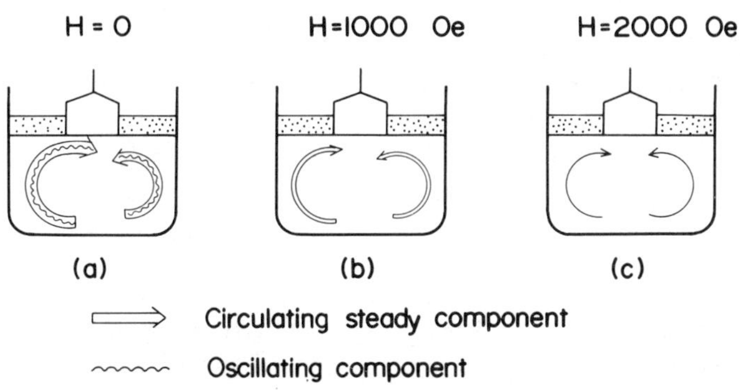

Figure 5 Schematic flow patterns of thermal convection in the melt with and without MF applied.

The axially asymmetric CSTC gives rise to rotational striations in a crystal. It is widely accepted that a crystal rotation is used to reduce an asymmetric melt temperature distribution around a growing crystal in the CZ technique. We have studied the dependence of macroscopic striation and impurity distribution on the crystal rotation. A crystal was grown under an MF of 1800 Oe with a crystal rotation that was changed in steps as 1, 6, 11, 1, 6 and 11 rpm as the crystal grew (fraction solidified $g = 0.2$–0.7), and with the crucible rotation at 1 rpm in the same direction as the crystal. A completely striation-free area was obtained when the crystal was rotated at 1 and 11 rpm. Rotational striations were present in the peripheral region when the crystal was rotated at 6 rpm. The macroscopic impurity distribution also changed as the crystal rotation was changed. The most uniform radial distribution of impurity concentration was obtained in the seed end side portion ($g = 0.4$) with the crystal rotation at 11 rpm. It has been suggested that these results are due to the balance between the CSTC and a forced convection generated by the crystal rotation [9]. It is readily understood that the optimum crystal rotation rate changes as a crystal grows because the strength of the CSTC should depend on the quantity of a residual melt.

As mentioned above, it is worth remarking that irregular striations can be suppressed by the application of an MF of at least 800 Oe but rotational striation cannot be suppressed unless the crystal rotation rate is appropriately chosen to cancel the influence of the residual CSTC.

Other advantages of the VMLEC technique found through this investigation are summarized as follows:

1. The yield of a single crystal increased. This may be associated with the convex growth interface caused by the melt temperature distribution as shown in Figure 4.
2. The controllability of a crystal diameter was improved. This may also be attributable to the radial melt temperature profile.
3. The crystal diameter could be controlled by an MF strength control. This is a result of the quick response of the melt temperature to any change in the MF strength as shown in Figure 3.

Properties of undoped semi-insulating VMLEC crystals

SI undoped crystals with resistivities of 5–$10 \times 10^7\ \Omega$ cm (mobility of about 6500 $cm^2\ V^{-1}\ s^{-1}$ at room temperature) were obtained by the VMLEC method. This result is different from the earlier published data reported by Terashima *et al.*, which were obtained with a horizontal MF LEC crystal [5]. To investigate the compensation mechanism in VMLEC crystals, three crystals, VMLEC undoped crystal (A), Se doped crystal (B) and a conventional LEC undoped crystal (C), of about 3 kg in weight were grown with rotation of the crystal and crucible at 11 and 1 rpm, respectively. A striation-free area of 60 mm in diameter at the centre of crystal B was obtained. The macroscopic carrier concentration radial profile in crystal B exhibited a W-shaped variation and its variation (maximum − minimum/average) was ±7% at the solidified fraction of g = 0.5–0.6.

Concentrations of chemical impurities determined by SIMS in the two undoped crystals A and C are summarized in Table 1. Both crystals were very high purity except for boron and oxygen. No distinct difference between the two crystals was recognized within experimental accuracy. The total concentration of shallow donor impurities (Si, S, Se and Te) was 3×10^{14} at. cm^{-3} and was one order of magnitude less than the concentration of the main shallow acceptor impurity carbon ($1–3 \times 10^{15}$ at. cm^{-3}). Carbon concentrations obtained by LVM (localized vibrational mode) measurements agreed well with those determined by SIMS. The concentrations of Mn and Fe (both acceptors), which are rather high, were below the background checked by using MBE (molecular beam epitaxy) GaAs. Therefore, it is assumed that the actual concentrations of Mn and Fe are far below the values shown in Table 1. Oxygen is believed to be a deep donor. Boron shows uncertain characteristics: Ta *et al.* reported that the concentration of a 0.073 eV acceptor level depends on boron content in Ga-rich and near stoichiometric crystals [10]. It was estimated from their results that the content of the 0.073 eV acceptor level depends on boron content in Ga-rich and near stoichiometric it can be assumed that the predominant electrically active background impurity is C in both crystals.

Table 1 *SIMS analysis of undoped crystals*

Sample	Si	S	Se	Te	C	Mg	Cr	Mn	Fe	O	B
A	2E14	1E14	6E12	1E13	1E15	1E14	4E14	8E14	3E15	1E16	1E17
C	1E14	1E14	2E13	3E13	3E15	1E14	3E14	8E14	2E15	2E16	8E16

The spectrum of the optical absorption around 1 μm exhibits the presence of a deep donor, generally referred to as EL2, in the two crystals. The EL2 concentration, determined by absorption at 1.1 μm using the capture cross-section reported by Martin [11], ranged from $1.5–2.2 \times 10^{16}$ cm^{-3} in the VMLEC crystal at g = 0.5–0.6, which was slightly less (~90%) than that in the conventional one. The radial distributions of the 1.1 μm and 2.0 μm absorptions in crystal B were measured in order to evaluate the total (occupied) EL2 concentration and its homogeneity in the VMLEC crystal. The spatial resolution was 400 μm^2. The result is shown in Figure 6 together with the result for crystal A. It is clear that the radial distribution is rather uniform. The spectrum of photoluminescence at 4.2 K showed a broad band around 0.8 eV.

It is concluded from these results that the SI behaviour in VMLEC crystals is attributable to the compensation of the shallow acceptor C by the deep donor EL2, quite similar to that in conventional LEC crystals. We believe that the discrepancy between our results and those by Terashima *et al.* [5] depends mainly on the difference in the melt composition, the quantity of residual impurity, growth rate and cooling rate rather than the difference of the applied MF direction.

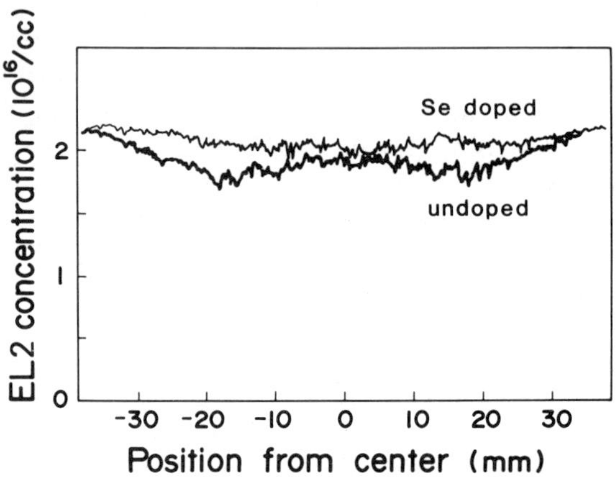

Figure 6 Radial EL2 concentration profile in undoped (bold line) and Se doped VMLEC crystals.

Summary

The effects of MF on melt stability and crystal properties of GaAs were investigated by using a newly developed VMLEC apparatus. It was reconfirmed that irregular striations which originated from the melt temperature fluctuation were suppressed by applying MF to the melt.

It was also found that the origin of the rotational striations generally present in MF applied crystals is axially asymmetric to the residual circulating steady thermal convection. The presence of this striation depends on the crystal rotation rate and other growth conditions.

The MF does not affect the SI mechanism in our striation-free undoped VMLEC crystals with homogeneous distribution of a deep donor EL2.

ACKNOWLEDGEMENTS

The authors are grateful to T. Izawa and S. Miyazawa for their continuous support and encouragement during the study. We are also indebted to H. Kohda and T. Kobayashi for their assistance in crystal growth and sample preparation, and K. Watanabe for his optical measurements.

References

1. Hoshi K., Suzuki T., Okubo Y. and Isawa N. (1980). In extended Abstracts, *Electrochem. Soc.*, Spring Meeting, Pennington
2. Suzuki T., Isawa N., Okubo Y. and Hoshi K. (1981). In *Semiconductor Silicon.* Electrochemical Society, Pennington, USA
3. Hoshikawa K., Kohda H. and Hirata H. (1984). *Jap. J. Appl. Phys.*, **23,** L37
4. Fiegel G. (1983). *Solid-State Technol.*, August, 121
5. Terashima K., Katsumata T., Orito F., Kikuta T. and Fukuda T. (1983). *Jap. J. Appl. Phys.*, **22,** L325
6. Witt A. F., Herman C. J. and Gatos H. C. (1970). *J. Mat. Science*, **5,** 882
7. Kobayashi T. and Osaka J. Unpublished
8. Abrahams M. S. and Buiocchi C. J. (1965). *J. Appl. Phys.*, **36,** 2855
9. Osaka J., Kohda H., Kobayashi T. and Hoshikawa K. To be published in *Jap. J. Appl. Phys.*
10. Ta L. B., Hobgood H. M. and Thomas R. N. (1982). *Appl. Phys. Lett.*, **41,** 1091
11. Martin G. M. (1981). *Appl. Phys. Lett.*, **39,** 747.

Growth and Some Characteristics of Low Dislocation Undoped SI GaAs Crystals

Mo Peigen, Wu Ju, Zou Yuanxi, Yang Jinhua and Li Shouchun

Shanghai Institute of Metallurgy, Academia Sinica, 865 Chang Ning Road, Shanghai 200050, China

Abstract

The LEF (liquid encapsulated freeze) technique is adopted for growing low dislocation undoped SI GaAs crystals of up to 60 mm in diameter from a quartz crucible. The best crystal, having a diameter of 30 mm, has an average EPD $< 7500\ cm^{-2}$, with that in one selected area as low as $10^2\ cm^{-2}$. Assessments of the crystal quality were made, including the thermal stability, the profiles of EPD (etch pit density), leakage current, I_L, and the sheet resistivity R_s. The preliminary results show that the electrical uniformity of the low EPD wafers is better than that of the high EPD ones. The complex $V_{Ga}O_iV_{Ga}$ acting as a deep donor is tentatively assigned to the 0.65 eV deep level in the low dislocation LEF crystals.

Introduction

Semi-insulating GaAs single crystals are widely used as substrates for the fabrication of GaAs ICs. The homogeneity of the available SI GaAs crystals grown by the LEC (liquid encapsulated Czochralski) technique is, however, not very satisfactory and dislocations are almost always present at a high density (10^4–$10^5\ cm^{-2}$). Major efforts have recently been concentrated on developing a reproducible technique for the growth of large and high quality undoped SI GaAs single crystals, but only a limited success has been reported so far.

In our recent work, a modified low pressure (<3 atm) LEC method, called the LEF technique [1], is employed for the growth of undoped SI GaAs crystals. The main advantage of the present technique is that the crystals can be grown under a reduced temperature gradient. Undoped SI crystals of 30 mm diameter with an average EPD $< 7500\ cm^{-2}$ have been successfully grown by this technique. The EPD, I_L and R_s profiles and the thermal stability of the resulting crystals as well as the main deep level have been studied. The electrical uniformity of the low dislocation crystals and the nature of the main deep level are discussed.

Experimental

In LEC crystals, dislocations are generated as a result of the stress set up by the temperature gradient existing between the crystal and its environment during growth [2]. The furnace temperature profile shows that the largest temperature gradient occurs between the B_2O_3 encapsulant and the Ar ambient. In view of this situation, a modified low pressure LEC method adopting certain features of the LEK (liquid encapsulated Kyropoulos) technique [3] is employed in the present experiments. The compound was synthesized through an As injection cell for *in situ* reaction between six-nines As and six-nines Ga in a quartz crucible. The B_2O_3 was dehydrated *in situ* at ~800°C for 1–2 h before Ga and As compounding. The B_2O_3 layer (20–30 mm) was thicker than that used in the conventional LEC method by a factor of two, so that a reduced temperature gradient can be maintained. The crystal was kept inside the B_2O_3 layer during the growth process, and was pulled out at a temperature in the range 700–800°C after solidification.

The resulting (100) wafers were subjected to the following tests along the ⟨110⟩ axis point by point: (a) EPD profiles were revealed by etching with molten KOH; (b) I_L profiles were measured at 1000 V in the dark by the two probe technique; and (c) R_s profiles were measured by the Barrera technique [4]. In addition, the thermal stability was assessed by thermal annealing under vacuum at 800°C for 30 min and the main deep level was determined by temperature dependent Hall measurements.

Results and discussion

The electrical resistivity and average EPD of a typical LEF crystal are illustrated in Figure 1. There are no difficulties in the growth of ⟨100⟩ single crystals. Some bubbles are usually observed at the end of the growth cycle indicating growth in an As-rich melt. Crystals of up to 60 mm diameter have been grown from a quartz crucible and the crystals exhibit a high resistivity ($>10^7$ Ω cm) from top to bottom under properly controlled conditions. The electrical properties of our crystals are somewhat different from those pre-

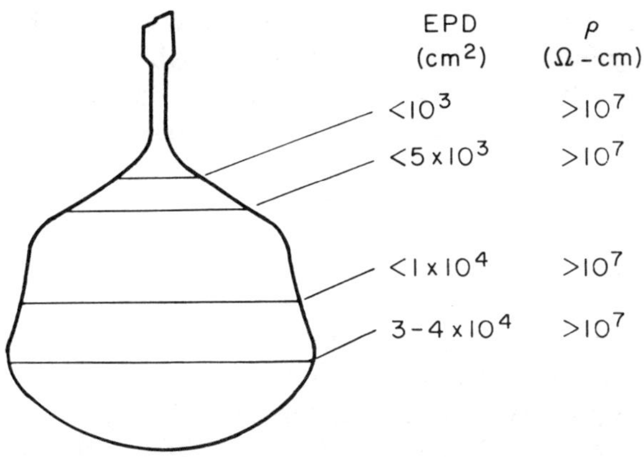

Figure 1 Typical data on EPD and resistivity in the LEF crystals.

pared by Jacob [3]. In our experiments the crystals are always n-type, even if the electrical resistivity is sometimes within the range of 10^4–10^2 Ω cm in the lower half of some crystals, whereas a change from n to p-type is observed in Jacob's work.

The average EPD of the resulting crystals with a diameter of 50–60 mm is about 10^4 cm^{-2}, lower than that of LEC crystals by one order of magnitude. The best crystal, having a diameter of 30 mm, has an average EPD of less than 7500 cm^{-2}, with that in selected areas as low as 10^2 cm^{-2}. This crystal is used in the following characterization experiments.

The wafers used for thermal annealing were placed in a closed box made of GaAs and heated under a vacuum of 8.5×10^{-5} torr to 800°C for 30 min in a sealed quartz ampoule. Electrical measurements were made on the samples both before and after annealing. The results on typical samples are given in Table 1. It is shown that the thermal stability of the low dislocation crystals meets the requirements for device fabrication fairly well.

Table 1 *Thermal stability of low dislocation undoped SI GaAs*

Sample	EPD (cm^{-2})	R_s or ρ	
		As grown	Post annealing
LSI-49	$\sim10^3$	4.8×10^{10} Ω/□	3.1×10^8 Ω/□
LSI-51	$\sim10^3$	3×10^7 Ω cm	3×10^7 Ω cm

The radial profiles of EPD and I_L on the (100) wafers along ⟨110⟩ axis show the W and M-shaped patterns respectively for the LEC wafers (EPD > 10^4 cm^{-2}), as shown in Figure 2, in agreement with those reported by Miyazawa *et al.* [5]. However, both the EPD and I_L profiles are approximately U-shaped and less prominent for the LEF wafers. The lower value of EPD obtained in LEF crystals compared with the LEC ones is remarkable. It may be due to the difference between the two processes in the thermal stresses induced in the growing crystal as it cools. The high I_L values shown in the periphery region of the low EPD wafer might, in our opinion, be accompanied by the low concentration of deep donors as a consequence of the loss of arsenic.

The distribution of R_s over a (100) wafer of low EPD is found to be quite uniform. The R_s along the ⟨110⟩ direction of a typical wafer is practically constant, and the largest difference in R_s over the wafer amounts to only 0.05×10^7 Ω/□ or about 3.8%, as shown in Figure 3. These results show that the lower value and more uniform distribution of EPD may help to improve the electrical uniformity of SI GaAs crystals.

The relationship between I_L and EPD is shown in Figure 4. It is clear that the I_L is inversely proportional to EPD for the high EPD wafers, whereas no such relationship exists when the latter is low enough. Since the parallel

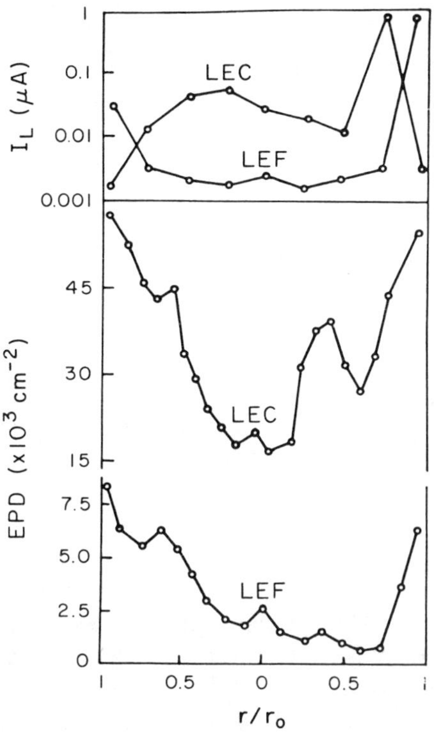

Figure 2 Radial profiles of EPD and I_L for the LEC and LEF crystals along the ⟨110⟩ axis.

relationship between EL2 and EPD has been well established [6], it is inferred that the deep donor in our low EPD wafers would be different from EL2. This seems to be verified by the results of Hall measurements at different temperatures showing that the activation energy for the LEF crystals is 0.65 eV.

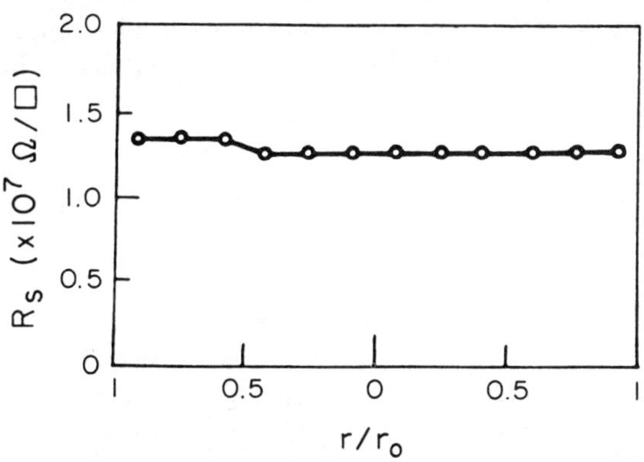

Figure 3 Radial profiles of R_s for the LEF crystals along the ⟨110⟩ axis.

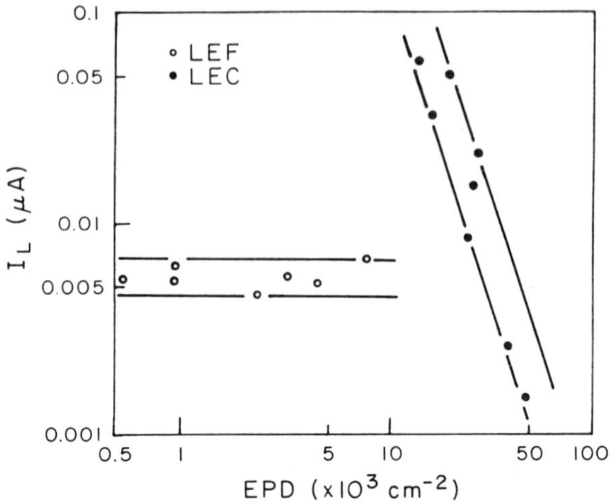

Figure 4 Relationship between EPD and I_L for the LEC (EPD > 10^4 cm^{-2}) and LEF (EPD < 10^4 cm^{-2}) crystals.

In a previous paper [7], we have pointed out that the formation of a complex will help to relieve the strain caused by plastic deformation and therefore the complex will be increased in the presence of stresses, just like dislocations. In this way, the parallel relationship between EPD and EL2 can be explained if a complex defect is assigned to EL2 such as $V_{Ga}As_{Ga}V_{As}$ proposed previously by one of the authors [8]. Now, our low dislocation LEF crystals are grown in a rather moderate temperature gradient and kept for a long time at a temperature near the melting point after solidification. As a result, the residual stress will be reduced to such an extent that it no longer controls the formation of the deep donor in question.

It is also our opinion that the deep level in LEF crystals should be related to oxygen which comes from the wet B_2O_3 used for growing SI GaAs crystals from a quartz crucible in agreement with the suggestions made by Ikoma [9], Yu and Walters [10] and Oliver *et al.* [11]. From a comparison of our results with those of Oliver *et al.* three different cases manifest themselves depending upon the crystal pulling conditions used, as shown in Table 2.

Table 2 *Main deep levels observed in LEC and LEF crystals*

EL2	0.65 eV	Crystal pulling conditions
Yes	Yes	Wet B_2O_3, LEC
Yes	No	Dry B_2O_3, LEC
No	Yes	Wet B_2O_3, LEF

It is conceivable, therefore, that the formation of EL2 and 0.65 eV levels compete with each other, favoured by high EPD and high oxygen content respectively. In a previous paper [8], the formation of EL2 is assumed to take place according to the following reaction:

$$As_{As}+(V_{Ga})_2^- = V_{Ga}As_{Ga}V_{As}+e^- \qquad (1)$$

We may assume, therefore, that the 0.65 eV level also originates from $(V_{Ga})_2^-$:

$$(V_{Ga})_2^- + O_i = V_{Ga}O_iV_{Ga}+e^- \qquad (2)$$

The complex $V_{Ga}O_iV_{Ga}$ may be assigned to the main electron trap (0.65 eV) in the low dislocation undoped LEF SI GaAs crystals. It is expected to play an important role in the compensating mechanism just like EL2 in the undoped LEC SI GaAs crystals. It is interesting to note that a similar complex defect of the configuration $V_{Ga}O_{As}V_{Ga}$ has been tentatively assigned by Zhou *et al.* [12] to the EL2-like deep level found in their LPE GaAs layers grown on high dislocation substrates. It may also be mentioned in passing that Martin and Makram-Ebeid [13] have questioned the justification of assigning EL2 to the level found by Ikoma *et al.* [9] in annealed LPE GaAs after electron beam-irradiation. Further work is in progress to shed more light on this problem.

Summary

A modified low pressure LEC method adopting certain features of the LEK technique, called LEF, has been employed for growing undoped SI GaAs crystals up to 60 mm diameter from a quartz crucible. The best crystal grown by this technique has an average EPD $<7500\ cm^{-2}$, with that in selected areas as low as $10^2\ cm^{-2}$. In comparison with the conventional LEC crystals, the radial variation of EPD along the wafer diameter for a low EPD wafer is seen to be less pronounced.

The I_L and R_s profiles show that the electrical uniformity of the low EPD wafers is better than that of the high EPD ones. It is found that I_L is independent of EPD when the latter is low enough. This implies that the deep donor in LEF wafers should be different from EL2 observed in the conventional LEC SI GaAs crystals. The complex $V_{Ga}O_iV_{Ga}$ is tentatively assigned to the 0.65 eV deep level in LEF crystals which has been found by Hall measurements at different temperatures. The formation of EL2 and the 0.65 eV level under different crystal pulling condition is discussed.

ACKNOWLEDGEMENTS

The authors would like to thank Lian Qi and Chen Zhengxin for the leakage current and temperature dependent Hall measurements.

References

1. Wu J., Yang J. H., Li S. H. and Mo P. G. To be published in *Rare Metals* (in Chinese)
2. Jordan A. S. (1980). *J. Appl. Phys.*, **51,** 2218
3. Jacob G. (1982). In *Semi-Insulating III–V Materials*, Evian, p. 2 (S. Makram-Ebeid and B. Tuck, eds.); Shiva Publishing, Nantwich, UK
4. Barrera J. (1975). Cornell Electrical Engineering Conference, 135
5. Miyazawa S., Mizutani T. and Yamazaki H. (1982). *Jap. J. Appl. Phys. Lett.*, **21,** L542
6. Martin G. M., Jacob G., Goltzene A., Schwab C. and Poiblaud G. (1981). *Inst. Phys. Conf. Ser.*, **59,** 281
7. Zou Y. X., Zhou J. C., Mo P. G., Lu F. Z., Li L. S., Shao J. A., Huang L., Sun H. H., Sheng C. and Sun Q. (1983). *Inst. Phys. Conf. Ser.*, **65,** 49
8. Zou Y. X. (1982). *Inst. Phys. Conf. Ser.*, **63,** 185
9. Ikoma T., Takikawa M. and Taniguchi M. (1982). *Inst. Phys. Conf. Ser.*, **63,** 191
10. Yu P. W. and Walters D. C. (1982). *Appl. Phys. Lett.*, **41,** 863
11. Oliver J. R., Fairman R. D. and Chen R. T. (1981). *Electron. Lett.*, **17,** 839
12. Zhou J. C., Li L. S., Luo Y., Zou Y. X., Li C. C., Lu P. F. and Sheng C. Presented at 1983 Nat. Symp. on GaAs and Related Compounds (Shanghai, China)
13. Martin G. M. and Makram-Ebeid S. (1983). *Physica B*, **116,** 371

Characterization of High-Resistivity Undoped GaAs VPE Layer Grown at Low Temperature

K. Kitahara, H. Itoh, J. Komeno and M. Ozeki

Fujitsu Laboratories Ltd., 1677 Ono, Atsugi 243-01, Japan

Abstract

High resistivity vapour-phase epitaxial layers were grown at temperatures (T_g) of 546–702°C without intentional doping, using the $AsCl_3/Ga/H_2$ system. The highest resistivity was obtained at $T_g = 650$°C which is slightly higher than the transition temperature from n to p-type conduction. From measurements of photoluminescence, secondary ion mass and deep level transient spectroscopies, it was deduced that these VPE (vapour-phase epitaxial) layers were of a high purity and that the electrical compensation between the Zn acceptor and deep-donor (EL2) contributed to the resistivity enhancement. Static characteristics of field effect transistors fabricated on ion-implanted vapour-phase epitaxial layers were better than those fabricated on directly ion-implanted LEC GaAs.

Introduction

For fabrication of GaAs MESFETs (metal-semiconductor field effect transistors) and ICs (integrated circuits), multiple VPE layers—high resistivity buffer layer and n-type active layer—are grown on SI substrates. The growth conditions of the high resistivity layers are adjusted for formation of the multilayers. Recently, the active layer has been fabricated by the ion-implantation process. Hence, the growth conditions of the high resistivity layers for ion-implantation must be reconsidered.

The high resistivity layers are usually grown by doping of impurities (Cr, Fe and V) [1–3], increasing $AsCl_3$ mole fraction [4], and using a dual bubbler [5]. For the ion-implantation material, the latter two non-doping methods are more suitable than the first one because of high purity. However, these non-doping methods require close control and careful selection of substrate [6]. As another non-doping method, we reported growth in a low-temperature region of 660–680°C [7]. Low-temperature growth (<700°C) has also been tried by other researchers [8, 9]. They obtained high purity and/or high resistivity layers. The characteristics of these VPE layers and devices fabricated by ion-implantation into them, however, have not yet been clarified.

This paper describes the electrical properties of the epilayers and ion-implanted layers, impurities and defects taking part in the electrical compensation, effects of inhomogeneity of the substrate on the epilayers and static characteristics of MESFETs fabricated on the ion implanted epilayers.

Samples

Epilayers were grown at temperatures (T_g) of 546–702°C using the $AsCl_3/Ga/H_2$ system. In our growth experiment, we did not use intentional doping of impurities. Various kinds of semi-insulating GaAs were used as substrates: these were heavily Cr-doped (~3 wt. ppm) GaAs grown by the HB (horizontal Bridgman) method and lightly Cr-doped (~0.3 wt. ppm) GaAs grown by the HB and the LEC (liquid encapsulated Czochralski) methods. Surface orientation of these substrates were 2 or 5 degrees off (100) towards (110). A mirror surface was obtained at $T_g \geq 580$°C with high reproducibility. In the region of $T_g < 580$°C the surface morphology tends to deteriorate with lowering of T_g.

The ion-implantation of Si^+ was done for five epilayers and an LEC GaAs wafer with an energy of 150 keV and a dose of 2.5×10^{12} cm^{-2}. Implanted layers were annealed at 800°C for 20 minutes with an AlN passivation. Epilayers without ion-implantation were also annealed to confirm a thermal conversion near the surface.

Simple-structure MESFETs were fabricated on ion-implanted layers for measuring their static characteristics. Their gate width and length were 3 and 50 μm respectively. Devices on the wafer were isolated from each other with a mesa-structure.

Results and discussion

ELECTRICAL PROPERTIES OF EPILAYERS

Hall-effect measurements were made at 300 K by the van der Pauw method. Figure 1 shows the relationship between the electrical properties and T_g for thirty three epilayers having thicknesses in the range of 3–6 μm. Sheet resistivities (ρ_s) greater than 3×10^7 Ω/□ were obtained with high reproducibility. Neglecting the parallel conduction in substrates whose sheet resistivities were 5×10^8–2×10^9 Ω/□ as illustrated with a shaded region in Figure 1, the resistivity of the epilayers was estimated to be greater than 1×10^4 Ω cm. Obvious dependence of sheet resistivity on Cr density and the substrate growth methods was not found for these epilayers. So, it was deduced that autodiffusion of Cr did not dominate the enhancement of resistivity.

Inversion of the conduction type was observed at $T_g \sim 640$°C. The samples were n-type at the lower T_g region and p-type at the higher T_g region. The highest sheet resistivity, $\rho_s = 8 \times 10^8$ Ω/□, was measured on a sample grown at $T_g = 650$°C, which is slightly higher than the inversion temperature from n to p-type conduction. This ρ_s value is comparable with that of the substrate. Electron mobilities (μ_e) of the n-type layers were 4500–8600 $cm^2\,V^{-1}\,s^{-1}$ at a T_g between 600 and 640°C. Their electron mobilities, in the range of $T_g <$ 600°C, decreased with lowering of T_g.

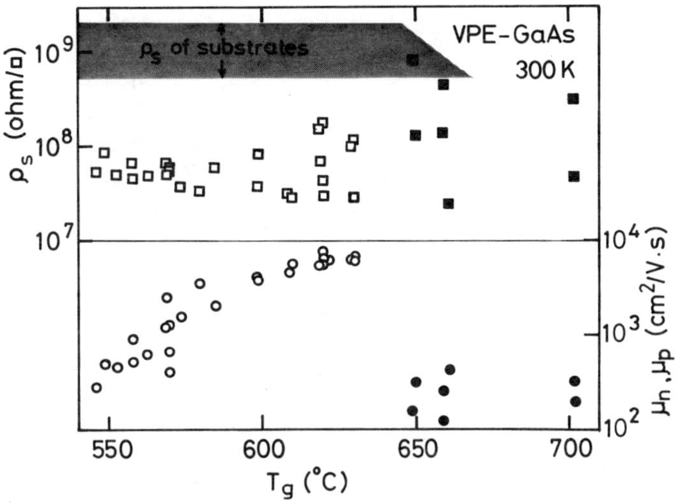

Figure 1 Growth temperature dependence of sheet resistivities (ρ_s) and mobilities (μ_n and μ_p) of epilayers having thicknesses of 3–6 μm.

The Fermi level energy in thermal equilibrium was estimated from the temperature gradient of the free electron density (n). The calculated values were $E_c - 0.49$ eV and $E_c - 0.62$ eV for $T_g = 558$ and 630°C respectively, where E_c is the lowest conduction band energy. The latter value is only 0.06 eV smaller than the activation energy of the deep-donor EL2 [10]. The contribution of EL2 to the electrical compensation is discussed later.

ELECTRICAL PROPERTIES OF ION-IMPLANTED LAYERS

Depth profiles of ion-implanted layers were monitored by differential capacitance-voltage measurements. Figure 2 shows typical results on epilayers grown at $T_g = 659$°C, and in LEC GaAs. Measurements were made at about twenty points between the arrow regions shown in the inset of Figure 2. Fluctuations along the solid and broken arrows are shown by shading in the profiles. The tails of the profiles are steeper in the VPE layer than in LEC GaAs. Activation rates of 62 to 67% (estimated from the peak value of n) and electron mobilities of 4720–4850 cm^2 V^{-1} s^{-1} were obtained for three VPE layers grown at $T_g = 619$, 659 and 699°C. These values were nearly equal to those of ion-implanted LEC GaAs. Poor activation rates of 21% and 0.4% (estimated from Hall measurements) were obtained for VPE layers grown at $T_g = 569$ and 552°C respectively.

IMPURITIES AND DEFECTS CONTRIBUTING TO COMPENSATION

For a quantitative analysis of the impurities, SIMS (secondary ion mass spectrometry) measurements were made for five samples grown at $T_g = 558$,

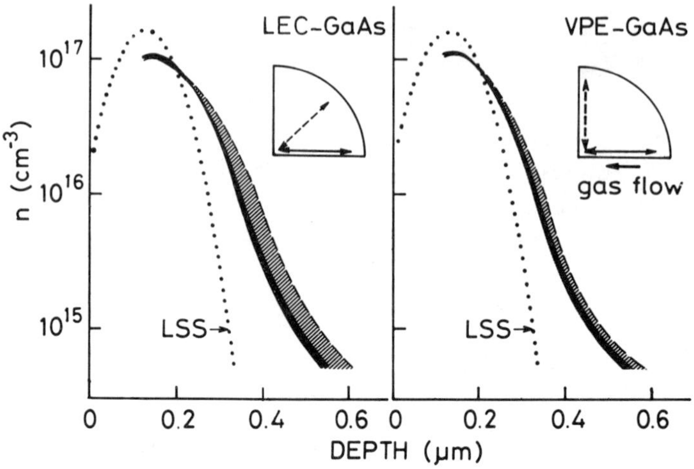

Figure 2 Depth profiles of carrier density (n) in ion-implanted layers. Dotted lines are LSS (Linhard, Scharff and Schiøtt) curves.

602, 610, 659 and 702°C. The measured densities of C, Si, S and Se atoms in all samples were nearly equal to, or lower than, the background level (the background levels were: C, 5×10^{16}; S, 5×10^{15}; Si, 2×10^{15}; and Se, $2\times10^{14}\ cm^{-3}$). The density of Cr in all samples was about $4\times10^{15}\ cm^{-3}$, while no dependence on T_g was found (the background level of Cr in this measurement is not clear). The present SIMS data show that the density of impurities in the epilayers were lower than their respective detection limits.

A more sensitive analysis was made by PL (photoluminescence) spectroscopy. The PL was excited by the 514.5 nm emission of an Ar^+ laser, analysed by a 750 mm double-pass or 250 mm spectrometer, and detected by a GaAs cathode photomultiplier or Si-photodiode. Also a few samples grown at $T_g > 702$°C were used for measurements. Figure 3 shows typical 4.2 K PL spectra of the epilayers grown at $T_g = 745$, 649 and 558°C. PL lines associated with excitons and the Zn acceptor were observed near the band-edge energy region. Such obvious splitting of exciton lines in the former two samples shows that these epilayers are high purity. A PL line associated with the Si acceptor was found for the sample grown at $T_g = 740$°C but not for the samples grown at lower temperatures. The disappearance of this line can be explained by the reduction of contamination from the growth system with lowering of T_g [11]. With a lowering of T_g, PL spectra showed the degradation of epilayers; that is, the near band-edge PL efficiency decreased, and a broad 1.2 eV line appeared at $T_g \leq 580$°C. This broad line has been attributed to a Ga vacancy-donor complex [12]. These degradation tendencies are consistent with the decrease of Hall mobilities shown in Figure 1.

Deep levels taking a part in the electrical compensation were studied by DLTS (deep level transient spectroscopy). Measurements were made on Schottky barrier diodes. Instead of as-grown epilayers, ion-implanted epilayers were used because quantitative measurement cannot be made for high

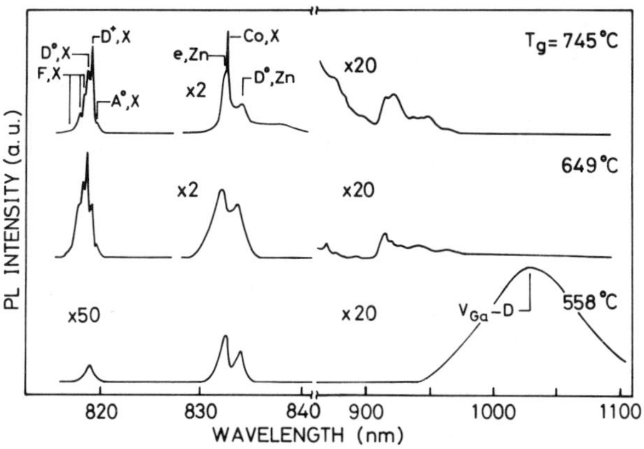

Figure 3 PL spectra at 4.2 K for epilayers grown at $T_g = 745, 649$ and 558°C. The lines appearing in the near-band-edge region are due to free excitons (F, X at 1.5181, 1.5161 and 1.5151 eV), excitons bound to neutral donors, ionized donors or neutral acceptors (D^0, X at 1.5145 eV; D^+, X at 1.5138 eV; A^0, X at 1.5129 eV).

resistivity thin layers. Only EL2 was detected for all samples. The densities of EL2 in samples grown at $T_g = 619, 659$ and 699°C were 2.4×10^{15}, 1.9×10^{15} and 9.8×10^{14} cm^{-3} respectively, and increased with lowering of T_g. One of the authors has reported that the density of electron traps in VPE layers, grown by the $AsCl_3/Ga/N_2$ system, increased with lowering of T_g in the region of 636–757°C [13]. So the increase of EL2 density with lowering of T_g seems to be a common characteristic of VPE layers grown in the low-temperature region.

The results of PL and DLTS measurements show that formation of an n-type high resistivity layer at $T_g < 650$°C can be explained by the compensation between the Zn acceptor and EL2. The presence of a deep acceptor, which was not detected in the present experiment, should be taken into account to explain the formation of p-type high resistivity layers at $T_g \geqslant 650$°C.

UNIFORMITY OF EPILAYERS

Comparison of uniformity between LEC GaAs and epilayers was made through distributions of PL intensity on wafers as shown in Figures 4a and 4b, respectively. Measurement was made at 4.2 K. The Ar^+ laser was focussed to a diameter of about 100 μm on the sample wafers. The PL was analysed by the 250 mm spectrometer.

For LEC GaAs, measurements were made along the ⟨100⟩ and ⟨110⟩ axes. Microscale severe inhomogeneity was found for the 1.515 eV line associated with excitons. The distributions of the 1.494 eV line associated with the C acceptor shows the well-known W-shape. The intensity at the peripheral region was greater along the ⟨100⟩ than the ⟨110⟩ axes.

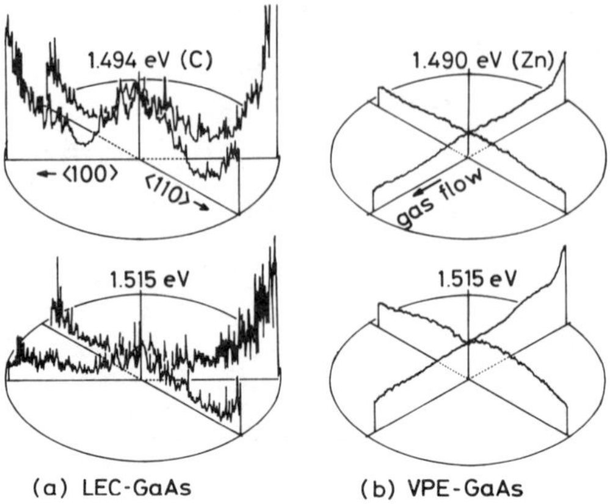

Figure 4 Distributions of PL intensity on (a) LEC GaAs and (b) VPE layers grown on LEC GaAs substrate.

PL distributions on the epilayer were measured along the diameter axes parallel and perpendicular to the gas flow in the furnace. This epilayer was grown at 659°C on a substrate cut from the same boule as the present LEC GaAs. The microscale and W-shaped inhomogeneities were not found for either the 1.490 eV line associated with the Zn acceptor or the 1.515 eV exciton line.

The microscale and W-shape distributions in the LEC GaAs are attributed to the gettering of nonradiative centres by dislocations [14] and thermal stress during the crystal growth [15], respectively. The improved uniformity of the VPE growth shows that the uniformity of the VPE layer is not dominated by the distribution of dislocations and impurities in the substrate.

STATIC CHARACTERISTICS OF MESFETS

Static characteristics of MESFETs, such as drain current looping and transconductance, were measured to compare the epilayers and LEC GaAs. The effects of materials on the device performances have frequently been masked by the fluctuation of device fabrication processes. So MESFETs with a simple mesa-structure were made on both materials by an identical process. Figures 5a and 5b show the drain current–voltage characteristics of MESFETs fabricated on the ion-implanted epilayers and those on the directly ion-implanted LEC GaAs, respectively. The obvious drain current looping appears in the latter, but not in the former. The transconductance of the epilayer is about 10% higher than that of LEC GaAs.

MESFETs arranged in ICs often have a planar structure. In this case, thermal conversion at the surface can degrade the isolation between devices; so, ρ_s of the epilayers was measured after annealing. The ρ_s values after annealing at 800, 825 and 850°C were 8×10^8, 3×10^9 and 4×10^9 Ω/□, respectively. These values were higher than those of the as-grown epilayers, $4\times 10^8\ \Omega/\square$; that is, no degradation of ρ_s by annealing was found.

The remaining problems for device fabrication are macroscale inhomogeneity, which was shown in Figure 4b, and resistivity, which is lower than that of a semi-insulating substrate. Inhomogeneity can probably be improved by controlling the gas flow velocity and $AsCl_3$ mole fractions [16]. Isolation between devices may possibly be improved by proton bombardment [17].

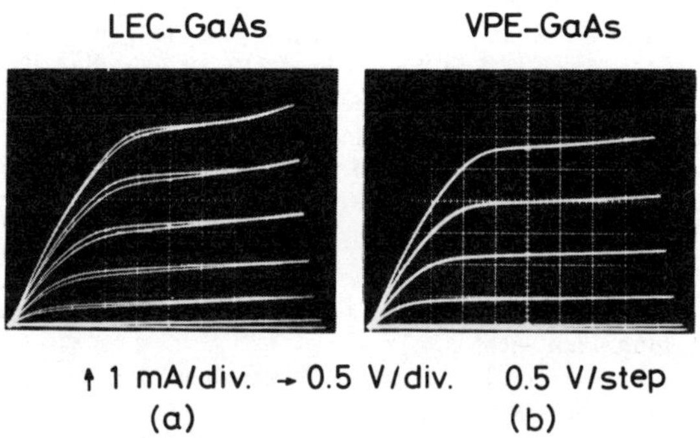

Figure 5 Drain current-voltage characteristics of MESFETs fabricated on (a) directly ion-implanted LEC GaAs and (b) ion-implanted epilayer.

Conclusions

Growth of high resistivity VPE layers was made at a temperature below 700°C without intentional doping of impurities. The highest sheet resistivity ($8\times 10^8\ \Omega/\square$) was obtained at 650°C for a VPE layer having a thickness of 3.8 μm. Excellent surface morphology, mobility, activation rate and luminescence efficiency were obtained in the temperature region of $T_g > 600$°C, while they tend to deteriorate with lowering of T_g in the range of $T_g < 600$°C. PL measurements showed that the VPE layers were of a high purity and that the dominant acceptor impurity was Zn. Resistivity is enhanced by the electrical compensation between Zn and the EL2 deep-donor. PL intensity profiles showed that the uniformity of the VPE layer was not dominated by the distribution of dislocations and impurities in the substrate. Static characteristics of MESFETs fabricated from the ion-implanted VPE layers are better than those of directly ion-implanted LEC GaAs.

ACKNOWLEDGEMENTS

The authors would like to thank S. Okamura, K. Kodama and E. Taguchi for their contributions to this work and K. Nakai, Y. Nakayama and H. Nishi for their helpful discussions. The authors are also grateful to T. Kotani and

A. Sibatomi for their encouragement. The SIMS measurement was carried out by C. Evans and Associates.

References

1. Mizuno O., Kikuchi S. and Seki Y. (1971). *Jap. J. Appl. Phys.*, **10,** 208
2. Nakai K., Kitahara K., Shibatomi A. and Ohkawa S. (1977). *J. Electrochem. Soc.*, **124,** 1635
3. Terao H., Sunakawa H., Ohata K. and Watanabe H. (1982). In *Semi-Insulating III–V Materials, Evian*, p. 54 (S. Makram-Ebeid and B. Tuck, eds.); Shiva Publishing, Nantwich, UK
4. Cairns B. and Fairman R. D. (1968). *J. Electrochem. Soc.*, **117,** 197c
5. Nozaki T., Ogawa M., Terao H. and Watanabe H. (1975). *Inst. Phys. Conf. Ser.*, **24,** 46
6. Cox H. M. and DiLorenzo J. V. (1980). In *Semi-Insulating III–V Materials, Nottingham*, p. 41 (G. J. Rees, ed.); Shiva Publishing, Nantwich, UK
7. Komeno J., Kitahara K. and Ohkawa S. (1979). *J. Cryst. Growth*, **47,** 601
8. Hallais J., Gibod G. B., Chane J. P., Durand J. M. and Hollan, L. (1977). *J. Electrochem. Soc.*, **124,** 1290
9. Shaw D. W. (1977). *152nd Electrochem. Soc. Meeting, Ext. Abstract No. 77*, 862
10. Martin G. M. (1980). In *Semi-Insulating III–V Materials, Nottingham*, p. 13 (G. J. Rees, ed.); Shiva Publishing, Nantwich, UK
11. DiLorenzo J. V. and Moore Jr G. E. (1971). *J. Electrochem. Soc.*, **118,** 1823
12. Williams E. W. (1968). *Phys. Rev.*, **168,** 922
13. Ozeki M., Komeno J., Shibatomi A. and Ohkawa S. (1979). *J. Appl. Phys.*, **50,** 4808
14. Kitahara K., Ozeki M. and Shibatomi A. (1983). *Appl. Phys. Lett.*, **42,** 188
15. Jordan A. S., Caruso R. and von Neida A. R. (1980). *Bell Syst. Tech. J.*, **59,** 593
16. Komeno J., Nogami M., Shibatomi A. and Ohkawa S. (1980). In *Inst. Phys. Conf. Ser.*, **56,** 9
17. Lee C. P. (1982). In *Semi-Insulating III–V Materials, Evian*, p. 324 (S. Makram-Ebeid and B. Tuck, eds.); Shiva Publishing, Nantwich, UK

Large Diameter, Low Dislocation In-Doped GaAs: Growth, Characterization and Implications for FET Fabrication

H. M. Hobgood, R. N. Thomas, D. L. Barrett, G. W. Eldridge, M. M. Sopira and M. C. Driver

Westinghouse R & D Center, 1310 Beulah Road, Pittsburgh, Pennsylvania 15235, USA

Abstract

Preliminary results are presented on the effectiveness of elemental melt doping with dilute concentrations of certain Group III and Group V impurities to prepare large diameter, SI GaAs crystals with low dislocation densities in a high pressure, Melbourn LEC (liquid encapsulated Czochralski) puller. In-doping is highly effective in reducing dislocations, and large central regions (75% diameter) which are virtually free of dislocations are observed in 50 mm diameter substrates. In-doped GaAs substrates are indistinguishable from undoped GaAs in most respects: thermally stable, high resistivities in the 10^7 to $10^8\ \Omega$ cm range and mobilities $>4500\ cm^2\ V^{-1}\ s^{-1}$. Under Si implantation, implant profiles, activation efficiencies and FET (field effect transistor) channel mobilities equivalent to those obtained with high purity, undoped GaAs substrates are observed. DC characteristics of directly implanted FETs fabricated on In-doped and undoped GaAs substrates are comparable, suggesting that the Schottky barrier height is unchanged by the In dopant.

Introduction

General concern of the adverse effects of the high dislocation densities ($>10\,000\ cm^{-2}$) in present day, large diameter GaAs substrates upon the performance of linear and digital ICs, coupled with recent direct correlations of poor FET characteristics to substrate dislocations [1], has renewed interest in reducing dislocation densities in GaAs crystal growth by reducing thermal gradients [2, 3] and/or using impurity hardening effects [4–6].

In this study, the effectiveness of melt doping with Sb, B and In to produce low dislocation density, SI GaAs crystals has been explored. A reduction in dislocation density was observed with all the dopants examined. However, Sb-doped GaAs crystals proved difficult to grow because of the multiple twinning which was observed—a consequence of the low effective segregation coefficient of Sb in GaAs. B-doped GaAs growths resulted in twin-free, SI crystals but with rather modest reductions in dislocations (a factor of six compared to undoped GaAs), presumably a consequence of the low solubility of B in GaAs. In contrast, indium is found to be readily incorporated into the solid GaAs by elemental melt dopant at concentrations up to $10^{20}\ cm^{-3}$ range. In-doping is found to be highly effective in reducing dislocation densities in large diameter GaAs crystals. Large central regions (75% diameter) in 50 mm diameter In-doped GaAs substrates are observed to be virtually free of dislocations. The growth, compositional purity and crystalline perfection, electrical properties and the potential of dislocation-free, In-doped GaAs for directly implanted FET fabrication are discussed here.

Growth of In-doped GaAs crystals

In-doped GaAs crystals up to 75 mm diameter were grown in the ⟨100⟩ orientation from 4 kg, stoichiometric or slightly As-rich melts contained in a PBN (pyrolytic boron nitride) crucible using a high pressure, Melbourn LEC puller. *In situ* melt synthesis from six-nines purity Ga and As charges and elemental In melt doping at concentrations up to about 1 atomic % were used. Crystal growth was initiated from selected undoped GaAs seeds under conditions of counter-rotation of seed (6 rpm) and crucible (15 rpm), pull rate of 6 mm h^{-1}, an axial gradient of 150°C cm^{-1} and an argon ambient pressure of 300 psi.

Crystal growth from In-doped melts proceeded in a straightforward manner similar to growths of conventional undoped crystals. In addition, achievement of the desired stoichiometric or As-rich starting melt composition, for reasons not understood at present, appeared much easier with the presence of the In dopant. The In doping concentration in the crystals was determined to be in the mid-10^{19} cm^{-3} to low-10^{20} cm^{-3} range by X-ray microprobe analysis. From analyses of the In content as a function of the melt fraction solidified, an effective segregation coefficient of 0.1 for In in GaAs is determined (Figure 1), which is consistent with a value of 0.2 predicted from the pseudobinary phase diagram for the GaAs–InAs system at dilute In concentrations [7].

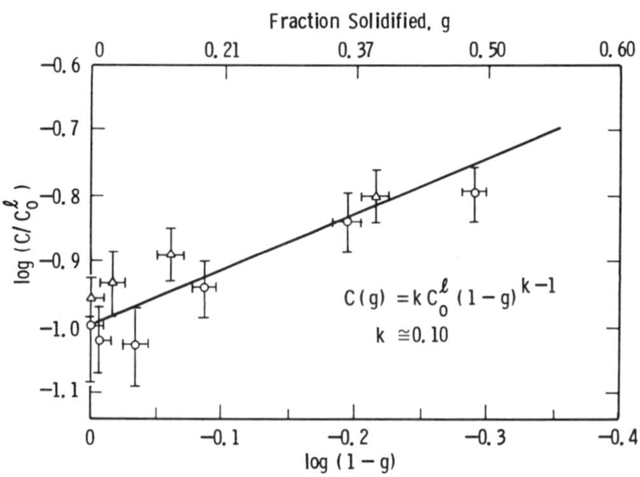

Figure 1 Effective segregation coefficient of In in GaAs derived from X-ray microanalysis of two In-doped GaAs crystals.

Secondary ion mass spectrometry (SIMS) analyses indicate no significant differences in the residual impurity content of In-doped and undoped GaAs samples.

Crystalline perfection

A comparison of In-doped and conventional undoped GaAs substrates by molten KOH etch-pit studies and reflection X-ray topography reveals striking differences in dislocation densities and their distribution. In contrast to the 10^4–10^5 cm^{-2} densities of large diameter, undoped GaAs which often exhibit lineage and polygonized dislocation networks (probably induced by the thermal history of the crystal), In-doped GaAs is remarkably free of these defects. The results of KOH etching of a typical 50 mm diameter In-doped GaAs wafer are shown in Figure 2. Over 75% of the diameter is virtually

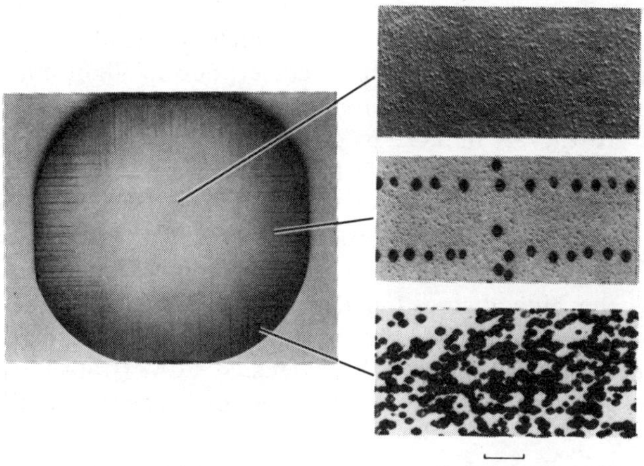

Figure 2 Dislocation EPD in 45 mm diameter, (100) In-doped GaAs wafer after KOH etching.

free of dislocations (100–400 cm^{-2}), while at the periphery dislocation densities are observed to increase, and are revealed as linear arrays aligned along the two orthogonal ⟨110⟩-glide directions lying in the (100) surface. Figure 3 illustrates the dramatic difference between the central regions of an In-doped and an undoped GaAs slice after KOH etching. It is interesting to observe that the undoped sample shows small dislocation-free areas of about 100 μm size which are enclosed by closed polygonized dislocation networks. In 75 mm diameter, In-doped GaAs crystals, the linear dislocation arrays now penetrate farther into the interior and orthogonally linear arrays are observed at the centre of the crystal.

Reflection X-ray topographic evaluations of In-doped GaAs wafers display prominent radial striations in the incorporation of the In dopant and almost

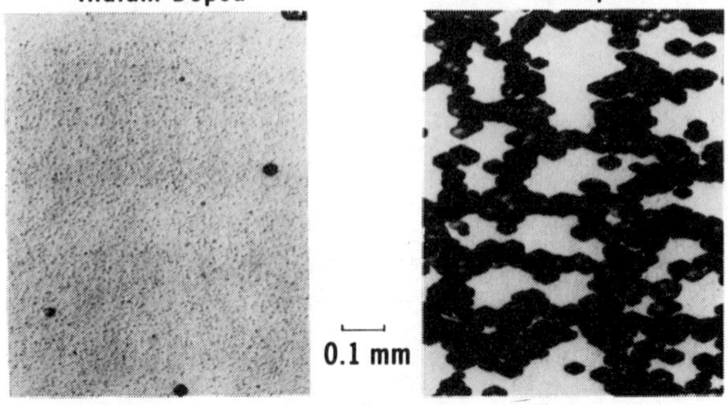

Figure 3 Comparison of EPD in central regions of (a) In-doped and (b) undoped GaAs wafers after KOH etching.

overshadow the linear dislocation arrays which are visible at the edge of the wafer (Figure 4). The striations which correspond to slight inhomogeneities of In incorporation (about 0.01 atomic %), are related to variations in the microscopic growth rate induced by the large thermal convective fluctuations (10°C) measured in the Melbourn LEC growth system [8]. While striated growth is a normal feature of Czochralski grown crystals, and striations have on occasion been observed in undoped and Si-doped GaAs crystals [8], their appearance by KOH etching or X-ray topography is usually masked or modified by the high dislocation densities. In contrast, In decoration of striations makes them clearly visible in In-doped GaAs samples. No evidence of In precipitation was revealed by extensive TEM (transmission electron microscopy) studies of dislocation-free regions. However, precipitates up to

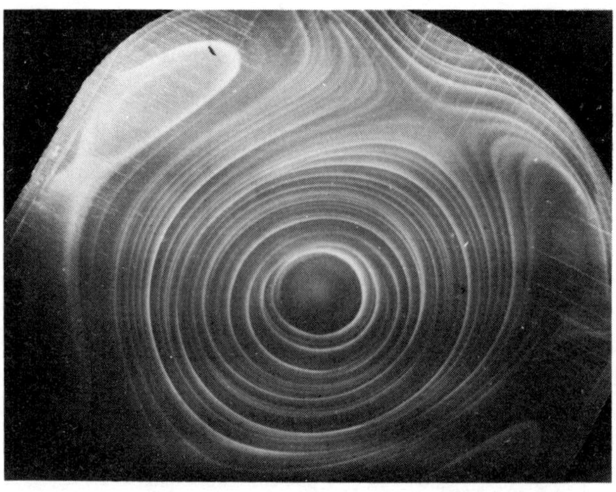

Figure 4 Reflection X-ray topograph of 45 mm diameter In-doped GaAs wafer.

5000 Å in size, which were shown to be highly As-rich by X-ray microanalysis, decorate the linear dislocation arrays near the edge of the crystal if grown from As-rich melts. Similar precipitation effects have been reported in As-rich undoped GaAs [8].

Electrical characterization

Group III and Group V impurity doping of LEC grown GaAs crystals pulled from stoichiometric melts affects the material resistivity quite differently. Sb-doped GaAs samples showed n-type, 100 Ω cm resistivities and mobilities as high as 7000 $cm^2 V^{-1} s^{-1}$ (we also note that similar results were also observed for undoped GaAs when growth was carried out under a nitrogen ambient). In contrast, B and In-doped GaAs showed high resistivity in the mid-10^7 Ω cm and higher, and the characteristic M-shaped resistivity distribution of undoped GaAs slices was noticeably absent in In-doped GaAs substrates. The measured mobility is about 5000 $cm^2 V^{-1} s^{-1}$ in the central dislocation-free region and decreases to 2500 $cm^2 V^{-1} s^{-1}$ at the dislocated edge regions. Substrates subjected to a standard implantation thermal annealing under As overpressure conditions at 860°C for 15 min maintain high resistivities $>10^7$ Ω cm, but some mobility degradation to 3000 $cm^2 V^{-1} s^{-1}$ was observed. No propagation of dislocations was observed following the thermal treatment.

Plots of log resistivity versus reciprocal temperature for undoped Group III and Group V doped GaAs samples are shown in Figure 5. An activation

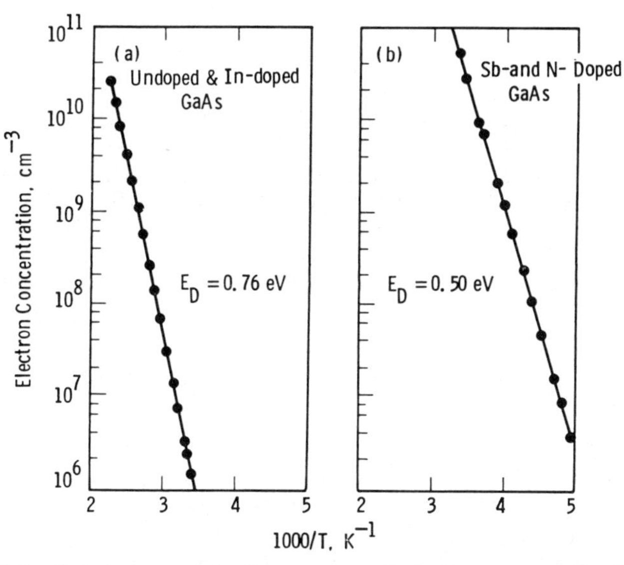

Figure 5 Plots of electron concentration versus reciprocal temperature for (a) In-doped and undoped GaAs and (b) Sb and N-doped GaAs samples.

energy of 0.76 eV is obtained for undoped (grown from either PBN or SiO_2 crucibles) and for In-doped GaAs, suggesting that the SI behaviour in In-doped GaAs is controlled by the same mechanism as in undoped GaAs, namely, compensation of residual acceptors by the deep donor EL2 defect level. However, N and Sb-doped GaAs samples yield an activation energy of 0.45 to 0.50 eV, and it is tempting to suggest that conduction in Group V-doped GaAs is controlled by a shallower 0.50 eV donor defect level.

Ion implantation and FET fabrication

Direct ^{29}Si implantation of dislocation-free, In-doped GaAs results in implant profiles, activation efficiencies and FET channel mobilities which are indistinguishable from those observed in undoped GaAs. Figure 6 shows a comparison of the differential efficiencies for donor activation (Figure 6a) and total ionized impurity activation (Figure 6b) for undoped and In-doped GaAs

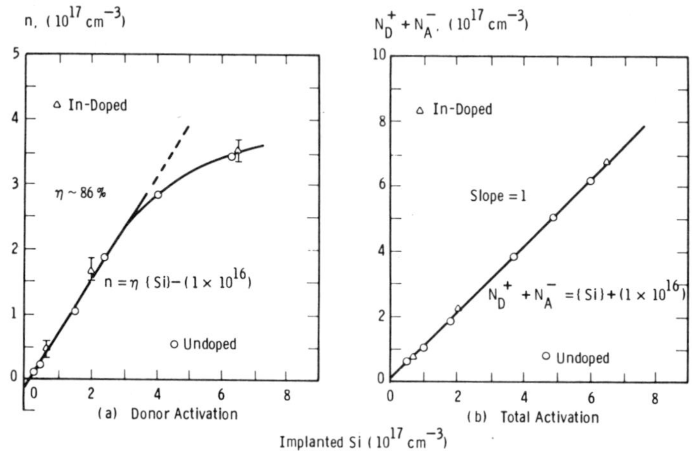

Figure 6 Silicon implantation activation characteristics of In-doped GaAs versus undoped GaAs: (a) donor activation; and (b) total ionized impurity activation.

samples. For implant concentrations below 4×10^{17} cm^{-3}, the donor differential activation efficiency is 86%, and the total ionized impurity activation efficiency is 100%. At the higher implant concentrations, the reduction in implant activation is attributed to the amphoteric nature of Si. A threshold concentration of less than 1×10^{16} cm^{-3} is determined from the intercept in Figure 6a. Measurements of the n-channel mobility as a function of the implanted donor density in In-doped and undoped GaAs are again equivalent: at 1×10^{17} cm^{-3} doping the mobility is 4900 cm^2 V^{-1} s^{-1} at 300 K. Low concentration implants (2×10^{16} cm^{-3}) into In-doped GaAs substrates exhibit mobilities in excess of 5600 cm^2 V^{-1} s^{-1} at 300 K and 12 000 cm^2 V^{-1} s^{-1} at 77 K.

Power FETs with approximately 1 μm gates and 100 μm gate peripheries were fabricated using planar, selective ^{29}Si implantation to form 3000 Å deep channels with 1.3×10^{17} cm^{-3} peak doping on dislocation-free In-doped GaAs substrates. Measurements of d.c. characteristics indicate source and drain contact resistance below 10^{-6} Ω cm^2, pinch-off of 6.5 V and g_m of 90 ms mm^{-1} corresponding to 420 mA mm^{-1} output current. These results are comparable to state-of-the-art FET measurements fabricated on high quality, undoped SI GaAs substrates.

Summary

These investigations indicate that isovalent In-doping of otherwise high purity crystals is highly effective in reducing dislocations in large diameter, SI GaAs crystals required for low-cost IC processing. With proper optimization of the In-doping density and rather modest reductions in the thermal gradients, it is anticipated that 75 mm diameter, SI GaAs crystals free of dislocations will be produced in the future. Indium appears to reside exclusively on the Ga sublattice in contrast to Group V dopants which suggest possible antisite occupancy. The mechanism of dislocation reduction in In-doped GaAs is not well understood, but appears to be associated with the strain-field which the lattice misfit of the In atoms introduces in the host GaAs lattice. This strain-field inhibits the propagation of dislocations through the lattice [9] and effectively increases the critically resolved shear stress of the In-doped lattice. The fact that high quality FETs can be fabricated on In-doped GaAs substrates indicates that the Schottky barrier height is not altered to any significant extent. The availability of dislocation-free, In-doped GaAs substrates may have important implications in the future, particularly in the development of ultra-uniform threshold voltage, digital GaAs ICs.

ACKNOWLEDGEMENTS

The authors gratefully acknowledge the valuable contributions of several of our colleagues to this study, especially Dr W. J. Takei for the X-ray topographic evaluations, Dr S. Wood for the TEM studies and Dr G. Wicks of Cornell University for the photoluminescence spectra. We personally thank Dr W. C. Mitchel of AFWAL for his support and interest in this work. This study was funded in part by AFWAL Contract F33601-83-C-0076.

References

1. Miyazawa S., Ishii Y., Ishida S. and Nanishi Y. (1983). *Appl. Phys. Lett.*, **43a,** 11
2. Jordan A. S., Caruso R. and von Neida A. R. (1980). *Bell Syst. Tech. J.*, **50(4),** 593
3. Fukuda T. (1983). Materials Research Society Meeting, Boston, Massachusetts

4. Seki Y., Watanabe H. and Matsui J. (1978). *J. Appl. Phys.*, **49(2),** 822
5. Jacob G. (1982). In *Semi-Insulating III–V Materials, Evian*, p. 2 (S. Makram-Ebeid and B. Tuck, eds.); Shiva Publishing, Nantwich, UK
6. Hobgood H. M., Barrett D. L., Ta L. B., Eldridge G. W. and Thomas R. N. (1983). Electronic Materials Conf., Burlington, VT
7. Stringfellow G. B. and Greene P. E. (1969). *J. Phys. Chem. Sol.*, **30,** 1779
8. Thomas R. N., Hobgood H. M., Eldridge G. W., Barrett D. L., Braggins T. T., Ta L. B. and Wang, S. K. (1984). In *Semiconductors and Semimetals 20*, p. 1 (R. K. Willardson and A. C. Beer, eds.); Academic Press, New York
9. See, for example, Cottrell A. H. (1953). In *Dislocations and Plastic Flow in Crystals*; Oxford Press, London

Effect of Argon Pressure on the Electrical Resistivity in LEC GaAs

Y. Kadota, M. Kakimoto, Y. Masa, A. Tanaka and T. Yokoyama

Electronics Materials Laboratory, Sumitomo Metal Mining Co. Ltd., 1-6-1, Seuhiro-cho, Ohme-shi, Tokyo, Japan

Abstract

In the undoped GaAs crystals grown by the LEC (liquid encapsulated Czochralski) method, the resistivity falls down drastically from 10^8 Ω cm to 10 Ω cm at a critical fraction of arsenic atom in the melt. This critical fraction changes with the pressure of the argon ambient during the crystal growth, such as 0.455 under 3.1 MPa and 0.477 under 2.0 MPa.

Argon pressure is one of the important parameters in the LEC method for the growth of GaAs single crystals. In this report we describe the effect of the argon pressure on the critical melt composition which gives the boundary between the semi-insulating and the conducting crystals.

The resistivity of GaAs single crystals depends strongly on the fraction of arsenic atoms [As]/([Ga]+[As]) in the melt during the growth as shown in Figure 1, where the arsenic fraction is evaluated by the method of Holmes *et al.* [1, 2]. The resistivity changes drastically from 10^8 Ω cm at an arsenic fraction in the melt of around 0.45. However, these changes occur at a different arsenic fraction in the melt under different argon pressures with the same temperature distribution in the B_2O_3 encapsulant.

Let us define the critical fraction of arsenic atoms as the fraction where the distribution of the resistivity shows very large fluctuations such as 10^2 Ω cm in a wafer and begins to show the so called W-shape. If the arsenic fraction is higher than the critical fraction, the resistivity is of the order of 10^8 Ω cm n-type, and shows a slight M-shape in a wafer. With a lower arsenic fraction, the resistivity decreases abruptly down to 10 Ω cm p-type. The critical fraction changes from 0.477 under an argon pressure of 2.0 MPa to 0.455 under 3.1 MPa.

In Figure 1, the resistivity means an average value over a wafer. From each wafer, 14 samples are cut along the diameter in the shape of parallelopipeds (4 mm × 4 mm × 0.4 mm), and the resistivity is measured by the van der Pauw method in each sample. It is assumed that the arsenic fraction in each wafer could be assigned a certain value because the principal variations generally occur along the growing axis of an ingot of a crystal.

Kirkpatrick *et al.* [3] determined the critical fraction from the concentrations of EL2 and carbon. When the concentration of EL2 decreases below

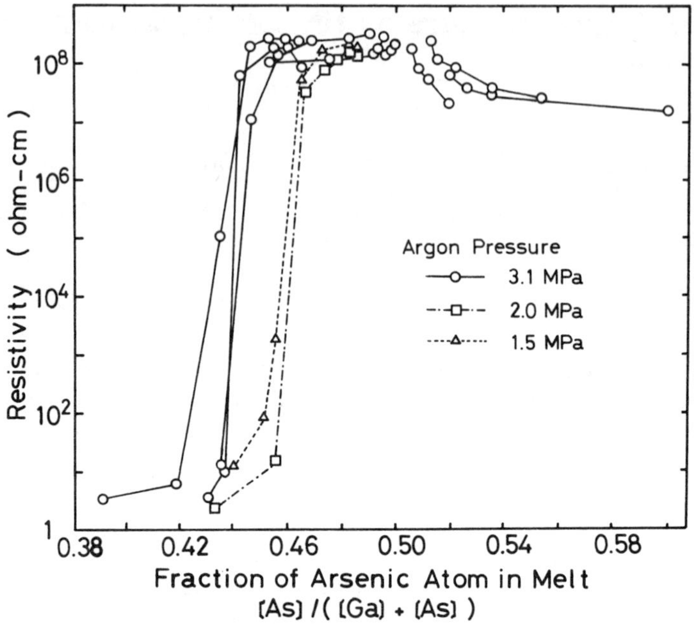

Figure 1 Dependence of the electrical resistivity of LEC GaAs on the fraction of arsenic atoms in the melt.

that of carbon, the hole concentration increases and the crystal shows p-type conduction. Concentrations of EL2 in the present sample are plotted in Figure 2 as a function of the fraction of arsenic atoms in the melt for the semi-insulating samples which are cut from the same ingots used in resistivity measurements shown in Figure 1.

The EL2 concentrations are determined by 300 K infrared absorption experiments [4]. The concentrations of carbon are also obtained from the absorption of the local vibrational mode at 582 cm^{-1} [5] and plotted partly in Figure 2. The carbon concentrations do not show any dependence on the arsenic fraction or on the argon pressure, and the values are very close to the observation of Kirkpatrick *et al.* [3]. In Figure 2, a full line (marked as A) shows the concentration of EL2 in the crystals grown in this work under an argon pressure of 3.1 MPa. A dotted line (marked as B) shows the concentration of EL2 of the crystal grown under an argon pressure of 2.1 MPa which was measured by a similar method and reported by Kirkpatrick *et al.* [2, 3].

The gradient of curve A is different from that of B depending on the argon pressures during the crystal growth. This indicates that the concentration of EL2 depends on the pressure of the argon ambient during the crystal growth. In the case of A, 0.455 is the critical fraction of arsenic atoms in the melt which is in good agreement with the fraction which is estimated from the intersection of the line of the carbon concentration with the extrapolated

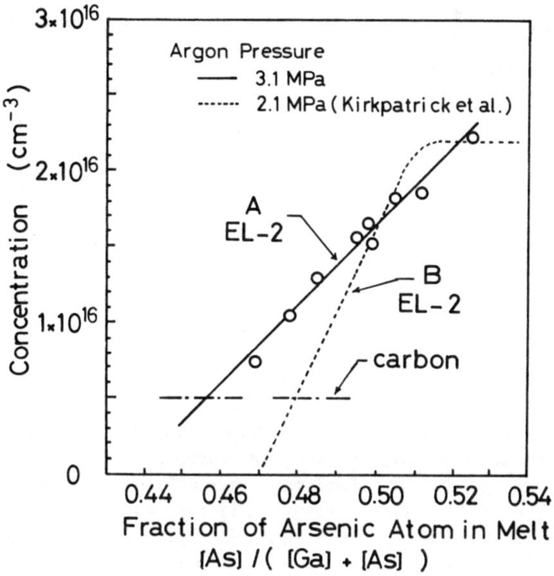

Figure 2 Concentrations of EL2 and carbon as functions of the fraction of arsenic atoms in the melt.

line of the EL2 concentration. In the case of B, a value 0.477 is obtained as the critical fraction from the similar intersection.

A GaAs single crystal grown under 3.1 MPa argon pressure would be more As-rich than the crystal under 2.0 MPa, and the critical fraction of arsenic atoms in the melt decreases from 0.477 to 0.455.

References

1. Holmes D. E., Chen R. T., Elliott K. R. and Kirkpatrick C. G. (1982). *Appl. Phys. Lett.*, **40,** 46
2. Holmes D. E., Chen R. T., Elliott K. R., Kirkpatrick C. G. and Yu P. W. (1983). *IEEE Trans. El. Dev.*, **ED-29,** 1045
3. Kirkpatrick C. G., Chen R. T., Holmes D. E., Elliott K. R. and Asbeck P. M. (1983). *Extended Abstracts of the 15th Conference on Solid State Devices and Materials, Tokyo,* 145
4. Martin G. M. (1981). *Appl. Phys. Lett.*, **39,** 747
5. Brozel M. R., Clegg J. B. and Newman R. C. (1978). *J. Phys. D: Appl. Phys.*, **11,** 1331

Homogeneity of LEC Gallium Arsenide Heavily Doped with Indium

E. J. Foulkes, M. R. Brozel, I. Grant*, P. Singer*, B. Waldock*
and R. M. Ware†

Department of Electrical Engineering, Trent Polytechnic, Nottingham, UK

Abstract

We have used radio-tracer techniques to study the incorporation of indium in LEC (liquid encapsulated Czochralski) GaAs. The value of the distribution coefficient obtained from the gradient of the Scheil plot is 3.6×10^{-2} (compared with the 'intercept' value at $g=0$ of 0.61 and values in the literature which range from unity to 10^{-3}). The radial distribution of the indium is very uniform with a standard deviation of 1.1% at the seed end increasing to 3.7% after 73% of the charge had been grown. The crystals are generally semi-insulating. Correlation of indium distribution with EL2 concentration as measured by infrared absorption, with dislocation densities and electrical data are reported.

Introduction

It has been shown by Jacob [1] that GaAs heavily doped with indium can be dislocation-free whilst retaining its SI properties. We have used radio-tracer, infrared absorption and electrical techniques to study the incorporation of indium in LEC GaAs. The indium dopant was augmented with a small fraction of ^{114}In, a gamma-emitter with a half-life of 50 days. We have compared the concentrations of indium in various parts of the crystals with both the concentrations of the deep donor level EL2 and electrical properties determined from van der Pauw measurements.

Experimental

A number of crystals have been grown by the standard LEC technique using 'dry' (about 300 ppm H_2O) boric oxide in PBN crucibles in a Malvern MSR6/R crystal puller. One per cent metallic indium, by weight, was added to the melt prior to synthesis. After growth the crystals were sectioned to allow radio-tracer measurements, autoradiographs, infrared absorption macroscopic analysis and van der Pauw measurements to be made. The first two analyses allow both quantitative and highly resolved qualitative determinations to be made of the concentrations of radio-tracer indium. Infrared

* Cambridge Instruments Ltd., Rustat Road, Cambridge CB1 3QH, UK.
† Rowland Ware Consultancy, 15 Telegraph Street, Cottenham, Cambridge, UK.

macroscopy has been used previously [2] to determine concentrations of EL2 as well as to detect second phases which can be present.

Figure 1a shows mappings of total indium content in slices taken from various sections of crystal A985/R. Figure 1b shows autoradiographs of adjacent sections. It is clear that a second indium-rich phase is present at values of g (melt fraction) greater than about 0.5, corresponding to an indium concentration of 3×10^{20} cm^{-3}. This also corresponds to the onset of polycrystallinity. However, we have also seen such precipitation in single crystal material.

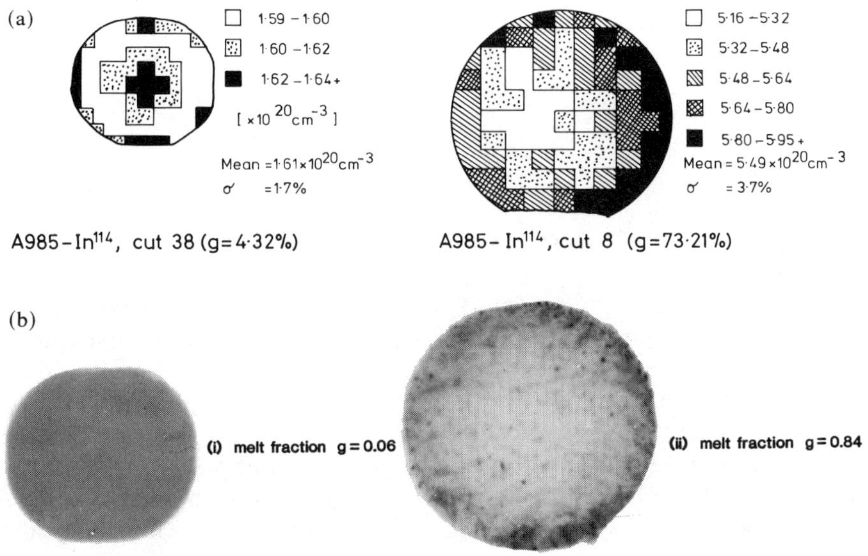

Figure 1 (a) In concentration mappings at seed and tail ends of radio-tracer doped crystal. (b) Autoradiographs of ingot A985 GaAs: In.

Quantitative measurements of [In] shown in Figure 1 have been used to derive a Scheil plot of the segregation behaviour of In. The graph shown in Figure 2 indicates that normal segregation is certainly not taking place. Indeed, values of k, the distribution coefficient, measured from the gradient and the intercept at $g = 0$ are 3.6×10^{-2} and 0.61, respectively. Such deviations from normal segregation have been discussed previously with regard to chromium in LEC GaAs [3] and to tin in LEC InP [4]. These anomalies appear to be closely associated with the non-stoichiometry of the melt and details of analysis in the present case will be presented elsewhere.

Van der Pauw measurements were made on a set of samples 5×5 mm square, cut along (110) diameters of slices taken from a non-radioactive crystal A987 grown under almost identical conditions. The resistivity and mobility are shown in Figure 3. These results are typical for SI undoped GaAs and indicate that the incorporation of In does not markedly affect the

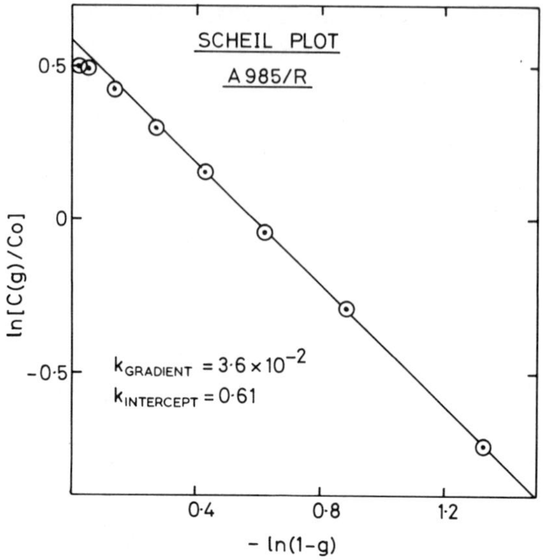

12017 *Figure 2* Indium segregation behaviour.

properties of these crystals. This is in good agreement with Jacob [1]. We have shown previously that the spatial variation of the electrical parameters of SI LEC GaAs can be accounted for by variations in the concentration of uncompensated EL2 centres [2]. We have presented a technique for imaging these EL2 distributions [5]. Such an infrared macrograph of the seed end of A987 is shown in Figure 4a. The contrast is due to variations in the concentration of un-ionized EL2 centres and we have found that the observed structure is directly associated with the presence of dislocations [6]. Thus the central

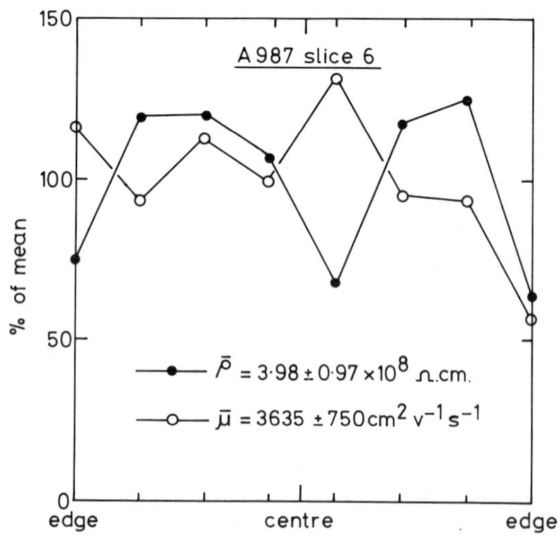

Figure 3 Electrical uniformity across (110) diameter of seed slice.

'square' region is thought to be due to dislocations propagated downwards from the seed (threading dislocations). The intermediate region has a very low dislocation density whilst the dislocations near the periphery appear to be generated by slip.

Figure 4b shows a micrograph of the central region. Note that the dimensions correspond well to the cross-sectional area of the seed. We believe that we are observing individual threading dislocations decorated with EL2. This is confirmed both by the bleaching behaviour at 77 K [7] and by the correspondence with features revealed by KOH etching.

(a)

(b)

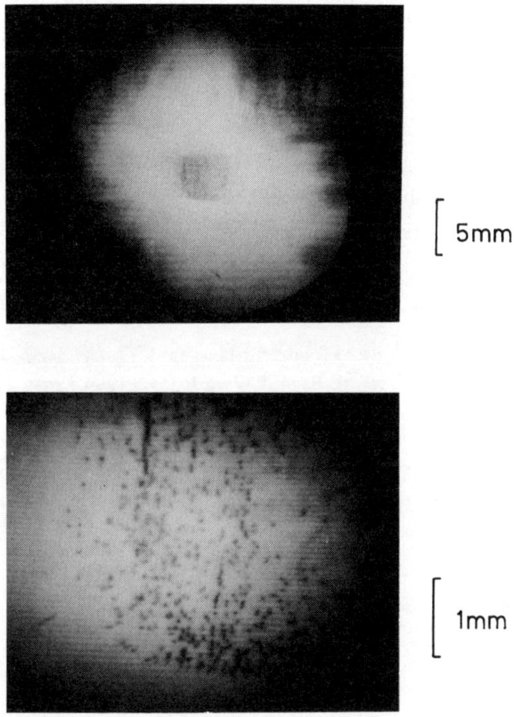

Figure 4 (a) Infrared transmission photograph of seed end of ingot A987. (b) Detail of the centre of the same slice.

Discussion and conclusions

The use of radio-tracer techniques has allowed us to investigate the segregation behaviour of In in LEC GaAs. Different values of the distribution coefficient derived in two ways from the Scheil plot indicate that freezing is not normal. This may explain the different values of k reported previously (Martin *et al.*[8], Willardson and Allred [9]). Autoradiography and radio-tracer mapping have shown that the radial In distribution is very uniform in the top half of the crystals. Precipitates of an In-rich second phase occur in the lower half of the boules when the In concentration is above 3×10^{20} cm^{-3}. Bleaching behaviour has given the first direct evidence of the presence of EL2 centres in highly In-doped GaAs.

ACKNOWLEDGEMENTS

Thanks are due to G. Davis for help in the growth of several In-doped crystals. This work was carried out with the support of Procurement Executive, Ministry of Defence, sponsored by DCVD.

References

1. Jacob G. (1982). In *Semi-Insulating III–V Materials, Evian*, p. 2 (S. Makram-Ebeid and B. Tuck, eds.); Shiva Publishing, Nantwich, UK
2. Skolnick M. S., Brozel M. R., Reed L. J., Grant I., Stirland D. J. and Ware R. M. (1984). *J. Electron. Mat.*, **13(1),** 107
3. Brozel M. R., Tuck B., Rumsby D. and Ware R. M. (1982). *J. Cryst. Growth*, **60(1),** 113
4. Brozel M. R., Foulkes E. J., Grant I. R., Hurle D. T. J., Li L. and Ware R. M. (1984). Unpublished
5. Brozel M. R., Grant I., Ware R. M. and Stirland D. J. (1983). *Appl. Phys. Lett.*, **42,** 610
6. Stirland D. J., Grant I., Brozel M. R. and Ware R. M. (1984). *Inst. Phys. Conf. Ser.*, **67,** 285
7. Skolnick M. S. and Reed L. J. (1984). *Appl. Phys. Lett.*, in press
8. Martin G. M., Duseaux M. and Farges J. P. (1983). IOOC 83, Tokyo
9. Willardson R. K. and Allred W. P. (1966). *Inst. Phys. Conf. Ser.*, **3,** 35

Electrical Behaviour of Annealed LEC GaAs

D. Rumsby, I. Grant, M. R. Brozel*, E. J. Foulkes* and R. M. Ware†

Cambridge Instruments Ltd., Rustat Road, Cambridge, UK

Abstract

Measurements of resistivity, mobility and EL2 concentration have been taken on annealed sections of undoped SI GaAs. A marked improvement in the radial uniformities of these properties has been found for anneals performed at temperatures above 650°C. A reduction in strain in the material, after annealing, was observed by XRT (X-ray topography) and was reflected in reduced wafer breakages during cutting.

The increasing use of GaAs for integrated circuits places ever more stringent demands on the quality and uniformity of substrates. However, it has been found that bulk electrical properties such as resistivity and mobility are very non-uniform in as-grown crystals. For instance, the radial variation in resistivity in the ⟨110⟩ direction shows, typically, an 'M' shaped distribution [1], whereas mobility and the concentration of the deep level EL2 typically show 'W' shaped distributions. The shape of the EL2 distribution has been correlated with the variations in dislocation density [2, 3]. In addition, crystals are often badly strained which can result in ingot cracking during cutting, or wafer breakage during processing.

We have shown previously [4], that annealing of LEC (liquid encapsulated Czochralski) GaAs ingots improves their homogeneity in terms of uniformity of resistivity, mobility and EL2 concentration. In this work, two 55 mm diameter crystals were grown in the ⟨001⟩ direction by the LEC technique, using a crystal puller with automatic diameter control. These crystals were subsequently cut into four 50 mm sections. Samples were taken for assessment before and after the crystal sections were annealed in evacuated, sealed quartz tubes. On completion of the anneal, the furnace was switched off and the sections left to cool to room temperature over several hours. Afterwards, sample slices were cut from one end of each section. The top slice of 1 mm thickness was discarded in order to avoid any surface effects. The remaining samples were used for measurement of infrared absorption, resistivity, mobility and etch pit densities. The technique of measuring EL2 concentrations has been described previously [3]. Resistivity and mobility were determined using van der Pauw measurements made on 5 mm square samples and etch

* Department of Electrical Engineering, Trent Polytechnic, Nottingham, UK.
† Rowland Ware Consultancy, 15 Telegraph Street, Cottenham, Cambridge, UK.

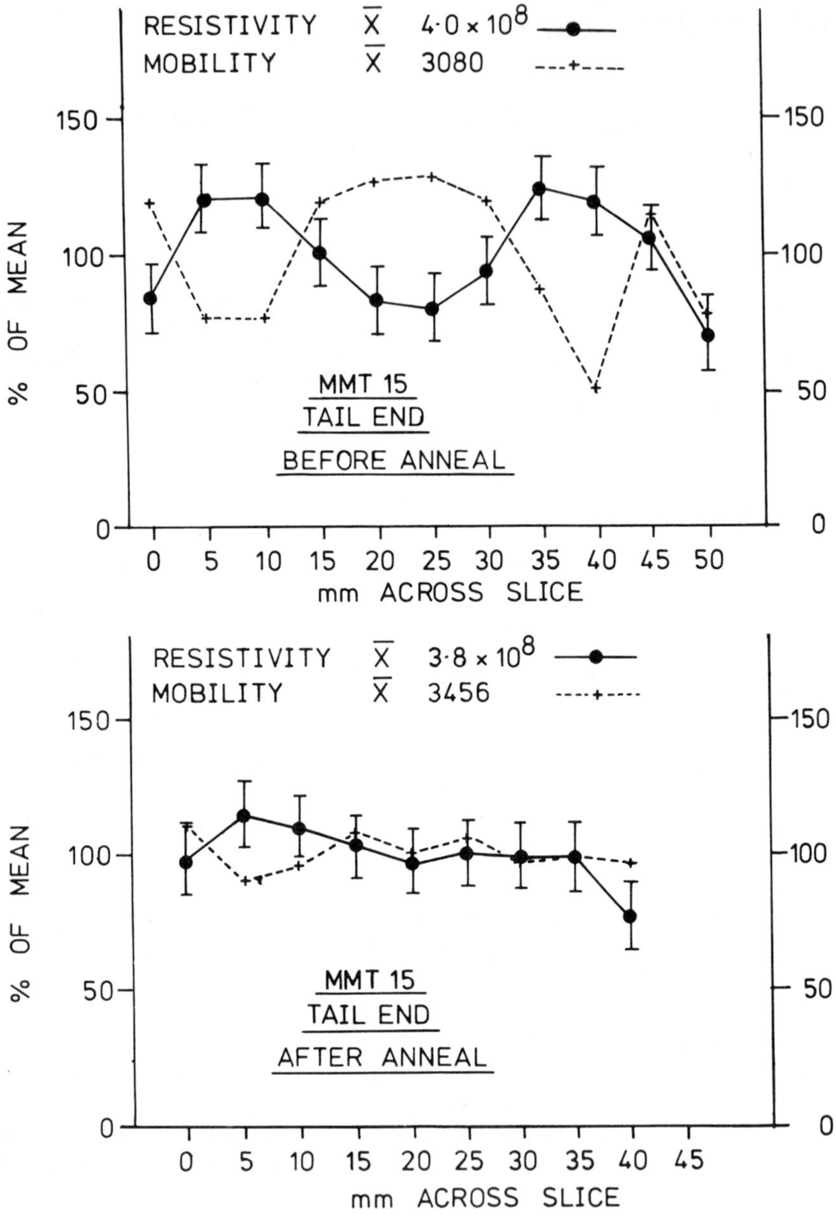

Figure 1 Two inch, ⟨100⟩ undoped GaAs before and after anneal at 950°C for 5 hours.

pit distributions revealed using the A–B etch. In addition, general strain was observed using double crystal XRT.

Changes in resistivity and mobility brought about by annealing at 950°C for 5 hours are shown graphically in Figure 1. It is clear that a substantial improvement in the uniformity of both these properties has taken place. More

Table 1 *Isochronal anneal (5 hours)*

Temperature (°C)	Mean resistivity (Ω cm)		Mean mobility ($cm^2 V^{-1} sec^{-1}$)	
	Before	After	Before	After
650	4.2×10^8 s.d. = 18%	4.2×10^8 s.d. = 22%	2091 s.d. = 43%	3102 s.d. = 17%
750	4.9×10^8 s.d. = 32%	5.4×10^8 s.d. = 12%	1999 s.d. = 37%	2550 s.d. = 14%
850	4.5×10^8 s.d. = 43%	6.3×10^8 s.d. = 10%	1392 s.d. = 51%	2368 s.d. = 12%
950	1.4×10^8 s.d. = 16%	3.8×10^8 s.d. = 6.4%	3079 s.d. = 23%	3456 s.d. = 5%

s.d.—standard deviation.

Table 2 *Isothermal anneal (950°C)*

Time (hours)	Mean resistivity (Ω cm)		Mean mobility ($cm^2 V^{-1} sec^{-1}$)	
	Before	After	Before	After
5	1.8×10^8 s.d. = 18%	1.5×10^8 s.d. = 14%	3760 s.d. = 16%	4450 s.d. = 3%
3	2.5×10^8 s.d. = 13%	2.1×10^8 s.d. = 7%	3348 s.d. = 12%	4263 s.d. = 3%
10	4.1×10^8 s.d. = 15%	3.3×10^8 s.d. = 10%	2564 s.d. = 32%	3885 s.d. = 8%
15	2.5×10^8 s.d. = 18%	2.2×10^8 s.d. = 6%	3289 s.d. = 27%	4207 s.d. = 7%

s.d.—standard deviation.

detailed results of the annealing experiments are given in Tables 1 and 2 for isochronal and isothermal anneals respectively. For isochronal annealing, it can be seen that even at 650°C there is an improvement in both the mean value and the uniformity of mobility, whilst the resistivity remains unchanged. At higher temperatures the uniformity of both mobility and resistivity improve. Note that the mean value of mobility is increased in all cases. The isothermal annealing experiments, carried out at 950°C, suggest that the annealing process is rapid. The electrical properties obtained after 15 hours anneal are not significantly different from the result of a 3 hour anneal.

Figure 2 shows infrared absorption measured at wavelengths of 1 μm, and 2 μm across a ⟨110⟩ diameter of a 55 mm wafer as grown, and after bulk section annealing at 950°C for 5 hours. The absorption curves are obtained by traversing the GaAs wafer across a narrow beam of infrared radiation [3]. Using the calibration due to Martin *et al.* [2], quantitative values of EL2 concentrations have been calculated from the measured absorption coefficients. The effect of annealing is to flatten the 'W' shaped distribution of EL2 and reduce the amplitude of the fine structure usually observed [3] in the absorption curve. In the sample measurements presented in Figure 2

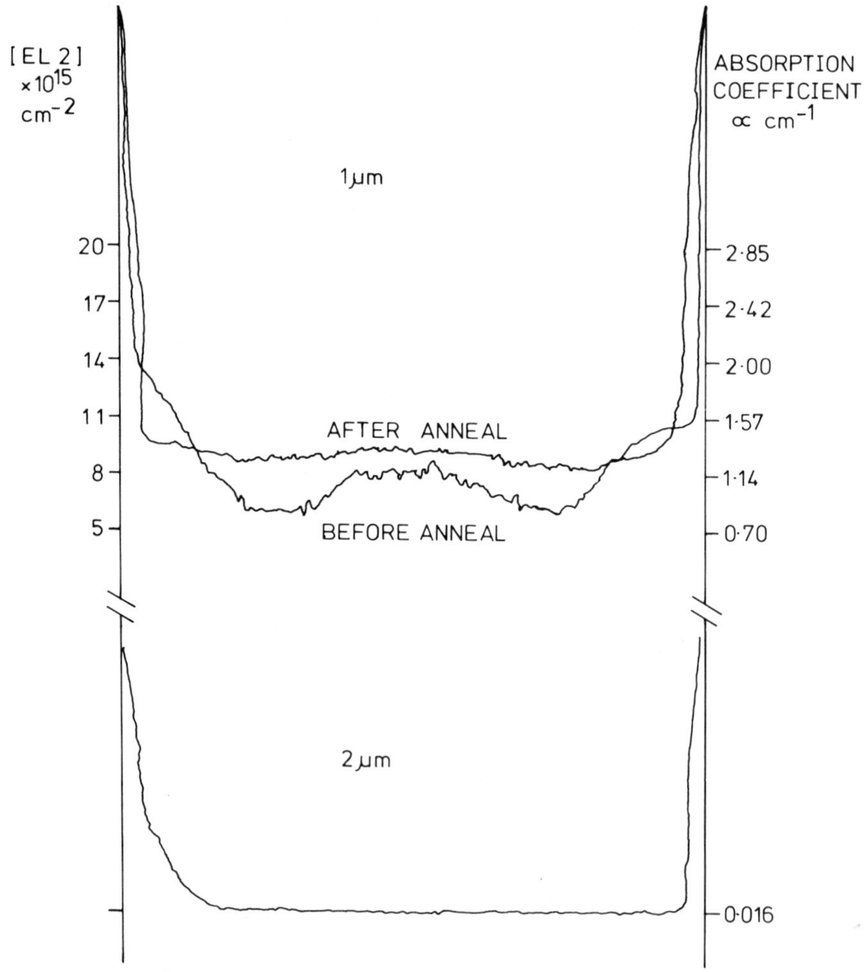

Figure 2 ⟨110⟩ radial EL2 variation before and after anneal at 950°C for 5 hours.

some vestige of the 'W' shaped distribution remains, though in another sample annealed at 1000°C, the EL2 distribution became completely uniform within the accuracy of the experiment.

In addition to the two crystals described, a large number of crystals of both two inch and three inch diameter have been annealed as whole boules. Measurements of resistivity, mobility and EL2 concentration after annealing give more uniform results with standard deviations in mobility and resistivity variations in the range 3–5%.

It is interesting to speculate on the nature of the annealing process, particularly on the redistribution of the deep level EL2. In the as-grown crystals, the distribution of EL2 closely follows that of the dislocations and it has been demonstrated that this association exists not only on the macroscopic scale of the 'W' shaped variation but also on the microscopic structural level of dislocation networks and low angle grain boundaries [5]. However, because the EL2 is redistributed during annealing without any noticeable change in the dislocation distributions or densities, it would appear that the presence of the EL2 is not simply governed by dislocations. To interpret these results, the rapidity of the redistribution has to be explained. The present consensus is that EL2 is related to the antisite defect As_{Ga}. On this basis the redistribution is unlikely to involve a simple diffusion mechanism, since a diffusivity of $>10^{-5}\ cm^2\ sec^{-1}$ would be required and self-diffusion in III–V compounds is known to be very slow [6]. It is possible that there is a thermodynamically determined equilibrium concentration of EL2 and that the levels in the crystal approach this value on annealing. Since the crystal is at a uniform temperature during annealing and during cooling, a uniform concentration would be expected to be frozen into the crystal. Van Vechten [7] has suggested that the concentration of antisite defects in GaAs, close to the melting point, should be about $10^{17}\ cm^{-3}$ and that on cooling these are likely to combine with gallium vacancies to form antisite-vacancy complexes. The measured value for the concentration of EL2 at around $1–3\times10^{16}\ cm^{-3}$ is not inconsistent with this mechanism.

Double crystal XRT is sensitive to crystal strain. Using a synchrotron radiation source, whole wafers up to 3 inch diameter can be imaged and strain mapping performed [8]. With this technique, overall strain reduction in annealed material was apparent and was reflected in the general reduction of slice cracking during cutting.

In conclusion, we have found that annealing GaAs ingots after growth reduces cracking of ingots on wafering, and greatly improves the uniformity of resistivity, mobility and EL2 concentration.

ACKNOWLEDGEMENTS

Thanks are due to Dr G. T. Brown for XRT results and to the Procurement Executive, Ministry of Defence (DCVD) for financial support.

References

1. Grant I., Rumsby D., Ware R. M., Brozel M. R. and Tuck B. (1982). In *Semi-Insulating III–V Materials, Evian* p. 98, (S. Makram-Ebeid and B. Tuck, eds.); Shiva Publishing, Nantwich, UK
2. Martin G. M., Jacob G. and Poiblaud G. (1980). *Acta Electron.*, **23(1)**, 37
3. Brozel M R., Grant I., Ware R. M. and Stirland D. J. (1983). *Appl. Phys. Lett.*, **42(7)**, 610
4. Rumsby D., Smith B., Ware R. M., Brozel M. R. and Foulkes E. J. (1983). GaAs IC Conf., Phoenix
5. Stirland D. J., Grant I., Brozel M. R. and Ware R. M. (1983). *Inst. Phys. Conf. Ser.*, **67**, 285
6. Palfrey H. D., Brown M. and Willoughby A. F. W. (1981). *J. Electrochem. Soc.*, **128(10)**, 2224
7. Van Vechten J. A. (1980). *Handbook on Semiconductors Vol. 3* (S. P. Keller, ed.); North Holland Publishing Company
8. Brown G. T., Skolnick M. S., Jones G. R., Tanner B. K. and Barnett S. J. This volume, p. 76

High-Pressure Direct Synthesis of Bulk Indium Phosphide

R. O. Savage, J. E. Anthony, T. R. Au Coin, R. L. Ross, W. Harsch* and H. E. Cantwell*

US Army Electronics Technology and Devices Laboratory (ERADCOM), Fort Monmouth, New Jersey 07703, USA

Abstract

Modified high-pressure synthesis techniques have been used successfully to compound InP *in situ*. External pressures up to 3000 psi were utilized to form polycrystalline charges in PBN crucibles contained inside sealed quartz ampoules. InP was also synthesized directly from the elements in unsealed crucibles at argon pressures up to 27 000 psi. In the latter method, all quartz components were eliminated from the compounding process. Using standard LEC growth techniques, crystals grown from these polycrystalline charges yielded free carrier concentrations as low as 1.5×10^{15} cm^{-3}, and mobilities up to 20 000 cm^2 V^{-1} sec^{-1} at 77 K. No attempt was made to purify the starting materials prior to compounding.

Single-crystal InP is an important semiconductor for FETs (field effect transistors), millimetre-wave devices, high-speed integrated circuits and electro-optic devices. However, broad use of InP is hindered by high substrate costs, a relatively immature processing technology and less than optimum polycrystalline material for LEC (liquid encapsulated Czochralski) crystal growth.

Various compounding methods have been used to synthesize InP, including direct reaction of the elements in horizontal quartz reaction tubes [1–4], phosphorus injection compounding in high-pressure crystal pullers [5–7] and reactions involving compounds of the elements [8]. Most of the methods utilize quartz components (crucibles, reaction tubes, boats, sealed ampoules, etc.) during the poly-InP synthesis phase. Since silicon is an electrically active impurity in InP, requiring intentional compensation to form semi-insulating material, methods which can reduce or eliminate silica from the synthesis process are highly desirable. After assessing the merits of the above methods, we concluded that a technique that could synthesize InP directly from high-purity elements, in a quartz-free ambient, would ultimately yield very pure material. Also, because of the high vapour pressure of phosphorus and the highly exothermic reaction, there is always the danger of an explosion due to the rapid build-up of pressure in sealed ampoules. By utilizing internally heated pressure vessels capable of sustaining high phosphorus

* Miami Research Laboratories, Eagle-Pitcher Industries, Miami, Oklahoma 74354, USA.

vapour pressures (≫27 atm), this explosion hazard can be greatly reduced, and *in situ* compounding from the elements can be accomplished.

In this study, we are investigating two approaches to the formation of high-purity poly-InP: synthesis in sealed quartz ampoules in a low-pressure autoclave; and reaction of the elements in unsealed PBN crucibles at argon pressures to 27 000 psi. In the first method (Figure 1a), a modified Bridgman–Stockbarger pressure furnace, designed by Medcalf and Fahrig [9], is used to crystallize the poly-ingot. Each synthesis run is performed in a Union Carbide PBN crucible which has been degreased, cleaned in aqua regia,

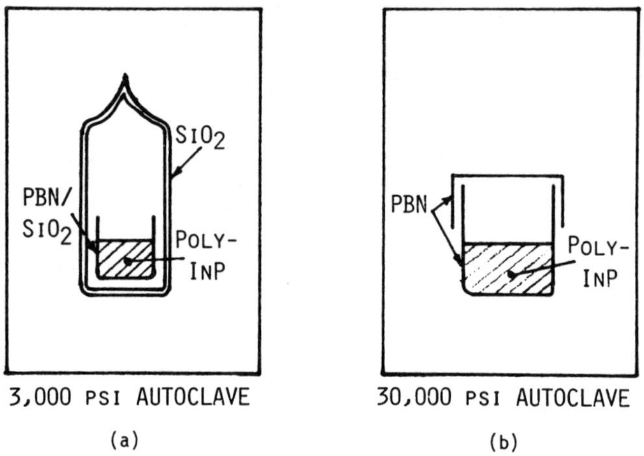

Figure 1 Direct synthesis of poly-InP in (a) sealed quartz ampoules and (b) unsealed PBN crucibles.

rinsed in DI (de-ionized) water and dried. The crucible is placed into a quartz ampoule, loaded with indium and phosphorus, vacuum dried and sealed. The ampoule is then placed into the autoclave and the temperature and pressure carefully ramped upward until approximately 3000 psi is reached at 1100°C. These conditions are held until the reaction and mixing are complete. The ampoule is then lowered through a temperature gradient to initiate freezing of the InP at the bottom of the melt. Solidification progresses upward providing one with a uniform polycrystalline ingot.

The second method under investigation (Figure 1b) employs a quartz-free *in situ* compounding technique using an internally heated Autoclave Engineers high-pressure vessel capable of achieving 30 000 psi at 2000°C in 1.5 litres working volume. Each synthesis run is conducted with stoichiometric quantities of indium and phosphorus plus excess phosphorus to compensate for the amount lost by diffusion during compounding (typically 5–10 wt. %). The crucible, with a loose-fitting PBN cover, is placed into the autoclave, and the desired pressure and heating ramp rates are programmed. Up to ten different heating/cooling/pressurization rates can be programmed into the autoclave system. A typical run-sequence is as follows: (a) the system is pressurized to approximately 12 000 psi argon and the temperature increased to 550°C at 10°C per minute; (b) a slow temperature ramp is maintained from 550°C to 700°C (2°C per minute) at approximately 27 000 psi; and (c) the temperature is rapidly increased to 1100°C (10°C per minute) and held for thirty minutes. The system is then cooled down and the poly-InP charge removed.

Using standard LEC growth techniques in Malvern MSR6-R and Varian HPCZ pressure-pullers, single crystals were grown from InP synthesized by both approaches. The measurements of carrier concentration and Hall mobility were performed using the van der Pauw method. In this work, the conventional 'clover-shaped' sample geometry is replaced by an equivalent Greek cross structure [10]. Shown in Table 1 are electrical characteristics of undoped crystals grown from poly-InP compounded in quartz and PBN crucibles.

Table 1 *Electrical properties of undoped crystals grown from InP synthesized in SiO_2 and PBN crucibles*

Sample	μ 300 K $cm^2\,V^{-1}\,sec^{-1}$	N_D-N_A $10^{15}\,cm^{-3}$	μ 77 K $cm^2\,V^{-1}\,sec^{-1}$	N_D-N_A $10^{15}\,cm^{-3}$
EP LEC-4 (SiO_2)	4000	17.8	11 300	11.9
EP LEC-3 (PBN)	4400	4.4	19 700	3.5

The above data were derived from InP samples that were compounded in sealed quartz ampoules in the low-pressure autoclave. Preliminary runs in unsealed PBN crucibles from the high-pressure system have been successful, yielding up to 200 g charges in 5 cm diameter crucibles. Although a lower purity grade of indium was used during the high-pressure feasibility studies, mobilities up to 9800 $cm^2\,V^{-1}\,sec^{-1}$ and free carrier concentrations as low as $1.2\times10^{16}\,cm^{-3}$ at 77 K were obtained.

Conclusions

We have successfully synthesized high-purity polycrystalline InP, using PBN crucibles in sealed silica ampoules and unsealed PBN crucibles at argon pressures to 30 000 psi. In the latter process, the poly InP is formed in a silica-free environment, thereby eliminating the possibility of silicon contamination from quartz components. Using vacuum and/or hydrogen baking of high-purity indium and phosphorus, in conjunction with optimized growth conditions, it is anticipated that a considerable improvement in background carrier concentration and electron mobility will be realized.

ACKNOWLEDGEMENTS

The authors gratefully acknowledge D. W. Eckart and S. Steele for their assistance in sample preparation, and to R. P. Moerkirk for electrical measurements.

References

1. Hilsum C. and Rees H. D. (1970). *Electron. Lett.*, **6,** 277
2. Fauth T. A. and Adamski J. A. (1979). *Rome Air Development Center Technical Report, RADC-TR-79-246*
3. Henry R. L. and Swiggard E. M. (1978). *J. Electron. Mat.*, **7,** 647
4. Iseler G. W. (1979). *Inst. Phys. Conf. Ser.*, **45,** 144
5. Farges J. P. (1982). *J. Cryst. Growth*, **59,** 665
6. Antypas G. A. and Hyder S. B. (1983). *Army Research Office Contract DAAG29-81-C0013*
7. Hyder S. B. (1982). *Army Research Office Contract DAAG-79-C-0009*
8. Effer D. and Antell G. R. (1960). *J. Electrochem. Soc.*, **107,** 252
9. Medcalf W. E. and Fahrig R. H. (1958). *J. Electrochem. Soc.*, **105,** 719
10. David J. M. and Buehler M. G. (1977). *Solid St. Electron.*, **20,** 539

LEC Growth of 3 Inch Diameter Undoped SI GaAs with Good Reproducibility Using an Improved Hot Zone

K. Matsumoto, H. Morishita, M. Sasaki, S. Nishine, M. Yokogawa, M. Sekinobu, K. Tada* and S. Akai

Sumitomo Electric Industries Ltd., Semiconductor Division, 1-1-1 Koyakita, Itami, Hyogo, 664 Japan

Abstract

An improved LEC (liquid encapsulated Czochralski) method of growing semi-insulating GaAs single crystals with low dislocation density has been established. It is based on an improvement of the hot zone in a Melbourn puller, where the axial temperature gradient in the B_2O_3 encapsulant was reduced to 30–60°C cm^{-1}, leading to an average EPD (etch pit density) as low as $1–4 \times 10^4$ cm^{-2} throughout the 3 inch diameter ⟨100⟩ single crystals. Average electrical resistivity was $2–6 \times 10^7$ Ω cm. Reproducibility of the above EPD and resistivity was confirmed to be excellent.

In order to make GaAs practical for IC applications, it is very important to obtain high-quality GaAs crystals grown by the LEC technique. In particular, a reduction in dislocation density is required. The principal origin of the dislocation is the stress induced by the steep temperature gradient during growth. A modification of the hot zone has been made to reduce the temperature gradient with very satisfactory results, as described below.

A high-pressure Melbourn puller made by Cambridge Instruments was used. The axial temperature gradient in the B_2O_3 encapsulant was as high as 100–150°C cm^{-1} before improvement. A computer analysis by the finite element method was applied to find out how improvements could be made. The calculated temperature contour lines and the gaseous convective flow in the conventional hot zone are shown in Figures 1a and 1b, respectively. By carefully studying the flow pattern, it was concluded that the principal origin of the steep temperature gradient was the flow of cooled gas on to the B_2O_3 encapsulant. It was thus decided to improve the hot zone by setting a roof made of graphite on the standard graphite heat shield to separate the downward flow of cold gas from the hot gas convection under the roof, as illustrated in Figures 1c and 1d. The temperature gradient in the B_2O_3 encapsulant can be changed by varying the thickness, shape and height of the roof. A temperature gradient of 30–60°C cm^{-1}, which is about one third of that observed with the standard heat shield, was obtained.

* Sumitomo Electric Industries Ltd., R&D Group, 1-1-3 Shimaya, Konohana-Ku, Osaka, 554 Japan.

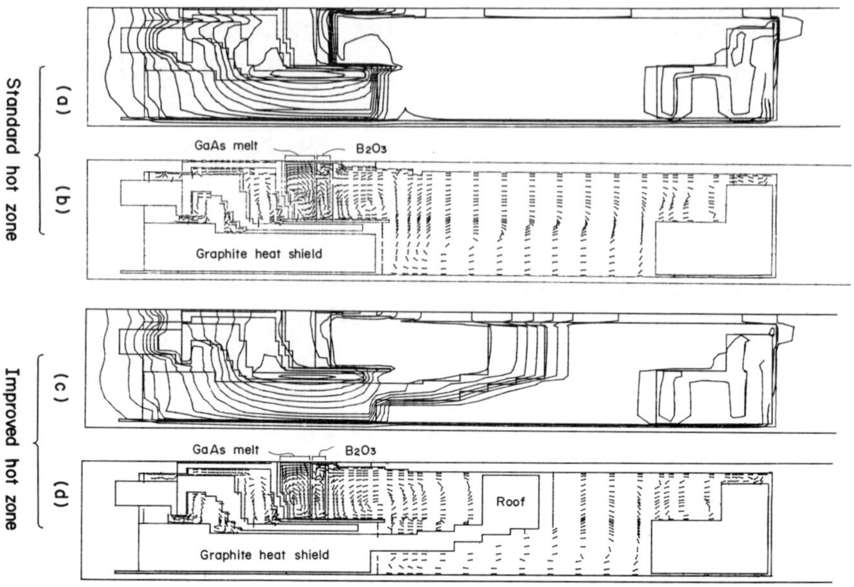

Figure 1 Computer analysis of thermal environments in half of the cross-section of the puller. (a) and (c) show temperature contours (with 100°C interval). (b) and (d) show gaseous convective flow.

Using an improved hot zone with a temperature gradient of 60°C cm^{-1}, undoped SI GaAs crystals of 3 inch diameter were grown in the ⟨100⟩ direction from a 6 inch diameter PBN crucible. The raw material was 4.5 kg of HB-grown polycrystalline GaAs. The diameter of the crystal was controlled by automatically adjusting the temperature in response to variations in the

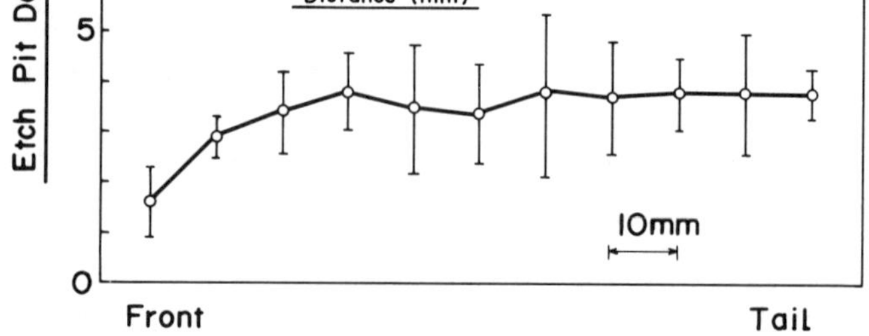

Figure 2 Longitudinal distribution of average EPD for ten crystals. Inset shows typical EPD profiles.

differential of crystal weight increase. The final crystals weighed from 4.0 to 4.4 kg.

Dislocations were revealed by etching (100) wafers from ten crystals with molten KOH. Though the EPD profiles on the (100) wafers had a 'W'-shaped distribution, as theoretically described by Jordan [1] for LEC-grown crystals, the average EPDs measured on the (100) wafers were within a range of $1–4 \times 10^4$ cm^{-2} from the front to the tail. The results are shown in Figure 2 together with their deviations.

Measurements of electrical resistivity were also carried out by using the two-electrode method. Average electrical resistivity was determined to be $2–6 \times 10^7$ Ω cm, with a standard deviation of 6–21% across the wafers. As shown in Figure 3, the radial variation has an 'M'-shaped distribution, which is clearly correlated to the 'W'-shaped distribution of dislocation density.

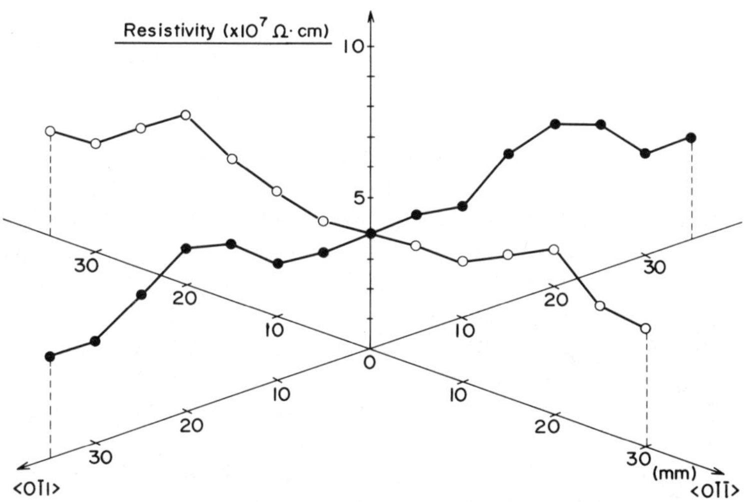

Figure 3 Resistivity variations along ⟨01̄1̄⟩ and ⟨01̄1⟩ directions on (100) wafer.

Electrical properties of an active layer formed in a 3 inch wafer by implanting 3.0×10^{12} cm^{-2} Si ions at 180 keV followed by annealing at 820°C for 20 min in nitrogen gas under face-to-face configuration were measured. Typical values for Hall mobility, sheet carrier concentration, sheet resistance and activation efficiency were found to be 4500 cm^2 V^{-1} s^{-1}, 1.4×10^{12} cm^{-2}, 990 Ω/□ and 47% respectively.

Reference

1. Jordan A. S. (1980). *J. Cryst. Growth*, **49**, 631

Residual Impurities in Boron Trioxide Encapsulant

M. R. Shropshall and P. E. Skinner

Johnson Matthey Chemicals Limited, Royston, UK

Abstract

The quality of the B_2O_3 encapsulant used in the LEC (liquid encapsulated Czochralski) crystal growth of III–V materials has a major influence on the performance of the resultant crystals. This B_2O_3 must be of the highest possible purity and have a closely controlled and quantified water level. The purity typically required is discussed, together with the theory of the infrared technique used for water level measurement. Some of the precautions necessary to maintain material quality are also mentioned.

When used as a liquid encapsulant in the Czochralski growth of III–V single crystals, B_2O_3 performs at least two distinct functions. The arsenic in GaAs and phosphorus in gallium and indium phosphides both have high vapour pressures at crystal growth temperatures, and so the encapsulant layer of B_2O_3 is needed to stabilize melt stoichiometry before and during crystal growth [1]. Secondly, the B_2O_3 layer provides a molecular environment more suitable for ionic species than does the III–V compound, ionic impurities therefore tending to migrate into this layer, leaving the III–V compound purer [2].

To perform well as an encapsulant the B_2O_3 should be of the highest possible purity with respect to metals and also be low in anionic or volatile impurities (e.g. H_2O/OH^-). Minimizing metallic impurities and reducing the water level are thus the areas of prime concern when producing B_2O_3 for LEC crystal growth. Unfortunately the reduction in water level leads to an increase in B_2O_3 melt viscosity which can lead to physical distortion of the crystal, especially with relatively soft systems such as indium phosphide [3]. It is therefore necessary to strike a compromise on water level to achieve the optimum encapsulant and melt viscosity characteristics for a specific application. This results in a situation where a wide range of water levels is required by the crystal growing industry. Thus, the B_2O_3 manufacturer should be able to control and measure water level over a wide range.

A B_2O_3 purity appropriate for III–V growth would have a total metallic impurity content of about 5 ppm maximum. Table 1 shows a typical analysis for JMC Puratronic® grade B_2O_3 which has these impurity levels. The rapid advance of III–V crystal growth technology probably means that higher and higher purity B_2O_3 encapsulant will be required.

The water level in B_2O_3 may be reduced by several methods, but the most important factor is the ability to control and monitor this process to give the

Table 1 *Typical purity of JMC Puratronic® B_2O_3*

Impurity	ppmw
Cd	<0.02
Co	0.02
Cr	0.02
Cu	0.04
C	<5
Fe	0.5
Mn	0.01
Ni	0.03
S	<0.2
Si	1–3
Zn	0.04

precise water level required by the individual crystal grower. We measure B_2O_3 water level by infrared spectroscopy, utilizing the Beer-Lambert law. Absorption due to the oscillation of the OH bond takes place in the infrared region and this absorption is directly proportional to the number of OH bonds and therefore the level of water present. The concentration C is given by:

$$C\ (\text{moles/litre}) = \frac{\log_{10}(I_0/I)}{\varepsilon L} \tag{1}$$

where I_0 is the intensity of incident radiation, I is the transmitted radiation, ε is the molar extinction coefficient (141 l mol^{-1} cm^{-1}) and L is the path length.

To convert the moles/litre figure into parts per million weight, we use the equation:

$$C\ (\text{ppmw}) = C\ (\text{moles/litre}) \times \frac{18\,000}{\rho L} \tag{2}$$

where ρ is the density of B_2O_3. This yields the final equation for water content:

$$C\ (\text{ppmw}) = \frac{18\,000 \times \log_{10}(I_0/I)}{\varepsilon L \rho} \tag{3}$$

The B_2O_3 spectra shown in Figures 1 and 2 can be used to illustrate the technique. Using the data from the spectrum in Figure 1: $I_0 = 65$; $I = 45$; $L = 0.043–0.033$ cm, mean $= 0.038$ cm; $\varepsilon = 141$ l mol^{-1} cm^{-1}; $\rho = 1.84$ g cm^{-3} and substituting into equation (3) gives a value of $C = 291.5$ ppmw. Using the data from the spectrum in Figure 2: $I_0 = 65$; $I = 14$; $L = 0.043–0.033$ cm,

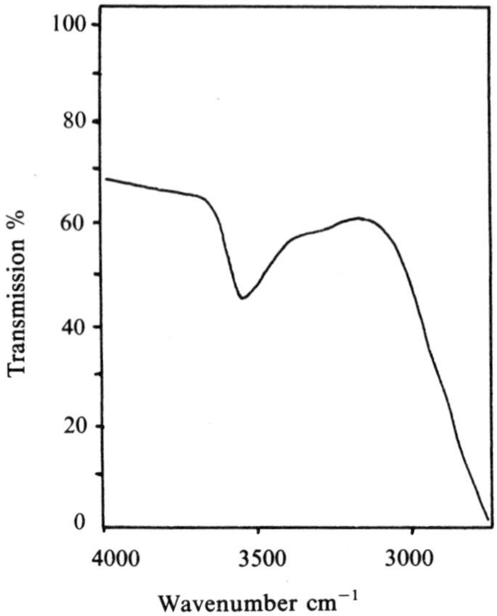

Figure 1 IR spectrum of freshly made sample of B_2O_3.

mean = 0.038 cm; $\varepsilon = 141$ l mol^{-1} cm^{-1}; $\rho = 1.84$ g cm^{-3} and once again substituting into equation (3) gives a value of C = 1217.3 ppmw.

The two spectra (Figures 1 and 2) illustrate the contrast between surface water contamination and internal water content. The same B_2O_3 sample was used for both spectra, but Figure 2 is the spectrum obtained after 2 minutes exposure to atmospheric moisture contamination. The peak at 3580 cm^{-1}

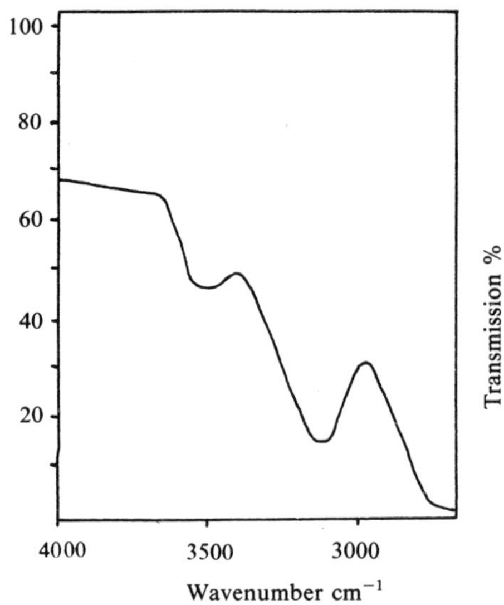

Figure 2 IR spectrum of the same B_2O_3 sample as Figure 1 after exposure to atmosphere for 2 min.

corresponds to the internal moisture content with the peak at 3220 cm^{-1} corresponding to surface moisture contamination; the calculation done for Figure 2 is a measurement of this surface water contamination. Slight errors are involved in using the same value of extinction coefficient at both wavelengths, but this does demonstrate the rapidity of water absorption at the B_2O_3 surface.

To achieve accurate results, care must be taken to ensure that the relevant parameters are measured with precision and that moisture contamination is prevented. The molar extinction coefficient is another obvious area where inaccuracy can arise; the figure of 141 1 mol^{-1} cm^{-1} has been determined and confirmed by several workers [4–7] and is now accepted as correct.

Having produced B_2O_3 to high purity and controlled water level, it needs to arrive at the crystal grower in the same condition. To ensure that this occurs, the material should be vacuum wrapped to provide an effective barrier to water vapour.

References

1. Metz E. P. A., Miller R. C. and Mazelsky J. (1962). *J. Appl. Phys.*, **33**, 2016
2. (1981). In *Gallium Arsenide and Related Compounds*, Oiso; Inst. Phys.
3. Cambridge Instruments Ltd. Private communication
4. Franz H. (1965). *Glastech Ber*, **38(2)**, 54
5. Pasteur G. A. (1973). *J. Am. Ceramic. Soc.*, **56(10)**, 548
6. Broadhead P. and Newman G. A. (1972). *Spectrotum Act. Part A*, **28(10)**, 1915
7. Pearson A. D., Pasteur G. A. and Northover W. R. (1979). *J. Mat. Sci.*, **14**, 869

Section 4:
EL2 and other point defects

Nature of EL2: the Main Native Midgap Electron Trap in VPE and Bulk GaAs

S. Makram-Ebeid, P. Langlade and G. M. Martin

Laboratoires d'Electronique et de Physique Appliquée, 3, avenue Descartes, 94450 Limeil-Brévannes, France

Abstract

The midgap donor level EL2 is of technological importance, especially because of the role it plays in the electrical compensation of high purity SI GaAs. The control of the concentration and uniformity of this deep level has made considerable progress, but the precise determination of its nature is still an unsolved problem.

Today, EL2 is known with certainty to be related to an intrinsic point defect of GaAs and to be associated with an excess arsenic stoichiometry. Strong arguments are in favour of its attribution to a defect complex including the As_{Ga} antisite defect.

The peculiar photoquenching behaviour of EL2 which is unique among GaAs defects is today the surest way of detecting its presence. The detailed study of this effect may, in the future, help to reveal the exact EL2 structure.

Introduction

Undoped SI GaAs is acquiring great importance because of the role it may play in the future development of the GaAs IC (integrated circuit) industry. The high resistivity of such material is due to an electrical compensation resulting in a Fermi level practically pinned to a midgap donor deep level labelled EL2. Shallower acceptor and donor levels are present in smaller concentrations and have a much less important effect on the Fermi level pinning position.

Considerable effort is currently devoted by different laboratories throughout the world to control the concentration and uniformity of this deep level donor with the purpose of obtaining high purity SI substrates of acceptable quality and homogeneity for industrial IC applications. Although very promising empirical rules have been discovered for achieving these goals and in spite of intensive characterization efforts, the problem of definitely establishing the nature of EL2 is not yet entirely resolved.

The role of oxygen in the early genesis of EL2

The early genesis of what is today known as EL2 dates back to the first attempts to dope GaAs materials with oxygen during crystal growth, which were reported as early as 1958 by Whelan and Wheatley [1]. Successful attempts to produce high resistivity materials in such a way were reported in 1961 by Hilsum and Rose-Innes [2]. Later on, SI materials were obtained by HB (horizontal Bridgman) growth either by introducing an oxygen partial pressure into the ampoule [3–5] or by adding Ga_2O_3 into the melt [5–8]. These high resistivity materials were mainly assessed by Hall effect measurements. The results showed that the activation energy of the Hall constant was close to 0.75 eV, and it was thus proposed that a deep level could actually lie at this energy. Since it was a drop of the electron concentration which was recorded as a function of the oxygen treatment, it was first proposed that the deep midgap level was of acceptor nature [9], compensating the shallow donors. But it was also suggested that it might well be a deep donor; the reduction of the free electron concentration could be explained by the selective chemical action of oxygen on shallow levels. This chemical action would leave us with an excess of shallow acceptors over shallow donors and the deep oxygen level would compensate the excess carrier, thus leading to the observed high resistivity [9–11]. This hypothesis would have been in favour of an assignment of the level to O_{As}, expected to be a donor. This midgap level was thought to be related to oxygen, but there was neither experimental evidence at that time for the existence of the level itself nor, of course, evidence for the increase of its concentration with oxygen treatments. It is only a few years later that the existence of that level was definitely established by spectroscopic experiments such as thermally stimulated current (TSC) measurements in the early stage [12–14] and later on, deep level transient spectroscopy (DLTS) [15, 16]. In the general classification of deep levels in GaAs [17], this level was labelled EL2, a label now generally accepted.

The direct involvement of oxygen in the formation of the deep level EL2 and in the compensation of SI materials was investigated later on by Martin *et al.* [18] by correlating electrical measurements with chemical analyses based on spark source mass spectrometry (SSMS) and the charged particle activation (CPA) method. They also studied the formation of the Si–O complex using the localized vibrational mode (LVM) infrared absorption method. These authors concluded that thermal equilibrium at high temperature, in the presence of oxygen vapour pressure, leads to the actual gettering of silicon present in the starting melt. This is a stronger and more complex effect than that previously suggested, i.e. the reduction of Si incorporation by preventing quartz wall dissociation. They also found that the correlation of oxygen with levels other than the Si (shallow donor) in the gap was questionable and that these levels, if any, were introduced in small amounts. They also concluded that the incorporation of Si–O complexes can be considered as negligible, i.e. in concentrations lower than 2×10^{15} cm^{-3}. No relation was found between the concentration of the deep donor EL2 and the concentration of Ga_2O_3 in the melt or the concentration of oxygen in the crystal. The EL2 concentration seemed to be at least 10 times larger than that of oxygen. This ruled out the possibility of this level being related to oxygen, a conclusion already suggested in a previous study of Huber *et al.* [19].

The results of a contemporary and independent study by the MIT material research team [20] yielded conclusions consistent with those stated above. Their chemical analyses were performed by C. Evans & Associates using secondary ion mass spectrometry (SIMS). Adding Ga_2O_3 did not result in a sizeable increase in the oxygen content of the GaAs crystals but mainly induced a substantial amount of silicon gettering. They reported an increase in the EL2 concentration with Ga_2O_3 doping which they attributed to the decrease in the silicon donor concentration and to the inverse correlation they observed between the EL2 and free electron concentrations.

Adding oxygen during VPE (vapour phase epitaxial) growth leads to a decrease in free electron concentration [21–23] as for Bridgman growth. Wallis *et al.* [23] reported that no correlation could be observed between the concentration of any deep level, including EL2, and oxygen injection during the VPE growth. Furthermore, EL2 is not detected in liquid phase epitaxial (LPE) layers [17, 24] even when Ga_2O_3 is added to the melt.

Electrical and optical characteristics of EL2

In order to remove any ambiguity as to the detection of EL2, it is important to know the different characteristics of this level that can be considered as a complete fingerprint. A detailed account of these characteristics is being published elsewhere [25]; we only give here an abridged account.

EL2 AS A DEEP DONOR

A clear demonstration of the donor nature of EL2 was provided by Mircea *et al.* [26]. Their experiment consisted in studying the EL2 outdiffusion after a 600–750°C annealing in VPE materials where the electron concentration (around 10^{15} cm^{-3}) was comparable with the EL2 concentration ($1–3 \times 10^{14}$ cm^{-3}). Profiling of EL2 was carried out by a differential DLTS technique. In-depth profiling was achieved by combining the above method with chemical etching. The electron concentration profile observed at low temperature, with all the EL2 centres occupied by electrons, proved flat and unaffected by the annealing process. The high temperature carrier profile (with all the EL2 centres ionized) closely followed the outdiffusion profile of EL2. This experiment thus clearly showed that the EL2 centres are electrically neutral when occupied by electrons and are positively charged when releasing these electrons. Unless two electrons can be simultaneously released or captured by an EL2 centre, this experiment also indicates the simple donor nature of this centre.

THERMAL ELECTRON EMISSION AND CAPTURE RATES OF EL2

The temperature dependence of the thermal electron emission rate e_n has been studied by many authors [27] and most of the data closely follow the law:

$$e_n = \gamma_n \sigma_{na} T^2 \exp(-E_n/kT) \quad (1)$$

where $\gamma_n = 2.28 \times 10^{20}\ cm^{-2}\ s^{-1}\ K^{-2}$ for GaAs, $\sigma_{na} = (1.5 \pm 0.5) \times 10^{-13}\ cm^2$, and $E_n = 0.825 \pm 0.010$ eV.

Considerably larger emission rates have been reported by a Tokyo University team [28] who also report a scatter in the photoionization thresholds of these levels when they are created by different methods. Lagowski *et al.* [29] report that two distinct levels with emission rate and electron capture cross-section differing by a factor of 4 are present in certain types of Bridgman materials. This type of scatter and plausible explanations will be discussed further on.

The electron capture cross-section of EL2, has been measured by Mitonneau *et al.* [30] in the temperature range of 50–273 K and their results can be put in the form:

$$\sigma_n = (\sigma_{n\infty} - \sigma_{n0}) \exp(-E_\sigma/kT) + \sigma_{n0} \quad (2)$$

where $\sigma_{n\infty} = 6 \times 10^{-15}\ cm^2$, $\sigma_{n0} = 5 \times 10^{-19}\ cm^2$ and $E_\sigma = 0.066$ eV.

Equations (1) and (2) yield the Fermi-Dirac electron occupation ratio for any temperature, provided the electron concentration (or Fermi level) of the host materials is known. The determination of the free energy of ionization, however, requires a knowledge of the degeneracy ratio g_1/g_0. Taking $g_1/g_0 = 1/2$ by using the theoretical results of Van Vechten and Thurmond [31] for a donor electron level, one obtains for the free energy of ionization E_T with respect to the conduction band minimum the expression:

$$E_T = (0.730 \pm 0.02) - \alpha T^2/(T + 204)$$

where $\alpha = (3.5 \pm 0.6) \times 10^{-4}\ eV\ K^{-1}$.

PHOTOIONIZATION PROPERTIES AND LATTICE CONFIGURATION DIAGRAM OF EL2

The photoionization cross-sections of EL2 with the centre initially empty (σ_p^0) and initially filled (σ_n^0) with an electron have been studied in detail by Chantre *et al.* [32] as a function of photon energy, as summarized in Figure 1. This type of study allowed an unambiguous fingerprinting of EL2 from its optical properties and also led to a direct demonstration of optical transition from the filled level to each of the Γ, L and X conduction band valleys. These authors also used their theory to deduce the Franck-Condon shift Δ_{FC} of EL2 which was found to be $\Delta_{FC} \cong 120$ meV. They then suggested a lattice-relaxation coordinate diagram to account for their σ_p^0 and σ_n^0 data.

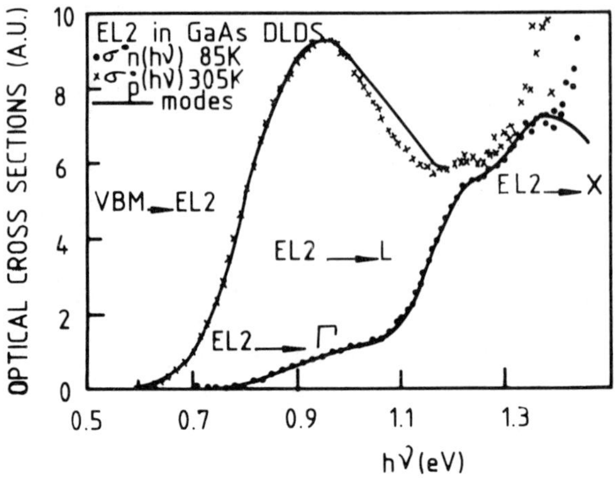

Figure 1 Spectral dependence of σ_n^0 (lower curve) and σ_p^0 (upper curve) after Ref. 32.

Another independent evaluation of the lattice relaxation parameter was deduced from detailed measurements of the high electric field ($F > 2.5 \times 10^5$ V cm^{-1}) and temperature dependence of the EL2 electron emission rate. A best fit with a multiphonon tunnelling emission theory [33] gave estimates for the Franck-Condon shift ($\Delta_{FC} = 130 \pm 10$ meV). The lattice configuration diagram agreed with that deduced by Chantre *et al.* [32] and is shown in Figure 2.

TRANSIENT PHOTOCAPACITANCE QUENCHING BEHAVIOUR OF EL2

Bois and Vincent [34] have reported a very peculiar transient photocapacitance behaviour associated with EL2. The typical behaviour is depicted in Figure 3. These photocapacitance transients can be observed in Schottky barrier diodes made on n-GaAs ($n = 10^{15}$ to 1.5×10^{17} cm^{-3}) which are cooled while forward biased from ambient temperature to below 140 K, then reverse biased and subjected to infrared illumination ($h\nu = 0.9$–1.5 eV). The observed photocapacitance transients consist of a fast increase of capacitance from the initial capacitance C_0 to a maximum value C_M followed by a decrease to a final value C_∞; the fall time is about 100 times larger than the rise time. For VPE samples where EL2 has a concentration larger than that of any other trap, the values of C_∞ and C_0 are nearly equal. This indicates that the phenomenon involved is EL2 related. Another proof of this assertion, which excludes, in particular, possible contributions from hole traps, is the experiment performed by Vincent *et al.* [35] who first reverse biased their diode at room temperature for a long enough time to empty EL2, then cooled it to 77 K; the photocapacitance transient observed was a decreasing one. Some thought shows that if a hole trap was involved in the process, an increasing

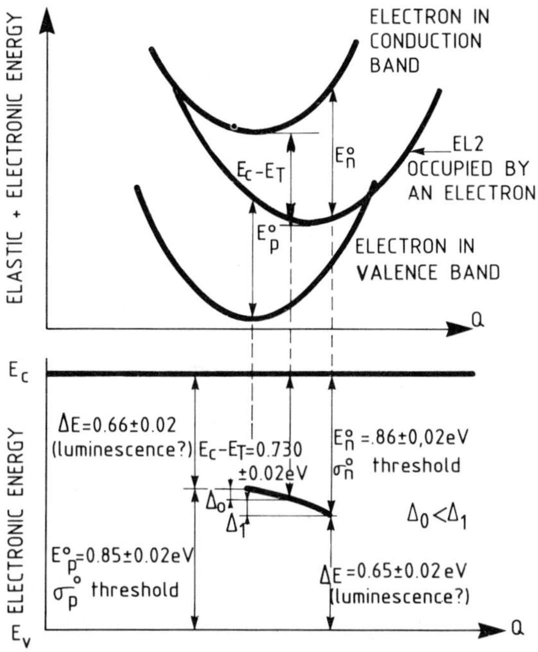

Figure 2 Lattice relaxation configuration diagram of EL2 as deduced from thermal and photoionization data. A non-linear coupling with the lattice is suggested by the fact that the thermal free energy of ionization is not midway between the ionization threshold energies for σ_n^0 and σ_p^0. All energies are referred to 0 K.

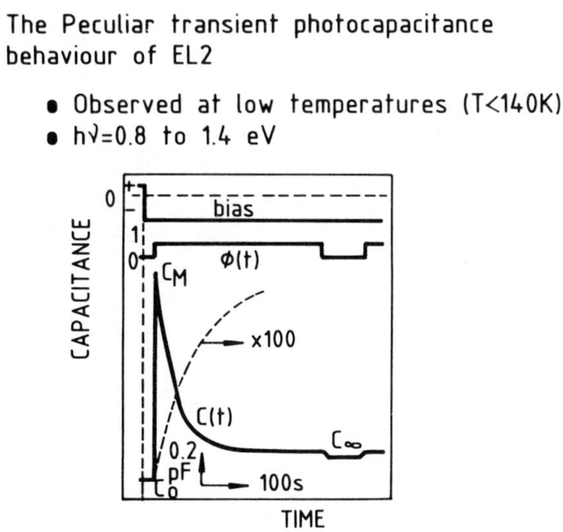

Figure 3 Transient photocapacitance observed in an n-GaAs Schottky barrier diode at 77 K after a direct electrical impulse and under an illumination of 1.17 eV photon energy. After Ref. 35.

photocapacitance transient would have been observed regardless of the bias sequence used.

After completion of the photocapacitance transient of Figure 3, switching the light off at 77 K and illuminating again does not result in any sizeable change in capacitance (except for small changes due to zone-edge effects, i.e. the non-zero free electron concentration at the depleted zone-neutral region boundary). The regeneration of the full overshoot behaviour visible in Figure 3 depends on the time spent in the dark and on the temperature. The return of the overshoot in the photocapacitance transient is exponentially related to the time spent in the dark and occurs at a rate r which has been determined independently by Mitonneau and Mircea [36] and by Vincent *et al.* [35] as a function of temperature. A phenomenological explanation of this behaviour which is unique among all known GaAs defects has been given by Bois and Vincent [34]. These authors were led to attribute to EL2 a state EL2M with an atomic configuration different from the fundamental normal state EL2F.

The de-excitation rate r (return from metastable to normal state) has been studied in detail by Mitonneau and Mircea [36]. They reported two different mechanisms: the first one being the above mentioned thermal de-excitation [35], the second one an Auger like process [36] in which the de-excitation rate is very much enhanced by the presence of a free electron density n. The total de-excitation rate r can, in general, be written as:

$$r = 2 \times 10^{11} \exp(-0.3\ \mathrm{eV}/kT) + 1.9 \times 10^{-14} \times n \times v_n \exp(-0.107\ \mathrm{eV}/kT)\ \mathrm{sec}^{-1}$$

where n is the free electron density in cm^{-3} and v_n the thermal electron velocity in cm sec^{-1} at the temperature T. In practice, above 140 K, the transfer to the normal state is very fast even in the absence of free carriers and one only observes the fundamental normal state above this temperature.

The only known way to transfer EL2 from normal to metastable configuration is by means of sub-bandgap light. The cross-section corresponding to this process was shown by Vincent *et al.* [35] to have the spectral dependence depicted in Figure 4 and peaking at 1.1 eV photon energy which is near that of Nd:YAG laser (1.17 eV).

The initial interpretation of Bois and Vincent [34] of the metastable behaviour of EL2 relied on an anomalously strong lattice relaxation effect. However, this is not the only possible interpretation. Levinson [37] proposes an explanation based on experiments performed by himself and other colleagues at Bell Laboratories on InP defects [38]. Very briefly, Levinson assumes EL2 to be composed of a shallow donor defect together with a deep donor defect. When both are positively charged, the transition to a metastable atomic configuration is made possible by the occurrence of the strong Coulombic repulsion. Return to equilibrium is favoured by Coulombic attraction between oppositely charged deep and shallow level components of EL2. This type of explanation appears more appealing and furthermore, Levinson

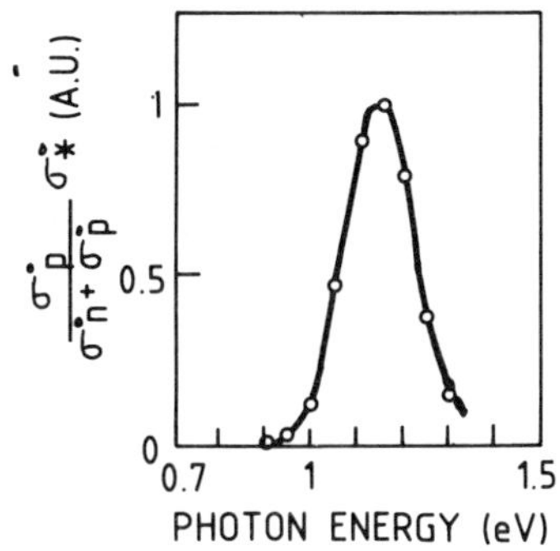

Figure 4 Spectral dependence of the cross-section σ^* corresponding to the optically induced transition rate from EL2F to EL2M. After Ref. 35.

explains the 'Auger like' effect for the return to equilibrium by an atomic rearrangement assisted by the release of phonons during the capture of an electron into the deep level component of EL2. Thus, for Levinson, the Auger like process is an instance of recombination enhanced defect reaction which frequently occurs in semiconductors.

EXTRINSIC OPTICAL ABSORPTION RELATED TO EL2

Martin *et al.* [39, 40] have identified an EL2 related extrinsic optical absorption band which is the dominant sub-bandgap near infrared absorption component in non chromium-doped SI and n-type bulk GaAs. The similarity between the spectra of the corresponding absorption band and that of σ_n^0 (see Figure 1) is quite remarkable. This attribution of the extrinsic absorption to EL2 was subsequently made more unequivocal [41] by observing a drastic change in the low temperature (10 K) extrinsic optical absorption spectrum of an undoped SI GaAs sample after a few minutes exposure to illumination from a tungsten filament lamp. Complete disappearance of the extrinsic optical absorption features was observed. This behaviour can be directly related to the photocapacitance quenching behaviour which is a unique fingerprint of EL2 as was previously discussed.

More recently, Kaminska *et al.* [42] have made an extremely interesting remark. They have noted a systematic difference (see Figure 5) between the shape of the electron *photoionization spectrum* of EL2 and that of the EL2 associated optical *absorption* in n-type materials. When properly normalizing the two curves so that they agree in the high and low energy parts of the spectrum, there appears in the optical absorption an extra band centred about 1.15 eV. The shape and position of this band coincides, within measurement accuracy, with the band depicted in Figure 4 and which represents the optical transition rates from the fundamental to the metastable state of EL2 [35]. Therefore, the optically induced transition to the metastable state is readily visible in the optical absorption data of GaAs. Although the photoionization

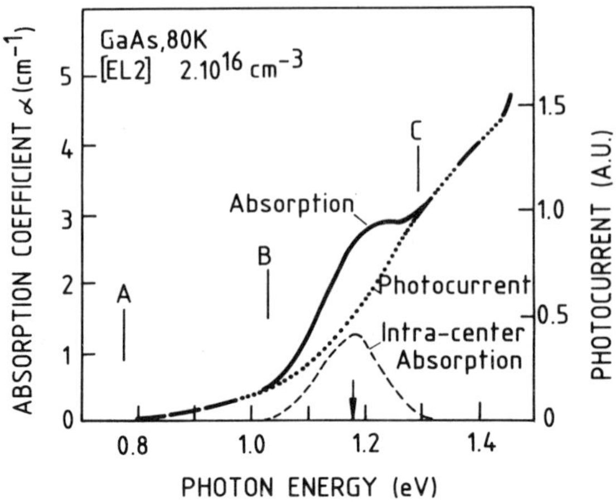

Figure 5 Optical absorption and photocurrent spectra measured on the same sample of melt-grown GaAs with EL2 concentration of about 2×10^{16} cm^{-3}. The difference between the two spectra defines the intra-centre absorption band. It is also similar to the shape of the optical transition from the fundamental to the metastable state (Figure 4). After Ref. 42.

and absorption data were available in the literature [39, 40], this was the first time that such a connection was made. Kaminska *et al.* [42] have examined in great detail and with a high resolution this EL2 intra-centre optical absorption in n-GaAs at 8 K. They revealed a fine structure in the absorption band (Figure 6). These authors interpret the fine structure as different optical transitions involving 0, 1, 2, 3, 4 or 5 phonons. The zero phonon line lies at 1.0395 eV and the associated phonon energy is 11 ± 1 meV.

The symmetry properties of the above zero phonon line were subsequently analysed in a uniaxial stress experiment by Skowronski *et al.* [43]. In spite of some problems with the interpretation (non-linear dependence with stress in certain directions), these authors prove that there is a strong influence of the trigonal field and almost no effect from the tetragonal field. They assume

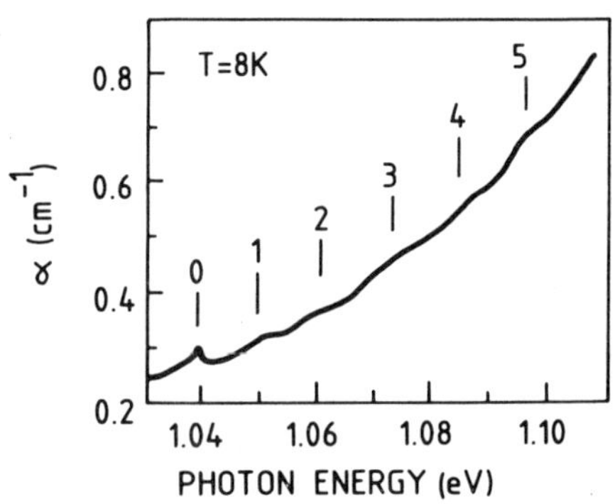

Figure 6 Fine structure in the 1.15 eV absorption band attributed to an EL2F to EL2M 'intra-centre' transition. After Ref. 42.

that the transition to the metastable state occurs via a strongly relaxed excited state having an electronic energy resonant in the conduction band and separated from this band by a purely vibronic barrier. The crystal field symmetry of this excited state, as revealed by their work is T_2 whereas that of the fundamental initial state is A_1.

PHOTOLUMINESCENCE BANDS ATTRIBUTED TO EL2

Wide luminescence bands at 0.63 and 0.68 eV have been studied by different authors [44–52]. Strong arguments have been presented to attribute to EL2 either the 0.64 eV band [44–46, 48] or the 0.68 eV band [50–52] or both [25]. One of the problems for the unequivocal identification of EL2 associated luminescence bands is that this level is the main source of extrinsic photo-electron-hole production in high purity materials. The fatiguing behaviour observed with YAG laser excitation ($h\nu = 1.17$ eV) for the 0.64 eV band [45, 48] and for the 0.68 eV band [52] is reminiscent of the EL2 metastable behaviour reported in the preceeding sections. This, however, cannot be considered as definite proof in favour of the association of these photo-luminescence bands with EL2. In fact, when this level is sent to the metastable configuration, it becomes optically inactive and the main mechanism for electron-hole pair production is switched off. One therefore expects *any* defect or impurity related luminescence to be reduced by this process as long as intrinsic photo-excitation is not used. Indeed, by using intrinsic light excitation Yu [52] has observed an enhancement of the C luminescence simultaneously with the fatiguing of the 0.68 eV band.

Factors affecting the EL2 concentration

As previously discussed, the EL2 donor level cannot be identified with oxygen and it is necessary to elucidate its origin. Considerable effort has been devoted to studying different factors affecting its incorporation during growth, its annealing behaviour and its creation using neutron irradiation, ion implantation or plastic deformation.

EFFECT OF STOICHIOMETRY

A systematic study of the correlation of EL2 with the stoichiometry of LEC (liquid encapsulated Czochralski) GaAs, has been conducted by Holmes *et al.* [53–55]. They report that the concentration of EL2 as deduced from optical absorption measurements increases markedly with the As/Ga stoichiometry ratio. This can explain the observed reduction of EL2 [56] from seed to tail side of LEC GaAs ingots as the As content decreases.

In Bridgman growth, the stoichiometry control is more accurate and is realized by adjusting the arsenic source temperature. This results in higher concentrations and more uniform distribution of EL2. Lagowski *et al.* [57] report that the concentration of this level increases with As partial pressure during Bridgman growth.

Similar observations hold for MOVPE (metalorganic vapour phase epitaxy) growth. The EL2 concentration has been reported by different groups [58–63] to increase with (As/Ga) source ratio according to a law of the form $(\mathrm{As/Ga})^p$. However, there is a discrepancy in the numerical value of the exponent p which is reported to be either 1 [59], 0.5 [62] or 0.25 [63]. Perhaps the most convincing data are those of Samuelson *et al.* [62] who kept all the growth parameters fixed except for the As mole fraction and accurately measured the EL2 concentration by DLTS with appropriate correction factors. The exponent they find is $p = 0.5$.

All these experiments show clearly that EL2 is systematically generated during growth under As-rich conditions. It is not surprising to note the absence of EL2 in thick LPE layers which are grown under Ga-rich conditions [17, 24, 64]. It is however more puzzling not to detect EL2 in thick molecular beam epitaxial (MBE) layers whatever the As/Ga source ratio used. Possible reasons for this are the low growth temperature (≤680°C) and also the very low growth rates as compared to those of the other growth methods mentioned above. In fact, increasing the growth rate does increase the EL2 concentration as was reported for VPE growth [64–66].

EFFECT OF THE FREE ELECTRON CONCENTRATION

Lagowski *et al.* [67] have measured by DLTS the concentration of EL2 in Bridgman crystals grown under identical conditions, but with a wide range of intentional donor dopant concentrations. The EL2 concentration was found to remain nearly constant and then very quickly dropped when the electron concentration increased beyond $2 \times 10^{17}\ \mathrm{cm}^{-3}$. This behaviour was independent of the donor dopant used and, furthermore, counter-doping with acceptors led to an increase in the EL2 concentration. In all cases, the reduction in the EL2 concentration with the net electron concentration n was reported to follow a law of the form n^{-4}. In order to avoid possible measurement errors related to high electric fields and to reverse-bias current electron refilling effects which plague capacitive measurements on strongly n-type materials [25], the same team [68] checked their results with optical absorption measurements. Extrinsic optical absorption causes less acute measurement problems when high electron concentrations are present. Similar observations regarding the decrease in the EL2 concentration with n have been reported by Zou *et al.* [69].

THE THERMAL ANNEALING BEHAVIOUR OF EL2

EL2 was first reported to be affected by high temperature (above 600°C) treatments by Hasegawa and Majerfeld [70]. This was later shown by Mircea *et al.* [26] to be due to an outdiffusion process in the case of VPE materials. Similar results were obtained by Okumara and Ikoma for bulk crystals [71]. A more quantitative and detailed investigation of EL2 outdiffusion in bulk GaAs was later reported [72] and it was shown that the capping conditions had a strong influence on the outdiffusion profile (departure from Fick's law) while the average outdiffusion coefficient mainly depended on the type of material used (different types of bulk and VPE materials). A more reduced influence of capping conditions was reported for LEC material by Hughes and Lee [73] while Li and Wang [74] observed a very limited outdiffusion in their HB materials. Despite the practical importance of the EL2 outdiffusion study to throw light on the device and circuit performances, it is difficult to draw definite conclusions from the above observations. Defects of both the Ga and As sublattice may play an important role in EL2 outdiffusion behaviour.

Lagowski *et al.* [75] reported that annealing in a hydrogen plasma at 300°C for 2 hours resulted in a decrease in the EL2 concentration of Bridgman n-GaAs. They explained this observation by the role of atomic hydrogen in saturating dangling bonds. Similarly, Pearton and Tavendale [76] show that all observed electron trapping levels existing in LPE layers decreased in concentration after a similar hydrogen plasma anneal. A subsequent vacuum anneal resulted in partial recovery of the previous deep level concentration.

Creation of EL2 by ion implantation

Jervis *et al.* [77] were the first to report an artificial creation of EL2 (in very small concentrations) after Si implantation of an LPE material through a Si_3N_4 cap and annealing at 860°C. One should remember here that the GaAs LPE materials are practically free of native electron traps (including EL2).

Ikoma *et al.* [78] and Taniguchi and Ikoma [28, 79] succeeded in creating levels having DLTS signatures similar to that of EL2 in very substantial concentrations. They achieved this by implanting thick LPE n-GaAs materials with As, Ga and O ions and subsequently annealed their samples for 15 min around 600°C. Later on, Martin *et al.* [80] gave evidence that the level thus created in bulk GaAs materials was indeed EL2, since it exhibited the photocapacitance quenching behaviour which is an unambiguous signature, as reported above. The authors did not observe such behaviour with LPE implanted layers. However, these last negative results were later found to be fraught with uncertainty because the LPE Schottky diodes used were leaky. However, the electron emission rate of the level created in LPE materials was found [81] to vary with electric field and temperature in a manner identical to that known for the EL2 level [34]. More recently, T. Ikoma of Tokyo University has privately communicated to the authors the occurrence of the transient photocapacitance overshoot behaviour typical of EL2 in LPE ion implanted layers. This clearly demonstrates the creation of EL2 by ion implantation and annealing in both bulk and LPE layers.

EPR observation of the arsenic antisite defect

Electron paramagnetic resonance (EPR) is a powerful tool, very frequently used by physicists to gain detailed information about defects in semiconductors. Wagner *et al.* [82] were the first to report an EPR 4 line spectrum which they attributed to the As_{Ga} antisite defect. This observation was made with undoped SI GaAs. A similar spectrum was reported for electron-irradiated material [83]. The same EPR spectrum was later created by fast neutron irradiation [84] and by applying uniaxial stress (4% deformation) at 400°C [85]. The EPR spectra are as depicted in Figure 7. The quadruplet (4 lines)

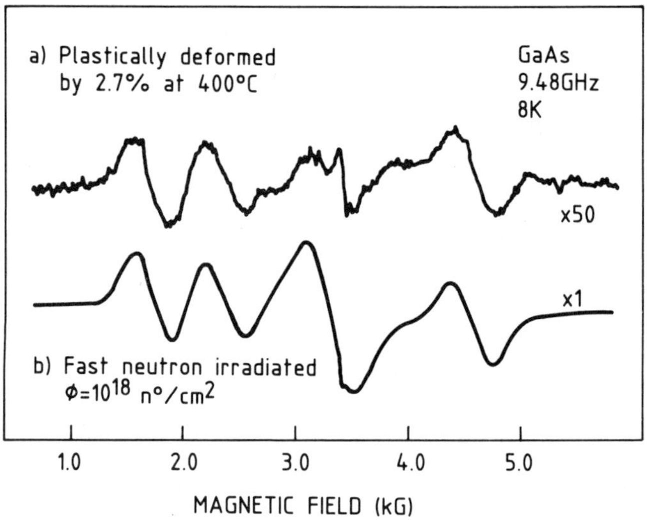

Figure 7 EPR spectrum of the As_{Ga} antisite defect in (a) plastically deformed and (b) fast neutron irradiated GaAs. After Ref. 84.

EPR spectrum is attributed to the As_{Ga} charge state of the antisite defect [84–89]. For fast neutron irradiations, this defect introduction rate was found to be 10 cm^{-1} for small neutron fluxes of about 10^{11} cm^{-2} s^{-1} [86] and to be about 3 times smaller for neutron fluxes larger than 10^{13} cm^{-2} s^{-1} [84]. For small neutron fluxes, Goltzene *et al.* [86] report that an additional singlet line can be extracted from the EPR spectrum. They found that the quadruplet and singlet components have closely correlated temperature dependences, with deviation in the same manner from Curie's law. The authors also report a decrease of the hyperfine structure separation with increasing temperature. They suggest that both the singlet and quadruplet lines are generated within the same defect complex of intrinsic nature and that the observed anomalies are due to closely separated defect interactions.

One of the problems with such EPR measurements is that they give wavefunction information on an atomic scale. In fact, different types of defect pairing or long range interactions which would yield very different electrical and optical deep level behaviours may yield identical EPR observations. One experimental way out of this difficulty is to perform photo-EPR measurements in an attempt to study possible changes in the defect charged states with different monochromatic light excitations. This type of experiment was performed by Weber *et al.* [85] who gave evidence for two photoionization thresholds in the As_{Ga}^{4+} EPR signal for photon energies $h\nu = 0.75$ and 1 eV which they attributed to the occurrence of 3 different charge states related to the As_{Ga} defect. The charge state As_{Ga}^{3+} should be neutral, the EPR active As_{Ga}^{4+} singly positively charged and the As_{Ga}^{5+} doubly positively charged. Attempts were also made by Weber and Schneider [89] to observe a photo-quenching behaviour for the As_{Ga}^{4+} EPR signal. They did in fact observe such a behaviour at 7 K, reminiscent of that of EL2, but the photoquenching was not complete.

Another possible criterion to compare defects observed by EPR with those observed by DLTS is to compare their thermal annealing behaviours. The EPR quadruplet signal was observed by Wörner *et al.* [84] to anneal as from a temperature of 450°C. This corresponds to the so-called stage 4 annealing occurring, in particular, for ion implantation and neutron irradiation defects [90, 91].

Correlation of EPR and DLTS measurements

The DLTS spectra related to electron traps created by ion implantation [90,92] or neutron irradiation [91] are very similar. Before annealing, both electron emission DLTS spectra are dominated by a wide peak near room temperature previously referred to as the U-band [91, 92]. The annealing behaviour of this peak was found to follow closely that of the As_{Ga}^{4+} EPR spectrum and it was therefore concluded that the U-band and not EL2 could be associated with the EPR defect [91]. In fact, the above conclusion does not take into account defect clustering effects which are known to be important for fast neutron irradiation and ion implantation [90]. Both types of defects are expected to be introduced by impinging neutron or ion particles which give up a significant portion of their kinetic energy to lattice atoms which in turn generate a cascade of closely spaced defects which are therefore expected to bunch into defect clusters.

In fact, the low electric field DLTS peak position is observed to gradually shift towards EL2 when one performs isochronal (30 min) anneals at increasingly higher temperatures [93, 94]. The DLTS U-peak positions become practically coincident with EL2 for annealing temperatures above 550°C [94]. This, in our view, is a convincing demonstration that defect clustering effects gradually disappear with annealing as the local defect density decreases. Furthermore, detailed observation of the annealing behaviour of the U-peak [95] shows that it is due to a long range defect migration behaviour. This is in sharp contrast with the annealing behaviour of shallower electron traps also produced by neutron irradiation and which seem to follow a local defect recombination scheme.

The above observations open up the possibility of reconciling the apparently conflicting DLTS and EPR data since it is not impossible that the U-band could be the EL2 level in disguise. Within the defect clusters created by fast neutron irradiation or ion implantation, an electron trapped in a (supposedly created) EL2 defect could either be directly emitted to the conduction band or transit first by a multiphonon tunnelling process to a nearby shallower electron trap before being emitted to the conduction band. For high density ($\gtrsim 10^{18}$ cm^{-3}) defect clusters, the indirect process is more probable because of the small GaAs Γ-conduction band effective mass which greatly enhances tunnelling probabilities [33, 94]. Furthermore, the wide distribution of electron emission rates (wide U-peak) could be explained by the random distribution of tunnelling lengths involved.

The possibility of associating the As_{Ga}^{4+} EPR signal with EL2 is also confirmed by plastic deformation experiments at 400°C which were shown to produce mainly a midgap electron trap [96, 97] which is presumably EL2 and with concentrations similar to that of the EPR As_{Ga}^{4+} signal generated by a similar deformation procedure [89]. Furthermore, a different annealing behaviour for the EPR signal is observed [89] and in particular a survival of this signal after annealing at temperatures as high as 550°C. This is not surprising since one expects defect clustering and long range defect recombination annealing to be less marked in plastically deformed samples as compared to neutron irradiated samples.

The identification problem

EL2 is clearly an intrinsic defect and not an impurity since, for example, it can be created by implantation of very different sorts of ions. All available data indicate that EL2 is systematically associated with an excess As stoichiometry during crystal growth. The arguments for associating EL2 with the As_{Ga} antisite defect seem today convincing. The dependence of the EL2 concentration on the free electron concentration, as well as the EPR and DLTS measurements performed on neutron irradiated or plastically deformed materials all converge to this conclusion. Of course, defect clustering effects frustrate a straight-forward interpretation of the observed DLTS data and annealing behaviours.

If EL2 is associated with the As_{Ga} defect, none of the above arguments prove or disprove possible associations with other defects. However, there are other reasons to associate another defect with the antisite to account for EL2:

1. Lagowski *et al.* [68] observed a correlation of the EL2 level with a shallow donor some 25 meV below the conduction band which occurs in their material with nearly the same concentration as EL2. They suggested that

EL2 is still positively charged when the deep level is occupied by an electron. This assertion, which is conflicting with the earlier data of Mircea *et al.* [26], lead them to propose for EL2 the defect complex $As_{Ga}V_{As}$.
2. The photoquenching behaviour of EL2 cannot be understood if one assumes EL2 to be composed exclusively of the As_{Ga} defect. It has recently been shown theoretically [98] that the isolated antisite defect cannot exhibit anomalously large lattice relaxation effects. It would therefore seem that the ideas of Levinson [37] are more appropriate than a lattice relaxation anomaly to explain the photoquenching; this type of explanation requires a more complex defect than the antisite.

One other question still remains: is there one EL2 level or is there sometimes confusion with another nearby midgap level, possibly related to oxygen? In fact, Taniguchi and Ikoma [28] report that an oxygen-related level may appear in their ion implantation experiments. They explain this by the differences between the thermal and photoionization behaviours of those levels introduced by oxygen ion implantation and those obtained by implanting other species. More recently, Lagowski *et al.* [29] have reported the simultaneous occurrence of two nearly coincident midgap levels in Bridgman samples. The shallower level seemed to be correlated with the oxygen content of their samples.

These observations are puzzling especially when one notices that the EL2 level created by an oxygen implantation followed by a 600°C anneal has the same type of photoquenching behaviour as the native levels and those introduced by implantation with other ionic species (Ga, As, B and Ne). It would be surprising if these two defects should have this very peculiar behaviour not observed for shallower levels. Of course, an alternative explanation to account for differences in photoionization, thermal ionization and photoquenching time constants would be some sort of long range defect interaction (tunnelling or Coulomb) of the sort suggested in the previous section.

Another observation which might suggest the involvement of oxygen in a midgap level near EL2 is that of Wolford *et al.* [99]. These authors observe a deep level luminescence in the $GaAs_{1-x}P_x$ ternary alloy system. They show that the peak energy of the luminescence band varies linearly with x, and that it coincides with O–Zn pair in GaP ($x = 1$) and with the 0.64 eV band probably related to EL2 in GaAs ($x = 0$).

Very recent contributions

This review would be incomplete without the mention of two recent contributions brought to our attention after the write-up of the present article.

Meyer *et al.* [100] have studied the magnetic circular dichroism (MCD) in the extrinsic absorption spectrum of GaAs. The MCD can be defined as the difference in absorption between left and right circular polarized light induced by a strong magnetic field. They thus put in evidence two new absorption bands centred about 1.05 and 1.29 eV photon energies. The origin of these bands were studied by monitoring the microwave (K-band) induced changes in MCD as a function of the magnetic field. They revealed the quadruplet signature of the As_{Ga} defect reported above. They were led to attribute the observed absorption bands to internal transitions of this antisite between a fundamental state in the band gap with A_1 symmetry and either one of two excited states having T_2 symmetry and lying at 1.05 and 1.29 eV above the fundamental state. Their investigation lead them to reject the identification of EL2 with the As_{Ga} antisite. Their arguments for this conclusion are:

1. The integrated MCD spectrum (for different photon energies) has a different shape from that related to the intra-centre absorption of EL2.
2. Their estimated optical cross-section for the intra-centre transitions in the As_{Ga} antisite (as deduced from the integrated MCD spectrum and EPR evaluation of the antisite concentration) is about a factor of 10 smaller than the corresponding EL2 cross-section (estimated from optical absorption and DLTS).
3. Their estimated EL2 concentration is about half that of the antisite defect.

In fact, these arguments of Meyer *et al.* do not definitely exclude the attribution of EL2 to a defect complex including the As_{Ga} antisite.

Another recent interesting contribution is that of Omling *et al.* [101]. These authors have made a detailed investigation of the EL2 photoquenching behaviour in the $GaAs_{1-x}P_x$ ternary alloy system (x ranging from 0 to 0.26). They report spectra for the optical transition cross-section from the stable to metastable configuration of EL2 with a shape practically independent of the phosphorus content (x). All the reported spectra are practically identical in shape to the spectrum of Figures 4 or 5. This observation, which confirms well the intra-centre character of such a transition rate, is in sharp contrast with the rapid changes reported by these authors for the activation energies related to ionization and thermal recovery to the normal state.

An important remark of the above authors [101] is the fact that the stable to metastable optical transition rate spectrum is temperature dependent (the peak energy decreases when the temperature increases). One of their proposed explanations is the occurrence of two excited intermediary states roughly 1.08 and 1.17 eV above the fundamental level.

Although the first of the two above contributions excludes the assignment of the isolated As_{Ga} antisite to EL2, they show similarities in the intra-centre transitions of these two types of defect. In both cases, the intra-centre transition can be attributed to an initial A_1 fundamental state or either one of two neighbouring excited states with T_2 symmetry.

Concluding remarks

The investigations of most research teams seem today to converge to the conclusion that EL2 is associated with the As_{Ga} antisite defect. More work has to be pursued to determine the nature of other defects (if any) associated with the antisite to form EL2. Very interesting possibilities have been demonstrated by a joint Warsaw University-MIT team who have demonstrated the possibility of studying the defect symmetry of EL2 by making use of the fine structure in the optical absorption band related to the phototransition rate between the stable and the metastable configuration of this defect. Further experimentation along those guidelines as well as a theoretical effort to understand these phenomena in detail seem to be necessary.

ACKNOWLEDGEMENTS

The authors are grateful to the many colleagues who have sent them preprints of their publications and in particular T. Ikoma, M. Kaminska, J. Lagowski, A. Goltzene, P. M. Omling and C. Schwab. Thanks are also due to G. Guillot, D. C. Look, A. Mircea, D. Paget, D. Pons, E. Weber and P. W. Yu for fruitful discussions.

References

1. Whelan J. W. and Wheatley G. H. (1958). *J. Phys. Chem. Sol.*, **6**, 169
2. Hilsum C. and Rose-Innes A. C. (1961). *Semiconducting III–V Compounds*, p. 142; Pergamon Press
3. Ainslie N. G., Blum S. E. and Woods J. F. (1962). *J. Appl. Phys.*, **33**, 2392
4. Woods J. F. and Ainslie N. G. (1963). *J. Appl. Phys.*, **34**, 1469
5. Fertin J. L., Lebailly J. and Deyris E. (1967). *Inst. Phys. Conf. Ser.*, **3**, 46
6. Woodall J. M. and Woods J. F. (1966). *Solid-State Commun.*, **4**, 33
7. Woodall J. M. (1967). *Trans. AIME*, **239**, 378
8. Shimoda T. and Akai S. (1969). *Jap. J. Appl. Phys.*, **8**, 1352
9. Allen J. W. (1960). *Nature*, **187**, 403
10. Blanc J. and Weisberg L. R. (1961). *Nature*, **192**, 155
11. Haisty R. W., Mehal E. W. and Stratton R. (1962). *J. Phys. Chem. Sol.*, **23**, 829
12. Blanc J., Bube R. H. and Weisberg L. R. (1964). *J. Phys. Chem. Sol.*, **25**, 225
13. Sze S. M. and Irvin J. C. (1968). *Solid St. Electron.*, **11**, 599
14. Martin G. M. and Bois D. (1978). *Proceedings of the Topical Conference on Characterization Techniques* (*Electrochem. Soc.*), **78–3**, 32
15. Williams R. (1966). *J. Appl. Phys.*, **37**, 3411
16. Mircea A. and Mitonneau A. (1975). *Appl. Phys.*, **8**, 15
17. Martin G. M., Mitonneau A. and Mircea A. (1977). *Electron. Lett.*, **13**, 191
18. Martin G. M., Jacob G., Hallais J. P., Grainger F., Roberts J. A., Clegg J. B., Blood P. and Poiblaud G. (1982). *J. Phys. C.*, **15**, 1841
19. Huber A. M., Linh N. T., Debrun J. C., Valladon M., Martin G. M., Mitonneau A. and Mircea A. (1979). *J. Appl. Phys.*, **50**, 4022
20. Kaminska M., Lagowski J., Parsey J., Wada K. and Gatos H. C. (1982). *Inst. Phys. Conf. Ser.*, **63**, 197
21. Wolfe C. M., Stillman G. E. and Korn D. M. (1977). *Inst. Phys. Conf. Ser.*, **33b**, 120
22. Palm L., Bruch H., Bachem K. H. and Balk P. (1979). *J. Electron. Mat.*, **8**, 555
23. Wallis R. H., Di Forte-Poisson M. A., Bonnet M., Beuchet G. and Duchemin J. P. (1981). *Inst. Phys. Conf. Ser.*, **56**, 73

24. Lang D. V. and Logan R. A. (1975). *J. Electron. Mat.*, **4,** 1053
25. Martin G. M. and Makram-Ebeid S. (1984). In *Deep Levels in Semiconductors* (S. Pantelides, ed.); Gordon and Breach
26. Mircea A., Mitonneau A., Hollan L. and Brière A. (1976). *Appl. Phys.*, **11,** 153
27. Martin G. M., Mitonneau A., Pons D., Mircea A. and Woodward D. W. (1980). *J. Phys. C.*, **13,** 3855
28. Taniguchi M. and Ikoma T. (1982). In *Semi-Insulating III-V Materials, Evian*, p. 283 (S. Makram-Ebeid and B. Tuck, eds.); Shiva Publishing, Nantwich, UK
29. Lagowski J., Lin S. G., Aogama T. and Gatos H. C. (1984). *Appl. Phys. Lett.*, **44,** 336
30. Mitonneau A., Mircea A., Martin G. M. and Pons D. (1979). *Rev. Phys. Appl.*, **14,** 853
31. Van Vechten J. A. and Thurmond C. D. (1976). *Phys. Rev. B*, **14,** 3539
32. Chantre A., Vincent G. and Bois D. (1981). *Phys. Rev. B*, **23,** 5335
33. Makram-Ebeid S. and Lannoo M. (1982). *Phys. Rev. B*, **25,** 6406
34. Bois D. and Vincent G. (1977). *J. Phys. Lett.* (*France*), **38,** 351
35. Vincent G., Bois D. and Chantre A. (1982). *J. Appl. Phys.*, **53,** 3643
36. Mitonneau A. and Mircea A. (1979). *Solid-State Commun.*, **30,** 157
37. Levinson M. (1983). *Phys. Rev. B*, **28,** 3660
38. Levinson M., Benton J. L. and Kimerling L. C. (1983). *Phys. Rev. B*, **27,** 6216
39. Martin G. M., Jacob G., Goltzene A., Schwab C. and Poiblaud G. (1981). *Inst. Phys. Conf. Ser.*, **59,** 281
40. Martin G. M. (1980). In *Semi-Insulating III–V Materials, Nottingham*, p. 13 (G. J. Rees, ed.); Shiva Publishing, Nantwich, UK
41. Martin G. M. (1981). *Appl. Phys. Lett.*, **39,** 747
42. Kaminska M., Skowronski M., Lagowski J., Parsey J. M. and Gatos H. C. (1983). *Appl. Phys. Lett.*, **43,** 302
43. Skowronski M., Kaminski M., Kuzko W. and Godlewski M. (1983). 4th Conf. on Deep Impurity Centers in Semiconductors, Eger, Hungary, and also Kaminska M., Skowronski M., Kusko W., Lagowski J., Parsey J. and Gatos H. C. (1984). *Czech Journal of Physics*, **334,** 409
44. Mircea-Roussel A. and Makram-Ebeid S. (1981). *Appl. Phys. Lett.*, **38,** 1007
45. Leyral P. and Guillot G. (1982). In *Semi-Insulating III–V Materials, Evian*, p. 166 (S. Makram-Ebeid and B. Tuck, eds.); Shiva Publishing, Nantwich, UK
46. Shanabrook B. V., Klein P. B., Swiggard E. M. and Bishop S. G. (1983). *J. Appl. Phys.*, **54,** 336
47. Yu P. W. (1979). *Solid-State Commun.*, **32,** 111
48. Leyral P., Vincent G., Nouailhat A. and Guillot G. (1982). *Solid-State Commun.*, **42,** 67
49. Shanabrook B. V., Klein P. B. and Bishop S. G. (1982). 15th Conference on the Physics of Semiconductors, Montpellier
50. Yu P. W., Holmes D. E. and Chen R. T. (1982). *Inst. Phys. Conf. Ser.*, **63,** 200
51. Yu P. W. (1982). *Solid-State Commun.*, **43,** 953
52. Yu P. W. (1984). *Appl. Phys. Lett.*, **44,** 330
53. Holmes D. E., Chen R. T., Elliott K. R., Kirkpatrick C. G. and Yu. P. W. (1982). *IEEE Trans El. Dev.*, **29,** 1045
54. Holmes D. E., Chen R. T., Elliott K. R. and Kirkpatrick C. G. (1982). *Appl. Phys. Lett.*, **40,** 46
55. Holmes D. E., Elliott K. R., Chen R. T. and Kirkpatrick C. G. (1982). In *Semi-Insulating III-V Materials, Evian*, p. 19 (S. Makram-Ebeid and B. Tuck, eds.); Shiva Publishing, Nantwich, UK
56. Martin G. M. (1982). *Physica Scripta*, **T1,** 38
57. Lagowski J., Gatos H. C., Parsey J. M., Wada K., Kaminska M. and Walukiewicz W. (1982). *Appl. Phys. Lett.*, **40,** 342
58. Miller M. D., Olsen G. H. and Ettenberg M. (1977). *Appl. Phys. Lett.*, **31,** 538
59. Bhattacharya P. K., Ku J. W., Owen S. J. T., Aebi V., Cooper C. B. and Moon R. L. (1980). *Appl. Phys. Lett.*, **36,** 304
60. Wagner E. E., Mars D. E., Hom G. and Stringfellow G. B. (1980). *J. Appl. Phys.*, **51,** 5434

61. Ozeki M., Komemo J. and Ohkawa S. (1979). *Fujitsu Sci. Tech. J.*, **15,** 83
62. Samuelson L., Omling P., Titze H. and Grimmeiss H. G. (1981). *J. Cryst. Growth*, **55,** 164
63. Watanabe M. O., Tanaka A., Nakanisi T. and Zohta Y. (1981). *Jap. J. Appl. Phys.*, **20,** L429
64. Subramanian S., Arora B. M. and Guha S. (1981). *Solid St. Electron.*, **24,** 287
65. Oseki M., Komemo J., Shibatomi A. and Ohkawa S. (1979). *J. Appl. Phys.*, **50,** 4808
66. Humbert A., Hollan L. and Bois D. (1976). *J. Appl. Phys.*, **47,** 4137
67. Lagowski J., Parsey J. M., Kaminska M., Wada K. and Gatos H. C. (1982). In *Semi-Insulating III-V Materials, Evian*, p. 154 (S. Makram-Ebeid and B. Tuck, eds.); Shiva Publishing, Nantwich, UK
68. Lagowski J., Kaminska M., Parsey J. M., Gatos H. C. and Walukiewicz W. (1983). *Inst. Phys. Conf. Ser.*, **65,** 41
69. Yuanxi Zou, Jicheng Zhou, Peigen Mo, Fengzmen Lu, Liansheng Li, Jiuan Shao and Lei Huang (1983). *Inst. Phys. Conf. Ser.*, **65,** 49
70. Hasegawa F. and Majerfeld A. (1976). *Electron. Lett.*, **12,** 52
71. Okumura T. and Ikoma T. (1977). Annual Research Report of the Solid State Electronics Laboratories (Institute of Industrial Science, Tokyo, Japan), Part C
72. Makram-Ebeid S., Gautard D., Devillard P. and Martin G. M. (1982). *Appl. Phys. Lett.*, **40,** 161
73. Hughes B. and Lee C. (1983). *Inst. Phys. Conf. Ser.*, **65,** 57
74. Li G. P. and Wang K. L. (1982). *J. Appl. Phys.*, **53,** 8653
75. Lagowski J., Kaminska M., Parsey Jr J. M., Gatos H. C. and Lichtensteiger M. (1982). *Appl. Phys. Lett.*, **41,** 1078
76. Pearton S. J. and Tavendale A. J. (1982). *Electron. Lett.*, **18,** 715
77. Jervis T. R., Woodward D. W. and Eastman L. F. (1979). *Electron. Lett.*, **20,** 620
78. Ikoma T., Takikawa M. and Taniguchi M. (1982). *Inst. Phys. Conf. Ser.*, **63,** 191
79. Taniguchi M. and Ikoma T. (1983). *Inst. Phys. Conf. Ser.*, **65,** 65
80. Martin G. M., Terriac P., Makram-Ebeid S., Guillot G. and Gavant M. (1983). *Appl. Phys. Lett.*, **42,** 61
81. Makram-Ebeid S. and Minondo P. Unpublished
82. Wagner J. R., Krebs J. J., Strauss G. H. and White A. M. (1980). *Solid-State Commun.*, **36,** 15
83. Goswami N. K., Newman R. C. and Whitehouse J. E. (1983). *Solid-State Commun.*, **46,** 473
84. Wörner R., Kaufmann U. and Schneider J. (1982). *Appl. Phys. Lett.*, **40,** 141
85. Weber E. R., Ennen H., Kaufmann U., Windscheif J., Schneider J. and Wosinski T. (1982). *J. Appl. Phys.*, **53,** 6140
86. Goltzene A., Meyer B. and Schwab C. (1983). *J. Appl. Phys.*, **54,** 3117
87. Goltzene A., Meyer B. and Schwab C. *Rev. Phys. Appl.*, in press
88. Schneider J. (1982). In *Semi-Insulating III–V Materials, Evian*, p. 144 (S. Makram-Ebeid and B. Tuck, eds.); Shiva Publishing, Nantwich, UK
89. Weber E. R. and Schneider J. (1982). Proc. of the 12th Int. Conf. on Defects in Semiconductors, North Holland, 398
90. Lang D. V. (1977). *Inst. Phys. Conf. Ser.*, **31,** 70
91. Martin G. M. and Makram-Ebeid S. (1982). Proc. of the 12th Int. Conf. on Defects in Semiconductors, North Holland, 371
92. Martin G. M., Secordel P. and Venger C. (1982). *J. Appl. Phys.*, **53,** 8706
93. Barnes C. E., Zipperrian T. E. and Dawson L. R. *J. Electron. Mat.*, in press
94. Langlade P. and Makram-Ebeid S. Unpublished
95. Martin G. M., Estève E., Makram-Ebeid S. and Langlade P. Unpublished
96. Ishida T., Maeda K. and Takeuchi S. (1980). *Appl. Phys.*, **21,** 257
97. Wosinski T., Morawski A. and Figielski T. (1983). *Appl. Phys. A*, **30,** 233
98. Scheffler M., Bachelet G. P., Meyer B. K. and Spaeth J. M. (1984). *Bull. Amer. Phys. Soc.*, **29,** 206
99. Wolford D. J., Modestie S. and Streetman B. G. (1983). *Inst. Phys. Conf. Ser.*, **65,** 501
100. Meyer B. K., Spaeth J. M. and Scheffler M. (1984). *Phys. Rev. Lett.*, **52,** 851
101. Omling P., Samuelson L. and Grimmeiss H. (1984). *Phys. Rev. B*, **29,** 15

Effect of Thermal History on Properties of LEC GaAs

D. E. Holmes, H. Kuwamoto, C. G. Kirkpatrick and R. T. Chen

Rockwell International Corporation, Microelectronics Research and Development Center, Thousand Oaks, California 91360, USA

Abstract

We have investigated the effects of thermal history of semi-insulating GaAs crystals grown by the LEC (liquid encapsulated Czochralski) technique on the distribution of the deep donor EL2. Several factors were evaluated, including the thermal stability of EL2 centres, variations of the temperature-time cycle along the growth axis of crystals during post-growth cooling, correlations of macroscopic and microscopic distributions of EL2 centres and dislocations, the stoichiometry of the melt during growth and the presence (or absence) of an As overpressure during annealing. Possible effects of solid–liquid interface shape on measured EL2 distributions were also taken into account. The results indicate that crystallographic symmetry of the EL2 distribution across LEC crystals arises from a combination of climb and gettering processes during postgrowth cooling. Furthermore, the formation of EL2 inhomogeneities by climb is reversible, and uniformity can be established after growth by a two-stage annealing process. Atomic mechanisms relating the formation of EL2 centres to dislocation dynamics are also discussed.

Introduction

The electrical properties of undoped semi-insulating GaAs grown by the LEC technique are normally controlled [1] by the balance between the deep donor EL2 and carbon acceptors. EL2 is believed to be a native defect, and there appears to be a growing consensus [1–3] that EL2 is the antisite defect As_{Ga} or a related defect complex. Spatial variations of the EL2 concentration result in corresponding variations of the electrical properties [4] of the material. Martin *et al.* [5] showed that linear profiles of the EL2 concentration across LEC crystals have a 'W' shape characteristic of the dislocation density. Two-dimensional contour maps of EL2 centres and the dislocation density published more recently [6] confirm the crystallographic nature of the EL2 distribution (four-fold pattern of EL2 in ⟨100⟩ crystals correlates with the dislocation pattern). High resolution infrared imaging studies [7] have also revealed a correlation between the pattern of dislocation networks and EL2 centres on the microscale. The relationship between the patterns of the line (dislocations) and point (EL2) defects, therefore, has great practical as well as fundamental importance.

The purpose of this investigation was to study the effect of the thermal history of LEC crystals on the development of crystallographic EL2 distributions, and to understand why EL2 distributions are invariably more uniform toward the tail of LEC crystals, as discussed in detail in Ref. 6. A thermal history of our crystals was first constructed in this study from available temperature profile data. The interface shape along the growth axis was also determined to evaluate changes in radial temperature gradients during growth. Finally, extensive thermal annealing experiments were conducted to study the thermal stability and mobility of EL2 centres in LEC material. The results of the thermal annealing and interface shape experiments were then evaluated in the light of the temperature-time cycle of the seed and tail ends of the crystal to describe the evolution of EL2 patterns.

Experimental

Undoped semi-insulating 3 inch ⟨100⟩ GaAs crystals characterized in this investigation were grown by the high pressure (MRC Melbourn system) LEC technique from 3 kg charges (see Ref. 1 for details). Thick slices were cut perpendicular to the growth direction from the seed and tail ends of the crystals. The thick slices were cut into bars for optical profiling of the EL2 concentration (see Ref. 6 for details of the optical absorption method). Dislocation density profiles were determined by preferential etching with KOH. Annealing experiments were conducted with these thick bars under a flowing argon atmosphere using proximity capping or under a 1 atm arsenic overpressure in a sealed quartz ampoule. Annealed samples were always polished prior to optical profiling to remove surface damage that could have interfered with the absorption measurements. Microscopic distributions of the EL2 concentration were determined by infrared microscopy. The shape of the solid–liquid interface was determined by optically enhanced preferential etching in a solution of ($15H_2SO_4 : 1H_2O_2 : 1H_2O$).

Results: thermal history

Idealized cooling rate curves for the seed and tail ends of LEC crystals grown in our system are shown in Figure 1. These curves were constructed from static thermal profiles [8] of the Cambridge high pressure Melbourn system and pulling rate parameters used in our process. Although the curves are not highly accurate, the fundamental differences between the seed and tail are qualitatively representative.

The cooling rate of the seed end of the crystal is first controlled by the vertical motion of the crystal through the boric oxide encapsulant. Since the thermal gradients in the encapsulant are high (100–200°C cm^{-1}), the seed end ('seed' end refers to the 3 inch portion of the crystal nearest the seed) in our crystals cooled at about 100°C h^{-1}. After emerging from the encapsulant, and probably after some transient cooling, the seed end cooled at a slower rate. At the termination of growth, the cool down to room temperature was programmably controlled.

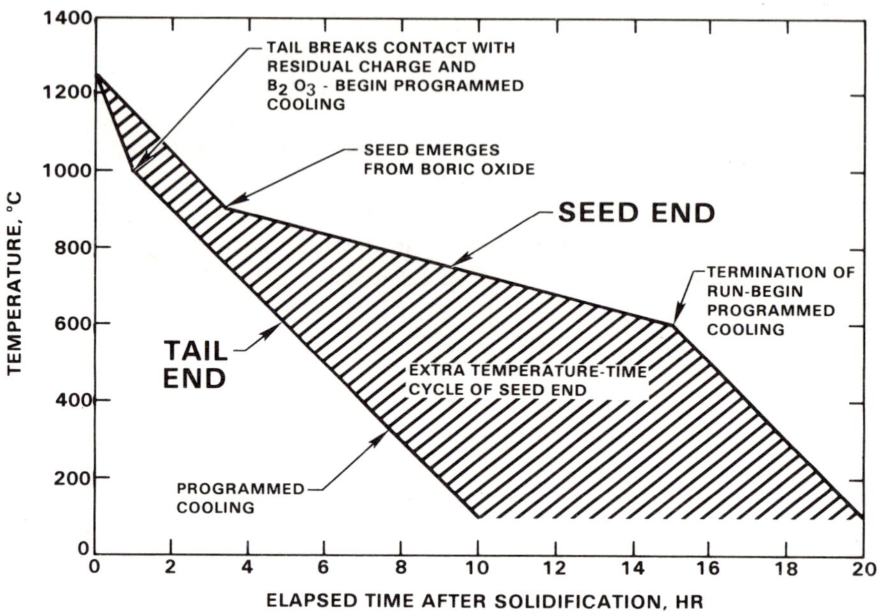

Figure 1 Temperature-time profiles for the seed and tail ends of a LEC crystal. The differences between the seed and tail ends are qualitatively represented. The tail end spends less time at higher temperature than the seed end.

The cooling rate of the tail of the crystals was generally determined by the exact method by which the crystal was separated from the remaining melt (if any) and the boric oxide encapsulant. After careful evaluation of our separation process, we estimated that the tail ends cooled to about 900–1100°C within about one hour, which is a much faster initial cooling rate than at the seed end. The cool down to room temperature was then programmably controlled.

Results: interface shape

The interface shapes at the seed end of the crystals were found to be convex with respect to the solid. The slope of the interface with respect to horizontal is maximum midway between the centre and edge. The vertical displacement of the centre and edge was about 5 mm and independent of cone angle. In contrast, the interface shape at the tail was concave with respect to the solid, and the vertical displacement varied up to 3.5 mm. The inversion of the interface shape at the seed and tail indicates that the interface was nearly flat at some point along the body of the crystal accompanied by a minimum radial temperature gradient. This finding is unexpected because the average dislocation density in our crystals always increases toward the tail of our crystals, indicating progressively higher stresses toward the tail of the crystals.

Higher stresses imply higher radial gradients and more pronounced interface curvature. Therefore, it is likely that the gradient-induced stress increased during postgrowth cooling when the material was within the encapsulant [9].

It is noted that since the 4 mm thick slices used for characterization were cut perpendicular to the growth direction, the edge of the slices cut from the seed end could have solidified up to 30 min after the centre as a result of the interface curvature. Segregation of arsenic in arsenic-rich melts, therefore, could have contributed to apparent (measured) radial EL2 variations. However, calculations showed that even in the most arsenic-rich crystals we have grown, this effect could only have caused an apparent radial concentration variation of 5×10^{14} cm^{-3}.

Results: thermal annealing

4 mm thick bars were annealed at temperatures ranging from 850 to 1100°C. The maximum annealing times were 110, 240 and 2 hours at 850, 950 and 1100°C, respectively. The EL2 centre, as indicated by the associated optical absorption, was found to be stable in all of the samples grown from As-rich and near stoichiometric melts for these annealing schedules. However, two distinct patterns of behaviour were observed for isothermal annealing depending on the annealing time. The results for 950°C anneals are discussed below.

For 'short' anneal cycles, corresponding to about 4–5 hours, the EL2 concentration increased by some 10–30%, as shown in Table 1 and Figure 2. The magnitude of the change in concentration across the crystal varied inversely to the dislocation density, i.e. the EL2 concentration increased more at the 'ring' region (between the centre and edge) than at the edge. The uniformity of the EL2 distribution also improved (see Table 1) as annealing

Table 1 *Annealing schedule of sample depicted in Figure 2*

Profile number*	Total annealing time (h)	Average EL2 concentration $\times 10^{-16}$ (cm^{-3})	Standard deviation (%)
1	None	1.70	13.4
2	0.33	1.87	8.0
3	0.67	1.97	8.6
4	1.0	2.00	6.3
5	1.67	2.01	6.2
6	7.67	1.95	4.1
7	22.67	1.89	5.1
8	122.67	1.70	3.4

* Profile numbers are keyed to Figure 2.

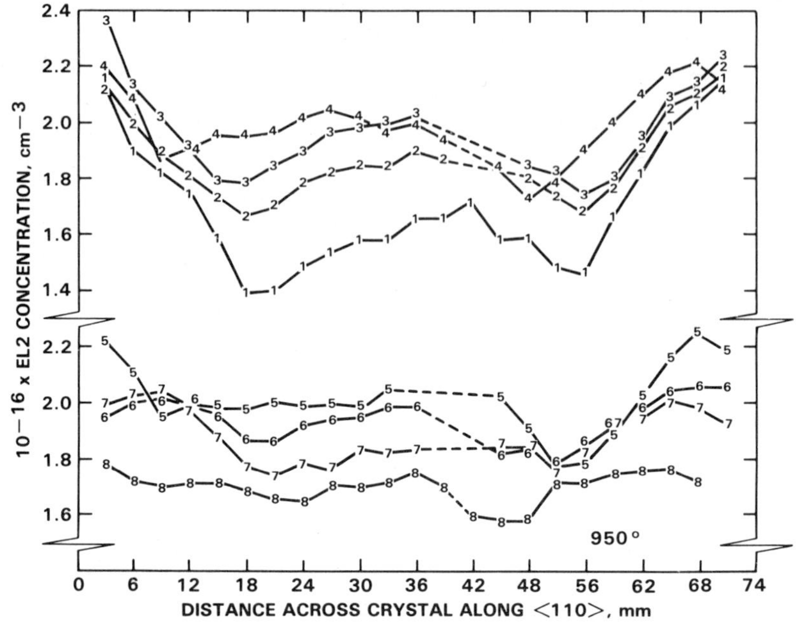

Figure 2 EL2 concentration profiles of a 4 mm thick bar sample annealed consecutively at 950°C. These profiles are keyed to the sample number in Table 1. The EL2 concentration increases during 'short-term' (4–5 hours) anneals, and then decreases, approaching 1.6×10^{16} cm^{-3}. The average concentration after 223 hours (not shown) was 1.67×10^{16} cm^{-3}. The uniformity of the profiles improves continuously during the entire annealing process.

time increased. The dislocation density was found not to change significantly during short-term annealing. After about 10–20 hours the EL2 concentration began decreasing, accompanied by continued macroscopic homogenization and a steadily decreasing rate of change of the EL2 concentration. After 'long' times (100–200 hours), the EL2 distribution was very uniform, with a typical standard deviation of 2–6%. The annealing experiments of Rumsby *et al.* [10], the first demonstration of improved uniformity upon annealing, fall into our 'short-term' stage.

The effects of long-term annealing on the two-dimensional EL2 distribution at the seed end of a crystal are shown in Figure 3. The four-fold symmetry of the pattern generally disappeared, leaving a more uniform but asymmetrical distribution. It was also found that, after long-term annealing at 950°C, the average EL2 concentration in all samples had the same end point value of $1.6 \times 10^{16} \pm 6\%$ cm^{-3}. This end point was found to be independent of the continuity of the sample (i.e. a 3 inch long, 4 mm thick bar would have the same average EL2 concentration after long term annealing if annealed in one piece or if cut and annealed as several pieces) and the stoichiometry of the melt from which the crystal was grown, as shown in Table 2. In addition, the annealing of a sample at 950°C for 100 h with a 1 atm overpressure of

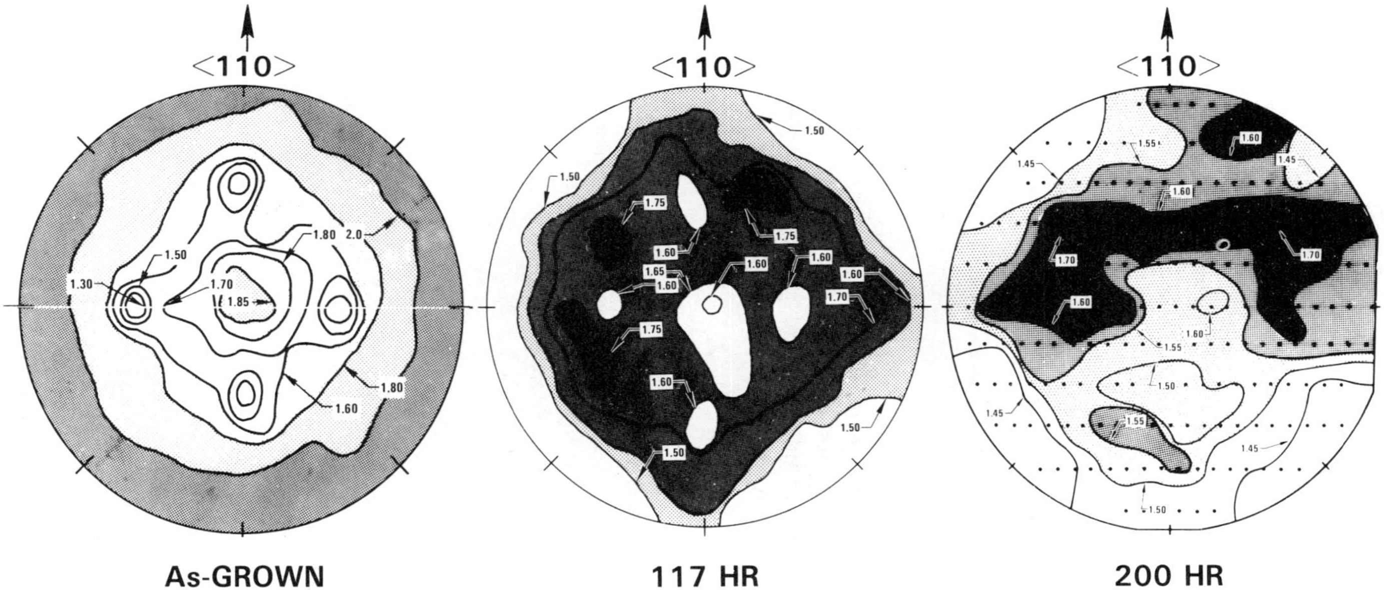

Figure 3 Two-dimensional contour maps of the EL2 concentration at the seed end of an undoped semi-insulating LEC crystal; (left) as-grown, (centre) after 117 hours at 950°C, and (right) after 200 hours at 950°C. Darker regions represent higher EL2 concentrations. The contour labels represent units of 10^{16} cm^{-3}. The average EL2 concentration and standard deviation are 1.86×10^{16} cm^{-3} ± 9.7% (left), 1.63×10^{16} cm^{-3} ± 7.4% (centre), and 1.53×10^{16} cm^{-3} ± 6.5% (right). The most pronounced change during annealing occurred at the edge along ⟨100⟩ where the EL2 concentration decreased up to 20% after 200 hours. The characteristic 'W' profiles along ⟨110⟩ and ⟨100⟩ became more uniform and 'M' shaped after annealing.

Table 2 *Comparison of annealing behaviour of material grown from As-rich and near-stoichiometric melts*

Sample	Initial EL2 concentration (cm^{-3})	Short-term anneal* Peak time (h)	Short-term anneal* EL2 concentration (cm^{-3})	Long-term anneal* Peak time (h)	Long-term anneal* EL2 concentration (cm^{-3})
Near-stoichiometric	1.9×10^{16}	3	2.4×10^{16}	240	1.7×10^{16}
As-rich	2.9×10^{16}	0.5	3.1×10^{16}	240	1.8×10^{16}

* 950°C.
Note: The as-grown As-rich material initially contains more EL2, but the concentrations after annealing converge toward the same level.

arsenic had no significant effect on the change of the EL2 pattern or end point concentration. This behaviour indicates that the EL2 concentration after long-term annealing is determined by local equilibrium conditions rather than a long-range redistribution process. Furthermore, short-term annealing produces a non-equilibrium excess of EL2 centres that can be reversibly annihilated by long-term annealing.

Samples were also characterized by infrared microscopy before and after long-term annealing. Relatively high absorption was observed in patterns corresponding to the microscopic dislocation patterns, as described in Refs. 7 and 11. These patterns were less apparent, although still distinguishable, after annealing. Our observations indicate that some redistribution of EL2 occurs over distances on the order of 100 μm during annealing.

Discussion

The results of the annealing studies and our evaluation of the thermal history of the crystals show that the improved uniformity of EL2 distributions at the tail of our LEC crystals is not due to effects of 'long-term' annealing or long-range diffusion. This is because the tail end simply had less time than the seed end at elevated temperatures for such processes to occur. This finding supports our earlier conclusion [6, 12] that macroscopic EL2 inhomogeneities with crystallographic distributions result from the superposition of two distributions. The first distribution is grown-in, and constitutes 80–90% of the EL2 concentration at the seed end of the crystal and greater than 95% at the tail. This distribution is macroscopically uniform across the crystal, and the average EL2 concentration is controlled by the melt stoichiometry during growth. The second distribution forms during postgrowth cooling, and the pattern corresponds to that of the dislocation density.

The general correspondence between the postgrowth cooling time of our crystals and 'short-term' annealing time described above is an indication that the increase of the EL2 concentration during each of these processes is controlled by the same mechanism. This mechanism could be dislocation climb-induced EL2 generation [13, 14] which is consistent with the present results for the following reasons.

The amount of EL2 centres produced by climb would generally increase as the dislocation density increases, as the distance increases over which a dislocation segment climbs, as time increases, and as the level of stress or stress gradient increases, among other factors. The distribution across the crystal of EL2 centres created by climb would therefore follow the pattern of the dislocation density, as observed experimentally [5, 6], because time (at temperatures where the material is plastic) is constant, and the dislocation density is generally proportional to the level of stress. However, the total EL2 concentration would not necessarily be quantitatively related to the total dislocation density after growth, as observed experimentally, because only 10–30% of the total EL2 centres are created by climb, and factors such as melt stoichiometry are more important. The distribution of EL2 centres becomes progressively more uniform [6] toward the tail of the crystals corresponding to the trend [9] of the distribution of dislocations. The climb model also explains the increase of the EL2 concentration during short-term annealing as this annealing undoubtedly involves the relief of grown-in stress by climb.

Unlike the EL2 distribution in the as-grown crystal, the EL2 concentration changes most during short-term annealing in regions of low dislocation density. However, these differences are probably related to differences in the structure of the dislocation network during postgrowth cooling and annealing. The observation of polygonized dislocation networks in the as-grown material implies that climb took place during postgrowth cooling. Therefore, EL2 formed while a more-or-less random network of dislocations aligned into cell networks during postgrowth cooling. During the annealing of as-grown material, on the other hand, the dislocation networks are already present. The network is especially tightly interlocked near the edge of the crystal, presenting more resistance to climb for stress relief.

Although the macroscopic EL2 pattern can be explained by climb, the clustering of EL2 to dislocations on the microscopic scale suggests that dislocation gettering is also important. This is because the distribution of dislocations at the solidification front is probably fairly uniform. The formation of dislocation networks occurs during postgrowth cooling. During this process an isolated dislocation could move several hundred microns before reaching a cell boundary. Unless the EL2 formed by the climb process subsequently gettered to the cell boundary, the material would contain tracks of EL2 inside the cell boundaries, which apparently does not occur.

The reversibility of climb-induced EL2 formation, and the independence of the end point EL2 concentration on melt stoichiometry after long-term annealing suggests that the EL2 concentration is self-adjusting. If we assume that EL2 is the antisite As_{Ga}, three self-adjusting mechanisms seem possible. First, the reduction of the EL2 concentration after 'long-term' annealing could be due to arsenic precipitation [15]. Second, it has been reported that sub-100 Å microdefects [15] occur in undoped semi-insulating LEC GaAs grown from non-stoichiometric melts. The nature of these defects has not yet been determined, but they could be loops corresponding to condensed vacancies or interstitials. These microdefects could react with and annihilate EL2 at elevated temperatures. However, microdefects are not observed [16] in material grown from near-stoichiometric melts. Therefore, the climb process itself may leave a residual point defect behind during postgrowth cooling that could recombine later during annealing. For example, Weber *et al.* [14] discuss how negative climb involving arsenic interstitials As_i can produce gallium vacancies V_{Ga}:

$$As_i - dc \rightarrow V_{Ga} \tag{1}$$

EL2 could then form during postgrowth cooling by the reaction of V_{Ga} and arsenic As_{As}:

$$V_{Ga} + As_{As} \rightarrow As_{Ga} + V_{As} \tag{2}$$

Reaction (2) leaves excess V_{As} behind, which could recombine with As_{Ga} during annealing. Both mechanisms are likely to be important in material grown from non-stoichiometric melts. Further understanding of the relationship between point defects (and defect complexes) and the motion of dislocations in GaAs is needed.

Summary and conclusions

We have shown that the thermal history of LEC GaAs crystals can have a measurable effect on the properties of the undoped semi-insulating material by affecting both the macroscopic and microscopic distribution of EL2 centres. The results suggest a two-stage model for the formation of EL2. The average EL2 concentration across LEC crystals is controlled by the melt stoichiometry. During postgrowth cooling, additional EL2 centres are formed by dislocation climb during polygonization of dislocations. This process imparts crystallographic symmetry to the EL2 distribution on the microscale following the pattern of dislocations. Clustering of EL2 centres to dislocations on the microscale probably results from a gettering process.

The results also show that material with exceptional uniformity can be produced by long-term annealing because the EL2 concentration varies asymptotically toward a single value in time. In addition, deviations of the EL2 concentration from this value in as-grown material produced by climb or excess As richness in the melt during growth are reversibly annihilated by a self-compensating recombination process. Further work is currently underway to better understand the relationship between dislocation dynamics and the formation of EL2.

ACKNOWLEDGEMENTS

The authors are grateful to Jeff Dreon and Shawn Johnston for their assistance throughout all aspects of this work.

References

1. Holmes D. E., Chen R. T., Elliott K. R. and Kirkpatrick C. G. (1982). *Appl. Phys. Lett.*, **40,** 46
2. Lagowski J., Gatos H. C., Parsey J. M., Wada K., Kaminska M. and Walukiewicz W. (1982). *Appl. Phys. Lett.*, **40,** 342
3. Elliott K. R., Chen R. T., Greenbaum S. G. and Wagner R. J. *Appl. Phys. Lett.*, in press
4. Holmes D. E., Elliott K. R., Chen R. T. and Kirkpatrick C. G. (1982). In *Semi-Insulating III–V Materials, Evian*, p. 19 (S. Makram-Ebeid and B. Tuck eds.); Shiva Publishing, Nantwich, UK
5. Martin G. M., Jacob G., Goltzene A. and Schwab C. (1980). In Proceedings of the 11th International Conference on Defects and Radiation Effects in Semi-Conductors, Oiso, Tokyo
6. Holmes D. E., Chen R. T., Elliott K. R. and Kirkpatrick C. G. (1983). *Appl. Phys. Lett.*, **43,** 305; Holmes D. E. and Chen R. T. (1984). *J. Appl. Phys.*, **55,** 3588
7. Stirland D. J., Grant I., Brozel M. R. and Ware R. M. Proc. 3rd Oxford Conf. Microscopy of Semiconductor Materials, *Inst. Phys. Conf. Ser.* (to be published); Brozel M. R., Grant I., Ware R. M. and Stirland D. J. (1983). *Appl. Phys. Lett.*, **42,** 610; Skolnick M. S., Brozel M. R., Grant I., Stirland D. J. and Ware R. M. 1983 Electronic Materials Conference, Burlington, Vermont
8. Hobgood H. M., Braggins T. T. and Thomas R. N. ONR/DARPA Contract No. N00014-80-C-0445, Status Report No. 3
9. Chen R. T. and Holmes D. E. (1983). *J. Cryst. Growth*, **61,** 111
10. Rumsby D., Ware R. M., Smith B., Tysberg M., Brozel M. R. and Foulkes E. J. (1983). *Digest GaAs IC Symposium, Phoenix*
11. Skolnick M. S. (1984). This volume, p. 446
12. Holmes D. E., Chen R. T. and Kirkpatrick C. G. 1983 Electronic Materials Conference, Burlington, Vermont
13. Ishida T., Maeda K. and Takeuki S. (1980). *Appl. Phys.*, **21,** 257
14. Weber E. R., Ennen H., Kaufmann U., Windschief J. and Schneider J. (1982). *J. Appl. Phys.*, **53,** 6140
15. Chen R. T., Holmes D. E. and Fairman R. D. (1981). 5th International Conference on Vapor Growth and Epitaxy/5th American Conference on Crystal Growth, Coronado, CA
16. Chen R. T. Unpublished data

The Fine Microstructure of Single Crystal LEC Undoped SI GaAs

P. A. Leigh, I. P. Hall, C. R. Elliott, B. Wakefield and M. H. Lyons

British Telecom Research Labs, Martlesham Heath, Ipswich, Suffolk IP5 7RE, UK

Abstract

Although undoped SI GaAs is required to be very uniform for integrated circuit production, currently available crystals have a high dislocation content. In the work reported here three types of ingots have been studied: standard high pressure LEC (liquid encapsulated Czochralski); annealed high pressure; and standard low pressure ingots. Wafers of standard thickness (300 μm) were assessed by X-ray topography, cathodoluminescence at 10 K and infrared imaging at $\lambda = 1$ μm. All techniques revealed a 300–1000 μm dimension cellular structure for all three samples, indicating that carbon and EL2 were similarly concentrated at the cell walls.

Introduction

Undoped SI (semi-insulating) GaAs, which is to be used for the production of GaAs integrated circuits, is required to be very uniform in its properties. The LEC technique for growing GaAs is currently receiving considerable attention. The technique appears capable of producing large diameter crystals with macroscopically uniform properties at a reasonable cost. The SI properties are believed to be caused by a deep donor, known as EL2, compensating residual carbon [1] acceptors. Therefore the uniformity of the SI properties depends on the uniformity of the distribution of the EL2 and carbon centres. Characterization of LEC GaAs by techniques such as X-ray topography [2] and defect etching [3] has revealed that the material contains a network of dislocations, arranged in a cellular structure. A cellular pattern has also been seen when similar material has been imaged using cathodoluminescence [4] and infrared transmission microscopy [5]. Variations of resistivity across wafers have been characterized and typically a W configuration is observed. It has been shown that heat treatment of the ingot at 950°C for 5 hours after growth results in a uniform resistivity across wafers and better uniformity of EL2 distribution [5]. To investigate the relationship between ingot treatment and physical properties, wafers were taken from three LEC ingots: (a) standard high pressure, (b) ingot-annealed high pressure and (c) standard low pressure. The wafers were assessed by X-ray topography, cathodoluminescence and infrared imaging at $\lambda = 1$ μm. A careful comparison using all techniques has been made for the same wafer from each of the three types of material in order to correlate the distributions observed.

Experimental

X-RAY TOPOGRAPHY

The X-ray topographs were taken using a Hirst topography camera which allows a complete standard wafer to be imaged. The camera design is based on the work of Wallace and Ward [6]. The technique depends upon flexing the wafer in such a way that all X-rays from a distant small source subtend the same angle of incidence at its surface. By this means whole wafer topographs can be obtained from a stationary wafer, with a consequent saving in time compared with the earlier scanning technique. Where the crystal lattice is distorted, for example by a dislocation, the intensity of the reflected X-rays will differ from that of the X-rays reflected by the perfect crystal. Dislocations, therefore, are represented by changes in contrast. In this work the camera was used in the transmission mode with silver $K\alpha$ radiation, and it was possible to penetrate standard 300 μm thickness wafers.

INFRARED ABSORPTION MICROSCOPE

The apparatus for studying infrared absorption of GaAs wafers is not complex. A converging beam of infrared radiation from an optically filtered incandescent source is formed by a pair of plano-convex condenser lenses and used to illuminate the wafer, which is viewed from the opposite side by a closed-circuit TV camera containing a silicon target Vidicon.

CATHODOLUMINESCENCE

The cathodoluminescence (CL) measurements were made on a Cambridge Instruments S180 scanning electron microscope (SEM), fitted with a liquid helium cooled cryostage to enable measurements to be made at any temperature between room temperature and about 4 K [7]. The detector used was a photomultiplier with an S1 photocathode, and the spectra were measured using a 0.5 m Spex monochromator. To obtain monochromatic images the CL was first passed through a Spex 'Minimate' monochromator before detection, the output of the photomultiplier then being used to brightness modulate the SEM display.

Results

X-RAY TOPOGRAPHY

The X-ray topographs displayed the dislocation distributions evident in the three types of material studied. Two main types of feature were observed—an alignment of high densities of dislocations along particular crystal planes (lineage) and a cellular structure where the dislocation bundles have formed cell walls leaving the centres of the cells relatively free of dislocations. The topograph from a wafer of the standard LEC high pressure grown ingot is shown in Figure 1a; clear lineage was observed in ⟨110⟩ directions in the form of a cross with branching occurring into other crystal directions. The

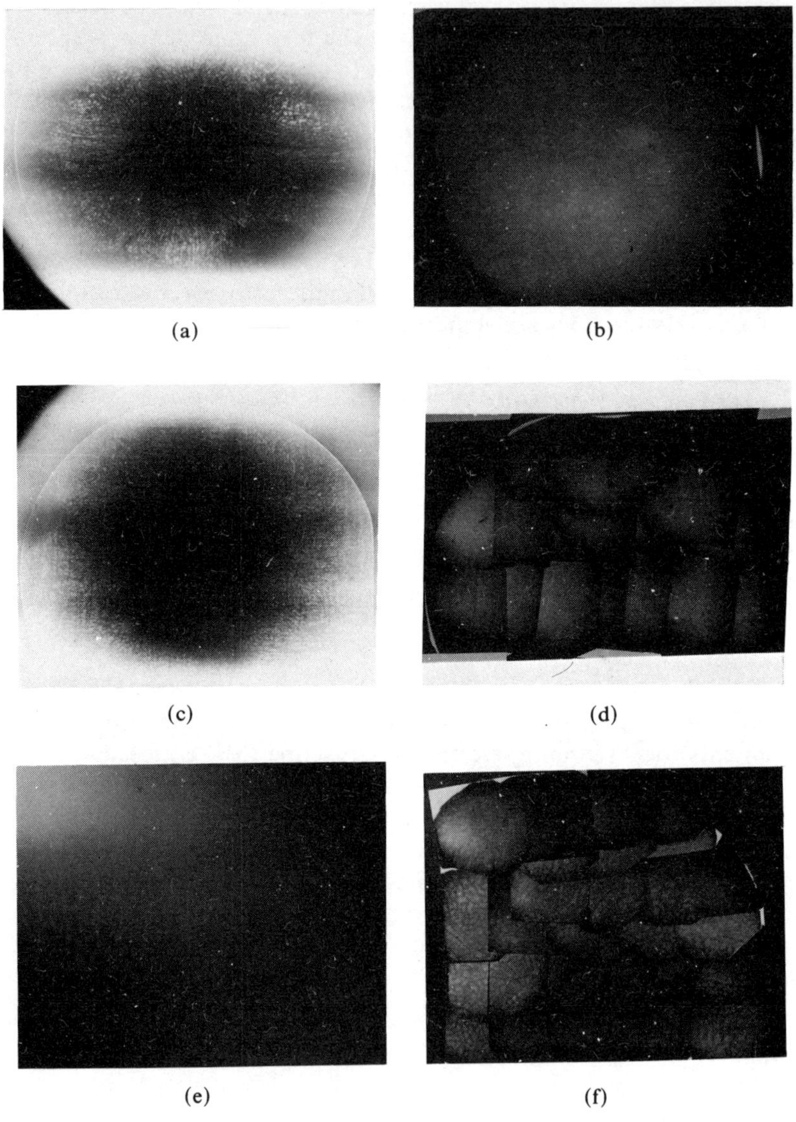

Figure 1 Comparison of the X-ray topograph and infrared absorption image for each of the three types of wafers studied: high pressure grown ingot (a) X-ray topograph and (b) infrared absorption at 1 μm; high pressure grown, annealed ingot (c) X-ray topograph and (d) infrared absorption at 1 μm (collage); low pressure grown ingot (e) X-ray topograph and (f) infrared absorption at 1 μm (collage).

cellular structure is well developed particularly in the centre region of the wafer with clusters of large and small cells. The cell structure becomes coarser and more fibrous towards the periphery of the wafer, intermingling with the branching of the lineage structure.

The standard material which had subsequently been annealed showed that considerable refinement of the dislocation structure had occurred. The strong lineage structure is hardly apparent although slight vestiges of its presence are apparent. The cellular pattern is considerably finer and movement of dislocations has resulted in alignment of the cells to form a pattern of 'woven' appearance in the topograph Figure 1c. The cell size is more uniform although it does coarsen from wafer centre to periphery, roughly doubling in area at the wafer edge. The very centre of the wafer retained a fine cellular structure with less alignment.

The low pressure grown ingot topograph is shown in Figure 1e and reveals a cellular structure similar to that of the standard high pressure ingot. However the direction of the lineage structure tended to ⟨100⟩ crystal directions rather than ⟨110⟩ as in the standard high pressure material.

INFRARED ABSORPTION AT 1 μM

The infrared absorption images are shown in Figures 1b, 1d and 1f (compared to the topograph of the identical area in Figures 1a, 1c and 1e). The imaging of infrared absorption at 1 μm is attributed to EL2. The EL2 distributions were identical to those observed for the dislocations imaged in the topographs for both the cellular and lineage structures. This is obvious for the standard high pressure ingot and low pressure grown ingot. The heat treated ingot has the finer structure and the correlation is less immediately obvious. However major features can be identified and careful study shows the distributions to be identical.

CATHODOLUMINESCENCE

Small samples only could be studied by CL (cathodoluminescence) at temperatures near 4 K in the SEM. Therefore areas showing typical structure were selected from the wafers of Figure 1 and samples 5×5 or 6×5 mm removed. To map out the variations in cathodoluminescence intensity the samples were imaged in the SEM as described previously. Cathodoluminescence micrographs were recorded using the light from the 1.494 eV peak emissions due to the presence of residual carbon acceptors. The results for these samples are shown in Figure 2. The comparison is made between the X-ray topograph, infrared absorption at 1 μm and the cathodoluminescence image.

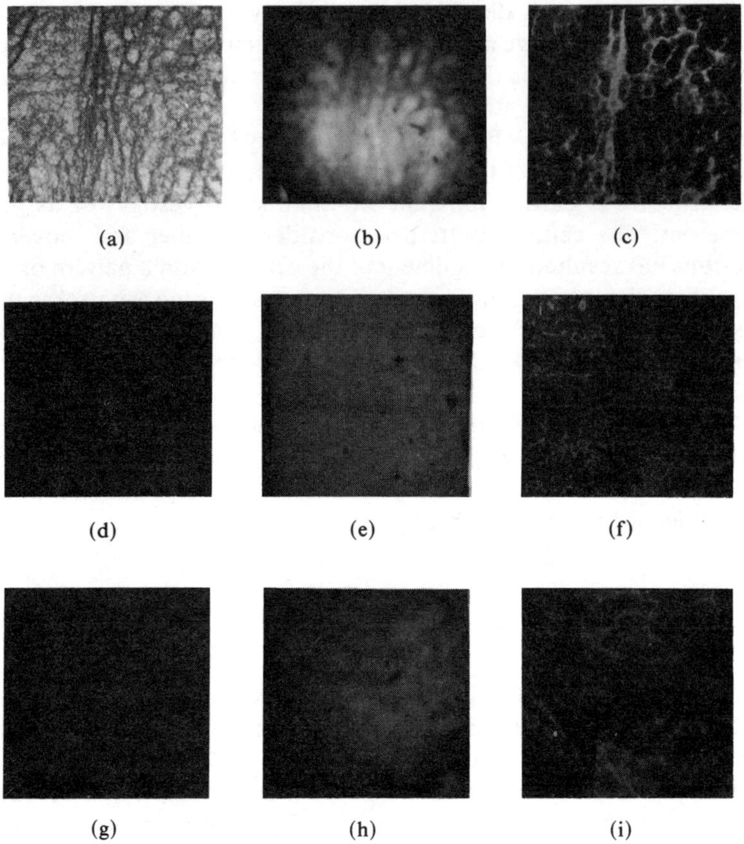

Figure 2 Comparison of the X-ray topograph, infrared absorption image (at 1 μm) and SEM cathodoluminescence (near 4 K) of carbon emission for the 5 mm chips taken from each wafer: high pressure grown ingot (a) X-ray topograph, (b) infrared and (c) SEM cathodoluminescence; high pressure grown, annealed ingot (d) X-ray topograph, (e) infrared absorption and (f) SEM cathodoluminescence; low pressure grown ingot (g) X-ray topograph, (h) infrared absorption and (i) SEM cathodoluminescence.

For each of the materials studied the correlation between the distributions is clear. Where the dislocation content is highest, i.e. in the cell walls or in the lineage structures, both the EL2 and carbon acceptors are concentrated. Concentrations of these two are relatively insignificant in the centre body region of the cells. The heat treated ingot shows the finer structure with alignment of the cells. Here some finer redistribution of the carbon is apparent away from the cell walls, Figure 2i.

An example of a cathodoluminescence spectrum recorded at a temperature near 4 K is shown in Figure 3a. It consists of two peaks; one at 1.514 eV associated with shallow donors or bound excitons, and one at 1.494 eV due

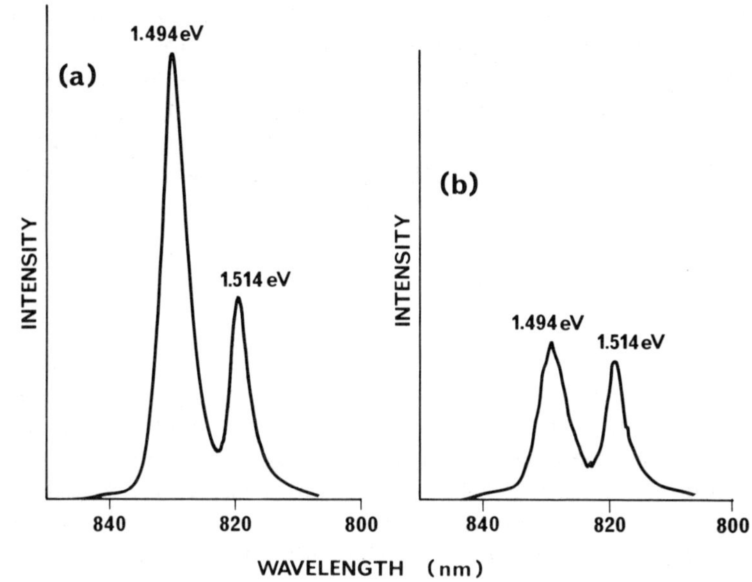

Figure 3 The cathodoluminescence spectra of the material comparing a bright region (a) to a cell interior (b).

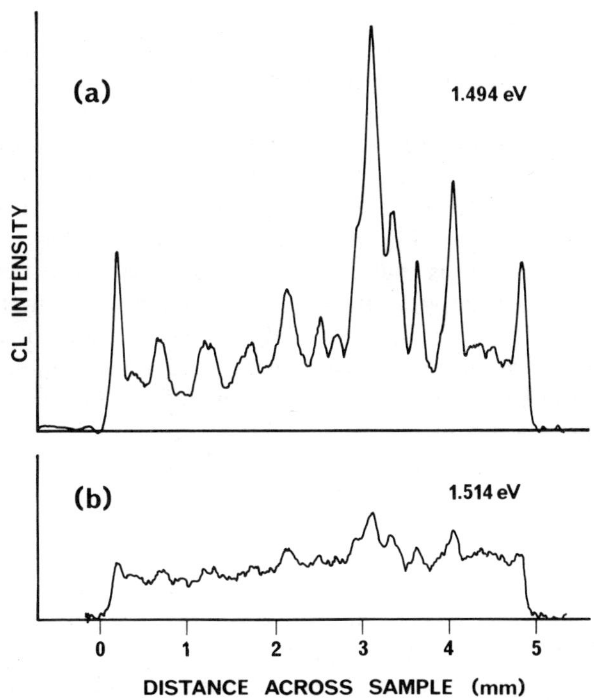

Figure 4 A scan across a 5 mm chip of the cathodoluminescence emission for the 1.494 eV carbon emission and the shallow donor 1.514 eV emission.

to the presence of the carbon acceptors. Figures 4a and 4b show plots of the CL intensity for these two peaks measured across a sample. For this experiment the sample was slowly scanned beneath an electron beam with a diameter of approximately 100 μm. The spectra measured at a bright area and at an adjacent darker area are plotted in Figures 3a and 3b. The results shown in these two figures clearly demonstrate that relative to the 1.514 eV emission, the emission due to the carbon impurities varies considerably more across the wafer.

Discussion

In the work presented here it is clear that all three ingots have a high dislocation content. The dislocations are arranged in a cellular system presumably to lower the free energy. In addition, considerable lineage is present which may lie along (110) or (100) planes, with considerable branching into other crystal directions. The heat treatment of such an ingot at high temperature does not *per se* reduce the dislocation content. However it is clear that the stored strain energy within the ingot is sufficient to cause dislocation movement and rearrangement of the cell structure. The centre region of the ingot retains the original cellular configuration, presumably because the thermal stresses are lower in this region. In addition, the stress fields around the dislocations result in both EL2 and carbon residing preferentially in the areas of the cell walls (the atmosphere effect). This is still apparent in the heat treated ingot—both EL2 and carbon still associate with the highly dislocated cell walls.

It is somewhat surprising that the CL intensity is greatest where the dislocation density is high, as dislocations are generally considered to be strong non-radiative centres. Nevertheless it is in agreement with the previously reported CL and PL (photoluminescence) studies [2, 4, 8]. The explanation given there is that dislocation cores can attract non-radiative impurities or non-radiative point defects with the result that the regions surrounding the dislocations, for a radius of several tens of micrometers, become denuded of non-radiative centres, and so will have a luminous efficiency greater than that in areas well away from dislocations. Regions where the dislocation density is high will therefore have higher than average luminescence. Only at the dislocation itself will the luminescence be reduced, as has been observed by Chin *et al.* [4] and Kamejima *et al.* [2], who found arrays of dark spots along the cell boundaries, which they attributed to individual dislocations.

If the gettering of non-radiative recombination centres were the only effect leading to an increase in CL intensity around dislocations, then it would apply equally to all parts of the spectrum. The relative intensities of the various emission peaks would not alter. But this is not what is observed. In this study, as can be seen in Figures 3 and 4, and from the observation that the CL cell pattern is more pronounced with the 1.494 eV emission than with the 1.514 eV emission, it is obvious that both the absolute intensity of the carbon acceptor peak (1.494 eV) and its intensity relative to the 1.514 eV peak, is greater in the vicinity of the dislocation network than in the dislocation-free regions. This suggests that the carbon impurity concentration is enhanced near to the dislocation network.

The repercussions of these results concerning the uniformity of the material for GaAs integrated circuit applications need to be considered. There is strong evidence to suspect that the electrical properties of the highly dislocated regions containing the majority of the EL2 and carbon impurities will be significantly different from the cell body regions of low dislocation content. Previous models generally consider a uniform distribution of these defects while it is clear from this work that this is not so. In conclusion, more investigation of the electrical properties of the cell walls compared with the cell body regions is called for. It will also be necessary to investigate LEC ingots of low dislocation content with similar stoichiometry and impurity content when these become available.

References

1. Holmes D. E., Elliott K. R., Chen R. T. and Kirkpatrick C. G. (1982). In *Semi-Insulating III–V Materials, Evian*, p. 19 (S. Makram-Ebeid and B. Tuck, eds.); Shiva Publishing, Nantwich, UK
2. Kamejima T., Shimura F., Matsumoto Y., Watanabe H. and Mitsui J. (1982). *Jap. J. Appl. Phys.*, **21**, L721
3. Clark S. and Stirland D. J. (1981). *Inst. Phys. Conf. Ser.*, **60**, 339
4. Chin A. K., Von Neida A. R. and Caruso R. (1982). *J. Electrochem. Soc.*, **129**, 2387
5. Brozel M. R., Grant I., Ware R. M. and Stirland D. J. (1983). *Appl. Phys. Lett.*, **42**, 610
6. Wallace C. A. and Ward R. C. C. (1975). *J. Appl. Cryst.*, **8**, 281
7. Wakefield B. (1983). *Inst. Phys. Conf. Ser.*, **67**, 315
8. Kitahara K., Ozeki M. and Shibatomi A. (1983). *Appl. Phys. Lett.*, **42**, 188

Oxygen-Related Midgap Level in GaAs

J. Lagowski, D. G. Lin, T. Aoyama* and H. C. Gatos

Massachusetts Institute of Technology, Cambridge, Massachusetts 02139, USA

Abstract

We established that in Bridgman as well as in LEC (liquid encapsulated Czochralski) grown GaAs there is a deep level related to oxygen, EL0, with essentially the same activation energy (825 ± 4 meV) as EL2; however, the electron capture cross section of EL0 is four times greater than that of EL2. In conjunction with measurements on 'oxygen-free crystals' containing only EL2 we have formulated practical procedures for the evaluation of the [EL2]/[EL0] concentration ratio from DLTS (deep level transient spectroscopy) measurements. The IR absorption spectra of GaAs containing only EL2 and EL2 + EL0 were found to be essentially the same indicating the need to re-examine the absorption characteristics of EL2. The present results answer the question of long standing as to whether or not oxygen is responsible for a deep level and confirm recent findings that EL2 is not the only midgap level in melt-grown GaAs.

Introduction

Deep energy levels in melt-grown GaAs have received a great deal of attention as compensating centres responsible for the high resistivity material required for IC applications. The complex electrical and optical properties of this material have been commonly interpreted in terms of 'the one deep level model' with only one midgap donor referred to as EL2. However, most recent discoveries of midgap levels other than EL2 [1], and especially of the oxygen-related level EL0, [2–4] have demonstrated that the simplistic one midgap level (EL2) concept of melt-grown GaAs cannot be valid. These findings have raised once again the question of long standing: is oxygen involved in midgap energy levels in melt-grown GaAs? We have addressed this question in our extensive experimental studies in the last three years. Our results [5] confirmed the earlier findings [6] of the oxygen-induced suppression of Si contamination. We have also identified indirect effects of oxygen doping [5] such as increase of the electron trap concentration, a decrease of the excess carrier lifetime and the suppression of the 1.2 eV luminescence band. These effects resulted from gettering the silicon impurity from the GaAs melt rather than from direct oxygen involvement in electrically active centres.

In our most recent study [3] we have succeeded in the identification of an oxygen-related midgap level, EL0, in Bridgman-grown GaAs which appears to be very similar to the midgap level in oxygen implanted GaAs [2] and in

* On leave from Hitachi Research Laboratories, Ibaraki, Japan.

LEC GaAs [1]. In the present communication we discuss the properties of EL0 and compare them with those of the native defect level EL2. We also present practical means for distinguishing between the two midgap levels.

Experimental

'Oxygen-doped' GaAs crystals were grown in a horizontal Bridgman-type (HB) apparatus utilizing quartz or PBN boats. Oxygen doping was realized by adding Ga_2O_3 to the GaAs charge material at a weight fraction ratio $W_{Ga_2O_3}/W_{GaAs}$ varying from zero ('oxygen free' crystals) up to 10^{-3}, or by filling the growth ampoule with oxygen gas to a pressure up to 300 Torr (at room temperature). The upper limits of added oxygen exceeded by about one order of magnitude the doping range utilized in the study of Ref. 7 which failed to recognize any midgap levels other than the EL2.

The oxygen concentration in our 'oxygen free crystals' was below 10^{16} cm^{-3}, and in the heavily doped ones it was $1–4 \times 10^{16}$ cm^{-3}, as determined by SIMS (secondary ion mass spectrometry) analysis. Our heavily oxygen-doped crystals remained conducting (electron concentration ranging from $0.5–2 \times 10^{16}$ cm^{-3}) due to the presence of sulphur impurity in the starting material.

Deep level characterization was carried out using transient capacitance measurements, photocapacitance transients (induced by monochromatic light) and optical absorption measurements. The features of our transient capacitance system included: (a) DLTS with variable duration filling pulses; (b) direct determination of the capacitance relaxation, C(t), by recording with a signal averaging technique; the various measurements could be electronically stored and directly compared; (c) precise temperature control and monitoring with an absolute accuracy and reproducibility better than 0.1°C.

Results and discussion

MIDGAP LEVEL PARAMETERS

The primary difficulty in studying EL0 stems from the striking similarity between its parameters and those of EL2. Accordingly, when both EL0 and EL2 are simultaneously present their unique separation is hardly possible without reference to standard material containing only one midgap level.

In our study the crystals grown without oxygen doping and containing only EL2 were used as reference. In Figure 1 we present Schottky barrier capacitance transients corresponding to electron emission from midgap levels in oxygen-free and in heavily oxygen-doped GaAs. The capacitance relaxation, C(t), in oxygen-free material (Figure 1a) is perfectly exponential consistent with the standard emission equation for a single deep level $\delta C = \delta C^{\circ} e^{-te_1}$, where t is time, δC° is the magnitude of the capacitance change, and e_1 is the emission rate. The oxygen-doped crystal (Figure 1b) exhibits a

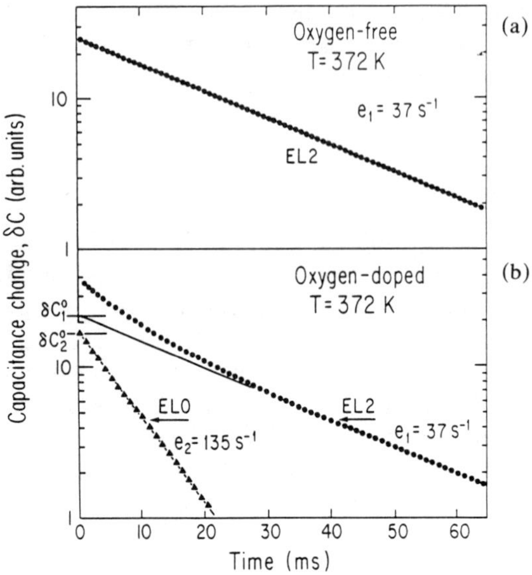

Figure 1 Capacitance transient corresponding to electron emission (a) from only one level EL2, (b) from two levels EL2 and EL0 simultaneously present.

two-component exponential transient: $\delta C = \delta C_1^\circ e^{-te_1} + \delta C_2^\circ e^{-te_2}$, with the slow component corresponding to EL2 emission as measured in oxygen-free material. By subtracting the EL2 component we separated the transient corresponding to EL0 and thus we were able to determine the EL0 emission constant e_2 and the EL0 concentration (from δC_2°). We have carried out such analysis of capacitance transients recorded over a wide temperature range 270–450 K. The results are summarized in Figure 2 in the form of thermal

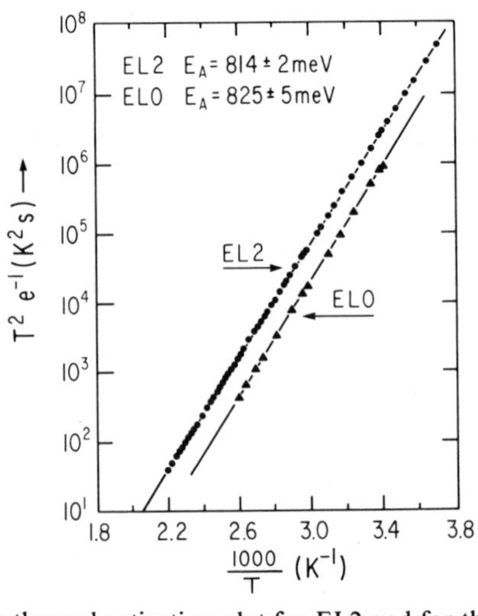

Figure 2 Emission rate thermal activation plot for EL2 and for the oxygen-related level EL0.

activation plots T^2e^{-1} versus $10^3/T$ for EL2 and EL0 levels. Least squares analysis yields for EL2 an activation energy $E_A = 814 \pm 2$ meV and an electron capture cross section $\sigma_{na} = (1.2 \pm 0.1) \times 10^{-13}$ cm^2; the corresponding values for EL0 are $E_A = 825 \pm 5$ meV and, $\sigma_{na} = (4.8 \pm 0.6) \times 10^{-13}$ cm^2. Empirical formulas for emission rates are:

$$\text{EL2: } e_1^{-1} = \frac{3.53 \times 10^{-8}}{T^2} \exp\left(\frac{9450}{T}\right) \text{ sec}^{-1} \qquad (1)$$

$$\text{EL0: } e_2^{-1} = \frac{6.84 \times 10^{-9}}{T^2} \exp\left(\frac{9580}{T}\right) \text{ sec}^{-1} \qquad (2)$$

where T is the absolute temperature in K.

We consider the above EL2 parameters more accurate than any previously published ones due to a much greater range of experimental data [8]. We propose that these values be used as standards for midgap level identification in GaAs.

It is not clear at present whether oxygen is directly and solely responsible for the midgap EL0 level. Theoretical study [9] and experimental data extrapolated from $GaAs_{1-x}P_x$ [10] indicated that an energy level of an oxygen atom on the arsenic site should be located in GaAs about 0.79 eV below the conduction band, i.e. very close to EL0 energy.

EL0 CONCENTRATION

As pointed out above, the activation energies of the electron emission rates are very similar for EL0 and EL2. Accordingly, both individual levels exhibit DLTS peaks of identical shape. However, the EL0 peak temperature T_{EL0} is shifted to a temperature lower than that of the EL2 peak, T_{EL2}, due to the larger value of the electron capture cross-section ($\sigma_{na}(EL0) \simeq 4\sigma_{na}(EL2)$). Using the emission rate formulas (1) and (2), the DLTS peaks can be exactly calculated for any given boxcar gate setting t_1/t_2 (or an emission rate window $(t_2 - t_1)/\ln(t_2/t_1)$). When EL0 and EL2 are simultaneously present at concentrations N_{EL0} and N_{EL2}, respectively, the resulting DLTS peak will occur at a temperature, T_{eff}, dependent on the concentration ratio of the two levels. As shown in Figure 3, the involvement of two levels in an overall DLTS peak is not by any means evident. The half width of the overall peak is only marginally larger than the half width of the component peaks corresponding to the individual levels. It is thus clear that the involvement of two levels of similar energies can be recognized more readily from direct capacitance transient measurements C(t) (see Figure 1) than from DLTS measurements. Nevertheless, when the sample temperature is precisely monitored, the ratio N_{EL0}/N_{EL2} can be accurately determined by deconvolution of a DLTS peak into components with emission rates described by equations (1) and (2) (as done in Figure 3).

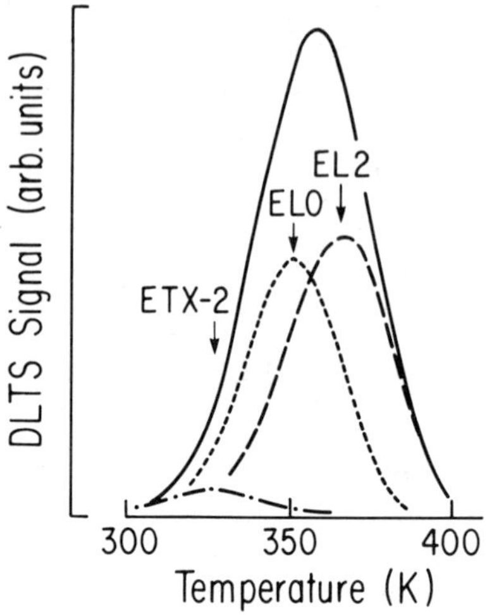

Figure 3 DLTS spectrum (rate window 25.6 s^{-1}) of heavily oxygen-doped GaAs deconvoluted using experimentally determined emission rates of EL2 and EL0 versus temperature. (Compare with Ref. 1.)

The relative concentration of midgap levels can be quickly evaluated from the position of an overall DLTS peak using the calculated values given in Figure 4. It is seen in that figure that increase of EL0 concentration gradually shifts T_{eff} from a temperature characteristic of the EL2 peak down to a temperature of the EL0 peak. The shift is linear with $X \equiv N_{EL0}/(N_{EL0} + N_{EL2})$ up to $X \simeq 0.6$. Furthermore, the shift is very similar for all rate windows. Thus, X can be very well approximated by a simple relation:

$$X = 0.05 \times \Delta T \tag{3}$$

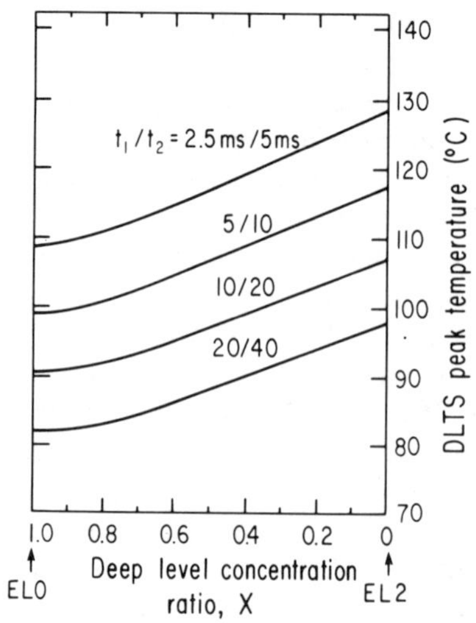

Figure 4 DLTS peak temperature versus deep level concentration ratio.

where ΔT (°C) is a difference between the actually measured peak position and the position corresponding to EL2 only, e.g. measured in oxygen-free GaAs. (Both measurements should be made with the same emission rate window.)

EL0 IDENTIFICATION USING SHORT FILLING PULSES

The above procedure assumes the identical initial occupation of EL0 and EL2 within a depletion layer. In our recent work on EL0 identification [3] we have discussed a very effective means for enhancement of the relative contribution of EL0 to the overall transient capacitance signal based on utilization of short duration filling pulses, which preferentially fill EL0 and not EL2. As shown in Figure 5a the magnitude of the capacitance transient corresponding to EL2 decreases when a short duration filling pulse is used. A corresponding decrease of the EL2 peak height is seen in the DLTS spectrum of Figure 5b. However, there is no shift in a peak position and no change in emission rate is observed in Figures 5b and 5a, respectively. In oxygen-doped GaAs (Figures 5c and 5d) a decrease of the filling pulse duration produces more significant changes. The increased contribution to EL0 in respect to that of EL2 (the effect is due to the larger capture

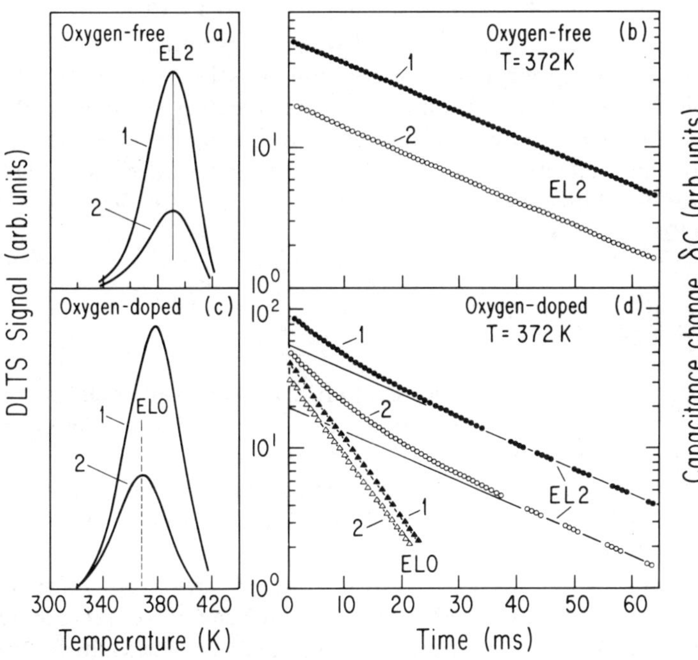

Figure 5 DLTS spectra and capacitance transient: 1—long duration; 2—short duration filling pulses.

cross-section of EL0 than that of EL2) is clearly seen in capacitance transients of Figure 5c. A corresponding shift of an overall DLTS peak toward lower temperature is visible in Figure 5d.

Using the procedures discussed above, we have identified EL0 not only in the heavily oxygen-doped HB GaAs grown in our laboratory, but also in commercial HB GaAs (typically $N_{EL0}/N_{EL2} \leqslant 0.3$) and in LEC grown GaAs ($N_{EL0}/N_{EL2} \sim 0.25$). The presence of EL0 in these crystals is not surprising, since they were grown from melts containing oxides (usually arsenic oxide added in HB melt) or encapsulated in B_2O_3 (LEC).

APPARENT EFFECTS OF ELECTRIC FIELD

Preferential filling of EL0 (as compared to that of EL2) is also observed for long duration filling pulses, when the magnitude of the employed pulses is smaller than the barrier height (i.e. insufficient to complete the filling of all levels within the depletion region). DLTS results obtained under such conditions could be easily misinterpreted as anomalous phenomena such as: an increase of the emission in low electric fields; non-uniform distribution of levels beneath the surface; or a spectrum of midgap levels rather than only two levels.

Interference in the determination of the EL0/EL2 ratio can also result from apparent effects of strong electric fields on the electron emission. We have investigated the effects of strong fields (up to 4×10^5 V cm^{-1}) using the double pulse technique of Ref. 11, and we have established that both levels exhibit a small field enhancement of electron emission, well within the range of the Poole-Frenkel effect. In contrast, we observed very pronounced electric field effects, however, only with diodes having large reverse bias currents. We refer to these effects as 'apparent' field effects because they originate in Schottky diodes with poor characteristics rather than in the nature of EL2 and EL0 [12].

OPTICAL ABSORPTION

It is evident that the one midgap level (EL2) concept of melt-grown GaAs cannot be valid and that the midgap level properties of GaAs must be re-examined in view of the presence of more than one level.

Re-examination of the midgap level properties is especially urgent with respect to the infrared absorption [13] commonly utilized for the determination of the EL2 concentration in semi-insulating GaAs. In Figure 6 we present optical absorption and photoionization capture cross-section spectra for oxygen-free GaAs containing EL2 only and for heavily oxygen-doped GaAs containing EL0 and EL2 in approximately equal concentrations. It is seen that the results for both crystals are almost the same. The most plausible explanation of these results is that the optical properties of EL0 and of EL2

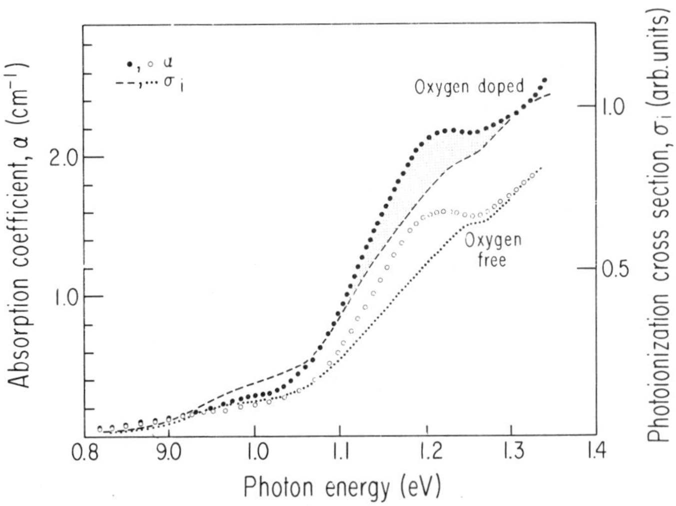

Figure 6 Absorption coefficient α; and photoionization cross-section σ_i (determined from capacitance transient) for oxygen doped GaAs containing EL0 and EL2 and for oxygen free GaAs containing only EL2.

are similar. Indeed both levels exhibit similar spectra of electron photoionization capture cross-section. What is even more surprising is that both levels exhibit quite similar intra-centre transitions in the region $1.05 < h\nu < 1.3$ eV manifested by the difference between the optical absorption and the photoionization cross-section (shaded areas in Figure 6). The similarities between the EL2 and the (EL0 + EL2) properties could be accounted for if the EL0 contribution to optical absorption is negligible in comparison with that of EL2. Perhaps very precise measurements on crystals containing different EL0/EL2 concentration ratios (currently in progress) should uniquely resolve this issue. In the meanwhile, we must conclude that infrared absorption measurements in melt-grown GaAs should be treated as a measure of midgap levels in general rather than as a unique measure of the EL2 only.

ACKNOWLEDGEMENTS

We would like to acknowledge financial support from the National Aeronautics and Space Administration.

References

1. Taniguchi M. and Ikoma T. (1983). *J. Appl. Phys.*, **54,** 6448
2. Taniguchi M. and Ikoma T. (1982). In *Semi-Insulating III–V Materials, Evian,* p. 283 (S. Makram-Ebeid and B. Tuck, eds.); Shiva Publishing, Nantwich, UK
3. Lagowski J., Lin D. G., Aoyama T. and Gatos H. C. (1984). *Appl. Phys. Lett.*, **44,** 336

4. Yu P. W. and Walters D. C. (1982). *Appl. Phys. Lett.*, **41,** 863
5. Kaminska M., Lagowski J., Parsey J., Wada K. and Gatos H. C. (1981). *Inst. Phys. Conf. Ser.*, **63,** 197
6. Cochran C. N. and Foster L. M. (1962). *J. Electrochem. Soc.*, **109,** 149; Woods J. F. and Ainsle N. G. (1963). *J. Appl. Phys.*, **34,** 1469
7. Martin G. M., Jacob G., Hallais J. P., Grainger F., Roberts J. A., Clegg J. B., Blood P. and Poiblaud G. (1982). *J. Phys.*, **C15,** 1841
8. Martin G. M., Mittonneau A. and Mircea A. (1977). *Electron. Lett.*, **13,** 191
9. Jaros M. (1980). *Adv. Phys.*, **29,** 409
10. Wolford D. J., Modesti S. and Streetman B. G. (1983). *Inst. Phys. Conf. Ser.*, **65,** 501
11. Makram-Ebeid S. (1980). *Appl. Phys. Lett.*, **37,** 464; Makram-Ebeid S. and Lannoo M. (1982). *Phys. Rev. B*, **25,** 6406
12. Lagowski J., Lin D. G., Gatos H. C., Parsey, Jr J. M. and Kaminska M. Submitted to *Appl. Phys. Lett.* (1984)
13. Martin G. M. (1981). *Appl. Phys. Lett.*, **39,** 747

Photoquenching Effect at the Midgap Electron Traps ('EL2 Family') in Different GaAs Crystals

M. Taniguchi*, Y. Mochizuki and T. Ikoma

Institute of Industrial Science, University of Tokyo, Roppongi, Minato-ku, Tokyo, Japan

Abstract

Transitions between normal and metastable states at various deep levels which belong to the 'EL2 family' have been characterized. As a consequence, we have found that the midgap levels which form the 'EL2 family' have various electron-lattice couplings and that their origins may be arsenic aggregates which have amorphous-like atomic structures.

Introduction

Since midgap electron traps act as main compensators in SI GaAs, understanding of these traps is of importance for achieving high speed GaAs integrated circuits. These traps are called 'EL2' after Martin *et al.* [1], but we have shown that these traps cannot be attributed to a single level [2]. We found [3] that there existed in LEC (liquid encapsulated Czochralski) GaAs a variety of midgap levels whose capture cross-sections were scattered in a wafer and were unstable for a thermal annealing. Furthermore, we reported [4] that all the levels showed the photoquenching effect [5] but their photoquenching efficiency spectra were clearly different. Therefore, we thought that these levels formed a group of similar deep states and introduced the concept of the 'EL2 family'.

In this paper, we discuss transitions between normal and metastable states at various deep levels of the 'EL2 family' in various GaAs crystals. The results indicate that the 'EL2 family' form a group of deep states which have different electron-lattice coupling strengths.

Samples and measurement method

Samples used in this study are undoped LEC, HB (horizontal Bridgman), VPE (vapour phase epitaxial), and O-implanted LPE (liquid phase epitaxial) GaAs. All these samples contain a midgap level as a main trap as described in Ref. 4. These traps are labelled as ETX-1 and ETX-2, found in the tail and front sections of LEC GaAs, respectively, 'EL2' in HB and VPE GaAs, and EL2-0 in O-implanted LPE GaAs. From thermal emission rates and

* Nippon Mining Co. Ltd., Toda-shi, Saitama, Japan.

photoquenching effects, these traps have been judged to belong to the 'EL2 family' [4].

In the present study, photocapacitance transients of Schottky barrier diodes were measured with a Boonton 72B capacitance meter under illumination of monochromatic light from the back of a sample.

When a photocapacitance change is large, a nonlinear relationship between space charge density and junction capacitance becomes non-negligible. In this case, we used a constant capacitance method where bias-voltage was changed and recorded by keeping the junction capacitance constant. A change of bias-voltage, in this case, is proportional to a change of space charge density. This constant capacitance technique also eliminates the edge effect that becomes important when a recovery from the metastable state to the normal state is fast in a neutral region.

Photocapacitance transients due to photoquenching

A photocapacitance transient was measured as follows: first, all traps in the depletion region of a Schottky barrier diode were filled with electrons and the diode was backward biased at −1 V. Then, the sample was illuminated with light having an energy of 1.17 eV and a photon flux of 2.1×10^{15} cm^{-2} sec^{-1} and the capacitance was measured. For a normal trap, the capacitance increases and saturates. However, for all the midgap levels measured here the capacitance first increased, passed a maximum and then decreased. This decrease of capacitance is due to the photoquenching effect [5], which is regarded as the fingerprint of the EL2 family. When this decrease of capacitance is plotted as a function of time for various midgap traps, we obtain Figure 1.

It should be noted first that photoquenching efficiencies defined by capacitance decreasing rates [6] as well as their spectra (which are not shown here but are reported in Ref. 4) are different at different traps. Second, *non-exponential* variations are remarkable, in particular for EL2-0 and ETX-1.

To study the non-exponential transients in more detail, the change of space charge density in a depletion region was measured for ETX-1 by the constant capacitance method. Figure 2a shows the result, where a bias-voltage change, $\Delta V(t)$, is plotted as a function of time for different photon energies. Here, a change of space charge density, $\Delta\rho(t)$, is given by:

$$\Delta\rho(t) = \frac{\Delta V(t)}{V_{bi} - V_{appl} - kT/q} N_d$$

where V_{bi}, V_{appl} and N_d represent built-in potential, initial reverse bias voltage and shallow donor density, respectively. It is clear that the transient from the normal state to the metastable state is non-exponential, indicating existence of multiple-metastable states [4].

To interpret the non-exponential transient, we assume that the photoquenching efficiency, $\sigma_p^o \sigma_*^o \phi/(\sigma_n^o + \sigma_p^o)$ [6] should have a distribution (here,

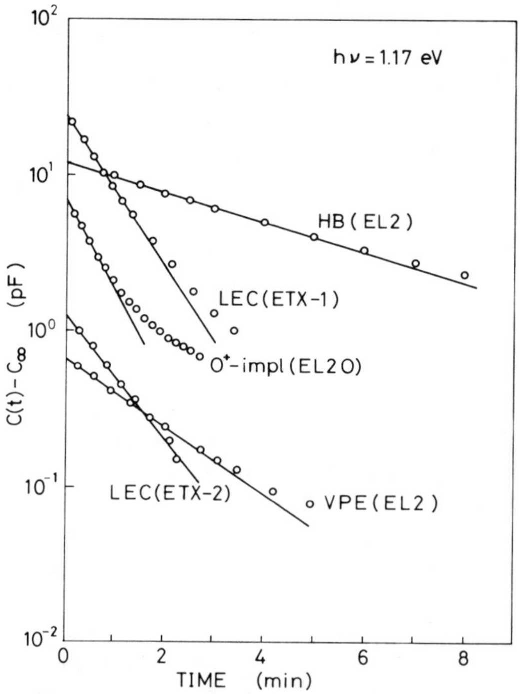

Figure 1 Semilogarithmic plots of $C(t)-C_{\infty}$ versus time for various midgap electron traps (ETX-1, ETX-2, 'EL2' in HB and VPE, and EL2-0).

σ_n^o and σ_p^o are photoionization cross-sections for electrons and holes, respectively, σ_*^o is the cross-section for transition from the normal state to the metastable state and ϕ is the photon flux). For the simplest case, a Gaussian distribution was assumed and a least squares fitting [7] was carried out.

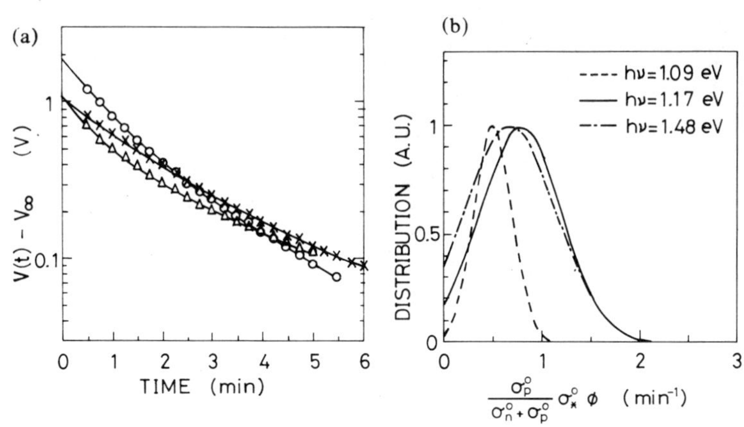

Figure 2 (a) Photoquenching transients measured by constant capacitance technique for ETX-1 in LEC-1(T). Excitation photon energies were 1.48 (Δ), 1.17 (○) and 1.09 eV (×). The solid lines are the calculated curves (see text). (b) Distributions of photoquenching efficiencies used to reproduce the solid lines in (a), which gave the best fit to the observed transients.

Distributions in the photoquenching efficiency thus obtained are shown in Figure 2b. The mean values are 0.50, 0.79 and 0.68 sec^{-1} for 1.09, 1.17 and 1.48 eV light energies, respectively, and the standard deviations are 0.19, 0.41 and 0.68 sec^{-1} for each photon energy. The fitting is excellent as shown by the solid lines in Figure 2a but the non-zero distribution in the negative region of the efficiency observed for the 1.17 and 1.48 eV energies in Figure 2b has no physical meaning. This suggests that *one* Gaussian distribution might not be appropriate for the present case.

Recovery from metastable state to normal state

The metastable state can be recovered to the normal state either by heating the sample above 120 K or by injecting free electrons [6, 8]. These recovery processes were studied for ETX-1. The measurement was carried out as follows. First, all traps were filled with electrons and the sample was illuminated with monochromatic light. The capacitance increased first and then decreased to reach a steady state. The magnitude of capacitance decrease, which is defined as the photoquenched capacitance, C_{PQ}, was measured. At the steady state, after illumination for a long time, all the 'EL2' levels were in the metastable state. Then, the diode was kept in the dark for a period, t_r, at a temperature T and again illuminated. After this, the amount of photoquenched capacitance was measured. The photoquenched capacitance measured at the second stage was smaller than at the first one because of a partial recovery during the period t_r. The measurement was repeated by changing t_r and the photoquenched capacitance, normalized to the initial value, was plotted against t_r. This curve gives a thermal recovery rate from the metastable state to the normal state.

Typical recovery curves are shown in Figure 3. In these transients, the non-exponentiality is clear. To interpret this non-exponentiality, we assumed a Gaussian distribution for the recovery efficiency and deduced effective

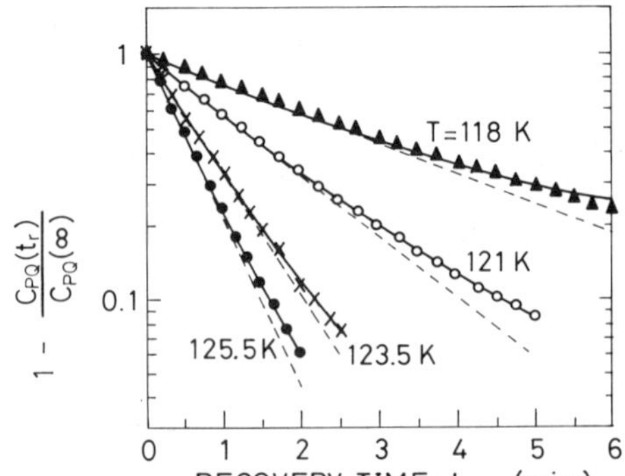

Figure 3 Transients of thermal recovery process measured for ETX-1 at four temperatures. The solid lines are the calculated curves (see text).

energy barrier distributions, by keeping the pre-exponential factor constant (4.8×10^{10} sec^{-1} in this case). The 'average' mean value of the barrier is 0.305 eV and the 'average' half width is 10 meV.

Vincent *et al.* [6], and Mitonneau and Mircea [8] reported the recovery rate to be $2 \times 10^{11} \exp(-0.30\ q/kT)\ sec^{-1}$ and $8 \times 10^{11} \exp(-0.34\ q/kT)\ sec^{-1}$, respectively. The present recovery efficiency is between these values.

The electron-injection recovery rate was also measured by using a method similar to that of Vincent *et al.* [6]. Here, we observed large non-exponentialities in the recovery transients. It is interesting to note that for EL2-0, no recovery was observed by injecting free electrons at 77 K. This also supports the existence of various kinds of metastable states for the 'EL2 family'.

Transition between bands and normal state

The photoquenching efficiency defined in the previous section includes photoionization cross-sections that correspond to the transitions from the valence band to the normal state, σ_p^o, and from the normal state to the conduction band, σ_n^o. Therefore, to elucidate effects of these transitions on the photoquenching efficiency, we measured the photoionization process for ETX-1 by avoiding effects of photoquenching and influences from other existing traps.

The measurement procedure was as follows. First, a sample was cooled down to 77 K and all the levels were made to be in the metastable state by

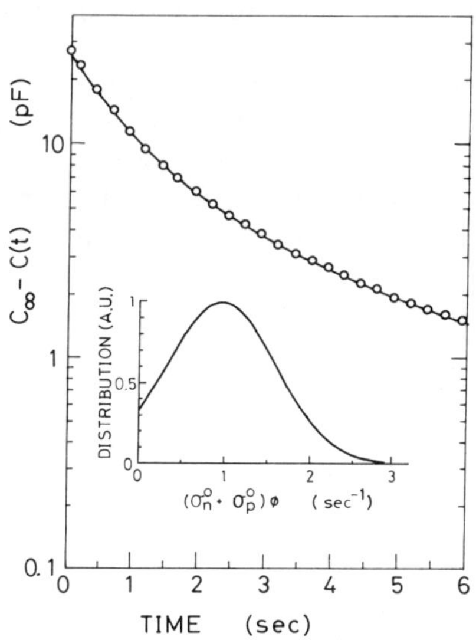

Figure 4 A photocapacitance transient corresponding to the photoionization process for ETX-1 measured at 140 K. The excitation photon energy was 1.17 eV. The solid line is the result of fitting, which was reproduced by assuming a distribution of $(\sigma_n^o + \sigma_p^o)\phi$ as shown in the inset.

illumination. The sample was heated up to 140 K with a reverse bias of −1 V in the dark, illuminated again, and the capacitance change was recorded. In this procedure, initial filling of the levels with electrons was achieved not by forward biasing, but by thermal recovery from the metastable state to the normal state. With this technique, electron occupations at other traps could be kept almost unchanged from the steady state values and influences from these traps were made almost negligible. Figure 4 shows such a capacitance transient, which exhibits a non-exponential transient. From this transient, the distribution of photoionization efficiency $(\sigma_n^o + \sigma_p^o)\phi$ is obtained as shown in the inset.

Discussion

In the previous paper [4], we have shown that the midgap levels found in LEC, HB, VPE and O-implanted LPE GaAs crystals exhibited different photoquenching efficiencies under illumination of the light with the same energy and photon flux, and that their spectra were different, although their thermal emission rates were very similar to each other. Thus, we introduced a concept of the ‘EL2 family’ for the midgap levels often detected in GaAs crystals.

As shown in previous sections, we have found that the optical transition from the normal state to the metastable state is non-exponential and this non-exponential transient can be interpreted by assuming a broadening of photoquenching efficiencies. Also, a thermal recovery transient from the metastable state to the normal state shows non-exponentiality which can be interpreted by an energy barrier broadening.

These broadenings suggest that these midgap levels consist of a group of distributed states. In particular, multiple-metastable states are suggested to exist for the ‘EL2 family’. One of the possible interpretations of these broadenings is that the states have different electron-lattice couplings as illustrated in Figure 5. Figure 5a schematically shows that the Huang-Rhys

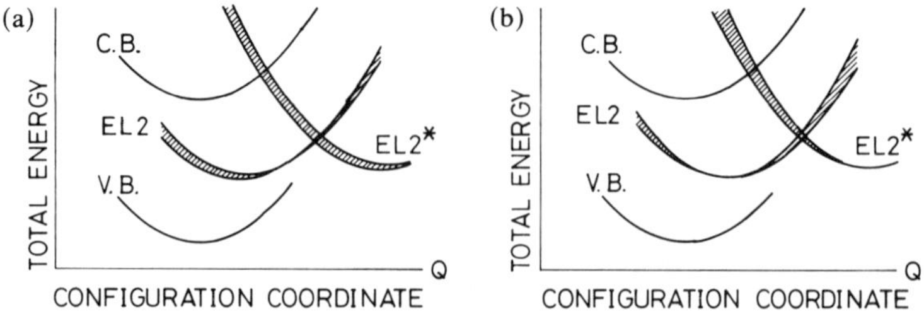

Figure 5 Configuration coordinate diagrams schematically illustrated for normal and metastable states of the ‘EL2 family’. Distributions in the trap parameters are due to those in (a) Huang-Rhys factors and (b) effective phonon frequencies.

factors are different among these states in the linear electron-lattice coupling approximation for both the normal and metastable states. Figure 5b illustrates a case in which the higher terms for the electron-lattice coupling are taken into account and hence effective phonon frequencies change.

We have already proposed that the 'EL2 family' might be arsenic-atom aggregates [4]. A variety of midgap levels in the 'EL2 family' existing in various GaAs crystals may be ascribed to variations in sizes and atomic configurations in arsenic aggregates. A transition from the normal state to the metastable state is regarded as an atomic configurational change upon a charge transfer [6]. This may happen readily when arsenic atoms in such aggregates are not located at normal lattice sites in GaAs, but have amorphous-like structures, in which bonding lengths and angles can be varied according to their charge states. The interface between the crystalline GaAs region and the amorphous arsenic region can introduce localized electronic states around the midgap, as inferred from the existence of surface states. Such amorphous-like arsenic aggregates can also explain existence of various electron-lattice coupling strengths observed in this study.

Recently, anion inclusions were reported to exist in GaAs, InP, GaP and GaSb [9]. Also, an anomalous amount of excess arsenic atoms or gallium-vacancies were detected by the X-ray reflection technique [10]. Such excess arsenic atoms are mobile and arsenic aggregates are therefore considered to be unstable under heat treatment. Thermally unstable natures of the ETX-1, ETX-2 and ETX-4 levels in LEC GaAs (all from the 'EL2 family') were observed [3]. The densities of these levels decreased rapidly after annealing around 400°C.

On the other hand, the 'EL2' in HB GaAs stayed unchanged after annealing at 700°C. The photoquenching efficiency spectra are also different from those of ETX-1 and ETX-2 [4]. All these characteristics suggest that the atomic structure of the 'EL2' in HB GaAs might be different from those of ETX-1, ETX-2 and ETX-4 in LEC GaAs. Therefore, not all deep levels in the 'EL2 family' have the same atomic structure. The model of arsenic aggregates mostly applies to the midgap levels found in LEC GaAs which are thermally unstable and have large fluctuations in density and properties such as capture cross-sections and optical transition rates [3].

Conclusions

The transitions between the normal state and the metastable state of the midgap level ETX-1 in LEC GaAs were studied by measuring photocapacitance transients. From the results obtained in this study together with the experimental results on photoquenching spectra previously reported [4], we have found that the midgap levels in LEC GaAs which form the 'EL2 family' have various electron-lattice couplings. From such characteristics, it is speculated that midgap levels found in LEC GaAs may be arsenic aggregates which have amorphous-like atomic structures.

ACKNOWLEDGEMENTS

The least squares fitting program, SALS, was provided by the Computer Center of the University of Tokyo. This work was partially supported by a Grant-in-aid for Scientific Research B and a Grant-in-aid for a Special Project Research from the Ministry of Education, Science and Culture.

References

1. Martin G. M., Mitonneau A. and Mircea A. (1977). *Electron. Lett.*, **13,** 191
2. Taniguchi M. and Ikoma T. (1982). *Inst. Phys. Conf. Ser.*, **65,** 65
3. Taniguchi M. and Ikoma T. (1983). *J. Appl. Phys.*, **54,** 6448
4. Taniguchi M. and Ikoma T. *Appl. Phys. Lett.*, in press
5. Vincent G. and Bois D. (1978). *Solid-State Commun.*, **27,** 431
6. Vincent G., Bois D. and Chantre A. (1982). *J. Appl. Phys.*, **53,** 3643
7. Computation was carried out using a program called SALS (Statistical Analysis by Least Square Fitting) in the program library of the Computer Center of the University of Tokyo
8. Mitonneau A. and Mircea A. (1979). *Solid-State Commun.*, **30,** 157
9. Gant H., Koenders L., Bartels F. and Monch W. (1983). *Appl. Phys. Lett.*, **43,** 1032
10. Fujimoto I. Submitted to *Jap. J. Appl. Phys.*

Identification of As_{Ga} Antisite Defects in LEC GaAs

K. R. Elliott, R. T. Chen, S. G. Greenbaum* and R. J. Wagner†

Rockwell International Corporation, Microelectronics Research and Development Center, Thousand Oaks, California 91360, USA

Abstract

We have studied the electron paramagnetic resonance of the As_{Ga} antisite defect in as-grown bulk undoped LEC (liquid encapsulated Czochralski) material in a series of samples. These samples have been characterized by SIMS (secondary ion mass spectrometry), local vibrational mode spectroscopy, photoluminescence and Hall measurements. Previous work has indicated that the compensation mechanism involves an unidentified native deep donor EL2, which has been compensated by residual carbon acceptors. In the present work we report the correlation of the EPR (electron paramagnetic resonance) signal with the concentration of residual carbon acceptors. Since the EPR sensitive charge state is As_{Ga}^{+} and since we have previously reported $|C| \approx EL2^{+}$, we can identify the defect As_{Ga} as the principal compensating species in the undoped, as-grown LEC material.

The development of high speed integrated circuits made from GaAs places stringent requirements on the quality of substrate material. An understanding of the defect species present in the material is crucial to such development. Deep donor centres referred to as EL2 in GaAs are responsible for the semi-insulating properties of undoped material grown by the LEC method. The identity of these deep donors has been controversial. A number of defect species may have properties which are consistent with those of EL2. These include substitutional oxygen atoms on arsenic lattice sites (O_{As}), substitutional arsenic atoms on gallium lattice sites (As_{Ga} antisites) and a number of complex defect species which may contain antisites, vacancies, interstitials and impurities.

The identity of these defects has been complicated by the anomalous behaviour of EL2 centres. The electrical properties of EL2 centres can be 'quenched' at temperatures below 120 K by photons near 1 eV in energy [1]. The centres exhibit a temperature activated electron capture cross-section [2]. It has been claimed that the centre has both a deep and shallow donor state [3]. The defect gives rise to a broad absorption band which may be

* Department of Physics and Astronomy, Hunter College of CUNY, New York 10021, USA. Work performed while S. Greenbaum was a National Research Council Postdoctoral Associate at the Naval Research Laboratory, 1981–1983.
† Naval Research Laboratory, Washington DC 20375, USA.

associated with intra-centre transitions [4]. DLTS (deep level transient spectroscopy) spectra give evidence that more than one defect may give rise to the features commonly associated with EL2 in such measurements [5, 6].

The only centre positively identified which might be associated with EL2 is the As_{Ga} antisite defect which has been observed by EPR [7, 8]. The identification of this defect has led to suggestions that the antisite and EL2 are one and the same. However, a positive correlation between the electrical properties of semi-insulating GaAs and the presence of antisites has heretofore not been reported.

The purpose of this work is to establish the relative importance of antisite defects in determining the electrical properties of semi-insulating undoped GaAs grown by high pressure LEC. We will show that in material grown at Rockwell, As_{Ga} antisites are the dominant deep donor species.

The samples used in this study have been characterized in great detail using a number of diagnostic techniques. These include Hall measurements, SIMS, photoluminescence, near and far infrared absorption and local vibrational mode spectroscopy (LVM) in the far infrared. The principal impurity species, as determined by LVM or SIMS were boron, which is present in a concentration range from 1×10^{16}–6×10^{17} cm^{-3}, and carbon, which ranges from 2×10^{15} to 1.3×10^{16} cm^{-3}. Silicon contamination can be eliminated as a problem by growing crystals in pyrolytic boron nitride crucibles. The main source of boron contamination is the B_2O_3 encapsulant.

The evaluation of the electrical and optical properties of the semi-insulating material indicates that the deep donor, commonly referred to as EL2, is the predominant deep centre. An optical absorption band between 1 and 1.4 μm previously identified with the EL2 centre was observed in all of the semi-insulating material [9]. The behaviour of the photoconductivity thresholds above and below 120 K was also found to be consistent with the presence of EL2. The concentration of EL2 as determined by optical absorption was found to depend on the melt stoichiometry, as shown in Figure 1, increasing from about 5×10^{15} to 1.7×10^{16} cm^{-3} as the As atom fraction increased from 0.48 to 0.51.

The concentration of a residual double acceptor observed in Hall measurements and also in the far infrared absorption spectra of the p-type material is also shown in Figure 1. It should be noted that the concentrations shown in Figure 1 represent absolute concentrations of the two defects as opposed to a measurement of the compensation of one defect by the other. The carrier concentrations obtained from the Hall measurements indicate compensation ratios for As_{Ga} (EL2) in the semi-insulating material of between 10 and 100 typically, indicating negligible compensation by residual acceptors. In the p-type Ga-rich material, Hall measurements indicate the presence of residual shallow carbon acceptors which have not been compensated by EL2 or other residual donors, placing an upper limit on the donor content at 2×10^{15} cm^{-3} in this material. The presence of the uncompensated carbon also guarantees that the deeper double acceptor is uncompensated, and that the infrared absorption measurements give an uncompensated measure of the concentra-

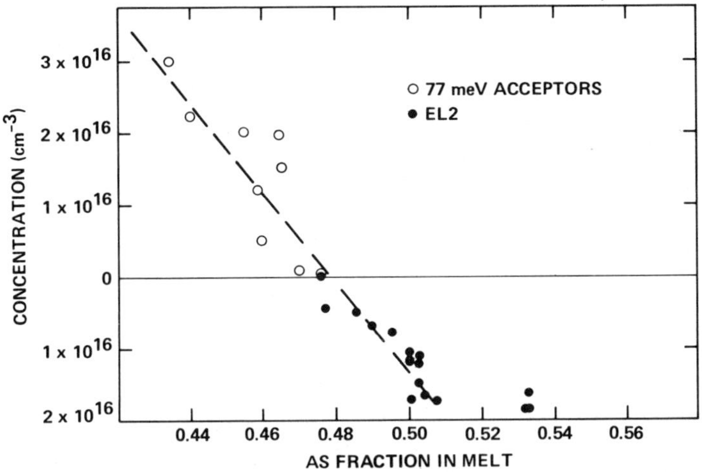

Figure 1 Stoichiometry dependence of EL2 (As_{Ga}) deep donors in LEC GaAs and residual double acceptors. The concentration of double acceptors is plotted on the positive axis. The concentration of EL2 deep donors is plotted on the negative axis.

tion. Photoluminescence measurements verify the absence of residual double acceptors in the semi-insulating material. The 78 meV double acceptor is probably due to either boron or gallium atoms substitutionally incorporated on As lattice sites (Ga_{As} or B_{As}). Both B_{As} and Ga_{As} defects are expected to have properties which can explain the behaviour observed for the double acceptor [10, 11].

The As_{Ga} EPR spectrum was identified in each of the semi-insulating samples grown from melts >0.47 As atom fraction. A typical spectrum is shown in Figure 2. P–type samples from Ga rich melts below the critical composition were also studied, but did not show As_{Ga} EPR spectra. These results indicate that the behaviour of As_{Ga} is generally consistent with the known behaviour of EL2. For instance, EL2 is only observed in LEC crystals grown above the critical melt composition of 0.47 As atom fraction. To identify EL2 with As_{Ga}, however, more definitive evidence is necessary. Unfortunately, a direct comparison between the optical absorption due to EL2 and the EPR signal of As_{Ga} is not the appropriate one. The optical absorption of the EL2 centre is thought to arise from photoionization of the neutral centre. The EPR signal, however, is only sensitive to the concentration of the singly ionized centre $As_{Ga}{}^{+}$. Our experiments substantiate this view. We have found no correlation between the EPR signal associated with As_{Ga} and near infrared absorption associated with EL2.

The appropriate comparison is between the concentration of ionized EL2 centres and the EPR signal. Previous studies on the semi-insulating crystals used in this work have established that the number of ionized centres is determined principally by compensation of EL2 by residual carbon acceptors

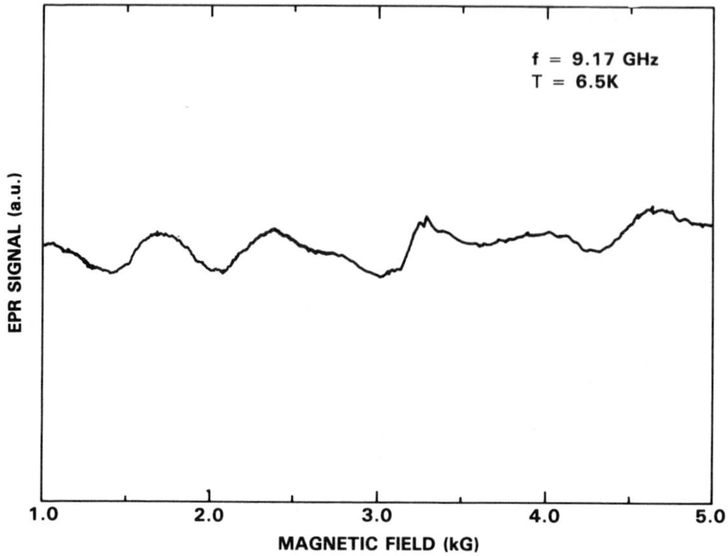

Figure 2 As_{Ga} EPR spectrum in GaAs. The spectrum consists of four closely spaced lines associated with the hyperfine splitting. The spectrum is isotropic.

[12, 13]. Hence, the net carbon content of these crystals can be used as a measure of the ionized EL2 concentration.

In Figure 3 we plot the relative EPR signal against the carbon concentration for the semi-insulating crystals used in this study. We find a linear correlation between the residual carbon content in this material and the EPR signal for the As_{Ga} defect. In addition, we have found through the study of other GaAs samples with both the present spectrometer and other high frequency spectrometers (35–337 GHz) that the lower limit of sensitivity is between 1×10^{15} cm^{-3} and 2×10^{15} cm^{-3} As_{Ga} spin density. By comparison with samples in this study, we conclude that the present samples yield absolute As_{Ga} densities which agree with the absolute carbon concentration. This agreement is well within the estimated uncertainty in the calibration procedure of ±50% for the absolute spin density.

The results presented in Figure 3 firmly establish that the occupation of As_{Ga} is controlled by the residual acceptor content. Our results also eliminate the possibility that As_{Ga} has a shallow donor level associated with the As_{Ga^+} and As_{Ga° charge states. In such a case the compensation would occur between As_{Ga^+} and $As_{Ga^{2+}}$. As a result, we would observe a decrease in the EPR signal with increasing carbon concentrations. Since we observe an increase in the EPR signal our results are inconsistent with such an interpretation. It may be possible to interpret previous data which suggest that EL2 can give rise to shallow donor levels as evidence that neutral As_{Ga° can bind an additional electron to form negatively charged As_{Ga^-} states. Such states are expected to be stable but shallow. Another possible explanation is that mixing between

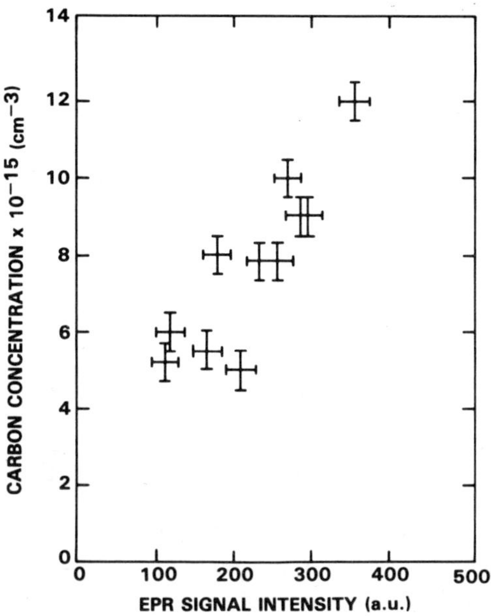

Figure 3 Relative EPR signal versus carbon concentration for the semi-insulating crystals used in this study. We find a linear increase in EPR signal with carbon concentration.

shallow donor states and EL2 deep donor states may lower the effective binding energy of the donor states. Such effects should be important when there is substantial overlap between shallow donor and As_{Ga} donor wavefunctions.

These results also place constraints on the density of other donor defects which may occur in the material. If these defects have ionization energies less than that of As_{Ga}, they will be compensated by carbon prior to any compensation of As_{Ga}. Hence in order to see any EPR signal, the carbon content must be higher than the residual donor content. For the material in this study, this puts limits on the residual donor content $<2\times10^{15}$ cm^{-3}. Although levels deeper than the (As_{Ga^0}/As_{Ga^+}) ionization could play a role, the concentration of such levels must be stoichiometry dependent to explain Hall measurements in undoped p-type material. It is unlikely that such levels exist in concentrations $>2\times10^{15}$ cm^{-3} if they are extrinsic in origin.

In order to clarify the role that such defects might play, DLTS measurements were made on samples intentionally doped with Se during growth. The growth procedure was identical in all other ways to the growth of the semi-insulating crystals used in this study. A typical DLTS spectrum is shown in Figure 4. The DLTS spectrum establishes that EL2 is the principal donor species. Analysis of the EL2 capacitance transient showed that the transient decays exponentially with an emission rate which agrees with that of other EL2 measurements [6]. We see no evidence for the presence of other 'EL2'

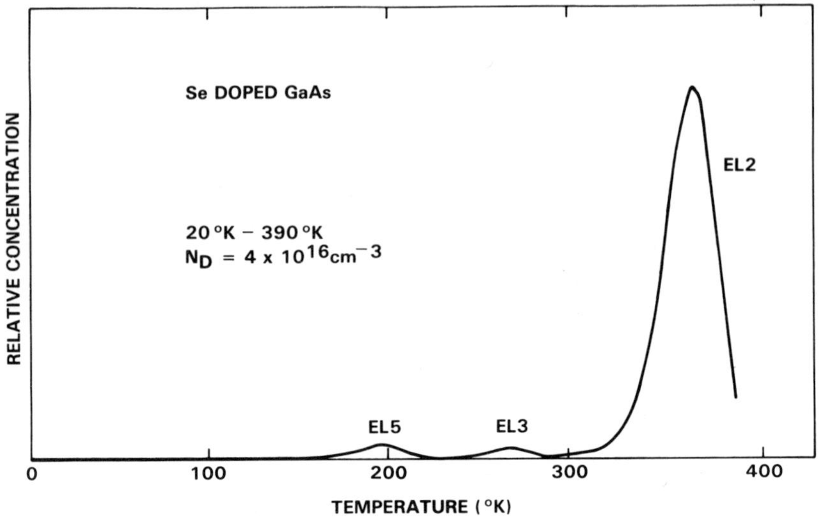

Figure 4 DLTS spectra of Se doped GaAs grown under As-rich conditions. The EL2 peak arises from a single exponential decay indicating that only a single defect contributes to this portion of the spectra.

defects and must conclude that this spectrum is associated with the antisite As_{Ga} defect.

We see evidence of two other defects in the DLTS spectrum which can be identified with EL5 (0.42 eV) and EL3 (0.57 eV) [14]. Both defects have relatively low concentrations ($\sim 5 \times 10^{14}$ cm^{-3}). EL5 has the same energy as the defect E3 observed in electron irradiated material. It has been established that E3 exhibits recombination enhanced defect motion and is associated with a primary defect formed by displacements on the As sublattice [15, 16]. In view of the As-rich nature of this material, and assuming that E3 is the same trap as observed here, identification of the defect as an As interstitial would seem reasonable. In material grown with very low carbon content it plays a role in determining residual carrier concentration.

In conclusion, we have correlated the presence of As_{Ga} as determined by EPR with compensating acceptors in bulk undoped LEC GaAs. This correlation identifies EL2 in this material as the antisite As_{Ga}. We also find that in this LEC material the compensation occurs between As_{Ga° and As_{Ga^+} charge states.

ACKNOWLEDGEMENTS

We acknowledge T. A. Kennedy for help in establishing the absolute spin density, T. A. Kennedy and B. V. Shanabrook for useful conversations. We acknowledge D. E. Holmes and C. G. Kirkpatrick for useful discussions and support.

References

1. Vincent G. and Bois D. (1978). *Solid-State Commun.*, **27,** 431
2. Lang D. V. and Logan R. A. (1975). *J. Electron. Mat.*, **4,** 1053
3. Walukiewicz W., Lagowski J. and Gatos H. C. (1983). *Appl. Phys. Lett.*, **43,** 112
4. Kaminska M., Skowronski M., Lagowski J., Parsey J. M. and Gatos H. C. (1983). *Appl. Phys. Lett.*, **43,** 302
5. Taniguchi M. and Ikoma T. (1983). *J. Appl. Phys.*, **54,** 6448
6. Lagowski J., Lin D. G., Aoyama T. and Gatos H. C. (1984). *Appl. Phys. Lett.*, **44,** 336
7. Wagner R. J., Krebs J. J., Stauss G. H. and White A. M. (1980). *Solid-State Commun.*, **36,** 15
8. Goswami N. K., Newman R. C. and Whitehouse J. E. (1981). *Solid-State Commun.*, **40,** 473
9. Martin G. M. (1981). *Appl. Phys. Lett.*, **39,** 747
10. Elliott K. R., Holmes D. E., Chen R. T. and Kirkpatrick C. G. (1982). *Appl. Phys. Lett.*, **40,** 898
11. Elliott K. R. (1983). *Appl. Phys. Lett.*, **42,** 274
12. Holmes D. E., Chen R. T., Elliott K. R. and Kirkpatrick C. G. (1982). *Appl. Phys. Lett.*, **40,** 46
13. Holmes D. E., Chen R. T., Elliott K. R., Kirkpatrick C. G. and Yu P. W. (1982). *IEEE Trans. Microwave Theory and Tech.*, **MTT-30,** 949
14. Martin G. M., Mitonneau A. and Mircea A. (1977). *Electron. Lett.*, **13,** 191
15. Lang D. V. (1977). *Inst. Phys. Conf. Ser.*, **31,** 70
16. Loualiche S., Guillot G. and Nouailhat A. (1982). *Phys. Rev.*, **B26,** 7090

Concentration and Thermal Stability of As_{Ga} in GaAs: Correlation with EL2

U. Kaufmann, J. Windscheif, M. Baeumler, J. Schneider and F. Köhl*

Fraunhofer-Institut für Angewandte Festkörperphysik, D-7800 Freiburg, West Germany

Abstract

A quantitative ESR (electron spin resonance) study of As_{Ga} antisite defects has been performed in ten different as-grown doped and undoped GaAs materials. The As_{Ga}^{4+} ESR is consistently observed in a concentration of $1\text{–}6 \times 10^{16}\ cm^{-3}$. It is found to be thermally stable up to annealing temperatures of 850°C in contrast to previous observations on neutron-irradiated material. Isochronal annealing studies indicate an increase in As_{Ga}^{4+} concentration near 900°C. Annealing of As_{Ga}^{4+} is observed at or above 950°C depending on the annealing atmosphere. Infrared images of entire 2 inch GaAs slices were obtained using a direct photographing technique. They reveal patterns which reflect the inhomogeneous EL2 distribution within a wafer. These patterns can be optically bleached at low temperature. They also disappear upon annealing at 950°C. The results are considered as further support for the As_{Ga} antisite model for EL2—but not as a proof.

Introduction

Although the first definite experimental confirmation for the existence of antisite defects in III–V compounds dates back to 1976 [1] the important influence of these defects on the material properties has only recently been established [2–4].

In 1980, Wagner *et al.* [5] positively identified the isolated As_{Ga} antisite in the 300 GHz ESR spectrum of Bridgman-grown (HB) SI GaAs:Cr. Although neither the ubiquitous nature nor other properties of As_{Ga} had been established, its ESR identification led to speculations [6–8] whether isolated As_{Ga} or As_{Ga} complexes could produce the midgap donor level EL2 which is considered to be responsible for the SI properties of undoped, LEC (liquid encapsulated Czochralski) PBN GaAs [9–11]. Experimental support for this view was provided by the observation that EL2 formation in bulk material is favoured by excess As [10–12]. More direct evidence linking EL2 and As_{Ga} was obtained from photo-ESR. These measurements indicated that As_{Ga} is a deep double donor with its first ionization energy close to midgap [13] and that the As_{Ga}^{4+} (D^+) ESR signal shows a persistent optical quenching behaviour [14] analogous to the persistent photocapacitance quenching of EL2 [15].

Although the above findings support the identification of As_{Ga} with EL2 one should keep in mind that there are serious objections against this model

* Wacker-Chemitronic, D-8263 Burghausen, West Germany.

[16, 17]. Two of these will be refuted in this paper [18]. It is first shown that As_{Ga} is *consistently* present in 8 SI LEC or HB GaAs samples investigated and not only sometimes as stated elsewhere [16]. It is further demonstrated that the isolated As_{Ga} defect in as-grown bulk material is thermally stable up to annealing temperatures of 900°C, in contrast to previous results for neutron irradiated crystals [19], consistent, however, with annealing data presented herein and those of other authors [11, 20] for the so called EL2 absorption in as-grown material.

The findings presented in this work add further support to the antisite model for EL2—although they do not prove it. However, if it turns out that the two defects should be different, the present results imply that As_{Ga} controls the compensation of SI GaAs in much the same way as EL2 does.

Experimental

The ESR measurements were performed with a commercial 9.7 GHz spectrometer at a temperature of 20 K or 5 K. *In situ* optical excitation was achieved through the slotted front wall of the ESR cavity. Absolute defect concentrations were determined using a calibrated GaP:S standard as described previously [21].

Near-infrared transmission photographs of 2 inch slices were obtained using either an IR-sensitive vidicon camera and a set-up similar to the one described recently [22, 23] or, simpler and more direct, a conventional 35 mm camera loaded with an IR sensitive film (Kodak 2481). The quality of the photographs in both cases were similar (e.g. contrast and resolution). We emphasize that this latter technique is the simplest, cheapest and most direct method for fast homogeneity assessment of thick SI GaAs wafers.

Table 1 *Specifications for the GaAs materials investigated by ESR*

No.	Label	Growth method	Crucible/ boat material	Doping	Carrier concentration (cm^{-3})	Supplier
1	Wa 48	LEC	PBN	—	*$n \approx 10^{15}$	Wacker
2	Wa 23	LEC	PBN	—	SI	Wacker
3	Wa 24	LEC	SiO_2	Cr	SI	Wacker
4	Wa 25	LEC	SiO_2	Ga_2O_3	SI	Wacker
5	Wa 26	LEC	SiO_2	Cr, Ga_2O_3	SI	Wacker
6	LC 259	LEC	SiO_2	Cr	SI	Siemens
7	Mil	HB	SiO_2	Cr	SI	Mitsubishi
8	Wa 41E	LEC	PBN	C	$p \approx 10^{12}$	Wacker
9	UE 14	†THM	SiO_2	—	*$p \approx 10^{16}$	Univ. Erlangen
10	MCP 4	HB	SiO_2	Zn	$p = 4 \times 10^{16}$	MCP Ltd.

*See Ref. 25; †Travelling heater method.

Ten different melt-grown GaAs materials have been investigated by ESR. Growth method, dopants and electrical properties as specified by the suppliers for the individual wafers/ingots are listed in Table 1. The sample position within the wafer was not especially selected. Several samples from wafer numbers 2–5 were subject to high-temperature heat-treatments, either under an As-atmosphere in sealed quartz ampoules or under an arsenic free (flowing N_2) atmosphere. Long time annealing resulted in strongly p-type converted surface layers which were removed to a depth of about 80 μm either by chemical etching or by mechanical grinding after each annealing step.

In addition to the materials listed in Table 1, six 3–9 mm thick 2 inch undoped LEC SI PBN GaAs slices were examined by NIR transmission imaging.

ESR results and discussion

In discussing quantitative ESR results one should be aware of the fact that ESR in general provides information about the concentration of a specific charge state only, rather than giving the total concentration of a defect. Only if the Fermi level (E_F) is suitably positioned will these two concentrations be the same [24].

AS-GROWN SI CRYSTALS

All as-grown SI samples (materials 2–7) have two ESR features in common at a measurement temperature of 20 K. First, under dark conditions they exclusively display the ESR spectrum of the singly ionized isolated As_{Ga} double donor, As_{Ga}^{4+} (see Figure 1 and Table 2). This signal cannot be enhanced optically. Second, after optical excitation with photon energies >1 eV the As_{Ga}^{4+} antisite is quenched below the detection limit and the ESR of the neutral isolated Fe acceptor (Fe^{3+}) appears. In addition, all Cr-doped samples show a light-induced Cr^{4+} signal. The As_{Ga}^{4+} concentration in *all* SI samples lies in the range $1–6\times10^{16}$ cm^{-3}. This fact is considered as convincing evidence for the omnipresence of As_{Ga} antisite defects in SI GaAs. The concentration falls in the concentration range quoted in the literature for the EL2 level (low 10^{16} cm^{-3}).

AS-GROWN p AND n-TYPE CRYSTALS

The p-type sample 10 turned out to be too lossy even at 5 K to give reliable results [25]. Sample 8, p-type due to intentional carbon-doping, revealed only the Fe^{3+} signal in the dark. After optical excitation (photon energy <0.8 eV) the As_{Ga}^{4+} signal appeared while Fe^{3+} remained unchanged. The failure to observe As_{Ga}^{4+} in the Ga rich THM sample (9) is not surprising, since As_{Ga}

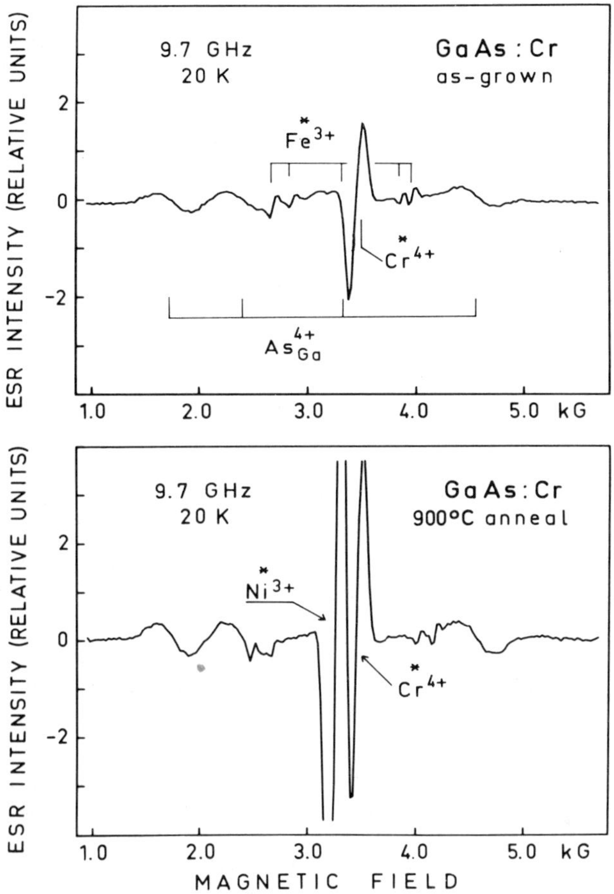

Figure 1 EPR spectra of SI GaAs:Cr (samples from 3 in Table 1) before and after heat treatment (900°C, 24 h, As atmosphere) with the magnetic field approximately along a ⟨100⟩ axis. Both traces correspond to 'dark minus light-on' difference spectra. Therefore all signals which appear only after *in situ* illumination (Fe^{3+} and Cr^{4+} upper trace; Ni^{3+}, Cr^{4+} lower trace) have inverted phase.

formation is not favoured under these growth conditions. The n-type material, sample 1 [25], shows a weak antisite signal (concentration about 1×10^{16} cm^{-3}) in the dark, which can be quenched by light with photon energies >1 eV. No enhancement by illumination with below-gap light and no Fe^{3+} signal was observed.

ANNEALED CRYSTALS

Annealed samples (900°C, 24 h, As) from wafers 2–5

Inspection of Table 2 shows that in these heat-treated samples the Fe^{3+} concentration is strongly enhanced and that the Fe^{3+} signal is observed

Table 2 *Concentration ($10^{16}\ cm^{-3}$) of transition metal acceptors and As_{Ga}^{4+} antisites in as-grown and heat-treated (900°C, 24 h, As atmosphere) SI GaAs samples from wafers 2–5 in Table 1 as determined by ESR*

Sample defect	PBN GaAs		GaAs:Cr		GaAs:Ga_2O_3		GaAs:Cr:Ga_2O_3	
	as-grown	900°C	as-grown	900°C	as-grown	900°C	as-grown	900°C
Fe^{3+}	0.007*	0.5	0.008*	002	0.004*	0.2	0.004*	0.3
Ni^{3+}	n.d.	2	n.d.	1*	n.d.	1*	n.d.	0.6*
Cr^{4+}	n.d.	n.d.	0.3*	1*	n.d.	n.d.	0.2*	2*
As_{Ga}^{4+}	1	5*	6	10	3	10	3	6

* Only observed or strongly enhanced after illumination; n.d.—not detected

already in the dark, together with the As_{Ga}^{4+} signal. The latter is also enhanced by a factor of 2–5 by the annealing. After optical excitation an enhanced Cr^{4+} signal and the newly appearing Ni^{3+} ESR are observed (see Figure 1).

The fact that the As_{Ga}^{4+} ESR is stable up to 850°C was surprising. Obviously, previous annealing data for As_{Ga} in neutron irradiated crystals [19] have to be reinterpreted: the 500°C annealing stage is not characteristic for the annealing of As_{Ga}. Rather, another neutron induced defect becomes mobile at 500°C and annihilates As_{Ga}.

The origin for As_{Ga}^{4+} enhancement under 900°C anneal is presently not clear. It may be an effect of Fermi level lowering but thermal generation and subsequent freeze-in of As_{Ga} cannot be excluded. Work to elucidate this question is in progress.

Isochronal and isothermal annealing of As_{Ga}

To investigate the thermal stability of As_{Ga} one sample from wafer 3 was subject to a 24 h isochronal anneal under flowing N_2. The result is shown in Figure 2. The As_{Ga}^{4+} concentration remains constant up to 850°C, it is enhanced at 900°C—consistent with the independent observations reported in Table 2—and drops below the detection limit $<10^{16}\ cm^{-3}$ at 950°C. If the annealing is performed in an As atmosphere the drop at 950°C does not occur. The effect of As_{Ga}^{4+} enhancement has been independently observed elsewhere [26].

A further sample from wafer 3 was subject to a 950°C isothermal anneal under N_2 flow (see Figure 2). Again, there is an increase in As_{Ga}^{4+} for the shorter annealing periods, but after a total annealing time of 15 h the signal has disappeared.

These results once again confirm the high thermal stability of As_{Ga} and its pile-up at 900°C, but they also demonstrate the disappearance of the As_{Ga} defect for a 950°C anneal under N_2 flow. The latter statement is worth a comment, since disappearance of the As_{Ga}^{4+} ESR could also be due to a drop of E_F below the $E_c - 1.0$ eV level of As_{Ga}. Samples with such a position of E_F are, for instance, those from wafer 8 in Table 1 and the strongly p-type

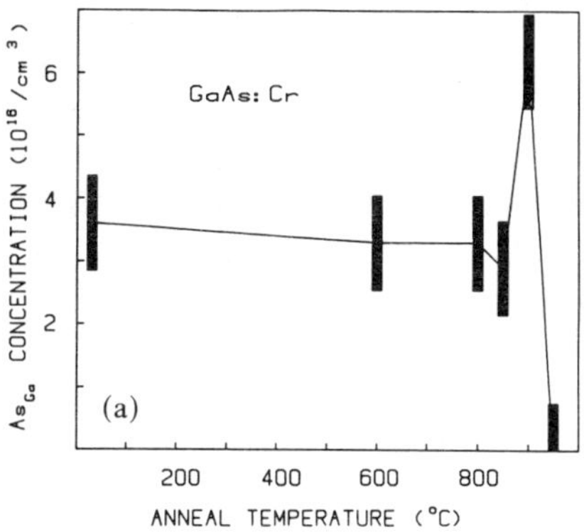

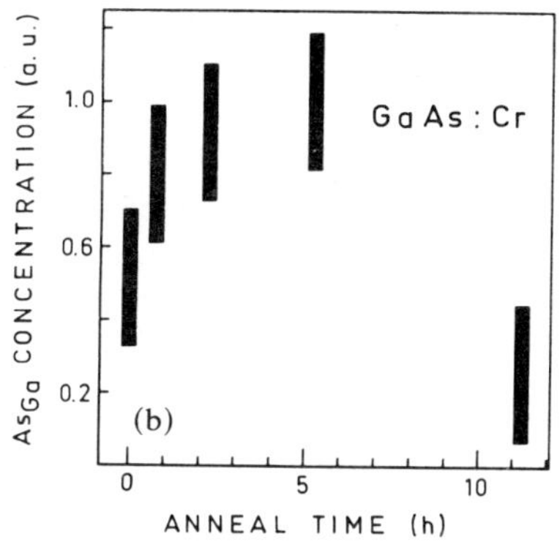

Figure 2 (a) Variation of As_{Ga}^{4+} concentration in LEC SI GaAs:Cr upon heat treatment (24 h, N_2 flow). (b) Isothermal anneal (950°C, N_2 flow) of As_{Ga}^{4+}.

converted undoped sample from Table 2. They show in fact no As_{Ga}^{4+} ESR in the dark but ESR can be optically excited. For the samples of Figure 2 annealed for 15 h or longer at 950°C this is no longer possible. We therefore conclude that the As_{Ga} defect has annealed. The fact that As_{Ga} does not disappear at 950°C if the annealing is performed under an As atmosphere demonstrates the, not unexpected, important influence of the ambient atmosphere on the minimum annealing temperature of As_{Ga} in as-grown GaAs.

NIR transmission images of 2 inch undoped SI GaAs slices

Figure 3 presents two-dimensional NIR transmission photographs of a ⟨100⟩ and a ⟨111⟩ oriented 2 inch diameter SI PBN GaAs slice, not listed in Table 1. The camera is viewing along the respective growth direction of the crystal. The photographs were obtained at room temperature with the *direct* photographing technique mentioned above.

The same structured patterns were observed with the IR vidicon set-up. If, however, a vidicon-camera with a sensitivity cut-off near 2 μm was used and only the transmitted light in the 1.6–2 μm range was monitored, the structures completely disappeared.

Infrared images with the wafer cooled to 80 K were also taken. No significant increase in either resolution or contrast was noticed. However, if the wafers were illuminated for several minutes with an intense white light source, the structure *completely vanished.*

One undoped SI wafer was subjected to a 15 h anneal, first at 850°C and then at 950°C in flowing N_2. The structures did not change for the 850°C treatment, while they had nearly vanished after the 950°C anneal. The disappearance of the structures in the IR photographs for a wavelength >1.6 μm and even more their low temperature quenching behaviour show without doubt that these structures arise from an inhomogeneous distribution and/or compensation of the defect(s) responsible for the absorption between the band edge and 0.8 eV. Thus, dark regions in Figure 3 correspond to regions of higher infrared absorption. Note that the overall azimuthal distribution reflects the symmetry of the growth direction and that the overall radial distribution in the ⟨100⟩ crystal has the well known 'W' shape [27]. If one trusts in the assignment of the below-gap absorption to the EL2 level, the images in Figure 3 can be considered as two-dimensional mappings of the *neutral* EL2 distribution. The NIR images presented here confirm the overall, inhomogeneous NIR absorption distribution in as-grown GaAs wafers very recently reported elsewhere [22, 23, 28].

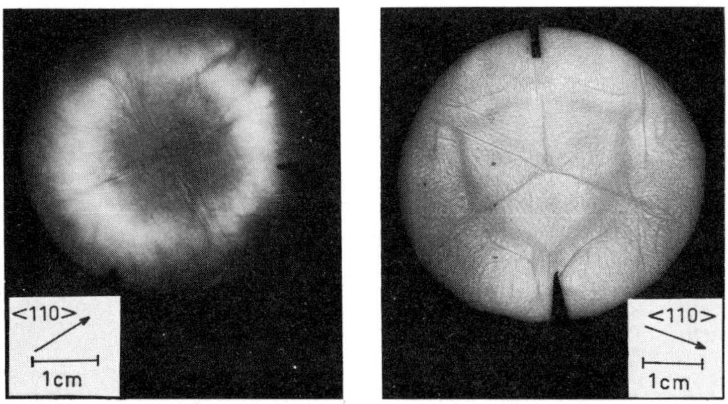

Figure 3 Two-dimensional NIR transmission images of undoped SI GaAs slices viewed along the growth direction of the ingot. Slice from ⟨100⟩ (left) and from ⟨111⟩ (right) grown ingot. Slice thickness 4 mm, diameter 2 inches. These images were taken at room temperature with the *direct* photographing technique.

A first attempt has been made to correlate the As_{Ga} antisite ESR with the images of Figure 3 by cutting samples from dark and bright regions in the wafer. The observed As_{Ga}^{4+} concentration shows an inverse 'W' shaped profile along the wafer diameter. This anticorrelation with the EL2 profile is not in conflict with the As_{Ga} model for EL2 as will be discussed elsewhere [29].

Our observation that the structured absorption pattern in the NIR images withstands a 15 h, 850°C anneal but disappears after annealing for another 15 h at 950°C is in line with earlier findings [11, 20] and supports the conclusion [20] that the defect responsible for the below-gap absorption begins to move freely in the bulk of very large crystals, thus also implying the possibility for its outdiffusion. This result for the EL2 absorption is in full agreement with the thermal stability of As_{Ga} as inferred from ESR. This striking coincidence gives further strong support to the view that EL2 is the isolated As_{Ga} antisite.

Conclusions

The isolated As_{Ga} antisite defect is a persistent defect in SI LEC and HB GaAs occurring in the same concentration range (low 10^{16} cm^{-3}) as the EL2 level. Thus As_{Ga} controls the compensation of SI PBN GaAs in much the same way as EL2 does. In as-grown materials it anneals at the same temperature (950°C) as the fluctuations of the EL2 absorption do. These findings are considered to be further strong support for the isolated As_{Ga} model for EL2 but not as a proof. The properties which the two defects do not seem to share have been pointed out. These issues have to be clarified before final conclusions can be reached.

ACKNOWLEDGEMENTS

We thank B. Zysset for the loan of the video cameras and H. Kniepkamp, H. Miki and R. Stein for supplying samples. We also thank E. Olander and B. Matthes for the preparation and the annealing of the samples. The technical assistance in taking the IR photographs by L. Marek is highly appreciated. This work was supported by the Bundesministerium für Forschung und Technologie of the Federal Republic of Germany.

References

1. Kaufmann U., Schneider J. and Räuber A. (1976). *Appl. Phys. Lett.*, **29,** 312
2. Kaufmann U., Schneider J., Wörner R., Kennedy T. A. and Wilsey N. D. (1981). *J. Phys.*, **C14,** L951
3. Killoran N., Cavenett B. C., Godlewski M., Kennedy T. A. and Wilsey N. D. (1983). *Physica*, **116B,** 425
4. O'Donnell K. P., Lee K. M. and Watkins G. D. (1982). *Solid-State Commun.*, **44,** 1015
5. Wagner J. R., Krebs J. J., Stauss G. H. and White A. M. (1980). *Solid-State Commun.*, **36,** 15
6. Zou Y. X. (1981). *Inst. Phys. Conf. Ser.*, **56,** 213
7. Kaminska M., Lagowski J., Parsey J., Wada K. and Gatos H. C. (1982). *Inst. Phys. Conf. Ser.*, **63,** 197
8. Kaufmann U. and Schneider J. (1982). In *Advances in Electronics and Electron Physics, 58,* p. 81 (C. Marton, ed.); Academic Press
9. Martin G. M., Farges J. P., Jacob G., Hallais J. P. and Poiblaud G. (1980). *J. Appl. Phys.*, **51,** 2840
10. Holmes D. E., Chen R. T., Elliott K. R. and Kirkpatrick C. G. (1982). *Appl. Phys. Lett.*, **40,** 46
11. Ta L. B., Hobgood H. M., Rohatgi A. and Thomas R. N. (1982). *J. Appl. Phys.*, **53,** 5771
12. Lagowski J., Gatos H. C., Parsey J. M., Wada K., Kaminska M. and Walukiewicz W. (1982). *Appl. Phys. Lett.*, **40,** 342
13. Weber E. R., Ennen H., Kaufmann U., Windscheif J., Schneider J. and Wosinski T. (1982). *J. Appl. Phys.*, **53,** 6140
14. Weber E. R. and Schneider J. (1983). *Physica*, **116B,** 398
15. Vincent G. and Bois D. (1978). *Solid-State Commun.*, **27,** 431
16. Martin G. M. and Makram-Ebeid S. (1983). *Physica*, **116B,** 371
17. Meyer B. K., Spaeth J. M. and Scheffler M. (1984). *Phys. Rev. Lett.*, **52,** 851

18. A preliminary account of part of this work was presented as a poster paper: J. Windscheif, U. Kaufmann, H. Ennen, T. Kimura and F. Köhl; *4th Lund Int. Conf. Deep Level Impurities in Semicond.*, Eger, Hungary (1983)
19. Wörner R., Kaufmann U. and Schneider J. (1982). *Appl. Phys. Lett.*, **40,** 141
20. Rumsby D., Ware R. M., Smith B., Tyjber M., Brozel M. R. and Foulkes E. J. (1983). *GaAs IC Symposium, Phoenix, Arizona Technical Digest*, p. 34
21. Kaufmann U. and Kennedy T. A. (1981). *J. Electron. Mat.*, **10,** 347
22. Stirland D. J., Grant I., Brozel M. R. and Ware M. R. *Inst. Phys. Conf. Ser.*, in press
23. Skolnick M. S., Brozel M. R., Reed L. J., Grant I., Stirland D. J. and Ware R. M. (1984). *J. Electron. Mat.*, **13,** 107
24. Walukiewicz W., Lagowski J., Gatos H. C. (1983). *Appl. Phys. Lett.*, **43,** 192
25. Our own electrical measurements on wafer 10 indicate a carrier concentration at least one order of magnitude higher than that specified by the supplier. For the n-type material 1 the room temperature conductivity was strongly inhomogeneous as judged by the Q of the microwave cavity for different samples from the wafer. Therefore the carrier concentrations of these two materials cited in Table 1 have to be considered with caution
26. E. R. Weber, private communication
27. Martin G., Jacob G., Poiblaud P., Goltzene A. and Schwab C. (1981). *Inst. Phys. Conf. Ser.*, **59,** 281
28. Holmes D. E., Chen R. T., Elliott K. R. and Kirkpatrick C. G. (1983). *Appl. Phys. Lett.*, **43,** 305
29. Windscheif J., Baeumler M., Kaufmann U. and Schneider J. Unpublished

Effect of Deep Level Ionization on Low Temperature Electron Mobility in SI GaAs

W. Walukiewicz*, Le Wang†, L. Pawlowicz, J. Lagowski and H. C. Gatos

Massachusetts Institute of Technology, Cambridge, Massachusetts 02139, USA

Abstract

Photoionization of midgap levels, by means of IR illumination, was used to increase the concentration of free carriers in SI GaAs and thus extend the Hall effect and conductivity measurements to low temperatures. We found, from electron mobility characteristics, that the optical transition of EL2 to a metastable state is most likely accompanied by a change of the charge state of this defect. We also established that the low temperature transport of SI GaAs is largely controlled by fluctuations of the electrostatic potential which lead to 'activated conductivity'. At elevated temperatures ($T > 300$ K) these fluctuations cause an effective electron scattering which limits the high temperature mobility values.

Introduction

Hall effect and conductivity measurements provide the means for the standard assessment of semiconducting materials. Application of these techniques to SI GaAs is limited to temperatures >300 K [1–5]. Low temperature measurements of the Hall effect and conductivity are possible only if additional electrons are excited to the conduction band [6]. In this paper we analyse low temperature measurements of the mobility and the concentration of photoelectrons in SI GaAs excited from the midgap levels EL2 and EL0 by IR radiation, including the spectral range of transitions to the metastable state.

Experimental

Conductivity and Hall effect were measured on 'undoped' SI GaAs in the temperature range of 70–200 K. Monochromatic sub-bandgap light of energy 0.9–1.3 eV was used for the illumination. Optical absorption in this spectral region is commonly assigned to the photoionization and intra-centre transitions of the midgap levels EL2 [7] and EL0 [8]. We found that the electron mobility behaviour above 100 K is distinctly different from that below 100 K.

* Permanent address: Institute of Physics, Polish Academy of Sciences, Warsaw, Poland.
† Permanent address: Shanghai Institute of Metallurgy, Academy of Sciences of China, Shanghai, China.

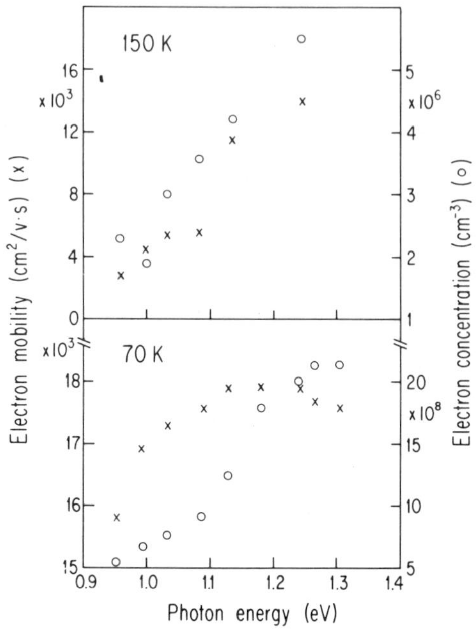

Figure 1 Electron mobility and Hall concentration versus photon energy at 150 and 70 K (×–electron mobility; O–electron concentration).

Thus, as seen in Figure 1, at higher temperatures (T>100 K) the mobility increases with increasing photon energy in a way similar to the increase of the electron concentration. In contrast, for temperatures lower than 100 K the mobility increases slightly at small photon energies, and saturates, or even decreases, at higher photon energies, although the electron concentration steadily increases in the entire spectral range.

The dependence of the electron concentration on the photon energy, shown in Figure 1, reflects the spectral dependence of the photoionization cross-section of the midgap levels EL0 and EL2. The saturation and decrease of the mobility at 70 K occurs at photon energies between 1.05 and 1.3 eV, i.e. exactly in the spectral range of the optical transitions of EL2. It is thus evident that the transition to the metastable state of EL2 substantially affects the electron mobility. Consistent with this finding was the measured time dependence of the electron concentration and mobility at two photon energies 1.01 eV and 1.24 eV, which lead to negligible and high efficiency transition to the mestable state, respectively. As seen in Figure 2, the electron mobility and the electron concentration decrease with time for the photon energy of 1.24 eV, whereas even prolonged illumination with photon energy of 1.01 eV does not affect either the mobility or the electron concentration values.

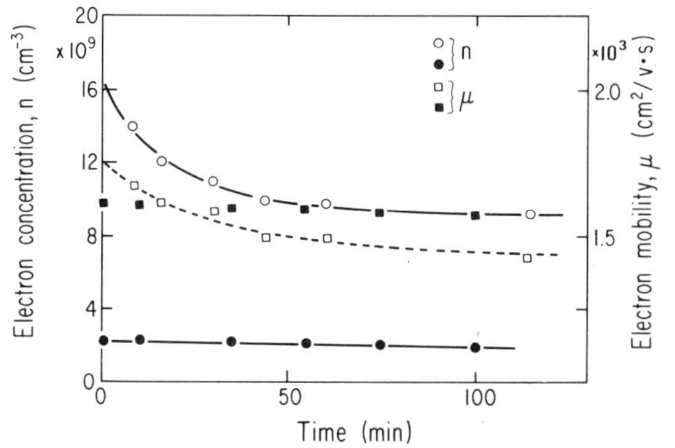

Figure 2 Time dependence of electron mobility and electron concentration upon illumination with photon energies, $h\nu = 1.24$ eV (□, ○) and $h\nu = 1.0$ eV (■, ●).

Theoretical analysis

ELECTRON MOBILITY AT LOW TEMPERATURES

The experimentally observed strong dependence of the mobility on the electron concentration in the temperature range of 100–200 K cannot be explained in terms of the classical electronic transport involving free carrier scattering by the lattice and by ionized impurities. In GaAs an increase of the Hall mobility with increasing carrier concentration can be due to enhanced screening effects which reduce scattering efficiency. However, the magnitude of this effect is too small to account for the present experimental results. Transition from a mixed conductivity to n-type conductivity can in general increase the Hall mobility in SI GaAs; however, this cannot be the case here, as seen from a simple comparison of the mobility and the conductivity characteristics.

Thus, it is apparent that an entirely different concept of electron transport is required to understand the low temperature electron mobility behaviour in SI GaAs. Spatial fluctuations of the electrostatic potential become a key element of such a concept. In 'undoped' SI GaAs there are shallow acceptors and donors in concentrations ranging from 10^{15}–10^{16} cm^{-3}. The net concentration of ionized acceptors $N_A^- - N_D^+$ is compensated by deep midgap donors, EL2 and EL0. This process leads to the localization of the Fermi energy close to the midgap and consequently to the high resistivity of the material. Spatial fluctuations of concentration of the ionized donors, N_D^+, and acceptors N_A^- will result in large fluctuations of the electrostatic potential. It has been proposed that the conductivity in such a solid system should have an activation character, σ given by [9, 10]:

$$\sigma \sim \exp(-E_1/kT) \qquad (1)$$

The activation energy E_l depends on the distribution of impurities in the sample. If the distribution of impurities is 'frozen' at a temperature T_0 in the process of the post-growth cooling, the characteristic length at which the distribution is correlated, R_0 is given by [9]:

$$R_0 = \left(\frac{\varepsilon_0 k T_0}{4\pi(N_D + N_A)e^2} \right)^{1/2} \tag{2}$$

where ε_0 is the dielectric constant and $N_D + N_A \simeq 2N_D$. Following Ref. 9, we find that:

$$E_l = CV(R_0) \left[\ln \left(\frac{n_t}{n} \right) \right]^{1/2} \tag{3}$$

where

$$V(R_0) = \frac{e^2 (N_D R_0^3)^{1/2}}{\varepsilon_0 R_0} \tag{4}$$

where $V(R_0)$ is the magnitude of a typical potential fluctuation, C is a constant, n is the average free electron concentration and n_t is given by:

$$n_t = \frac{m^* V(R_0)}{8\pi^2 h^3}$$

It should be noted that n is not equal, but only proportional, to the electron concentration determined from Hall effect measurements in our experiments.

According to equation (1) the activated conductivity is an exponential function of the variable $[\ln(n_t/n)]^{1/2}$. The experimental results, shown in Figure 3, are in good agreement with this dependence. It is also seen that the slopes of the straight lines in Figure 3 are inversely proportional to temperature, consistent with the dependence $\sigma \sim \exp(-E_l/kT)$.

The completely different behaviour of the electron mobility observed at 70 K can be explained on the basis of equations (2) and (3) by noting that the activation energy increases with increasing concentration of the ionized donors. The finding that the saturation of the mobility occurs in the spectral range in which the EL2 is optically excited to the metastable state indicates that the transition to the metastable state is accompanied by a change in the charge state of EL2. This change takes place when the transition to the metastable state consists of two steps: intra-centre excitation of an electron within the EL2 defect, followed by a transition to an acceptor state. Such processes are known to be responsible for bleaching of photochromic materials [11]. Thus, the activation energy at 70 K increases with increasing free electron concentration n and decreases with ionized donor concentration $N_D^+ = N_D + N_{dD}^+$, where N_{dD}^+ is the concentration of optically ionized deep donors. Since, as seen in Figure 1, those two contributions almost cancel

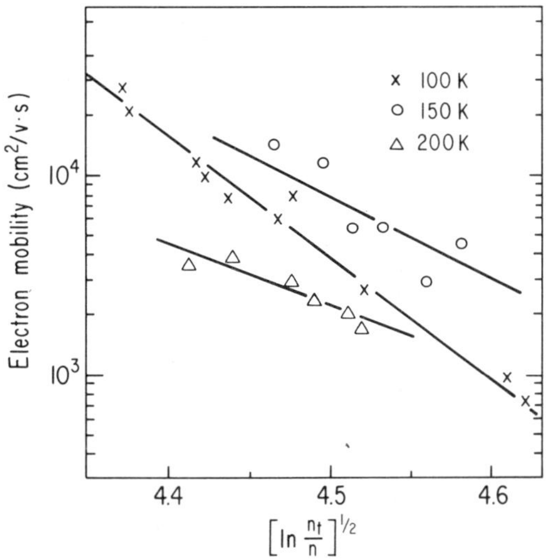

Figure 3 Electron mobility as a function of the variable $[\ln(n_t/n)]^{1/2}$ for various temperatures. The slope of the straight line is inversely proportional to the temperature.

each other, we can estimate the change of the ionized donor concentration N_{dD}^+. For the case given in Figure 1, it is found that the required optically induced change of N_{dD}^+ for the photon energy 1.24 eV is approximately equal to 0.16 N_D. Since the shallow donor concentration is assumed to be of the order of $\sim 10^{16}$ cm^{-3}, we obtain a value of 1.6×10^{15} cm^{-3} for the concentration of ionized deep donors. It was assumed in the present analysis that the typical size of the electrostatic potential fluctuations is determined by the distribution of donors and acceptors at the temperature T_0. This is valid only if the concentration of the intrinsic carrier at T_0 is lower than the concentration of donors and acceptors. When the intrinsic concentration is greater than that of the donors and acceptors, there is no correlation of the donor and acceptor distributions, and the typical magnitude of the potential fluctuations is determined by the concentration of free carriers [10]:

$$R_1 \approx (N_D/n^2)^{1/3} \qquad (5)$$

The transport of electrons in this case can also be described in terms of activated conductivity with the activation energy given by the expression:

$$E_1 = C_1 \frac{e^2(N_D R_1^3)^{1/2}}{\varepsilon_0 R_1} \qquad (6)$$

where C_1 is a constant.

ELECTRON MOBILITY AT ELEVATED TEMPERATURES

All of the above considerations clearly indicate that there are large fluctuations of the electrostatic potential in SI GaAs. This finding has important consequences for high temperature electron transport. In a series of papers on high temperature carrier transport in SI GaAs, it was assumed that ionized impurity scattering is the important mechanism limiting electron mobility [3, 4, 12]. However, this scattering process is based on the assumption that all ionized impurities act as individual scattering centres. In the light of the present analysis it is obvious that the effect of ionized impurities on high temperature electron mobility in SI GaAs should be reconsidered. At temperatures exceeding room temperature, the conductivity ceases to have an activated character and the treatment of electron conductivity in terms of the Boltzmann equation is applicable. However, the fluctuations of the electrostatic potential can act as efficient scattering centres, although they cannot localize electrons. In order to estimate the effect of such fluctuations on the electron mobility, we assume that the scattering rate is given by the expression:

$$\frac{1}{\tau_f} = \frac{v}{R_0} P(R_0) \tag{7}$$

where v is the electron velocity, R_0 is the electron free path, equal to the typical size of the fluctuation in our case, $P(R_0)$ is the scattering probability of a given fluctuation which can be taken as proportional to the square of the potential $[V(R_0)]^2 \sim [N_D^{1/2}]^2 = N_D$. Thus:

$$\frac{1}{\tau_f} = PN_D \left(\frac{x}{m^*}\right)^{1/2} (kT)^{1/2} \tag{8}$$

$x = E/kT$, E is the electron energy and P is a constant. We have included this scattering process in mobility calculations at high temperatures ($T \geq 300$ K). The mobility was calculated using the variational procedure [13]. Polar optical phonon, piezoelectric and deformation potential acoustic scattering were also included. The results of the calculations for two different samples, along with experimental data in the temperature range 300–550 K, are given in Figure 4. The scattering due to potential fluctuations, together with phonon scattering, account very well for the temperature dependence of the electron mobility. Combining phonon scattering with scattering on isolated ionized impurities, a much weaker dependence on temperature is obtained.

In view of the present results the validity of the commonly used methods for the characterization of SI GaAs must be reconsidered [4, 5, 12]. In such methods, the concentration of ionized impurities is determined by comparing the experimental mobility with that calculated theoretically, assuming ionized impurity scattering as one of the dominant scattering mechanisms. Using this procedure for the results presented in Figure 4 one would obtain an apparent

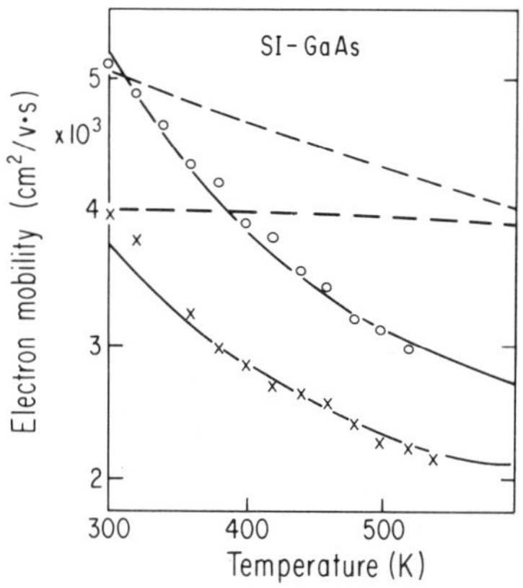

Figure 4 Temperature dependence of the electron mobility of undoped SI GaAs. The solid curves represent the theoretical calculations including scattering due to potential fluctuations with parameter $PN_D = 1.4 \times 10^5$ cm^{-1} (○), $PN_D = 2 \times 10^5$ cm^{-1} (×). The dashed curves represent the results of mobility calculations in which scattering takes place on isolated ionized impurities. The assumed ionized impurity concentrations are: $N_i = 10^{16}$ cm^{-3} (upper curve) and $N_i = 3 \times 10^{16}$ cm^{-3} (lower curve).

ionized impurity concentration strongly dependent on temperature. Since the parameter PN_D, describing the strength of the potential fluctuation scattering, is proportional to the donor concentration, the values of electron mobility can still be used for a qualitative assessment of SI GaAs. More quantitative assessments, however, would require a more detailed formulation of the electron scattering by the fluctuations of the electrostatic potential.

Conclusions

In this paper we have considered 'undoped' SI GaAs with electrons as the majority carrier; however, the present findings are also applicable to SI GaAs exhibiting mixed type conductivity. In this case, the effect of the potential fluctuations on the hole as well as on the electron mobility has to be considered. In particular, the potential fluctuation scattering will affect the electron-to-hole ratio, which is an essential parameter in the phenomenological description of the Hall effect mobility in SI GaAs exhibiting mixed conductivity. It should be noted that, owing to screening effects, the efficiency of the potential fluctuation scattering decreases rapidly with increasing free carrier (electron or hole) concentration. Therefore, this mechanism does not play a prominent role in conducting GaAs.

ACKNOWLEDGEMENTS

The authors are grateful to the Office of Naval Research for financial support.

References

1. Martin G. M., Verheijke M. L., Jansen J. A. J. and Poiblaud G. (1979). *J. Appl. Phys.*, **50,** 467
2. Look D. C. (1977). *J. Appl. Phys.*, **48,** 5141
3. Winter J. J., Leupold H. A., Ross R. L. and Ballato A. (1982). In *Semi-Insulating III–V Materials, Evian*, p. 134 (S. Makram-Ebeid and B. Tuck, eds.); Shiva Publishing, Nantwich, UK
4. Martin G. M., Farges J. P., Jacob G. and Hallais J. P. (1980). *J. Appl. Phys.*, **51,** 2840
5. Lee H. J. and Park Y. S. (1984). *J. Appl. Phys.*, **55,** 402
6. Lin A. L., Omel'yanovski E. and Bube R. H. (1975). *J. Appl. Phys.*, **47,** 1852
7. Martin G. M. (1981). *Appl. Phys. Lett.*, **39,** 747
8. Lagowski J., Gatos H. C., Aoyama T. and Lin D. G. (1984). This volume, p. 60
9. Halpern Y. S. and Efros A. L. (1969). *Fiz. Iverd. Tela.*, **11,** 2301
10. Efros A. L., Halpern Y. S. and Shlovsky B. I. (1972). *Proceedings of the XI Int. Conf. Phys. Semicon., Warsaw*, p. 126; edited by Polish Scientific Publishers
11. Markham J. (1966). *F Centres in Alkali Halides*; Academic Press
12. Walukiewicz W., Pawlowicz L., Lagowski J. and Gatos H. C. (1982). In *Semi-Insulating III–V Materials, Evian*, p. 121 (S. Makram-Ebeid and B. Tuck, eds.); Shiva Publishing, Nantwich, UK
13. Walukiewicz W., Lagowski J., Jastrebski L., Lichtensteiger M. and Gatos H. C. (1978). *J. Appl. Phys.*, **50,** 899

Defect Thermodynamics in Melt-Grown GaAs and Semi-Insulating Behaviour

H. E. Ruda*, J. Lagowski, H. C. Gatos and W. Walukiewicz†

Massachusetts Institute of Technology, Cambridge, Massachusetts 02139, USA

Abstract

We present a thermodynamic treatment of defects in melt-grown GaAs which provides a framework for understanding semi-insulating behaviour. Two defect schemes are considered: in one only native point defects and their complexes are considered, while in the other chemical impurities are also included. The present approach takes into account the defect state of the crystal at the solidification temperature and the evolution of defects upon cooling. By solving the partial electroneutrality conditions in conjunction with mass balance criteria as a function of thermodynamic system parameters, the conditions under which primary defect equilibria occurred are established. It is shown that both schemes predict a set of possible SI regimes. By comparing the two schemes, the influence of impurities on SI behaviour is made evident; at the same time both schemes can lead to stoichiometry-controlled SI to p-type transitions. This transition takes place at a critical arsenic activity (or arsenic fraction) in the growth melt which depends on the microscopic origin of the deep donor and also on the impurity concentration.

It is now widely recognized that the very high resistivity of 'undoped' melt-grown GaAs is attributable to the balance between the net concentration of shallow acceptors and that of deep donors. This balance is upset when the arsenic fraction in the growth melt falls below a certain critical value [1, 2]. It is thus evident that point defects controlled by the arsenic at or near the melting point play an important role in achieving SI GaAs.

In this work we consider the evolution of defects during the post-solidification cooling of the crystal. Using defect thermodynamics, we address the fundamental question: can SI GaAs be obtained as a result of a given balance among native defects and/or their complexes; or is it necessary to involve impurity related centres? We pursue two defect schemes: the native defect scheme (ND) and the defect-impurity scheme (DI), respectively.

Due to the finite existence region the composition of melt-grown GaAs can deviate from the ideal 1 : 1 arsenic-to-gallium concentration ratio. This deviation is controlled by the arsenic activity in the melt, a_{As}, (or the melt composition) and is accommodated in the solidifying crystal in the form of native point defects. The arsenic-to-gallium ratio remains unchanged during

* Present address: University of Toronto, Downsview, Ontario, Canada.
† Permanent address: Institute of Physics, Polish Academy of Sciences, Warsaw, Poland.

the post-solidification cooling of the crystal. However, the types of defects and their concentrations change in order to approach thermodynamic equilibrium at lower temperatures.

The corresponding defect interactions are analysed using mass balance criteria and the electroneutrality condition. In the present treatment we approximate the total set of these conditions by partial conditions (PC) in which two defects are considered to predominate. The choice of defect pairs determining a given PC was done by taking into account the compensation mechanism in SI GaAs and the conversion to p-type.

In the native defect scheme we have considered the following ionized donor type defects: arsenic vacancy V_{As}^{+}; arsenic interstitial As_i^{+}; the antisite defect As_{Ga}^{2+} and two antisite complexes $(V_{As}As_{Ga})^{2+}$ and $(V_{As}As_{Ga}V_{Ga})^{2+}$. The antisite defect As_{Ga} [3] and related complexes [4, 5] are likely candidates for the microscopic model of the native midgap donor EL2. The following acceptor type defects were treated: gallium vacancy V_{Ga}^{-}; gallium interstitial Ga_i^{-} and the antisite defect Ga_{As}^{2-}. A discussion of the defect-impurity scheme will be limited to the role of the substitutional acceptor dopants.

The partial neutrality conditions in the ND scheme and their evolution upon transition from gallium-rich to arsenic-rich growth melts are shown in Figure 1. Acceptor type native defects dominate on the gallium-rich side, which is consistent with the inherent p-type behaviour of undoped crystals grown from Ga-rich melt [1, 2]. The predominant native acceptor defects are Ga_{As} and Ga_i. Increase of the arsenic activity increases the concentration of donor type defects. Close to ideal stoichiometry compensation of GaAs can be maintained due to the balance between the acceptor type, Ga_{As}, and donor type, As_{Ga} antisite defects. A balance between gallium and arsenic interstitials or between gallium and arsenic vacancies, although possible, must be very unlikely. Simultaneous incorporation of excess gallium and arsenic atoms

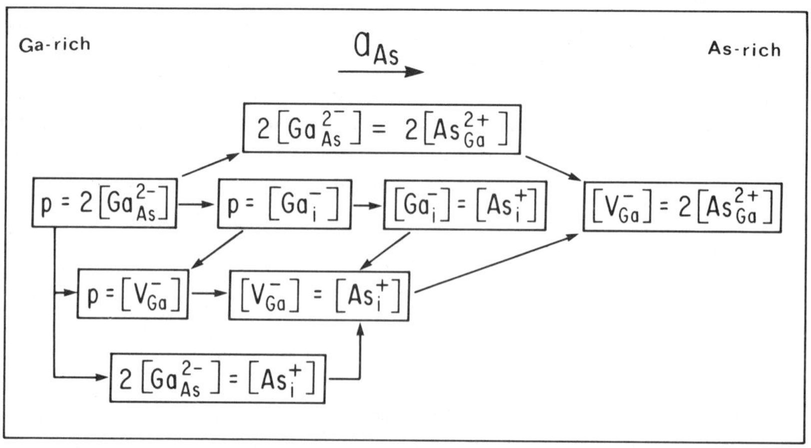

Figure 1 Change of partial electroneutrality conditions with increasing arsenic activity in the growth melt.

(as interstitials) during solidification is very unlikely. On the other hand, the incorporation of large concentrations of both gallium and arsenic vacancies would undoubtedly lead to vacancy condensation and thus to their subsequent removal from the lattice upon formation of dislocation loops [6]. Growth from arsenic-rich melts favours the formation of the gallium vacancies and the antisite defect As_{Ga}. Therefore, the ND scheme, at high a_{As}, can be approximated by only one partial condition: $[V_{Ga}^{-}] = 2[As_{Ga}^{2+}]$.

The incorporation of defect complexes increases significantly the number of PCs. In the present brief report we combine a discussion of defect complexes with the role of impurities.

The significant effects of impurities on defect interactions in GaAs have been demonstrated by our experimental results on EL2 suppression by donor impurities [4]. Most recently we have discovered that dislocations formed upon vacancy condensation are suppressed by donor impurities and enhanced by acceptor impurities [7]. For SI GaAs the most critical effects of impurities are those associated with carbon. These effects calculated from our model are illustrated in Figures 2 and 3 for three different microscopic models of the deep donors: As_{Ga}, $As_{Ga}V_{As}$ and $V_{Ga}As_{Ga}V_{As}$, respectively.

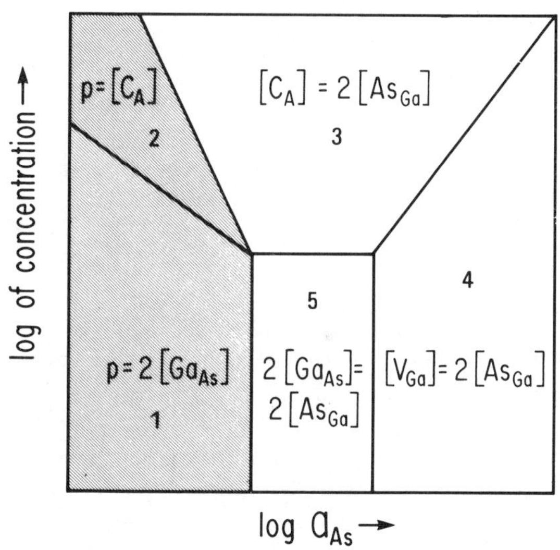

Figure 2 Regions of different partial electroneutrality conditions taking As_{Ga} as the EL2 defect. The vertical axis is the log of the substitutional acceptor concentration. Shaded areas denote p-type GaAs and plain areas correspond to SI GaAs.

The logarithm of the impurity concentration and the logarithm of the arsenic activity in the melt are two independent variables in these figures. Regions corresponding to a given PC condition are enclosed by boundary lines with slopes calculated from the present model. Shaded areas denote p-type material, while the plain areas correspond to semi-insulating material.

It is evident from Figure 2 that the transition from p-type to semi-insulating GaAs, p→SI, can be achieved by increasing the arsenic activity. It is also seen in this figure that if the midgap donor level As_{Ga} is a dominant donor defect, then an addition of acceptor impurity above a critical level can shift the transition point p→SI towards lower a_{As} values, thus extending the melt compositional range yielding semi-insulating material.

This phenomenon is opposite to the EL2 annihilation by shallow donors discussed in Ref. 4. It can be explained in terms of a general thermodynamic trend according to which addition of acceptors enhances the concentration of donors and vice versa [8]. A similar but quantitatively more significant tendency is observed for the other EL2 models (Figure 3). These antisite complex models of the deep donor give even stronger dependence of a_{As}^{crit} on the acceptor concentration extending SI behaviour to lower arsenic activities.

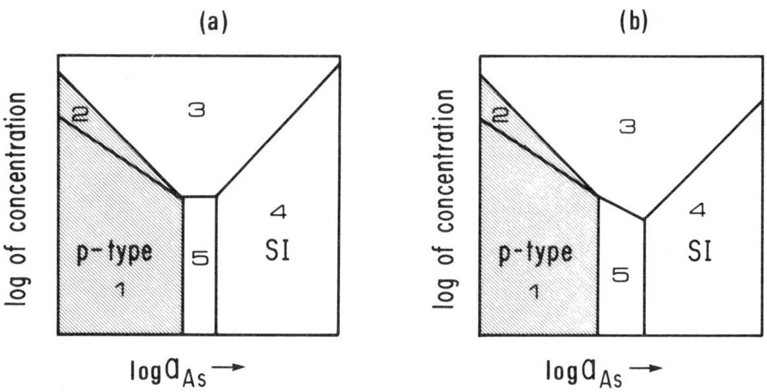

Figure 3 Diagram of partial electroneutrality conditions taking EL2 as (a) $As_{Ga}V_{As}$ and (b) $V_{As}As_{Ga}V_{Ga}$. The numbers correspond to the same neutrality conditions as in Figure 2.

The present method was also used to analyse defect thermodynamics in a more complicated situation in which complexes containing boron are included [9]. In this case the critical arsenic activity depends more strongly on the boron concentration than on the impurity concentration shown in Figures 2 and 3. However, at high a_{As} we invariably revert to the same defect-controlled SI behaviour.

ACKNOWLEDGEMENTS

The authors are grateful to the International Business Machines Corporation and the Air Force Office of Scientific Research for financial support.

References

1. Ta L. B., Hobgood H. M., Rohatgi A. and Thomas R. N. (1982). *J. Appl. Phys.*, **53,** 5771
2. Holmes D. E., Chen R. T., Elliott K. R., Kirkpatrick C. G. and Yu P. W. (1982). *IEEE Trans. El Dev.*, **ED-29,** 1045
3. Lagowski J., Gatos H. C., Parsey J. M., Wada K., Kaminska M. and Walukiewicz W. (1982). *Appl. Phys. Lett.*, **40,** 342
4. Lagowski J., Kaminska M., Parsey J. M., Gatos H. C. and Walukiewicz W. (1982). *Inst. Phys. Conf. Ser.*, **65,** 41
5. Chou Y. (1981). *Inst. Phys. Conf. Ser.*, **63,** 185
6. Brice J. (1970). *J. Cryst. Growth,* **7,** 9
7. Lagowski J., Gatos H. C., Aoyama T. and Ling D. G. (1984). This volume, p. 60.
8. Kröger F. A. (1964). *The Chemistry of Imperfect Crystals*; North-Holland Publishing Company, Amsterdam
9. Hobgood H. M., Eldridge G. L., Barrett D. L. and Thomas R. N. (1981). *IEEE Trans. El. Dev.*, **ED-28,** 140

Optical Properties of EL2 in GaAs

L. Samuelson, P. Omling, E. R. Weber* and H. G. Grimmeiss

Department of Solid State Physics, University of Lund, Box 725, S-220 07 Lund, Sweden

Abstract

The merits of absorption measurements for characterization of semi-insulating GaAs have been extended to allow selective measurement of EL2 absorption, which differs from previously reported absorption spectra. Using this technique it is shown that one can determine the EL2 equilibrium occupancy which is of importance for devices which use SI GaAs substrates. The EL2 absorption is also determined to be only a minor portion of the total absorption in plastically deformed samples. Comparison with EPR (electron paramagnetic resonance) measurements seems to exclude the direct EL2 – As_{Ga} correlation. Analyses of photoluminescence and photoluminescence excitation spectra of material with different quasi Fermi level positions suggest that the EL2 luminescence is due to electron capture into the EL2 ground state.

Introduction

The technical importance and remarkable physical properties of the GaAs: EL2 defect has made it one of the most investigated during recent years [1]. From a technical point of view it is important to obtain techniques which allow control of the incorporation and properties of the midgap centre EL2 in GaAs [2]. In this paper will be shown how an extension of conventional low temperature absorption measurements considerably increases the usefulness of such techniques. This will be exemplified by data from different types of bulk crystals. Photoluminescence techniques combined with absorption measurements form the basis for a discussion about the radiative processes in SI GaAs.

Experimental

The GaAs material used for this investigation was bulk GaAs grown by the LEC (liquid encapsulated Czochralski) technique [3]. EL2-containing material has been especially grown, selected and treated in order to obtain material with different concentrations of EL2 and different positions of the Fermi level [4]. Plastically deformed (PD) samples (2.5% deformation) were obtained by static compression at 400°C in an atmosphere of 8% H_2 and 92% N_2 [5]. These bulk samples have been used for EPR (electron para-

* Permanent address: Department of Materials Science, UC Berkeley, California 94720, USA.

magnetic resonance), PL (photoluminescence) and PLE (photoluminescence excitation) and absorption measurements.

PL was excited by 10–200 mW of 5145 Å Ar ion laser light focussed on samples immersed in superfluid He ($T \simeq 2$ K). PL was detected by a 0.75 m double grating monochromator (Jarrell-Ash) equipped with a LN_2 cooled InAs photodiode. The scanning light source for PLE measurements was a 250 W tungsten lamp filtered through a Zeiss MM3 prism monochromator. In this case the PL signal was detected broad-banded using a Ge filter in front of the InAs detector. A similar arrangement as for PLE was employed for absorption measurements.

Results and discussion

As was shown previously by Martin [6], conventional absorption measurements of GaAs bulk material can yield information about the concentration of EL2, [EL2], in the material. However, this type of investigation is unsuitable for material with contributions to the absorption e.g. from other deep level defects. There is, however, a very simple solution to the problem of how to selectively measure the EL2 component of absorption spectra, since the EL2 absorption (α_{EL2}) can be bleached by transformation of a normal, neutral EL2 state ($EL2^0$) to its metastable state ($EL2^*$). A straightforward measurement of the transmission spectra before ($T(h\nu)$) and after ($T_Q(h\nu)$) pumping in the absorption band which induces the transition $EL2^0 \rightarrow EL2^*$ ($1.0\text{ eV} \leqslant h\nu \leqslant 1.3\text{ eV}$) directly allows the determination of α_{EL2} by:

$$\alpha_{EL2}(h\nu) = \frac{1}{d} \ln\left(\frac{T_Q}{T}\right) \qquad (1)$$

Note that this measurement of the EL2 absorption coefficient is independent of assumptions about reflection coefficients, internal reflections etc. ($d =$ sample thickness). Figure 1 shows that this method for determination of α_{EL2} works for widely different types of material, ranging from n-type to different SI as well as plastically deformed SI material. The spectra in Figure 1 (obtained at $T \simeq 2$ K) thus show, for the n-type material (W900, Si doped, $n \simeq 10^{17}\text{ cm}^{-3}$) the true spectrum for $\sigma_n^0(h\nu)$ plus contributions from the cross section leading to quenching of EL2, $\sigma^*(h\nu)$, i.e. $\alpha_{EL2} = [EL2] \times (\sigma_n^0 + \sigma^*)$. This selectivity in the measurement of α_{EL2} is responsible for the resolution of the peak at $h\nu \simeq 1.2$ eV which was not seen in the low temperature spectra of Martin [6]. It is also hidden in the background of EL2-unrelated absorption bands in our data if we plot the total absorption spectra. This well-resolved absorption band supports the recent suggestion that the 1.2 eV peak is due to an internal EL2 excitation [7, 8] but is in contradiction with earlier interpretations in terms of thresholds in the σ_n^0 photoionization cross-section [9].

In SI material we have observed two types of α_{EL2} behaviour. Many types of materials behave very similar to n-type material, i.e. the absorption reflects

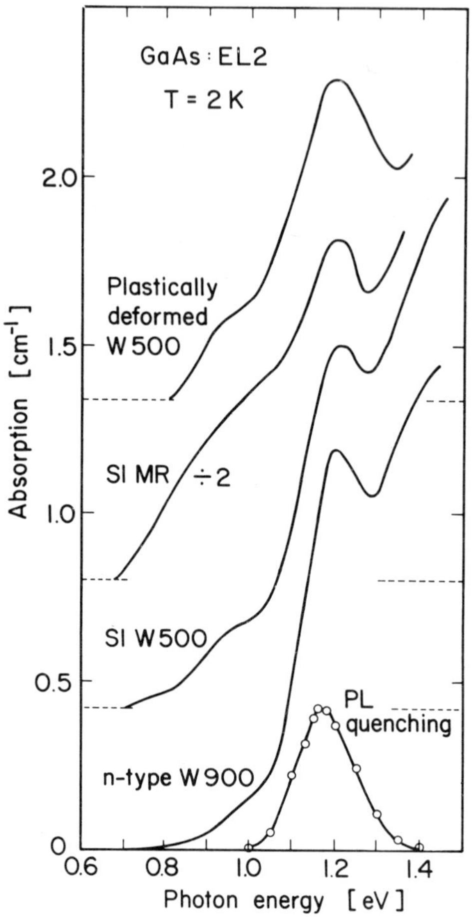

Figure 1 Absorption spectra of EL2 in different GaAs materials using the new measurement technique described in the text. The spectral distribution of the $EL2^0 \rightarrow EL2^*$ transition (from Ref. 15) is also included.

excitations from the filled EL2 donor level. This is exemplified by sample W500. Other materials, like sample MR in Figure 1, show a different spectral behaviour, especially for low photon energies. We identify this difference as being due to contributions from σ_p^0 of EL2, which at 300 K is known to have a peak around 0.9 eV. Separating the total α_{EL2} into its contributions involving filled and unfilled EL2 centres will thus allow determination of the fractional occupancy (f) of EL2, by $\alpha_{EL2} = [EL2]\ (f(\sigma_n^0 + \sigma^*) + (1-f)\sigma_p^0)$. Optical, non-destructive characterization of SI GaAs material can thus determine not only the total EL2 concentration but also the actual occupancy of the level. Acquiring this information is important for the understanding of interactions between free carriers in devices and the deep EL2 centres in the substrate [2]. Knowing the actual occupancy of EL2 is, furthermore, essential for evaluation of how a substrate is prone to thermal conversion during e.g.

post-implantation thermal annealing, since the degree of compensation due to acceptors accumulated close to the surface (e.g. Mn) will depend on the equilibrium EL2 occupancy [2, 10].

Plastic deformation of SI GaAs material has been shown to produce high concentrations of the arsenic antisite (As_{Ga}) as determined by EPR [5]. Since plastic deformation also increases the total level of extrinsic absorption in the material it is crucial for investigation of any correlation between EL2 and As_{Ga} to determine the portion of the absorption which is due to EL2. Using the method of selective EL2 absorption, as in equation (1), we find that the EL2 absorption (see Figure 1) is almost identical in its spectral dependence as well as the absolute magnitude in undeformed (W500) and plastically deformed material (PD-W500). In these materials the As_{Ga} concentrations have, from the As_{Ga}^{4+} resonance, been determined to be 5×10^{15} cm^{-3} and 3×10^{16} cm^{-3}, respectively. We thus conclude that the induced increase in the concentration of As_{Ga} defects does not correspond to a change in the EL2 concentration but to other, non-quenchable, deep levels. However, one can not exclude the possibility that EL2 comprises a subset of As antisites which include As_{Ga} in a complex.

Numerous papers have dealt with the luminescence of bulk GaAs which in most cases consists of a band peaking at 0.63 eV, another at 0.68 eV or

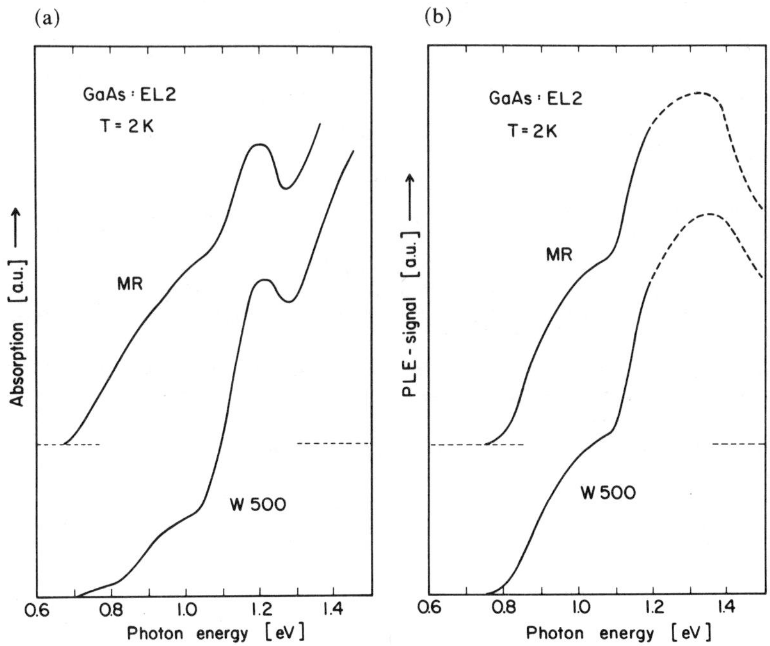

Figure 2 Comparison of (a) absorption and (b) PLE measurements of two different SI GaAs crystals. The spectral shapes of PLE above 1.2 eV (dashed lines in (b)) are probably distorted due to self-absorption, which is more severe for the MR sample where the EL2 concentration is higher (see Figure 1).

combinations of both [11–14]. PLE spectra of these bands detected broad-banded show structure in the intrinsic and near-intrinsic spectral regions [14, 15]. In one paper [15] an extrinsic PLE spectrum was identified as σ_n^0 of EL2. Since EL2 is the dominant deep level available for two-step excitations in this type of material, care must be taken in the interpretation of such spectra. The EL2 absorption spectra of the two type cases of SI GaAs are plotted in Figure 2a and the corresponding PLE spectra in Figure 2b. Evidently, the extrinsic PLE spectra are very similar in shape even when the absorption spectra display differences due to different equilibrium occupations. Both spectra in Figure 2b resemble the σ_n^0 spectrum of EL2 with a characteristic second step at $h\nu \simeq 1.1$ eV. Therefore, we believe that this is a measurement of the σ_n^0 spectrum without contributions from σ^*, since σ^* excitations do not lead to PL. The increased importance of σ_n^0 in PLE as compared with absorption spectra suggests that the σ_n^0 process directly excites the luminescent electronic states. This interpretation leads to the capture of electrons into EL2 as being responsible for the 0.68 eV PL band in SI GaAs.

References

1. Martin G. M. and Makram-Ebeid S. (1983). *Physica*, **116B,** 371
2. Zucca R. (1980). In *Semi-Insulating III–V Materials, Nottingham*, p. 358 (G. J. Rees, ed.); Shiva Publishing, Nantwich, UK
3. Metz E. P. A., Miller R. C. and Mazelsky R. (1962). *J. Appl. Phys.*, **33,** 782
4. Special thanks to Dr F. Köhl (Wacker-Chemitronic) for growth of specific material for this investigation
5. Weber E. R., Ennen H., Kaufmann U., Windscheif J., Schneider J. and Wosinski T. (1982). *J. Appl. Phys.*, **53,** 6140
6. Martin G. M. (1981). *Appl. Phys. Lett.*, **39,** 747
7. Kaminska M., Skowronski M., Lagowski J., Parsey J. M. and Gatos H. C. (1983). *Appl. Phys. Lett.*, **43,** 302
8. Omling P., Samuelson L. and Grimmeiss H. G. (1984). *Phys. Rev.*, **B29,** in press
9. Chantre A., Vincent G. and Bois D. (1981). *Phys. Rev.*, **B23,** 5335
10. Nordquist P. E. R., Klein P. B., Bishop S. G. and Siebenman P. G. (1981). *Inst. Phys. Conf. Ser.*, **56,** 569
11. Mircea-Roussel A. and Makram-Ebeid S. (1981). *Appl. Phys. Lett.*, **38,** 1007
12. Leyral P., Vincent G., Nouailhat A. and Guillot G. (1982). *Solid-State Commun.*, **42,** 67
13. Yu P. W. (1984). *Appl. Phys. Lett.*, **44,** 330
14. Shanabrook B. V., Klein P. B., Swiggard E. M. and Bishop S. G. (1983). *J. Appl. Phys.*, **54,** 336
15. Leyral P. and Guillot G. (1982). In *Semi-Insulating III–V Materials, Evian*, p. 166 (S. Makram-Ebeid and B. Tuck, eds.); Shiva Publishing, Nantwich, UK

Complications in Interpreting Extrinsic IR Photoconductivity for Semi-Insulating GaAs

S. Rahimi*, S. G. Johnson†, W. B. Leigh, C-K. Chau‡ and J. S. Blakemore

Oregon Graduate Center, Beaverton, Oregon 97006, USA

Abstract

There has been a substantial literature on the analysis of the spectral features of extrinsic photoconductivity for semi-insulating GaAs, mostly on the Cr midgap acceptor and on the EL2 defect (or family of defects). Such analyses are complicated by the mutual activity of processes that can generate holes and electrons, thus the problem is ambipolar. We comment here on the effect of compensation on the spectrum for SI GaAs: Cr, and also note for undoped (EL2 containing) SI GaAs the effects of compensation, and of low temperature photoquenching.

Introduction

Compared with some other experimental techniques for the study of deep-level centres, one complication with extrinsic photoconductivity is that the photon energy is often able to excite valence band holes as well as conduction band electrons. Thus the non-equilibrium situation in photo-excited SI GaAs is apt to be an ambipolar one.

Nevertheless, extrinsic photoconductivity is one of the few methods that can be used with SI material, with no further treatment of the material than the application of 'ohmic' contacts. The provision of such contacts has been, and continues to be, one of the major experimental headaches. However, making such contacts (such as evaporated and alloyed Au–Ge) should produce much less irreversible change than the preparation methods for many other kinds of investigation—such as the consequences of donor implantation so that DLTS (deep level transient spectroscopy) can be used in the region made conducting.

Accordingly, there has been a substantial volume of literature on extrinsic photoconductivity analysis for SI GaAs, the early and continuing efforts on spectral response [1–4], and more recently the transient photoconductivity 'PITS' analysis [5–8]. A paper by some of us in the last volume of this series [9] dealt with both of these aspects apropos chromium in GaAs, but here we restrict ourselves to the problem of spectral (photon energy) analysis. We wish to comment on the ambipolar presence for Cr and for EL2 in SI GaAs.

* Present Address: Sonoma State University, Rohnert Park, California, USA.

† Present Address: Intel Corporation, Aloha, Oregon, USA.

‡ Permanent Address: Chico State University, Chico, California, USA.

Chromium

The key equations here are:

For photoionization: $$Cr^{3+} + h\nu \rightarrow Cr^{2+} + h^{+} \tag{1}$$

For photoneutralization: $$Cr^{2+} + h\nu \rightarrow (Cr^{2+})^{*} \tag{2a}$$

$$\rightarrow Cr^{3+} + e^{-} \tag{2b}$$

Most workers now accept the model, first proposed by Kolchanova *et al.* [1], that photoneutralization takes place in two stages, with the $(Cr^{2+})^{*}$ excited state resonant in energy with the conduction band.

Photons exceeding some 0.8 eV can, in principle, ionize neutral Cr^{3+} (producing holes of modest mobility) or excite (leading to photoneutralization of) ionized Cr^{2+} acceptors. The latter, while of lower optical efficiency, provides carriers of much higher mobility, an important consideration in the analysis of photoconductive and photo-Hall data. Thus the photoconductive responsivity spectrum in the range 0.8–1.4 eV includes contributions from electrons and holes, to an extent that depends on the Cr^{2+}/Cr^{3+} compensation, as well as on mobilities and lifetimes of carriers of the two kinds. From the optical studies of Martinez *et al.* [10] and Hennel *et al.* [11], we can see that the optical cross-section for equation (2) maximizes at $\sigma_E \simeq 1.2 \times 10^{-17}\ \mathrm{cm}^2$ for $h\nu \simeq 0.9$ eV; whereas the cross-section σ_p^0 for equation (1) has become

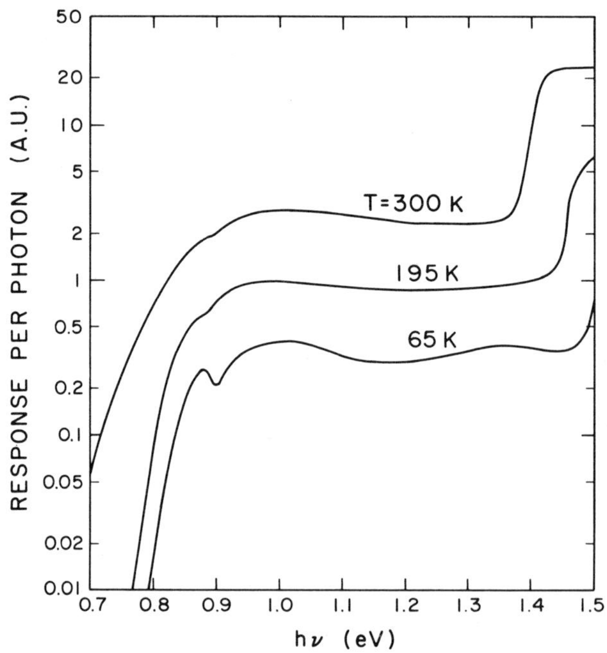

Figure 1 Extrinsic photoconductive response per photon, for a Cr-doped SI GaAs sample. Spectra at three temperatures as indicated. (A. U.—arbitrary units.)

three times larger for that energy, and continues to rise to $\sim 9 \times 10^{-17}$ cm^2 for $h\nu \sim 1.1$ eV.

Figure 1 shows the form of the photoconductive responsivity per photon, for a Cr-doped SI GaAs sample (Cambridge Instruments material). This shows *slight* compensation of the Cr^{2+}/Cr^{3+} level, from ambipolar transport and magnetotransport analysis, and as confirmed by the weakness of the 0.9 eV 'bump' resulting from equation (2). Examples of GaAs:Cr crystals with a much stronger peak at that energy were reported in the work of Kolchanova *et al.* [1] and of Look [3]. Despite all that we do know about the various processes that produce electrons and holes, it is still difficult to make a quantitative comparison between photoconductive spectra such as Figure 1, and the optical absorption characteristics underlying equations (1) and (2). A principal reason for this is that SI GaAs (whether Cr-doped or not) is usually grown under conditions that allow EL2 to be present. On the assumption (still not as fully established as one would like) that the EL2 donor ground state is slightly closer to the conduction band than the Cr^{2+}/Cr^{3+} level, then thermal equilibrium is a situation with ionized EL2 having ionized an equivalent number of Cr acceptors. All four of the optical processes then possible have comparable threshold energies!

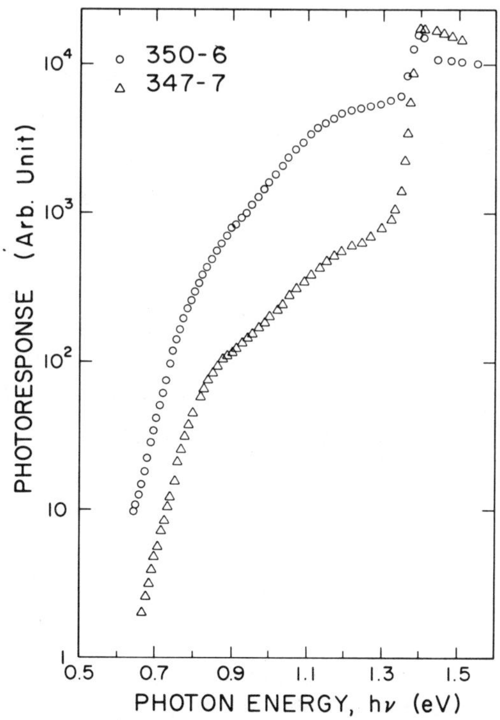

Figure 2 Photoconductive response per photon at 300 K for two undoped LEC SI GaAs samples, these coming from crystals from two different suppliers.

It is thus slightly more straightforward to examine SI material which is supposed to contain EL2, but for which the accidental Cr concentration is hoped to be in the $<10^{15}$ cm^{-3} range. That type of sample is discussed next.

The EL2 mid-gap donor

Among the interesting properties of EL2 is the photoquenching, caused by photoconversion of neutral EL2 into the state EL2* which is metastable at low temperatures [12, 13]. However, the recovery time from the EL2* state is rapid even in the dark above ~140 K, and photoquenching does not affect 300 K photoconductive spectra such as those of Figure 2. Both of these samples are of undoped LEC (liquid encapsulated Czochralski) SI material, sample 347-7 being of Cambridge Instruments material, and sample 350-6 of Cominco material.

Two pieces of evidence in Figure 2 show that the EL2 compensation is larger for Sample 347-7 (and this is supported by transport analysis and photo-Hall data). One clue is the lower extrinsic sensitivity, sign of a shorter electron lifetime (more $EL2^+$ capture sites). The other is a slight 'bulge' superimposed on the falling curve below 1 eV, with its maximum effect near

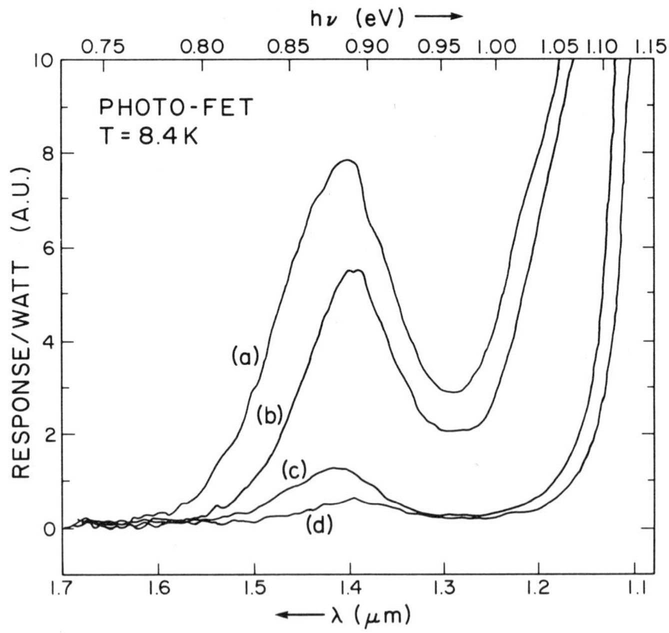

Figure 3 Photoconductive response (per watt) for a GaAs MESFET made by implanting a N^- doped channel on to a SI substrate; showing the spectral region of EL2 response and quenching. Curve (a) obtained with an initial low-intensity scan, and curves (b), (c) and (d) with illumination by 1 eV photons for periods of 1, 5 and 20 minutes, respectively. (A.U.—arbitrary units.)

0.85 eV, apparent for 347-7 but not for 350-6. This comes from the greater opportunity for photoholes to make a contribution, resulting from the $EL2^+ + h\nu \rightarrow EL2^0 + h^+$ photoneutralization reaction.

Lower temperature photoconductive data must be acquired and analysed with even greater care, because of the photoquenching properties of EL2, as exemplified by the succession from curves (a) through (d) in Figure 3. These data are taken using a GaAs MESFET (metal-semiconductor field effect transistor) just below pinch-off, with the channel formed by silicon ion direct implant. The sample has some interesting and much stronger responses above 1 eV (associated with shallow complexes) which we do not show. However, the EL2 response is easily seen at the longer wavelengths, to an extent that can be eliminated with a few minutes of 1 eV illumination. And so photoconductive analysis is, once again, a complicated matter—despite the simplicity of the experimental arrangement.

ACKNOWLEDGEMENTS

This research has been supported in part by the National Science Foundation, through Grant DMR 8305731. We wish to thank R. Y. Koyama of Tektronix Inc., for the GaAs MESFETs, and also for providing several 'virgin' SI wafers that have provided bulk samples.

References

1. Kolchanova N. M., Nasledov D. N., Mirdzhalilova M. A. and Ibragimov V. Yu. (1970). *Sov. Phys. Semicond.*, **4,** 294
2. Lin A. L. and Bube R. H. (1976). *J. Appl. Phys.*, **47,** 1859
3. Look D. C. (1977). *Solid-State Commun.*, **24,** 825
4. Ridley B. K., Arikan M. C., Bishop P. J., Hassan M. F. M. and Machado W. V. (1982). *J. Phys. C*, **15,** 6865
5. Martin G. M. and Bois D. (1978). *Electrochem. Soc. Electron. Div.*, **78-3,** 32
6. Hurtes Ch., Boulou M., Mitonneau A. and Bois D. (1978). *Appl. Phys. Lett.*, **32,** 821
7. Yoshie O. and Kamihara M. (1983). *Jap. J. Appl. Phys.*, **22,** 621
8. Kremer R. E., Abele J. C. and Arikan M. C. (1984). This volume, p. 480
9. Blakemore J. S., Johnson S. G. and Rahimi S. (1982). In *Semi-Insulating III–V Materials, Evian*, p. 172 (S. Makram-Ebeid and B. Tuck, eds.); Shiva Publishing, Nantwich, UK
10. Martinez G., Hennel A. M., Szuszkiewicz W., Balkanski M. and Clerjaud B. (1981). *Phys. Rev. B*, **23,** 3920
11. Hennel A. M., Szuszkiewicz W., Balkanski M., Martinez G. and Clerjaud B. (1981). *Phys. Rev. B*, **23,** 3933
12. Vincent G. and Bois D. (1978). *Solid-State Commun.*, **27,** 431
13. Mitonneau A. and Mircea A. (1979). *Solid-State Commun.*, **30,** 157

Photoluminescence Excitation of 0.68 eV Emission in Undoped Semi-Insulating GaAs

P. W. Yu

University Research Center, Wright State University, Dayton, Ohio 45435, USA

Abstract

Photoluminescence excitation spectra have been obtained at 4.2 K for the 0.68 eV EL2 emission present in undoped SI (semi-insulating) GaAs. The excitation spectrum consists of oscillatory structures at the above-gap energy, a threshold at ~1.49 eV and the subsequent peak at ~1.50 eV. The oscillatory structure results from the energy relaxation of the conduction band electron through the successive longitudinal phonon emission. The 1.50 eV peak increases with the shallow acceptor carbon-related emission intensity.

Characterization of undoped SI GaAs is of great importance in view of its technological applications. Midgap defect levels play an important role in achieving SI materials through compensation mechanisms [1]. One of the deep centres is the intrinsic defect donor, EL2, present at ~0.75 eV from the conduction band edge. The commonly known compensating shallow acceptor in SI materials is the residual carbon on the As site.

We have recently shown [2–4] that three broad featureless PL (photoluminescence bands at 0.63, 0.68 and 0.77 eV are present at 4.2 K in nominally undoped SI materials. The 0.63 eV band [3] was assigned to a centre related, directly or indirectly, with oxygen. The 0.68 eV and 0.77 eV bands are usually present together [4, 5] with a different relative strength in undoped SI materials. Temperature-dependent measurements [2] showed that the 0.68 eV band is due to a radiative transition between EL2 and the valence band. The band has a large Frank–Condon shift $\Delta_{FC} = 110$ meV, and the thermal energy E_{th} from the conduction band edge follows the expression:

$$E_{th} = 0.73 - \frac{5.4 \times 10^{-4}\, T^2}{T + 204}\ \text{eV}$$

Also, the 0.68 eV band exhibits a persistent photoluminescence quenching effect [6] caused by near infrared light at ~0.95–1.40 eV. In this work, we report the photoluminescence excitation (PLE) study made of the 0.68 eV emission.

The photoluminescence excitation spectra were obtained at 4.2 K. The details of the experiment are given elsewhere [7]. Undoped n-type SI crystals grown by the LEC (liquid encapsulated Czochralski) method were used. Typical room temperature electron concentrations are 4×10^7–2×10^8 cm^{-3}.

Figure 1 shows a typical 4.2 K PL spectrum of the 0.68 eV emission present in the crystals chosen for the PLE experiment. Samples having a small component of the 0.77 eV band were chosen. Overall PL in the samples consists of the 0.68 eV and C_{AS}-related emissions. The C_{AS}-related emissions include the neutral donor-to-acceptor pair and the conduction band electron-to-neutral acceptor transitions.

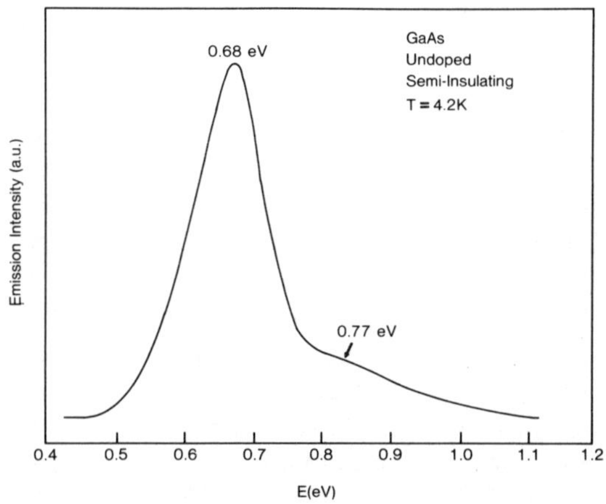

Figure 1 Typical 0.68 eV emission spectrum of a SI crystal used for the PLE experiment.

Figure 2 shows the two types of PLE spectra obtained at 4.2 K. The PLE spectra in Figures 2b and 2a, respectively, exhibit an oscillating behaviour and almost flat response at energies higher than the bandgap. One common feature of the two types of PLE spectra is the presence of a large peak at ~1.50 eV and the PLE threshold at ~1.49 eV. The period of this oscillating

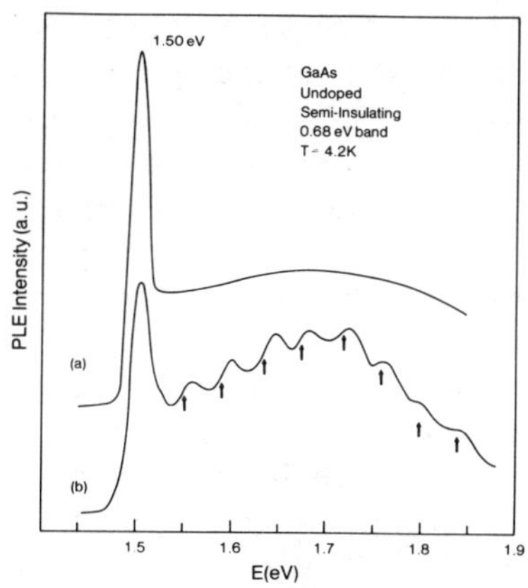

Figure 2 Two different PLE spectra obtained at 4.2 K.

structure is 41.4±0.4 meV and can be described by $\Delta E = \hbar\omega_{LO}(1 + m_e/m_{hh})$, where $\hbar\omega_{LO} = 36.7$ meV is the longitudinal–optical (LO) phonon energy at Γ, m_e is the electron effective-mass and m_{hh} is the heavy-hole effective mass.

Oscillating structures of PLE spectra have recently been observed in deep-centre PL bands [7, 8] at 0.63 and 0.77 eV. Such oscillatory structure is due to the energy relaxation of conduction-band electrons by successive emission of LO phonons. The energy-dependent capture of a conduction-band electron into a shallow ionized donor is essential for the observation of the oscillatory structures. In particular, the maxima and minima of the oscillatory structures give information on the type of oscillatory structure involved in radiative recombination. The two possible types of oscillatory structure were discussed [7, 8] in order to identify the oscillatory structures of the 0.63 and 0.77 eV emission PLE spectra. In this experiment, the maxima and minima of the oscillatory structure of the 0.68 eV emission PLE spectrum exhibit the same maxima and minima obtained from the PLE oscillatory structure of the 0.77 eV emission [7]. This shows that (a) the energy relation of the oscillation of the 0.68 eV PLE spectrum can be represented by the relation $E_n = E_g + (n\hbar\omega_{LO} - E_D)(1 + m_e/m_{hh})$, (where the symbols have the usual meanings) and (b) the conduction-band electron capture into EL2 is made through the shallow donor state. The arrows in Figure 2b show the energy relation given by the above equation. The parameters used are $E_g = 1.5196$ eV, $E_D = 5.8$ meV, $\hbar\omega_{LO} = 36.7$ meV and $m_e/m_{hh} = 0.12$. It may be noted that the arrows would lie approximately at the maxima of the structures if one neglected the donor term in the above relationship. A detailed discussion concerning which of these two models is correct is outside the scope of the present paper.

The non-existence of the pronounced oscillating structure, as shown in Figure 2a, is due to a violation of the necessary condition for energy-dependent capture of the conduction-band electron after successive emissions of the LO phonons. This condition depends on (a) the relation between the photogenerated carrier lifetime and other lifetimes for energy-loss mechanisms and (b) other factors influencing the probability of capture of low kinetic-energy conduction-band electrons into the shallow donor state.

The PLE spectrum in the extrinsic energy range consists of a threshold at ~1.49 eV and the subsequent PLE peak at ~1.50 eV. We observed that the 1.50 eV PLE peak increases proportionally to the relative PL strength of the C_{AS}-related emissions. The crystals used for this experiment, as described earlier, contain carbon as a main shallow acceptor, which is usual in the Czochralski-grown undoped SI materials. The extrinsic PLE spectrum can be attributed to the C_{AS}.

In conclusion, we have observed oscillatory structures in the PLE spectrum of the 0.68 eV EL2 related band present in undoped SI GaAs materials. The characteristics of the structures are the same as those of the 0.77 eV band. A shallow donor state plays an important role in the formation of the oscillation. The extrinsic PLE peak at ~1.50 eV increases with the C_{AS}-related emission intensity.

ACKNOWLEDGEMENTS

This work was performed at the Avionics Laboratory, Wright-Patterson Air Force Base, under Contract No. F33615-84-C-1423.

References

1. Holmes D. E., Chen R. T., Elliott K. R., Kirkpatrick C. G. and Yu P. W. (1982). *IEEE Trans. El. Dev.*, **ED-29**, 1045
2. Yu P. W. (1982). *Solid-State Commun.*, **43**, 953
3. Yu P. W. and Walters D. C. (1982). *Appl. Phys. Lett.*, **41**, 863
4. Yu P. W., Holmes D. E. and Chen R. T. (1981). *IOP Conf. Proc.*, **63**, 209
5. Tajima M. (1982). *Jap. J. Appl. Phys.* **21**, L227
6. Yu P. W. (1984). *Appl. Phys. Lett.*, **44**, 330
7. Yu P. W. (1984). *Phys. Rev. B*, **29**, 2283
8. Shanabrook B. V., Klein P. B., Swiggard E. M. and Bishop S. G. (1983). *J. Appl. Phys.*, **54**, 336

Quantitative Mapping of EL2 Concentration Across SI GaAs Wafers from Infrared Transmittance Data

P. Dobrilla, J. S. Blakemore and R. Y. Koyama*

Oregon Graduate Center, Beaverton, Oregon 97006, USA

Abstract

Measurements are described which seek to determine the IR transmittance of polished undoped SI GaAs wafers with maximum accuracy and stability, in order to deduce thereby the concentration of neutral EL2 over the wafer's area. The measurements reported are at $\lambda = 1.10\ \mu m$, with a 300 K sample temperature. Results are given as grey-scale mosaics (with 1 mm^2 pixels), showing the areas of larger and smaller EL2 concentration. Together with results for wafers of ~5 mm thickness (used in most previous reports of such studies because this gives an optical density $\alpha t \sim 1$), results are also shown here obtained with a wafer only 0.5 mm thick—as typically used in IC (integrated circuit) fabrication. Data for a sample with such a small optical density places maximum requirements on the precision and stability of the transmittance measurements during a lengthy scanning sequence.

Introduction

A knowledge of how various attributes of a grown ingot vary through the volume can assist progress towards more homogeneous and satisfactory SI GaAs for future generations of ICs. Wafer mapping can most easily provide such knowledge. The Evian volume in this series reported on wafer mapping for chromium [1], resistivity [1, 2], dislocation density [1–3] and concentrations of carbon and neutral EL2 [4]. An optical absorption calibration by Martin [5] for EL2 has since been used in other studies [6, 7] and is used here.

Brozel *et al.* [7] studied the fine scale of the EL2 spatial distribution. It was further investigated, qualitatively, by displaying infrared transmittance as detected with a silicon vidicon [8, 9]. The EL2 concentration microstructure, and its relation to dislocation density inhomogeneity, is certainly significant for research into how EL2 arises and can be controlled.

Our report here is more in the tradition of the report by Holmes *et al.* [6], mapping N_{EL} with less spatial resolution (~0.5 mm), but as nearly quantitatively as possible over the wafer's area. This is of interest for the evaluation of as-grown material, and potentially for tracking EL2 redistribution caused by bulk thermal annealing [10]. A particularly demanding aspect is how to do this reliably for a wafer of normal IC thickness (~0.5 mm).

* Tektronix Inc., Beaverton, Oregon 97077, USA.

Experimental arrangement

Figure 1 shows how we measure transmittance, at present at 300 K, and with wafers of 2 inch (5 cm) diameter. The radiation must be reasonably monochromatic if transmittance is to be convertible into a bulk absorption coefficient, and hence to a value for neutral N_{EL}. While stronger EL2 absorption is available at shorter wavelengths, we have chosen (so far) to use a silicon filter to block grating second-order output, and to set the monochromator at a wavelength of 1.10 μm ($h\nu = 1.13$ eV).

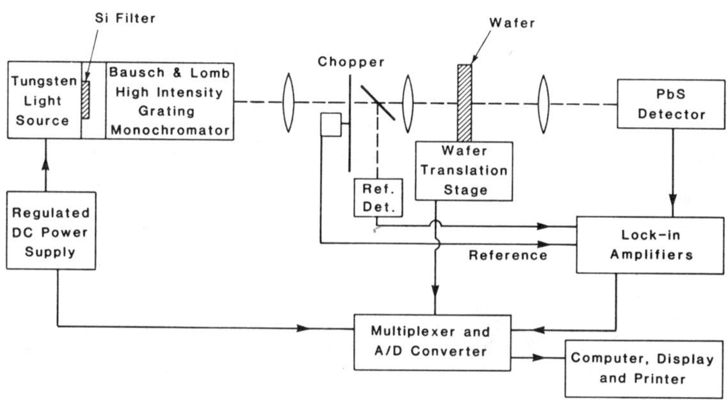

Figure 1 A schematic of the experimental arrangement.

Martin's calibration [5] gives a 300 K absorption cross-section of $\sigma_A = 8.5 \times 10^{-17}$ cm^2 for that wavelength, which we use while recognizing that the room temperature values used by ourselves and others [4, 6, 7] might eventually need some modest readjustment. For, Kaminska *et al.* [11] report at liquid nitrogen temperatures an absorption strength some 15% larger than Martin does for comparable temperatures. Based on refractive index data [12], we also assume a polished surface normal reflectance $R = 0.303$ at 1.10 μm and $T = 300$ K.

Figure 1 shows that our beam of radiation is chopped, focussed on to the wafer (to an area 0.25 mm^2), and refocussed on a detector. A beam-splitter takes part of the beam to a reference detector. Lock-in amplifiers pass on the detector chopped signal components to digitizers and a minicomputer. Up to 30 s of signal averaging may be used in determining the transmittance at each location on the wafer. This information, together with wafer thickness t, and σ_A as noted above, yields a number for neutral N_{EL}. The regulated power supplies for light source, detectors, etc. are in turn fed from a power conditioner to maximize stability during a 400-measurement run that may take up to four hours. Stability to a few parts in 10^4 is thus obtainable, as is necessary for good N_{EL} resolution across the area of a *thin* SI GaAs wafer (i.e. one with $t \sim 0.5$ mm). This is much less critical for the higher optical density of a *thick* wafer ($t \sim 2$–5 mm).

We initially measured along various wafer diameter paths, looking for the W-shaped profile and four-fold symmetry described by others [4, 6, 7]. The data now described use a wafer translation stage providing X–Y coverage of (most of) a wafer's area. The Apple computer interpolates between measured locations in creating a grey-scale plot of neutral N_{EL} so deduced, for a mosaic display.

Results

Figure 2 shows such plots for two thick (t ~ 5 mm) wafer slabs from a 2 inch diameter undoped SI LEC (liquid encapsulated Czochralski) GaAs Cominco crystal. Sample (a) is from the seed end and (b) from the tail of the same ingot. N_{EL} does increase almost 20% from seed to tail in this case—more or less as reported by Holmes *et al.* [6]. However, crystals grown to a 2 inch

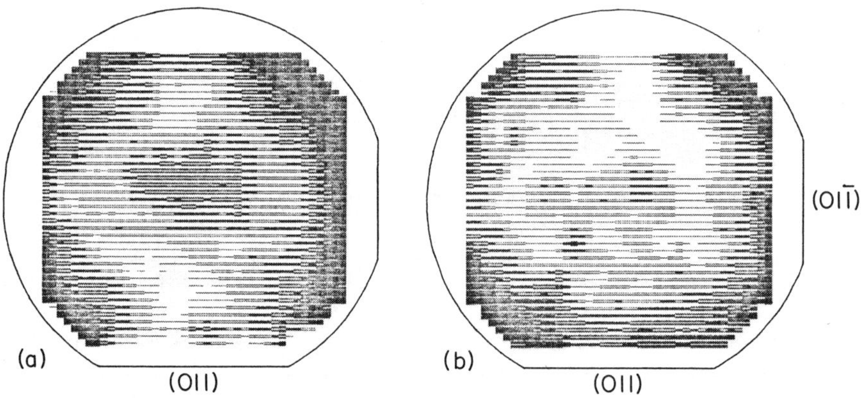

Figure 2 Two examples of mosaic grey-scale plots of N_{EL} over the area of 2 inch diameter thick wafer slabs ($t \simeq 5$ mm), as deduced from the 300 K transmittance at 1.10 μm. Both samples are from the same undoped SI LEC Cominco crystal, with sample (a) from the seed end and sample (b) from the tail end. With units of 10^{16} cm^{-3} for N_{EL}: $\bar{N}_{EL} = 1.71$ for sample (a), with maximum shading for $N_{EL} > 1.925$, no shading for $N_{EL} < 1.475$ and six intermediate steps of 0.075. With tail-end sample (b), $\bar{N}_{EL} = 1.97$, and here maximum shading indicates $N_{EL} > 2.15$, no shading $N_{EL} < 1.85$, and six intermediate steps of 0.05.

size do not necessarily have the same thermal and defect generation history as the 3 inch material reported on by Holmes; and in fact we have compared seed/tail samples from another 2 inch Cominco ingot which showed a slight *decrease* of N_{EL} as the tail region was reached.

The seed end sample of Figure 2a does show a central maximum, but the surrounding ring of reduced N_{EL} shows only two of the four (011) minima that might have been expected from other reports. The tail end sample (Figure 2b) shows neither mirror nor rotational symmetry in its N_{EL} distribution, though it does retain a weak central maximum and evidence of a rise of N_{EL} again towards the periphery.

Figure 3 compares tail end samples from two other Cominco ingots, and neither of these shows either a central maximum or any sign of 4-fold or 2-fold symmetries. The sample of Figure 3a does show a broad stripe of near-constant N_{EL} of (001) inclination, but no symmetry bordering that band!

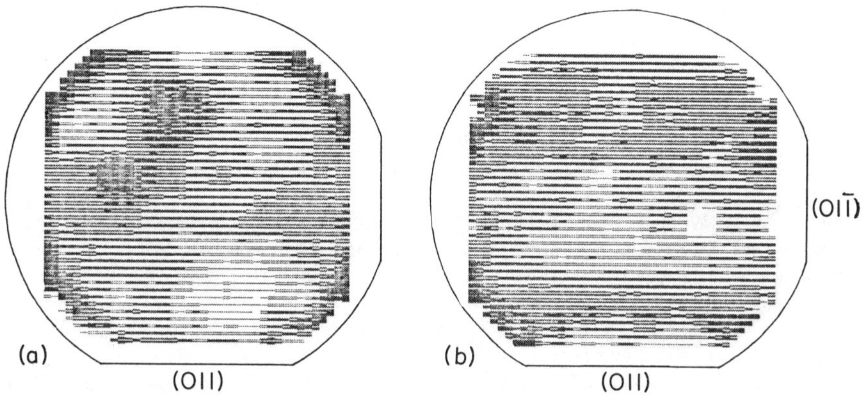

Figure 3 Grey-scale mosaic plots of N_{EL} for tail end wafers from two different undoped SI LEC Cominco crystals, of 2 inch diameter. Sample (a) here is 5 mm thick, as with the samples of Figure 2, and has $\bar{N}_{EL} = 2.25$ (in units of 10^{16} cm^{-3}), with maximum shading for $N_{EL} > 2.40$, no shading for $N_{EL} < 2.10$, and six intermediate steps of 0.05. Sample (b) is much thinner, t = 0.5 mm; and the transmittance data here indicate $\bar{N}_{EL} = 1.84$, with maximum shading for $N_{EL} > 2.00$, no shading for $N_{EL} < 1.70$ and six intermediate steps of 0.05. The 'empty square box' just below right centre is not an area of missing data, but all pixels within that 'box' show N_{EL} to be just less than 1.70×10^{16} cm^{-3}.

The data for the sample of Figure 3b are pleasing to report in that these were obtained using a *thin* wafer, t = 0.5 mm. Thus for the average neutral $\bar{N}_{EL} = 1.84 \times 10^{16}$ cm^{-3}, the optical density $(\sigma_A \bar{N}_{EL} t) = 0.078$, for a transmittance of 48.7%, only 5% smaller than for a sample with no internal absorption. (Indeed, the entire range of the grey-scale in Figure 3b corresponds to a transmittance change of only 0.8%.) Good instrumental stability is thus necessary for data acquisition with a thin wafer, but this is feasible. We hope soon to attempt these quantitative measurements with cooled wafers, recognizing the EL2 absorption quenching possibilities [5] this can present.

ACKNOWLEDGEMENTS

This work is supported in part by the National Science Foundation, through Grant DMR 830-5731. We wish to thank G. Needham of Cominco Ltd. for the 'thick' wafer slabs and J. F. Holmes for valuable discussions on measurement precision.

References

1. Grant I., Rumsby D., Ware R. M., Brozel M. R. and Tuck B. (1982). In *Semi-Insulating III–V Materials, Evian*, p. 98 (S. Makram-Ebeid and B. Tuck, eds.); Shiva Publishing, Nantwich, UK
2. Blunt R. T. (1982). In *Semi-Insulating III–V Materials, Evian*, p. 107 (S. Makram-Ebeid and B. Tuck, eds.); Shiva Publishing, Nantwich, UK
3. Bonnet M., Lent B., Visentin N., Duchemin J. P. and Oberti J. D. (1982). In *Semi-Insulating III–V Materials, Evian*, p. 92 (S. Makram-Ebeid and B. Tuck, eds.); Shiva Publishing, Nantwich, UK
4. Holmes D. E., Elliott K. R., Chen R. T. and Kirkpatrick C. G. (1982). In *Semi-Insulating III–V Materials, Evian*, p. 19 (S. Makram-Ebeid and B. Tuck, eds.); Shiva Publishing, Nantwich, UK
5. Martin G. M. (1981). *Appl. Phys. Lett.*, **39,** 747
6. Holmes D. E., Chen R. T., Elliott K. R. and Kirkpatrick C. G. (1983). *Appl. Phys. Lett.*, **43,** 305
7. Brozel M. R., Grant I., Ware R. M. and Stirland D. J. (1983). *Appl. Phys. Lett.*, **42,** 610
8. Skolnick M. S., Reed L. J. and Pitt A. D. (1984). *Appl. Phys. Lett.*, **44,** 447
9. Skolnick M. S., Brozel M. R., Reed L. J., Grant I., Stirland D. J. and Ware R. M. (1984). *J. Electron. Mat.*, **13,** 107
10. Rumsby D., Ware R. M., Smith B., Tyjberg M., Brozel M. R. and Foulkes E. J. (1983). Digest 1983 GaAs IC Symposium; IEEE
11. Kaminska M., Skowronski M., Lagowski J., Parsey J. M. and Gatos H. C. (1983). *Appl. Phys. Lett.*, **43,** 302
12. Blakemore J. S. (1982). *J. Appl. Phys.*, **53,** R123

Optical, EPR and Radio-tracer Assessment of the Distribution of Cr and EL2 in GaAs

M. R. Brozel, L. Eaves* and J. P. J. McCann*

Department of Electrical and Electronic Engineering, Trent Polytechnic, Burton Street, Nottingham NG7 4BU, UK

Abstract

Cr-doped LEC (liquid encapsulated Czochralski) GaAs has a higher resistivity than undoped SI material grown by the same technique. In order to establish the compensation mechanism in LEC GaAs when both Cr and EL2 are present, we have measured the spatial distribution of Cr^{2+} by EPR (electron paramagnetic resonance) mapping in material uniformly doped with Cr. The distribution of Cr^{2+} follows that of EL2 in undoped samples. We conclude that in heavily Cr-doped GaAs, the Fermi level is pinned at Cr^{2+}/Cr^{3+} and that this level is lower in the gap than the $EL2^0/EL2^+$ level.

The introduction of Cr into otherwise undoped GaAs grown by the LEC method leads to material of resistivity up to 6×10^8 Ω cm [1]. Undoped GaAs is p-type due to residual carbon, but a deep-donor Cr defect is not found in Cr-doped GaAs because the intrinsic deep-donor level, EL2, acts as a compensating level for the shallow C-acceptors. We show here that the role of Cr in this material is to fully ionize those EL2 centres remaining after compensating shallow acceptor species and to pin the Fermi level at the Cr^{2+}/Cr^{3+} level which must lie deeper in the forbidden band gap than $EL2^0/EL2^+$. The distribution of Cr^{2+} centres in Cr-doped LEC GaAs is found to be very similar to that of neutral EL2 centres in undoped GaAs even though the Cr concentration, [Cr], is practically uniform.

Several GaAs crystals doped with radio-tracer ^{51}Cr were grown by high pressure LEC in a Metals Research MSR6/R puller. After growth, slices from various sections of the crystals were analysed by radio-tracer gamma-counting for total [Cr]. Figure 1 indicates the cutting procedure, a typical [Cr] analysis and the cutting of samples for EPR. Thermally detected EPR (TDEPR) was used because it is sensitive, reproducible and particularly suitable for broad resonances [2]. Due to the sensitivity of the Cr^{3+} EPR signal to residual and unintentional mounting stress, its concentrations were not measured. However, the Cr^{2+} signature does not suffer from these difficulties and can be reproducibly measured. The Cr^{2+} lines were identified by their Zeeman anisotropy. The absorbed microwave power which is proportional to $[Cr^{2+}]$ is shown in Figure 2 across the diameter indicated in Figure

* Department of Physics, University of Nottingham, Nottingham NG7 2RD, UK.

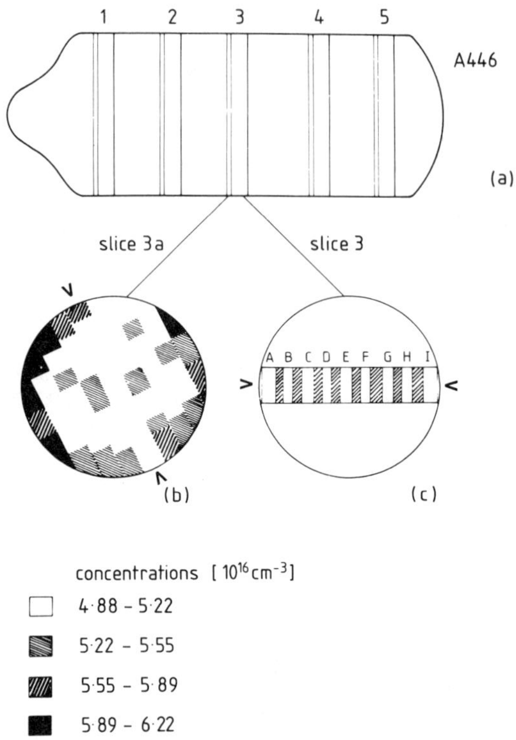

Figure 1 (a) The sectioning of the crystal. Sections 1 to 5 consist of pairs of slices of thickness 1 mm and 5 mm. The 1 mm slices are analysed for chromium content; the 5 mm slices are used for EPR assessment. (b) [Cr] assessments made by radio-tracer counting of the ^{51}Cr isotope. The distribution of [Cr] is relatively uniform but [Cr] generally increases near the periphery of the slice. The samples for EPR measurements are cut from the diameter marked with the arrows. (c) The nine EPR samples cut from the diametric cut as marked A to I.

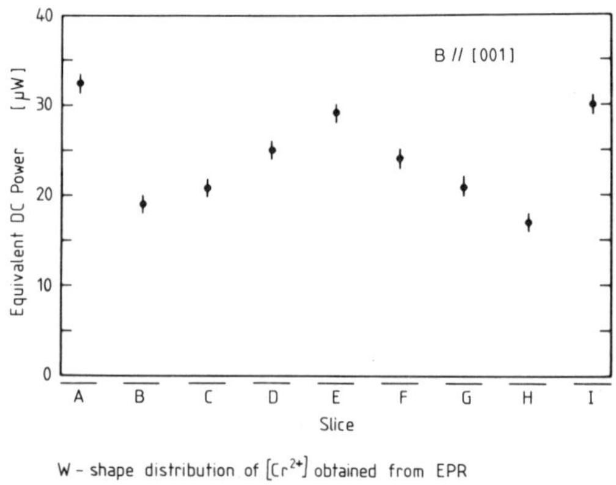

Figure 2 The variation of the magnitude of the Cr^{2+} EPR signal as a function of position across a ⟨011⟩ diameter showing a W-shaped variation.

1. A clear W shape is obtained. Although TDEPR is not strictly quantitative, an equivalent d.c. power of 20 μW is consistent with $[Cr^{2+}] \sim 10^{16}\ cm^{-3}$.

The presence of Cr^{2+} in our samples at a concentration of $\sim 10^{16}\ cm^{-3}$ means that compensating donors must also exist in this concentration. Chemical and physical analysis have shown that no chemical donors are present in such concentrations and that the only intrinsic donor of such concentration is EL2. The distribution of neutral EL2 across a similar diameter of undoped LEC GaAs is shown in Figure 3 as measured by optical absorption at 1 μm. Once again a clear W shape is seen. We have shown previously [1, 3] that such a W shape for [EL2] in undoped GaAs corresponds to an M shape in resistivity. Addition of Cr increases the resistivity generally and flattens the M-shaped resistivity map.

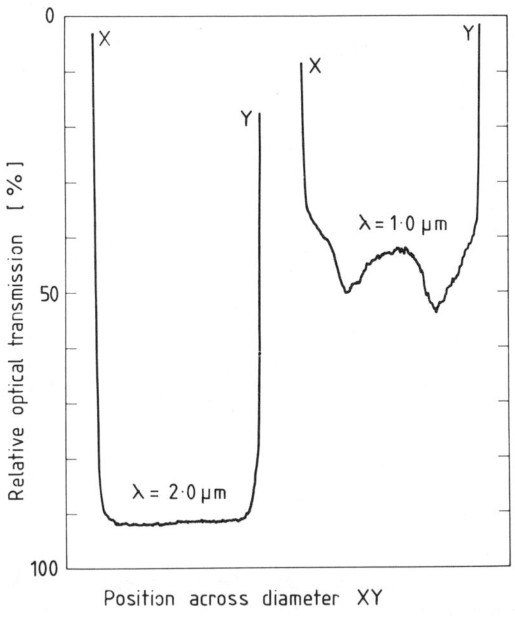

Figure 3 Spatial scans of the relative infra-red transmission across an ⟨011⟩ diameter of an undoped semi-insulating (100) slice at $\lambda = 1$ μm, in the middle of the $EL2^0$ absorption band and at 2 μm where there is no defect absorption. The zero transmission levels are the same for both scans. The W-shaped absorption reflects the concentration of the $EL2^0$ defects and are typical of undoped GaAs grown by the LEC method. The fine structure in the absorption at $\lambda = 1$ μm is genuine and has been discussed in [4].

Our results suggest that Cr introduced during growth can fully ionize native EL2 by:

$$Cr^{3+} + EL2^0 \rightleftarrows EL2^+ + Cr^{2+}$$

If [Cr] > [EL2], the distribution of $[Cr^{2+}]$ will reflect the intrinsic distribution of [EL2]. Thus the Cr^{2+}/Cr^{3+} level must lie deeper in the gap than $EL2^0/EL2^+$ and the Fermi level is pinned at Cr^{2+}/Cr^{3+} if $[Cr] > [EL2^0]$. This explains previous electrical measurements because the position of the Fermi level is not now controlled by concentrations of EL2 across a slice. If shallow C-acceptors are present, some EL2 will be ionized, even if no Cr is present. Under these circumstances the addition of Cr results in a change of the ratio $[EL2^0]/[EL2^+]$ until $[Cr]>[EL2]-[C]$ for which $[EL2^0]=0$. This behaviour is shown in Figure 4 which represents the relevant defect concentrations as a function of [Cr].

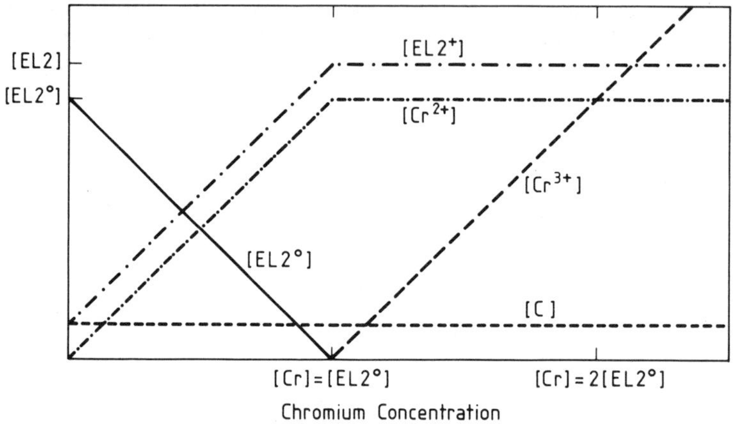

Figure 4 The variation of concentration of the various defect charge states as a function of increasing Cr concentration. In this figure, the presence of residual shallow donors is neglected for simplicity. The effect of their presence is to replace [C] by the net shallow acceptor concentration $[C]-N_D$. For $Cr<[EL2]-[C]$, the Fermi level is pinned at $EL2^0/EL2^+$ and for $Cr>[EL2]-[C]$ the Fermi level is pinned at Cr^{2+}/Cr^{3+}.

ACKNOWLEDGEMENTS

The authors acknowledge the help and encouragement of Dr R. W. Burns. The work is supported in part by SERC.

References

1. Grant I., Rumsby D., Ware R. M., Brozel M. R. and Tuck B. (1982). In *Semi-Insulating III–V Materials, Evian*, p. 98 (S. Makram-Ebeid and B. Tuck, eds.); Shiva Publishing, Nantwich, UK
2. Moore W. S. and Al-Sharbati T. M. (1973). *J. Phys. D., Appl. Phys.*, **6**, 367
3. Brozel M. R., Tuck B., Rumsby D. and Ware R. M. (1982). *J. Cryst. Growth*, **60**, 113
4. Brozel M. R., Grant I., Ware R. M. and Stirland D. J. (1983). *Appl. Phys. Lett.*, **42**, 610

Electronic Structures of Fast Neutron Generated Defects in SI GaAs

A. Goltzene, B. Meyer and C. Schwab

Laboratoire de Spectroscopie et d'Optique du Corps Solide Associé au CNRS No. 232, Université Louis Pasteur, 5 rue de l'Université, 67000 Strasbourg, France

Abstract

Fast neutron irradiation of SI GaAs induces two intensity correlated defects, whose EPR (electron paramagnetic resonance) spectra are attributed to the anionic antisite and the cation vacancy. The temperature dependence suggests that they are associated in a complex of intrinsic nature $As_{Ga}^{4+} - V_{Ga}^{2-}$.

Despite the growing importance of GaAs, there is only a limited knowledge of the intrinsic defects which are liable to occur in this compound semiconductor. Hence the determination of their electronic structures represents a fundamental challenge.

High energy particle irradiation is a convenient tool for generating such defects in the bulk of a sample at concentrations required by EPR techniques [1–3]. Fast neutron irradiation has the further advantage that it operates under conditions similar to those of ion implantation, since the primary knock-on kinetic energy is of the order of 50 keV for 1–2 MeV neutrons. In this context, a detailed study of fast neutron irradiated undoped SI GaAs has been performed at 9 GHz and over the 4.2–300 K temperature range [4].

Whatever the neutron fluence, the spectra consist of a quadruplet, previously ascribed to the anionic antisite As_{Ga}^{4+} [5], but the existence of one stronger component is not fully compatible with the former model. In fact, the experimental spectra can be decomposed into a quadruplet Q, constrained to the theoretical model of As_{Ga}^{4+}, and a singlet S of a mainly Lorentzian profile [6]. The intensities of Q and S remain correlated both as a function of the neutron fluence and of the recording temperature. However, the temperature dependence presents several peculiarities. Both Q and S show a similar non paramagnetic behaviour for temperatures $T > 100$ K (Figure 1). Another unexpected result is the decrease of the hyperfine structure constant A of Q with increasing temperature (Figure 2). Finally, an important shift of the position of S results in a thermal variation of its g value, concomitant with an anomalous line-width variation (Figure 3).

Assuming the validity of the As_{Ga}^{4+} model for Q requires an explanation for the discrepancy with the expected Curie law. Trapping effects can be excluded since the thermal dependence is fully reversible. The similar thermal

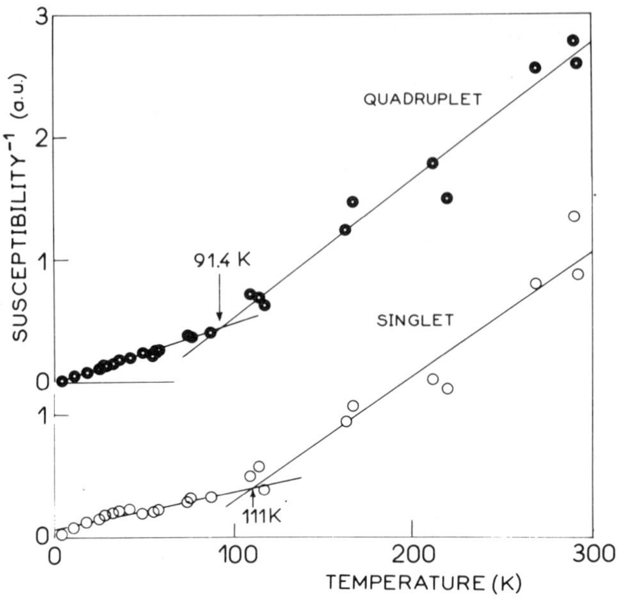

Figure 1 Inverse of the paramagnetic susceptibility of the quadruplet, assigned to As_{Ga}^{4+} and the superimposing singlet, assigned to V_{Ga}^{2-}, as a function of the temperature.

behaviours of Q and S may be a hint that they are linked in a kind of associated defect. For Q, this idea is also supported by the anomalous dependence of A(T) which corresponds to a thermal variation of the coupling between the electronic and nuclear magnetic moments of the core As^{4+} ion, i.e. a variation of the local electron density as induced, for example, by a strong lattice relaxation, already starting at the lowest T. In fact, such a relaxation could result from a Jahn-Teller effect, resulting from the intrinsic

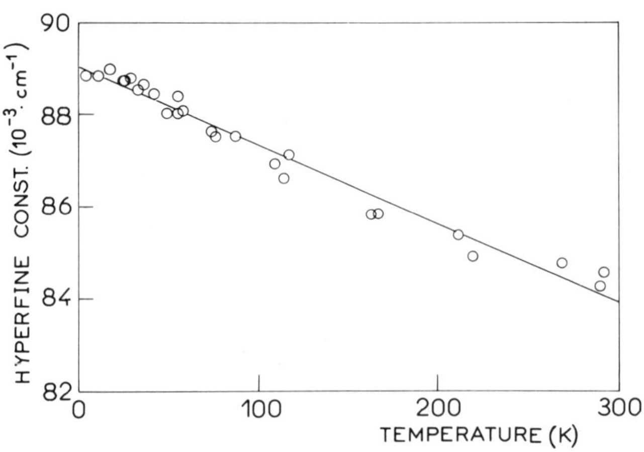

Figure 2 Thermal dependence of the quadruplet hyperfine structure constant.

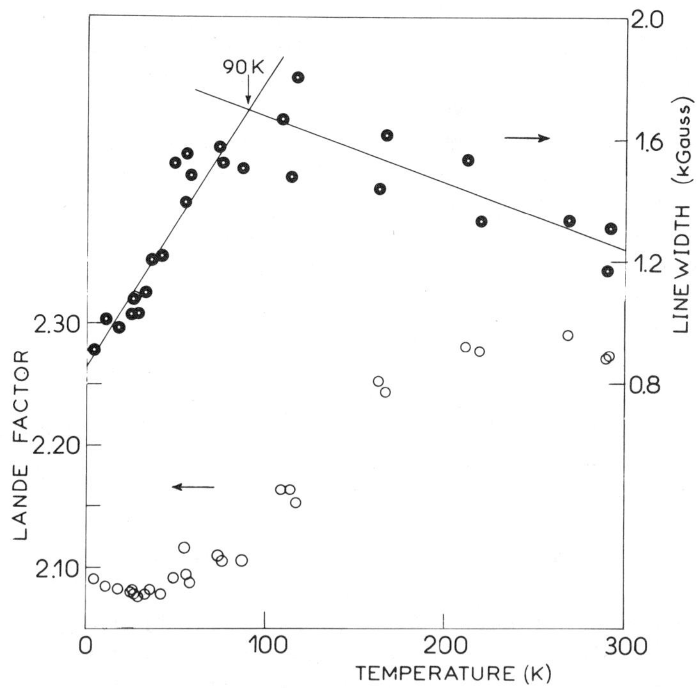

Figure 3 Thermal variations of the Landé factor and linewidth of the singlet.

instability of any electronically degenerate state and inducing either static or dynamic distortions according to their coupling with the vibrational modes.

Dealing with S, we note its analogy with the two distinct paramagnetic states observed in amorphous As and referred to as the optically induced response at 4.2 K and the thermally generated response [7]. However, both are ascribed to unpaired spins localized predominantly on As p-orbitals extending over less than three As atoms, i.e. the electronic configuration of a hole trapped at a cation vacancy [V_{Ga}^{2-}] in GaAs. Although this attribution is only based on a hyperfine interaction computation, we point out that the low temperature g value of S may be extrapolated from the values of the cation vacancies in ZnS [8] and ZnSe [9] if it is assumed that the mean g shifts:

$$\Delta\bar{g} = \bar{g} - g_e \sim \frac{2}{3}\frac{\lambda}{\Delta E}$$

are roughly proportional to the spin orbit coupling constant λ of the anion. These V_{Zn} centres precisely undergo a Jahn-Teller transition from a dynamic regime at high T to a static one at low T, where the hyperfine structures are actually resolved. For GaAs, these structures cannot be resolved, but the anomalous thermal behaviours of the Landé factor and linewidth of S are consistent with a transition from a static to a dynamic Jahn Teller effect, modifying considerably the EPR lineshapes although the mean g value is unaffected. Within this hypothesis, it is gratifying to note that derivation of the high temperature g value from that at low temperature leads to $g \sim 2.3$ in agreement with the 300 K value.

To sum up, although the similar thermal variations of the intensities of both Q and S could be described adequately by some Fermi-Dirac distributions, we believe that the A(T) variation of Q has to be explained in terms of a lattice relaxation effect, meaning that the As_{Ga}^{4+} centre senses a local electron density change, a precursor state for an associated defect. Indeed a Jahn-Teller effect on the As_{Ga}^{4+} centre itself is excluded on the basis of its fully symmetrical S state configuration.

In addition, there are concordant indications that the singlet S is related to a V_{Ga}^{2-} centre which likely presents different Jahn-Teller regimes, like the similar cation vacancies in other $A^{N}B^{8-N}$ compounds. This situation suggests that both centres may be associated in a complex of intrinsic nature. Hence the rapid decrease in the overall magnetic susceptibility of each centre above 100 K would result from antiferromagnetic coupling between the unpaired As_{Ga}^{4+} electron and the trapped V_{Ga}^{2-} hole or from their recombination, both phenomena having an increasing probability as T increases. However, the first hypothesis would require that the exchange integral increases more rapidly with T than the thermal decoupling of the spins. The second assumption relies on a thermally activated recombination process that could be directly linked to a configurational change of the $As_{Ga}^{4+}-V_{Ga}^{2-}$ complex. In any case the corresponding activation would be about 0.03 eV.

The tendency to form complex defects like antisite-vacancy associates has been predicted some time ago to be a predominant feature in III–V compound semiconductors [10]. Finally, recent experiments have shown that the quadruplet and the singlet could also be present in as-grown SI material [11].

ACKNOWLEDGEMENTS

Part of this work has been supported by Centre National d'Etudes des Télécommunications.

References

1. Goswami N. K., Newman R. C. and Whitehouse J. E. (1981). *Solid-State Commun.*, **40,** 473
2. Wörner R., Kaufmann U. and Schneider J. (1982). *Appl. Phys. Lett.*, **40,** 141
3. Goltzene A., Meyer B. and Schwab C. (1983). *J. Appl. Phys.*, **54,** 3117
4. Goltzene A., Meyer B. and Schwab C. (1983). *Rev. Phys. Appl.*, **18,** 703
5. Wagner R. J., Krebs J. J., Stauss G. H. and White A. M. (1980). *Solid-State Commun.*, **36,** 15
6. Goltzene A., Meyer B., Schwab C., Greenbaum S. G., Wagner R. J. and Kennedy T. A. Unpublished
7. Bishop S. G., Strom U. and Taylor P. C. (1976). *Solid-State Commun.*, **18,** 573

8. Watkins G. D. (1974). *Phys. Rev. Lett.*, **33,** 223
9. Title R. S. (1967). In *Physics and Chemistry of II–VI Compounds,* p. 265 (M. Aven and J. S. Prener, eds.); North-Holland Publishing Company, Amsterdam
10. Van Vechten J. A. (1975). *J. Electrochem. Soc.*, **122,** 419 and 423
11. Goltzene A., Meyer B. and Schwab C. Unpublished

EPR Analysis of Anion Antisite Defects in GaAs

E. R. Weber

Center for Advanced Materials, Lawrence Berkeley Laboratory, University of California, Berkeley, California 94720, USA

Abstract

Quantitative EPR (electron paramagnetic resonance) measurements of the annealing behaviour of As_{Ga} antisite defects in bulk GaAs show the thermal stability of these defects up to more than 900°C. Optical absorption shows an increase of the 'EL2' absorption in plastically deformed and n-irradiated GaAs, corresponding to the increase of As_{Ga} as measured by EPR. With both these findings, the identification of EL2 as an As_{Ga} related defect gains further support. Various recent results indicate, however, the existence of a family of midgap traps, for which a model of defect complexes, involving As_{Ga} antisite defects, is proposed.

Introduction

Anion antisite defects are the first directly identified intrinsic lattice defects in GaAs. Their EPR spectrum has been investigated, first by Wosinski, using plastically deformed GaAs [1], but not been correctly interpreted. Wagner *et al.* [2] identified the characteristic four-line spectrum as being due to As_{Ga}^{+} (in the ionic notation As_{Ga}^{4+}), in analogy to the well-known P_{Ga} antisite defect in GaP [3]. Photo-EPR experiments showed two distinct quenching steps of this defect. These were interpreted [4] as the two donor levels of As_{Ga} at $E_c - 0.75$ eV (0/+) and $E_v + 0.52$ eV (+/++). Recent Green's function [5] as well as tight-binding calculations [6] support this model.

The relation of As_{Ga} to EL2

The first donor level of As_{Ga} at midgap is at the same energy (within experimental uncertainty) as the dominant midgap electron trap 'EL2' found in LEC (liquid encapsulated Czochralski), HB (horizontal Bridgman), VPE (vapour phase epitaxial) and MOVPE (metalorganic VPE) GaAs grown under As-rich growth conditions, suggesting a correlation of these centres [4, 7]. In fact, even before the As_{Ga} energy levels were determined, various authors had already discussed an identification of EL2 with As_{Ga} antisite defects, using mainly the results of stoichiometry variation experiments [8–10]. Today, the arguments in favour of this identification can be summarized as follows:

1. The As_{Ga} defect shows in photo-EPR a midgap donor level, like EL2.
2. EL2 and As_{Ga} appear in as-grown LEC GaAs in about the same concentration, around $5–10\times10^{15}$ cm^{-3}.
3. EL2 and As_{Ga} show a persistent photoquenching effect with the same characteristic barrier of 0.35 eV for thermal de-quenching [7].
4. The relationship of EL2 concentrations to the As/Ga stoichiometry ratio has been clearly established, for example Refs. 8–11.

As recently discussed [12], the major points of disagreement concerning an EL2 As_{Ga} identification appear to be the different thermal stabilities reported for EL2 [13] and As_{Ga} [14], and the question of EL2 formation during n-irradiation. The formation of As_{Ga} during n-irradiation has been clearly shown [14]. In the next two paragraphs, these questions will be addressed.

Thermal stability of As_{Ga}

Heat treatment of LEC-grown SI GaAs samples ($3.5\times3.5\times15$ mm^3) was carried out up to 600°C in a flushing atmosphere of reducing gas (8% H_2, 92% N_2), and at higher temperatures in an evacuated quartz ampoule containing crushed GaAs to minimize As losses. After annealing, a surface layer of at least 100 μm was mechanically and chemically removed. Quantitative EPR measurements were performed by comparing the spectra with a NBS ruby standard sample. Because of the large linewidth of the As_{Ga}^+ spectrum, in comparison with the standard, the EPR concentrations are accurate only within an absolute uncertainty of a factor of two, whereas the relative error of the various As_{Ga}^+ concentrations are less than 10%.

Results of the thermal annealing experiments with As_{Ga} in as-grown GaAs are shown in Figure 1. Undoubtedly, As_{Ga} defects in *bulk* GaAs are thermally stable up to more than 900°C, comparable to the results of DLTS (deep level transient spectroscopy) measurements with EL2 [13]. The increase in As_{Ga}^+ concentration evident after high-temperature annealing (Figure 1) might be due to a high-temperature point defect reaction resulting in additional As_{Ga} formation, or a small downwards shift of the Fermi level. The thermal stability of As_{Ga} in as-grown GaAs sheds some doubt on the identification of the 0.8 eV photoluminescence band with As_{Ga} [15], as that identification primarily was based upon the observation of thermal annealing of this emission around 450°C.

It is worthwhile discussing the clear annealing step of As_{Ga} antisite defects produced by plastic deformation [7], or n-irradiation [14] (see Figure 1). One possible explanation has been given previously [7]: this annealing step might be indicative of indirect annealing by the movement of another point defect. An example for this would be the movement of As vacancies, so that the reaction:

$$As_{Ga}+V_{As}\rightarrow V_{Ga}+As_{As}$$

would result in disappearance of part of the As_{Ga} concentration.

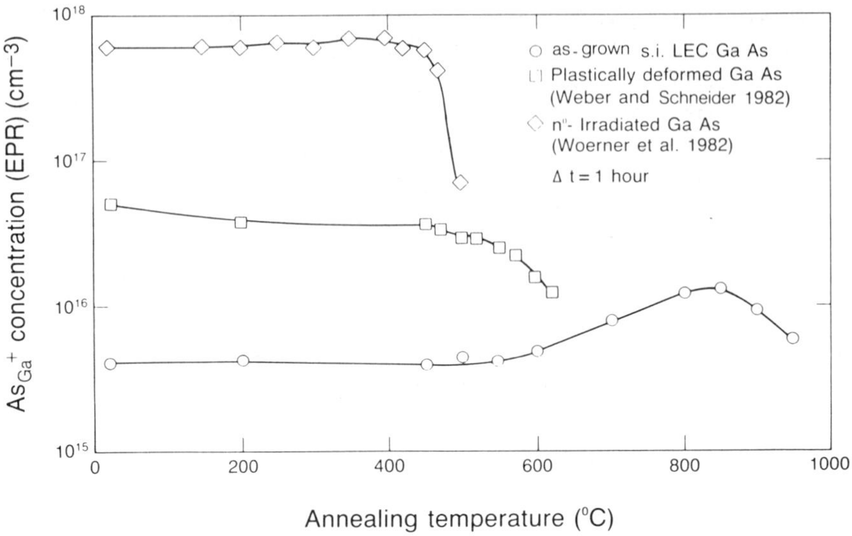

Figure 1 EPR measurements of As_{Ga} thermal annealing in as-grown bulk GaAs, compared with results from As_{Ga} defects formed in plastically deformed [7] and n-irradiated GaAs [14]. Each data point corresponds to one hour annealing time.

A second, new model questions the identity of all defects contributing to the As_{Ga}^+ EPR signal: the more thermally stable defects might be microscopically different from those annealing at 450°C. Thus a family of defects should be considered, involving isolated As_{Ga} defects as well as complexes with other defects, for example V_{Ga} or impurities like oxygen. These complexes might still show the same EPR spectrum, as the large linewidth of the As_{Ga}^+ spectrum inhibits the detection of *small* distortions of the full tetrahedral symmetry, for example by an associated defect on the next-nearest neighbour shell. An association with a defect on the adjacent As-shell seems to be irreconcilable with the As_{Ga}^+ EPR spectrum [14].

EL2 in n-irradiated GaAs

In order to treat the second question, the concentration of EL2 in n-irradiated SI GaAs has to be measured. The only quantitative technique available for EL2 in SI material is optical absorption at 1.13 and 1.24 eV, as described by Martin [16]. These measurements were carried out at 77 K, taking into account reflection and other losses by comparison with the transmission around 0.5 eV.

In Figure 2, the 'EL2 concentrations' derived from these measurements are plotted versus the As_{Ga}^+ concentrations established by EPR of the same samples. Both concentrations appear to be closely related, over more than two orders of magnitude in concentration. Deviations from the 45° line might

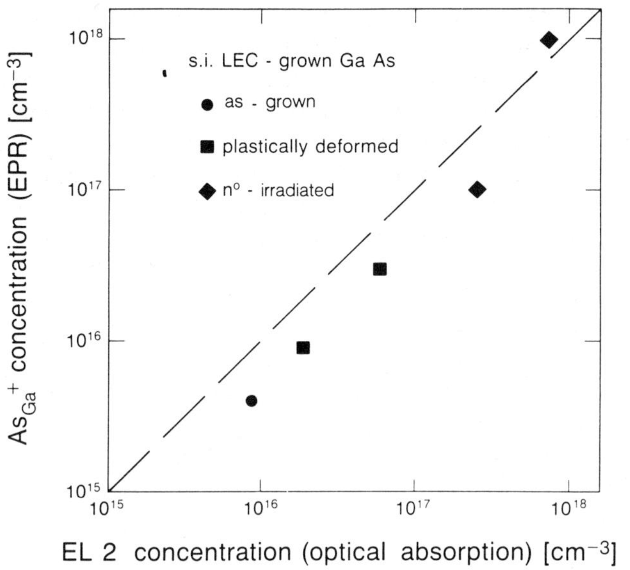

Figure 2 'EL2' concentrations in various GaAs samples, measured by optical absorption at 1.13 and 1.24 eV [16], plotted versus As_{Ga}^{+} concentrations determined by EPR with the same samples.

be due to the absolute errors of these methods and to slight shifts of the Fermi level during irradiation.

Evidence for more than one midgap trap

Recent investigations of the persistent photoquenching of the optical absorption in plastically deformed GaAs show that only a small fraction of the 'EL2 absorption' near 1.2 eV can be quenched by 1.0–1.3 eV light [17]. Moreover, this fraction corresponds to the concentration of EL2 and As_{Ga} in the as-grown samples, suggesting, in agreement with the discussion above, at least two microscopically different midgap traps. It should be noted that recent photoluminescence measurements [18–20] and DLTS results [20–23] also support the existence of at least two midgap defects with very similar, yet distinguishably different properties.

Conclusions

The thermal stability found for As_{Ga} antisite defects in as-grown SI GaAs, as well as optical absorption measurements of n-irradiated SI GaAs, support the correlation of anion antisite defects with the midgap traps commonly labelled EL2. However, there is strong evidence that the As_{Ga}^{+} EPR spectrum as well as the midgap traps in GaAs have to be due to a family of quite similar defects. The name 'EL2' should be restricted in the future to the defect showing the characteristic photoquenching, and no impurity relationship. The well-established stoichiometry dependence as well as the EPR results discussed here, suggest that all, or at least most, of these defects have a As_{Ga} antisite defect constituent. These midgap defects can be distinguished by their formation process, for example temperature during MOVPE growth [21], oxygen concentration in the melt [23], n-irradiation or plastic deformation, energy levels (photoluminescence [18–20] and DLTS [20–23]), thermal stability and persistent photoquenching behaviour [17].

In the near future, correlations between the different properties detected for this family of midgap traps will have to be established. It is hoped that detailed atomic models for those midgap traps which are related to As_{Ga} antisite defects might be determined by electron-nuclear double resonance (ENDOR) experiments.

ACKNOWLEDGEMENTS

The experiments described in this paper were performed during the author's stays in Cologne and Lund. For support and stimulating discussions during this time he is grateful to H. Alexander, H. G. Grimmeiss, P. Omling and L. Samuelson. For the GaAs samples, thanks are due to F. Koehl, Wacker-Chemitronic. This work was supported in part by the Division of Materials Science, US Department of Energy under Contract No. DE-AC03-765SF00098.

References

1. Wosinski T. (1980). *Phys. Stat. Solidi*, **A60,** K149
2. Wagner R. J., Krebs J. J., Stauss G. H. and White A. M. (1980). *Solid-State Commun.*, **36,** 15
3. Kaufmann U., Schneider J. and Raeuber A. (1976). *Appl. Phys. Lett.*, **29,** 312
4. Weber E. R., Ennen H., Kaufmann U., Windscheif J., Schneider J. and Wosinski T. (1982). *J. Appl. Phys.*, **53,** 6140
5. Bachelet G. B., Schlüter M. and Baraff G. A. (1983). *Phys. Rev.*, **B27,** 2545
6. Van der Rest J. and Pecheur P. (1983). *J. Physique*, **44,** 1297
7. Weber E. R. and Schneider J. (1983). *Physica*, **116B,** 398
8. Yuanxi, Z. (1981). *Inst. Phys. Conf. Ser.*, **63,** 185
9. Yu P. W., Holmes D. E. and Chen R. T. (1981). *Inst. Phys. Conf. Ser.*, **63,** 209
10. Lagowski J., Gatos H. C., Parsey J. M., Wada K., Kaminska M. and Walukiewicz W. (1982). *Appl. Phys. Lett.*, **40,** 342
11. Samuelson L., Omling P., Titze H. and Grimmeiss H. G. (1981). *J. Cryst. Growth*, **55,** 164
12. Martin G. M. and Makram-Ebeid S. (1983). *Physica*, **116B,** 371
13. Makram-Ebeid S., Gautard D., Devillard P. and Martin G. M. (1982). *Appl. Phys. Lett.*, **40,** 161
14. Woerner R., Kaufmann U. and Schneider J. (1982). *Appl. Phys. Lett.*, **40,** 141
15. Windscheif J., Ennen H., Kaufmann U., Schneider J. and Kimura T. (1983). *Appl. Phys.*, **A30,** 47
16. Martin G. M. (1981). *Appl. Phys. Lett.*, **39,** 747

17. Samuelson L., Omling P., Weber E. R. and Grimmeiss H. G. (1984). This volume, p. 268
18. Yu P. W. and Walters D. C. (1982). *Appl. Phys. Lett.*, **41,** 863
19. Shanabrook B. V., Klein P. B., Swiggard E. M. and Bishop S. G. (1983). *J. Appl. Phys.*, **54,** 336
20. Taniguchi M. and Ikoma T. (1982). *Inst. Phys. Conf. Ser.*, **65,** 65
21. Watanabe M. O., Tanaka A., Udagawa T., Nakanisi T. and Zohta Y. (1983). *Jap. J. Appl. Phys.*, **22,** 923
22. Taniguchi M. and Ikoma T. (1983). *Appl. Phys. Lett.*, **54,** 6448
23. Lagowski J., Lin D. G., Aoyama T. and Gatos H. C. (1984). *Appl. Phys. Lett.*, **44,** 336

Photoluminescence Study of the 78 meV Double Acceptor in GaAs

S. G. Bishop and B. V. Shanabrook

Naval Research Laboratory, Washington DC 20375, USA

Abstract

Time resolved and CW PL (photoluminescence) measurements have been performed on LEC grown GaAs crystals which exhibit the 78 meV double acceptor. In addition to a PL band at 1.44 eV (associated with the neutral charge state of the 78 meV double acceptor), these crystals exhibit two other bands at 1.283 eV and 1.316 eV. The difference between these two bands, which exhibit very different energies, temperature dependences and time decay characteristics, has previously been overlooked, and *both* have been related by independent authors to the negative charge state of the 78 meV double acceptor. The combination of our PL and far-infrared intra-centre absorption measurements provides rather strong evidence for associating only the 1.316 eV PL band with the negative charge state of a double acceptor.

Previous studies [1, 2] have reported a 77–78 meV acceptor level in GaAs crystals pulled from a Ga-rich melt by the LEC (liquid encapsulated Czochralski) technique. This level has been attributed to the neutral charge state of the Ga antisite double acceptor defect (Ga_{As}) [1, 2] on the basis of the following points: (a) crystals grown from Ga-rich melts favour the formation of Ga_{As} defects; and (b) that SIMS (secondary ion mass spectrometry) and LVM (localized vibrational mode) indicate that the only impurity present in sufficient concentrations to account for the 78 meV level is boron and a poor [1] or nonlinear [3] correlation has been reported between the boron content of the crystals and the concentration of the 78 meV defect level. (Another related model attributes the 78 meV acceptor level to a complex consisting of a gallium antisite and one or two boron atoms [3].) This acceptor level is manifested in PL measurements [1, 2, 4] by a donor-acceptor pair band occurring at 1.44 eV. In addition to the 78 meV acceptor, a level (or levels) with a binding energy of about 200 meV have been reported in crystals grown from Ga-rich melts. Specifically, Yu and co-workers [1] have reported a PL band at 1.284 eV and, subsequently, Elliott [2] observed a band at 1.316 eV. Apparently, the difference between the peak energies of these PL bands was unnoticed because, in both instances, these emission bands were associated with Ga_{As}^{-}/Ga_{As}^{2-} transistions. In the context of this model, the energies of the 1.284 eV and 1.316 eV PL bands imply that the negative charge state of Ga_{As} exhibits a binding energy of either 236 meV or 203 meV, respectively.

We have performed CW and time resolved PL measurements of bulk GaAs crystals grown in several different laboratories. Our measurements indicate that all of the emission bands reported in the energy range of 1.438 eV to 1.443 eV arise from donor-acceptor pair transitions involving the 78 meV acceptor level. The variety of peak positions observed in these samples is a result of differing final state Coulomb interactions involved in the recombination process. The results of this study will be reported in detail elsewhere [4]. In this article, we wish to emphasize (a) that two different deep levels, with binding energies of 203 meV and 236 meV, exist in GaAs, (b) that only the 203 meV level is related to the negative charge state of a double acceptor and (c) that the concentration of the defect which gives rise to the 236 meV level is related in an as yet unknown way to the concentration of the 78 meV double acceptor.

Shown in Figure 1 are 4.2 K, CW PL spectra of the (a) 1.283 eV and (b) the 1.316 eV emission bands that are characteristic of GaAs crystals grown from Ga-rich melts. The relative intensities of these two bands are sample

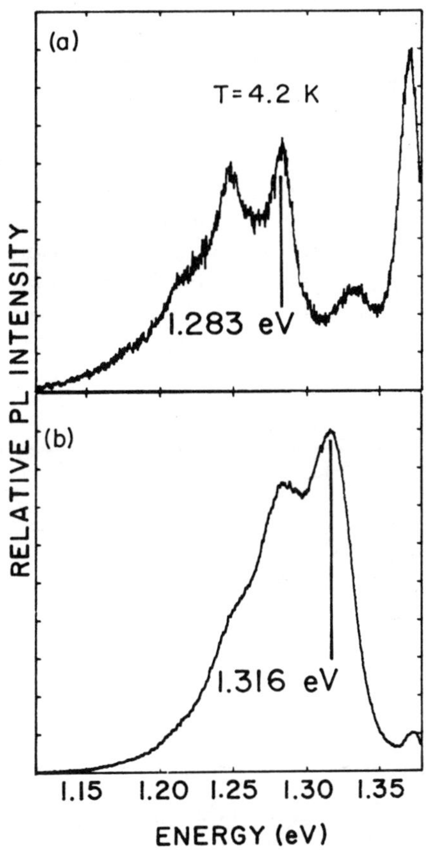

Figure 1 The CW PL spectra of (a) the 1.283 eV and (b) the 1.316 eV emission.

dependent and it appears that the 1.316 eV PL band is dominant when the Fermi-level of the sample is near the 78 meV level. The peaks observed at 1.37 eV and 1.34 eV are the 2 LO and 3 LO phonon replicas of the 1.443 eV (78 meV acceptor) PL bands. The peak at 1.283 eV apparently corresponds to the feature observed at 1.284 eV by Yu *et al.* [1] and which they attributed to an acceptor with a binding energy of 230 meV. In agreement with Yu *et al.* [1], we find that the intensity of the 1.283 eV PL band is always much smaller than but correlated in intensity with the 1.443 eV emission. This observation, which is true for samples from *different* suppliers, implies a fixed but unidentified relationship between the concentrations of the entities which give rise to the 1.443 eV and 1.283 eV emission bands. The PL spectrum shown in Figure 1b was strongest in a Si-compensated sample grown by Rockwell (R41-T). If one assumes that the 1.316 eV emission arises from a free-to-bound hole recombination mechanism, the corresponding acceptor binding energy is 204 meV.

We have also performed time-resolved PL measurements on the 1.283 eV and 1.316 eV emission bands. These measurements were performed by recording the time evolution of the selected emission band with a transient digitizer after the excitation of the sample by an 8 nsec wide laser pulse. While both the 1.283 eV and 1.316 eV emission bands exhibit time-decays which are non-exponential (see Figure 2), the 1.283 eV PL is characterized by a much larger rate of decay (about a factor of ten). Just as there is a one to one correspondence between the intensities of the 1.443 eV and 1.283 eV emission bands observed in the CW PL measurements, the rates of decay observed in the time-resolved measurements are similar for these two bands [4].

A detailed study of the infrared intra-centre absorption of the 78 meV double acceptor is made in Ref. 5. This investigation has shown that the neutral and negatively charged states of this centre exhibit binding energies of 78 meV and 203 meV, respectively. This implies, in contrast to what had

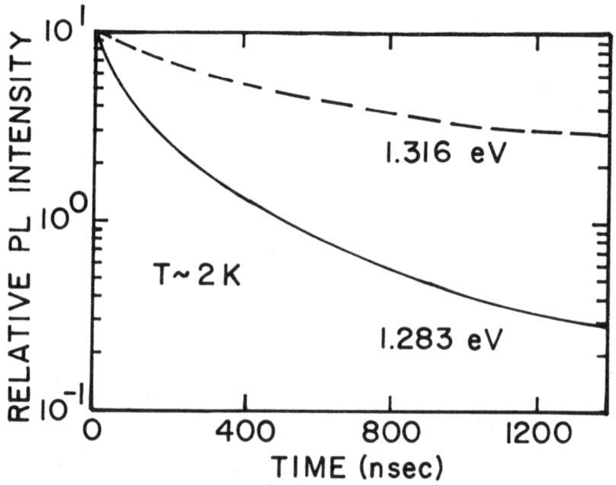

Figure 2 A comparison between the temporal decays of the PL occurring at 1.316 eV and 1.283 eV.

previously been suggested by Yu *et al.* [1], that the 1.316 eV PL band (not the 1.283 eV band) arises from a recombination mechanism which involves the negative charge state of the 78 meV double acceptor. This assignment allows two key features of the PL data to be explained easily. First, in the context of the double acceptor model, the slow decay of the 1.316 eV emission relative to that of the 1.443 eV PL arises because the former relies on the relatively slow capture of an electron by a centre which is already negatively charged, while the latter involves the more rapid capture of an electron by the neutral centre. Second, the observation that the 1.316 eV PL is strongest in samples where the 78 meV level is partially compensated is understandable because the compensation provides a large steady state population of the negatively charged double acceptor required for this recombination process. However, the nature of the relationship between the 78 meV level and the centre which is responsible for the 1.283 eV emission remains unexplained.

ACKNOWLEDGEMENTS

This work was partially supported by the Office of Naval Research. The authors wish to thank E. M. Swiggard of NRL, K. R. Elliott of Rockwell International and H. M. Hobgood of Westinghouse for generously supplying the samples examined in this study.

References

1. Yu P. W., Mitchel W. C., Mier M. G., Li S. S. and Wang W. L. (1982). *Appl. Phys. Lett.*, **41,** 532 and references therein
2. Elliott K. R. (1983). *Appl. Phys. Lett.*, **42,** 274
3. Ta L. B., Hobgood H. M. and Thomas R. N. (1982). *Appl. Phys. Lett.*, **41,** 1091
4. Bishop S. G., Shanabrook B. V. and Moore W. J. *J. Appl. Phys.* In press
5. Moore W. J., Shanabrook B. V. and Kennedy T. A. This volume, p. 453

Section 5:
Devices as affected by the substrate

Substrate Quality: How Important for Future GaAs ICs?

J. V. DiLorenzo, A. S. Jordan, A. R. Von Neida and P. O'Connor

AT & T Bell Laboratories, Murray Hill, New Jersey 07974, USA

Abstract

The increasing worldwide interest in GaAs ICs fabricated by direct ion implantation in SI substrates has led to a re-examination of the effect of dislocations on the uniformity of the activated implanted dopant. This paper reviews the reported correlations between some substrate properties (EL2 and carrier concentration), FET (field effect transistor) device characteristics (V_{th}, I_{DS}), and dislocation distribution. A particularly alarming observation is that device parameters of FETs can be influenced by a dislocation perhaps as far as 50 μm away. Thus, the theory and practice of dislocation reduction in GaAs has become a prominent issue and will be briefly surveyed. Finally, we discuss IC performance in light of the adverse effect of dislocations.

Introduction

In the last few years, there has been a rapidly growing interest in ICs fabricated in GaAs, manifested by the extensive activity in research laboratories throughout the world. In addition, there is a rush to develop GaAs ICs into commercial products, and in 1984, small scale circuits are available commercially. Why all the interest? Simply put, GaAs ICs will be able to operate at higher clock rates than Si. The attractiveness of GaAs based devices for high speed logic is summarized in Table 1 and Figure 1. High electron mobility and low saturation field compared to Si is evident from the plot of electron velocity versus electric field in Figure 1. In addition, as shown in Figure 2, the phenomenon of velocity overshoot has been theoretically predicted to happen as device dimensions approach the region of <0.5 μm gate lengths. As stated in Table 1, heterojunctions are possible, leading to exciting new devices such as the HEMT or SDHT and wide bandgap emitter bipolar transistors. Lastly, and in some respects, most importantly, SI substrates are available in GaAs which make device isolation simple, give low interconnect capacitance and allow via connections through the substrate.

The quality of SI substrates and its effect on IC performance has become a subject of much interest over the last few years. This paper will review some of the issues related to the device and circuit implications of substrate quality. However, in order to adequately describe the effect of substrates on GaAs ICs, the device and circuit concepts used presently in this technology are first discussed. Second, the substrate properties with respect to shape, SI

Table 1 *Attractiveness of GaAs for high-speed logic*

- High electron mobility and low saturation field
  - High velocity transport
  - Low power dissipation
  - Low parasitic resistances
- Enhanced velocity over short distances
  - Ballistic transport
  - Velocity overshoot
- Heterojunctions
  - Wide-gap emitter bipolar
  - Heterojunction-Gate FETs
  - Buffer layers
  - Mobility enhancement by selective doping
- Semi-insulating substrate
  - Device isolation
  - Low interconnect capacitance
  - 'Via' connections

behaviour, conversion, and doping uniformity are considered, where the latter appears to be associated with dislocation density. Subsequently, important recent correlations between some FET characteristics and dislocation pattern are presented. Then, methods to reduce the defect content of GaAs are surveyed. Finally, we offer highly speculative statistical and physical arguments to assess IC performance in light of the purported dislocation effect.

Device and circuit considerations

ENHANCEMENT AND DEPLETION MODE FETs

At the present time, GaAs is being pursued for monolithic microwave integrated circuits (MMIC), analog ICs and digital ICs. It is our opinion that high quality substrates will be most crucial to the drive to realize digital GaAs LSI ICs. Thus, in this paper we will focus entirely on digital technology. The cross-section of a typical GaAs MESFET (metal-semiconductor FET) used in IC technology is shown in Figure 3. Typically, n channel layers and n^+ source and drain regions are made by ion implantation directly into SI substrates. Two types of FET designs are used: the enhancement mode FET has an n channel layer thickness equal to that just pinched off by the built in Schottky barrier potential at zero applied gate bias. For example, using

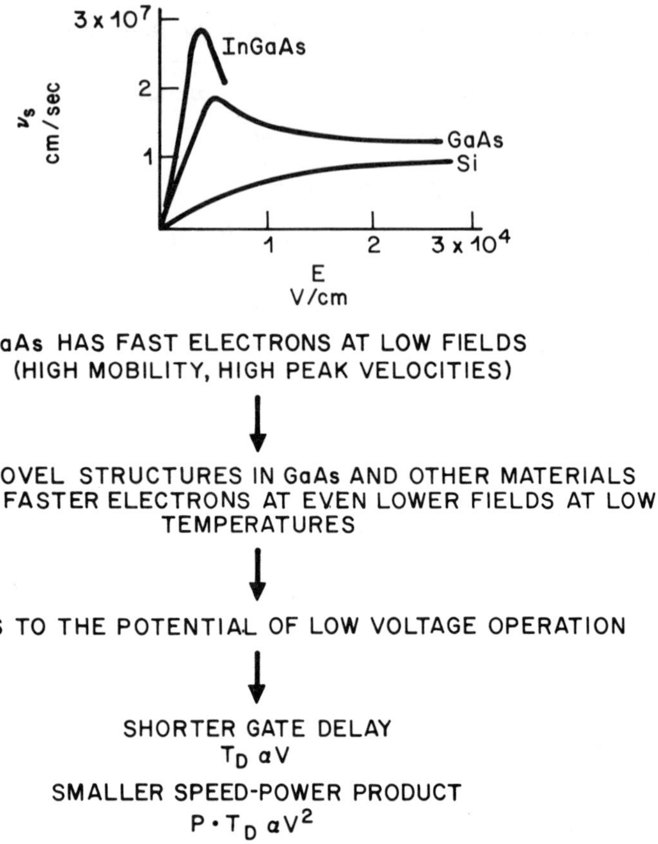

Figure 1 Why GaAs based devices for high speed.

$V_p = V_o + V_B$, where V_p = pinchoff voltage, V_B = built in Schottky barrier voltage (0.8 V) and $V_o \simeq 7\ Na^2$, where N is the doping level in units of $10^{16}\ cm^{-3}$ and a is the channel thickness in μm, a layer thickness of 0.11 μm at a doping of $10^{17}\ cm^{-3}$ is just pinched off by the built-in Schottky barrier potential, (0.8 V) with an applied gate source potential V_{gs}, equal to zero. In Figure 3, this is schematically shown by the depletion layer pinching off the n layer at $V_{gs} = 0$. It should be noted that $V_{gs} > 0$ is required to open the channel of an enhancement mode FET.

Depletion mode FETs are devices which require $V_{gs} < 0$ to pinch off the device, typically in the range −1 to −2.5 V. Using the above example of a channel doped at $10^{17}\ cm^{-3}$, a 0.15 μm channel layer would have a pinch off voltage of −1.6 V and would require a bias of −0.8 V applied to the gate to pinch off channel conduction.

Enhancement mode FETs, as will be discussed later, are the best choice for fabrication of complex LSI GaAs IC circuits requiring low power consumption. But the enhancement mode FET places severe requirements on

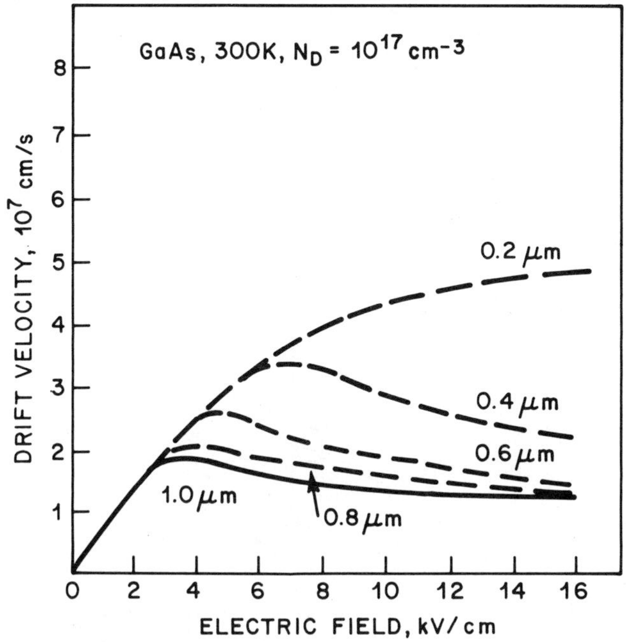

Figure 2 Drift velocity versus electric field as a function of gate length for GaAs.

the uniformity of channel doping and thickness and is the device structure that is most severely influenced by substrate quality.

Armed with the possibility of both enhancement and depletion mode FETs circuit designers have come up with an impressive array of logic circuit designs. In Figure 4, the most commonly used digital logic concepts are

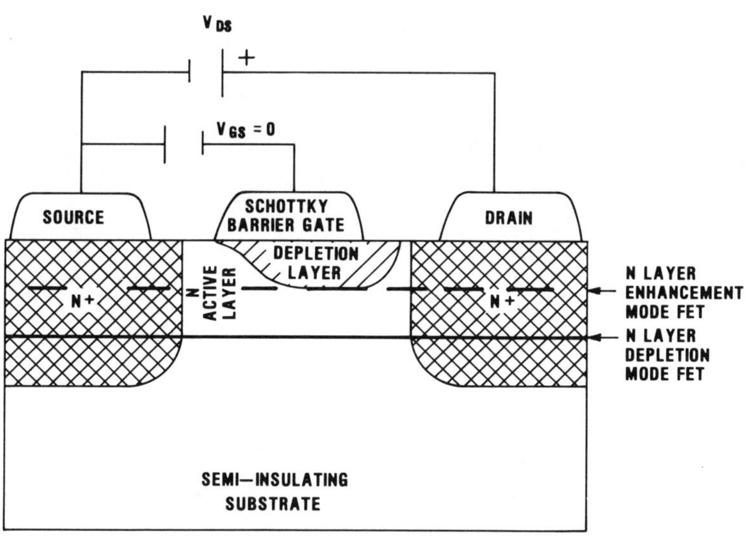

Figure 3 Cross-section of an enhancement mode and depletion mode GaAs MESFET.

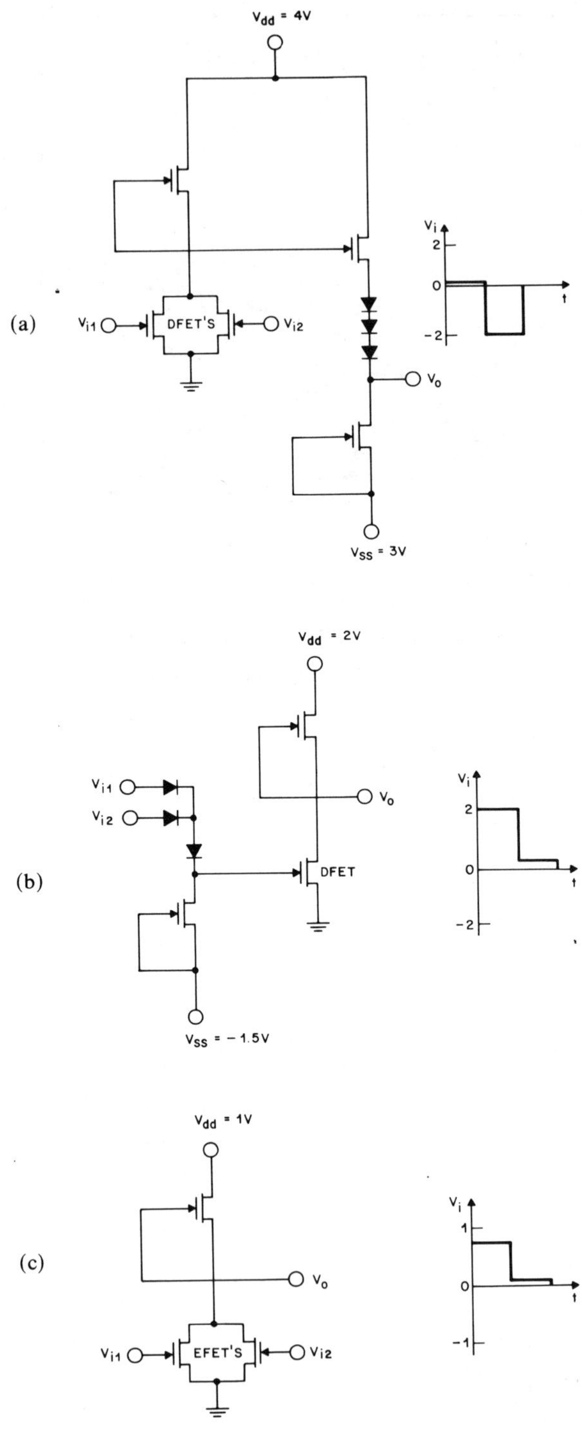

Figure 4 (a) Buffered FET logic; (b) Schottky diode FET logic; (c) direct coupled FET logic.

Table 2 *Technology comparisons*

	V_p (V)	Power/transistor (mW)	Area/gate (μm^2)	V_{bias} (V)	τ^*_{PD} (GHz)	ETC†
BFL	~ −2.5	4–5	5000	+5 and −3	5–6	4000
SDFL	−1.0	1	2000	+2.0 and −1.5	1–2	20 000
DCFL	0.0	0.2	1200	+1.0	2–3	100 000

* 1 μ gate length. τ_{PD} refers to the maximum flip flop toggle frequency of the master slave ÷2 circuits for BFL, and type D÷2 circuits for SDFL and DCFL.
† Estimated Transistor Count.

shown. Figure 4a shows a schematic for a two input NOR gate in buffered FET logic (BFL)[1]. This design uses depletion mode FET devices and loads, Schottky barrier level shifting diodes and requires two power buses for drain voltage, V_{dd}, and source voltage, V_{ss}. Although BFL has been widely used, it has the disadvantages of increased circuit area and power consumption relative to other designs and thus currently is not a prime candidate for LSI.

In Figure 4b, a two input NOR gate in Schottky diode FET logic (SDFL)[2] is schematically shown in which Schottky diodes are used as input switching elements and DFETs are used as inverters and amplifiers. This approach is less power consumptive than BFL, uses somewhat less area and is a better choice than BFL for larger scales of integration.

In Figure 4c, a two input NOR direct coupled FET logic (DCFL)[3] gate is schematically shown. This circuit design is the simplest, has the fewest number of elements, lowest power consumption and is the best choice for VLSI circuits in GaAs. In Table 2, a technology comparison is made for BFL, SDFL and DCFL showing DCFL as the clear choice for VLSI applications. A trend toward DCFL for LSI GaAs IC has been apparent in the advances made in GaAs IC complexity in various laboratories in recent years. Figure 5 plots GaAs IC complexity versus calendar year. To be noted is that BFL has reached an FET density of ~1000/chip, SDFL ~5000 FETs/chip and DCFL has reached 26 000 FETs/chip with the announcement by Fujitsu in 1984 of a 4K SRAM. The obvious choice for VLSI is DCFL which has the most demanding process and material uniformity of any of the circuit designs.

Substrate requirements

SHAPE AND DIAMETER

In general, boules grown by the horizontal Bridgman (HB) and gradient freeze techniques (HGF) are D-shaped, while those pulled by the liquid encapsulated Czochralski (LEC) method are nearly circular in cross-section.

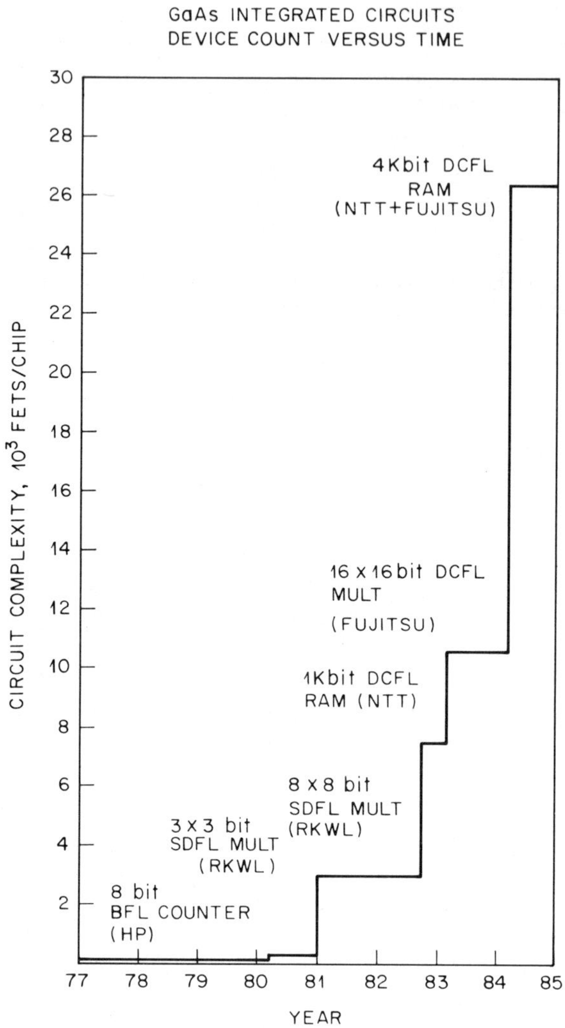

Figure 5 Time-evolution of GaAs complexity.

Confinement by the walls of the vessel produces crystals of uniform shape in horizontal systems. To preclude undulations in diameter common to LEC pulling a high sensitivity load cell is usually incorporated in the growth chamber to continuously monitor the weight signal. This has led to recent advances in diameter control. A complete analysis of the signal, including the effects of the B_2O_3 encapsulant and meniscus, has been performed [4]. Lately, a computer-automated scheme capable of regulating within $\pm 1.5\%$ the diameter of a nominally 2 inch crystal was described by Fukuda [5]. Although details of the method were not revealed, it appears that only a buoyancy correction was applied. The control loop was closed by means of empirical calibrations.

Regardless of the growth technique, grinding the boule to the desired diameter is unavoidable. Clearly, on account of its D-shape as well as the need to slice a ⟨111⟩ boule into (100) wafers, horizontal techniques suffer a greater material loss than LEC, especially when grown in the ⟨100⟩ direction. Currently, an informal consensus has been reached among GaAs IC fabricators to standardize initially on 3 inch diameter round substrates. This objective can be easily met by modern LEC pullers. Pioneering work by Akai and collaborators has led to the development of 3 inch diameter circular wafers cut from D-shaped HB grown boules [6]. Typical boule lengths are 20 and 40 cm for the LEC and HB techniques respectively. Indeed Katsumata *et al.* [7] have demonstrated by the LEC pulling of 4 inch diameter crystals that there are no fundamental obstacles in further increasing the size of GaAs when normal gradients are employed.

ASSURANCE OF SI BEHAVIOUR

In contrast to elemental semiconductors, GaAs, a binary compound with minuscule intrinsic electron concentration at room temperature usually cannot be prepared with high resistivity by means of thorough purification procedures alone. The required high resistivity material is always SI, a condition that reflects a delicate balance of deep and shallow donors and acceptors. Through much of the history of GaAs growth the bane of obtaining SI crystals has been Si donors originating from the quartz apparatus. Although the addition of Ga_2O_3 [8], in particular in HB, suppressed Si contamination, perfect compensation with a small concentration of shallow acceptors which shift the Fermi energy to the position of the intrinsic level is practically impossible. Consequently, the deep acceptor Cr has been universally employed to balance the uncompensated donors.

The present state of the art in a high pressure LEC puller using SiO_2 crucible and direct *in situ* synthesis guarantees resistivities between $\sim 10^8$ and 10^9 Ω cm for Cr concentrations ranging from ~0.1 to 2 ppm, respectively [9]. Recently, the HB growth in SiO_2 vessels of 3 inch diameter GaAs boules with greater than 10^8 Ω cm resistivity has been achieved by incorporating 0.26 ppm Cr in the crystal and adding oxygen in the form of As_2O_3 to the melt [6, 10]. In our laboratory, 2 inch wide D-shaped boules have been prepared by HGF without Cr that are SI with better than 10^7 Ω cm resistivity [11]. Key features of this technique are the replacement of the SiO_2 boat with BN and the incorporation of 10^{-4} wt.% Ga_2O_3 in the liquid. The etch pit densities for the above SI HB and HGF substrates were given as an average of 5000 and 3000 cm^{-2}, respectively.

The growth of 3 inch diameter undoped SI substrates in BN crucibles by direct *in situ* synthesis using modern high pressure LEC pullers represents a major breakthrough in supplying high purity substrates for FETs. The SI property is a result of EL2 deep donors [12] compensating carbon acceptors [13]. It has been shown that a prerequisite to attain the desired high resistivity

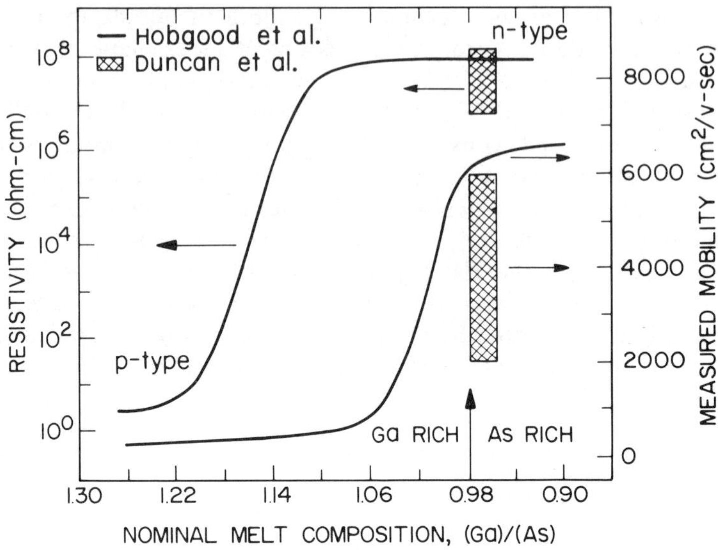

Figure 6 The variation of resistivity and mobility on melt composition for GaAs grown by low and high pressure direct *in situ* synthesis.

is an As-rich melt (~51%). This effect and other careful experiments have led to the tentative identification of EL2 as the antisite defect As_{Ga} [14].

In Figure 6 we present the resistivity and mobility of undoped SI crystals pulled by Hobgood *et al.* in a high pressure LEC system at various Ga/As ratios [15]. It can be readily seen that for As-rich compositions the resistivity and room temperature mobility approach 10^8 Ω cm and 7000 cm^2 V^{-1} sec^{-1}, respectively. Moreover, the high mobility following ^{29}Si implantation closely follows the theoretical bulk mobility as a function of peak donor concentration [16]. We have also superimposed in Figure 6 similar data by Duncan *et al.* obtained on boules grown by direct synthesis in SiO_2 crucibles employing a low pressure LEC system [17]. Although the scatter appears to be greater, the low pressure technique is only marginally inferior to the high pressure one. The mechanism of SI behaviour in this material is as yet unresolved.

CONVERSION AND ITS PREVENTION

Semi-insulating substrates used in GaAs ICs made by direct ion implantation must maintain their high resistivity in electrically inactive regions throughout all processing steps. A critical procedure is the post-implantation activation anneal that requires heating the material at ~800°C for ~30 min or longer. An undesirable aspect of this treatment is the frequently observed thermal conversion of GaAs at elevated temperatures. The converted wafer exhibits an increase in conductivity accompanied by a change to n or p-type behaviour.

Many experiments, including SIMS (secondary ion mass spectrometry) analysis, photoluminescence profiling, etc., have shown that type-conversion can be interpreted as anomalous out-diffusion of an intentional dopant or a background impurity to an internal or external surface of the substrate.

In particular, SIMS profiles of Cr-containing ($\sim 2\times 10^{17}$ cm^{-3}) SI substrates annealed above 800°C display a steep 'uphill' Cr diffusion front at the surface followed by a pronounced minimum ~ 0.2 μm therefrom [18]. Eventually, between 1 and 2 μm from the external boundary the Cr concentration merges with the bulk level. Due to the deficit of Cr in the surface domain the slice converts to n-type. In contrast, upon annealing the unexpexted presence and migration of the residual impurity manganese in a nominally undoped n-type

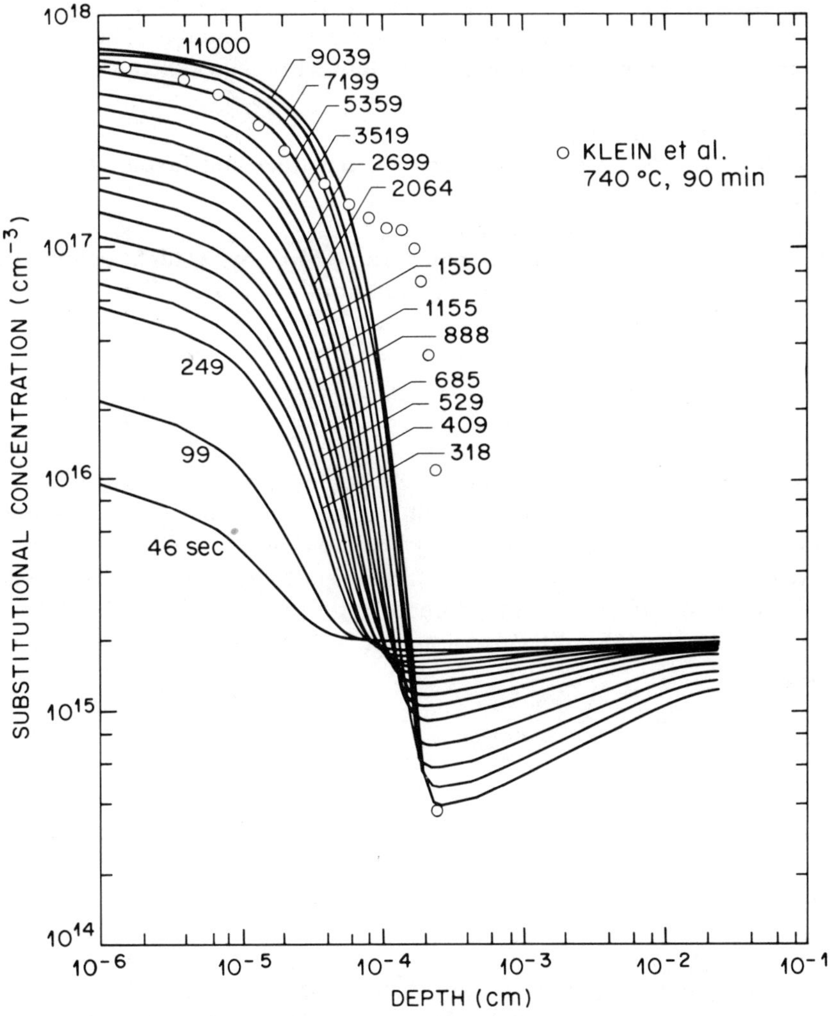

Figure 7 Mn out-diffusion profile in a GaAs slice annealed at 740°C for 90 min. The calculated curves are based on a substitutional-interstitial model.

or SI substrate leads to p-type conversion [19]. In a convincing set of experiments using photoluminescence profiling and SIMS, Klein *et al.* have shown that after 90 min annealing at 740°C in H_2 or vacuum, Mn accumulates within a 2–3 μm skin with a surface concentration climbing to a maximum value of 6×10^{17} cm^{-3} [19]. The Mn redistribution data of Klein *et al.* are reproduced in Figure 7 together with our recent theoretical analysis based on an interstitial-substitutional diffusion model [20, 21]. P-type conversion as a consequence of Mn out-diffusion was also reported in Si and Se^+ implanted GaAs substrates after annealing at 900°C. Kamber *et al.* found that only boules grown in a PBN crucible using *in situ* direct synthesis preceding LEC pulling were immune to conversion [22]. An important factor here may be the purification of the initial As charge.

Though type conversion is a prevalent feature of the high temperature processing of GaAs substrates, it is by no means unavoidable. The following list offers several remedies to reduce or eliminate this problem:

1. Avoid or minimize Cr doping.
2. Use high purity starting materials, BN crucibles and favour *in situ* synthesis.
3. Attempt using flash annealing.
4. Remove background contaminants (e.g. Mn) by a high temperature anneal/etch cycle prior to ion implantation.

DOPING UNIFORMITY

Axial variation in resisitivity

According to Thomas *et al.* undoped SI boules grown in BN crucibles by the high pressure LEC technique exhibit a seed to tail increase in resistivity between $\sim4\times10^7$ and 10^8 Ωcm [16]. Fukuda recently reported that As evaporation causes a several order of magnitude drop in resistivity from a plateau value of 10^8 Ω cm with a spread better than a factor of 2 [5]. As yet there is no evidence that FET performance is compromised by axial changes in resistivity.

Manifestations of radial inhomogeneity: EL2

Martin *et al.* measured the radial distribution of EL2 concentration in LEC wafers by infrared absorption [23]. As shown in Figure 8 the characteristic 'W'-shape is observed, resembling the radial dislocation density pattern. In later work Holmes *et al.* confirmed the W-shape of the EL2 profile in the seed sections of 3 inch diameter boules [14]. However, as the tail region is approached the distribution becomes more diffuse and uniform with a narrow dynamic range.

Holmes *et al.* have attributed the W-shape to convective flow patterns in the GaAs melt and questioned the significance of correlation with dislocation density [14]. But recently Brozel *et al.*, employing a high spatial resolution

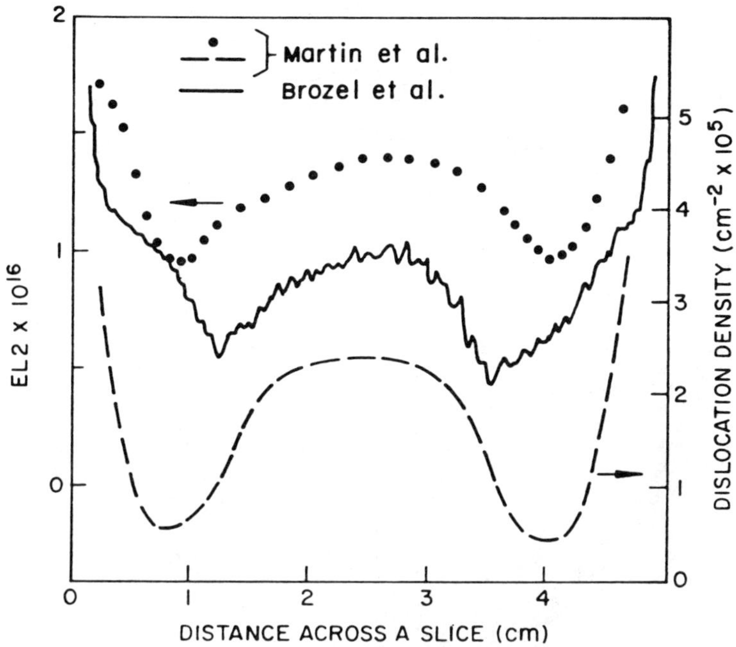

Figure 8 The radial distribution of EL2 and dislocation density in a GaAs slice.

(1 and 2 μm scans) near infrared absorption method, provided additional evidence connecting EL2 concentration to dislocation density (see Figure 8) [24]. Combining the fine structure of the optical spectrum with an infrared transmission micrograph and an A/B etch pattern, they have shown in detail on a microscopic scale that increased EL2 concentrations are associated with regions of high dislocation density [24].

To explain this relationship one may speculate that dislocations in GaAs serve as sinks for Ga atoms, creating Ga vacancies which in turn absorb As atoms to form the antisite defect As_{Ga}, which presumably is associated with EL2. Independent of detailed mechanisms in undoped SI material EL2 is the compensating donor responsible for the precarious balance assuring SI behaviour, and thus a radial variation in its concentration may diminish the high resistivity of a slice in some regions.

Resistivity

A ±25% variation in sheet resistance is typical in SI GaAs. Blunt *et al.* have recently developed a novel 'dark spot' non-destructive wafer scanning technique to map the sheet resistance [25, 26]. Briefly, in an illuminated wafer a strip of material becomes ohmic by the creation of numerous current carriers, providing a low resistance path everywhere except in a high resistance region protected by a movable mask. In this manner the sheet resistance of Cr-doped material near the seed, reproduced in Figure 9, was determined. The dislocation density distribution revealed by the A/B etch on the same

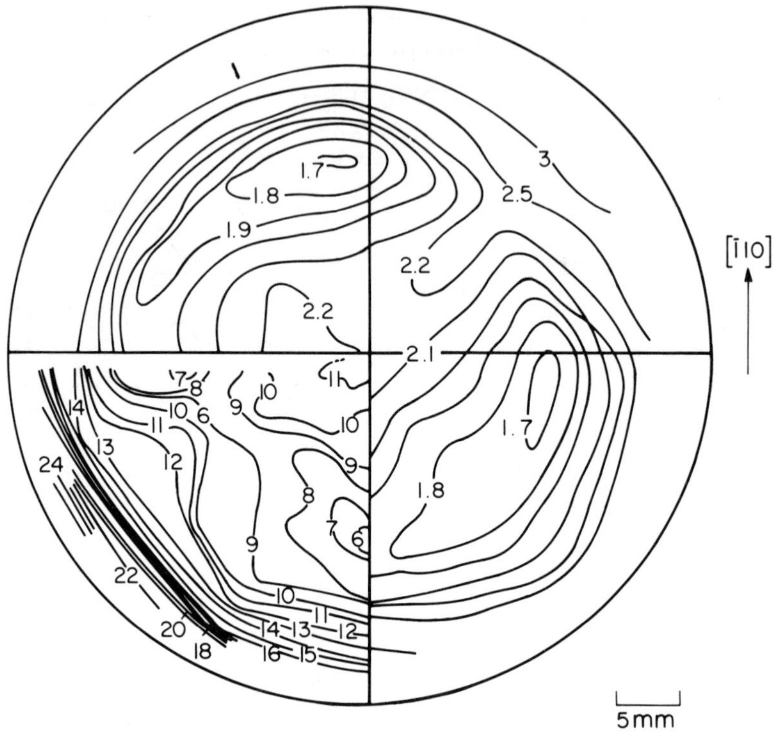

Figure 9 Dark spot sheet resistance contour lines for a GaAs wafer. The lower left quadrant depicts the experimental dislocation map corresponding to the resistance shown in the top right quadrant. The units for dislocation density and resistance are 10^4 cm^{-2} and 10^7 $\Omega/\square$, respectively.

sample is shown in the third quadrant. There is a clear correlation between the sheet resistance and dislocation contour lines [26].

In Figure 10 we display a 'grey scale' representation of the dislocation map for the top wafer of a ⟨001⟩ boule obtained from the quasi-steady state heat transfer/thermal stress model of dislocation generation in GaAs [27]. Accordingly, there is convincing evidence that thermal-stress induced slip dislocations are responsible for the observed etch pit pattern. Moreover, here we can readily visualize the origin of the often-quoted W-shape, already mentioned with regard to the EL2 distribution.

As yet no plausible interpretation has emerged connecting the W-shapes in sheet resistance and dislocation density. However, in applying the van der Pauw technique to measure the bulk resistivity of SI LEC substrates, Grant *et al.* [9] and Bonnet et al. [28] have found an M-shaped (inverse W-shaped) variation across their slices. Excluding the possibility that either experimental method is in error, the source of this opposite behaviour may be a boule-to-boule (or even wafer-to-wafer) difference in the impurity balance that is responsible for the SI character of the material. In this connection, it deserves mentioning that according to Blunt *et al.* the W-shape seen in the seed region

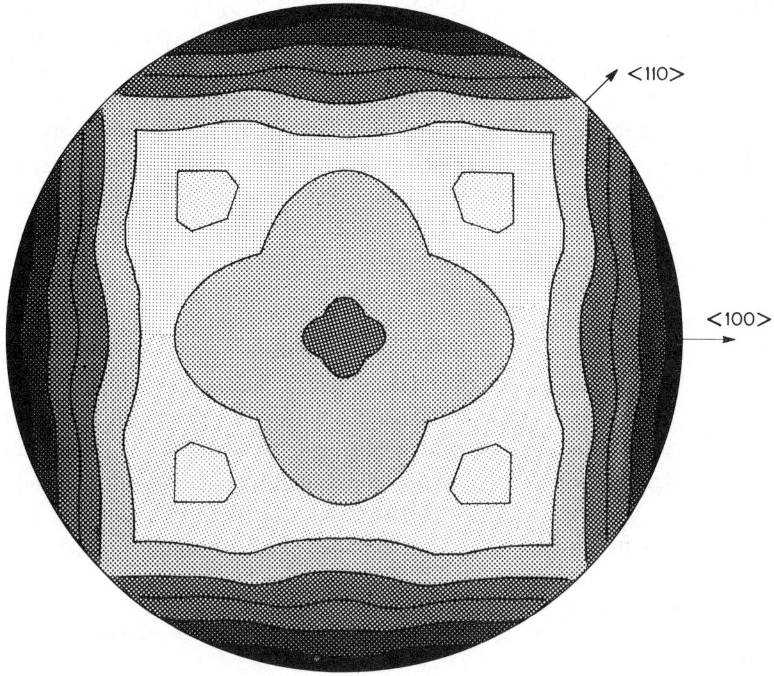

Figure 10 Calculated dislocation density contour lines for a ⟨100⟩ GaAs wafer.

becomes less pronounced as the tail section is reached [26]. In fact, one tail-end slice exhibits an M-shape in the ⟨110⟩ direction.

The inhomogeneity in resistivity persists even after Si ion implantation into Cr-doped SI substrate. The van der Pauw sheet resistance profile indicates an M-shaped distribution between 3 and $3.5 \times 10^3\ \Omega/\square$ in both ⟨100⟩ and ⟨110⟩ directions while the dislocation density displays the usual W-shape of LEC wafers [29].

At present, we can give only a tentative interpretation of the discrepant results on the course of the resistivity variation across SI substrates. One may speculate that under some conditions, even when some Cr is added, the deep donor EL2 governs the charge balance. It is then not surprising that sheet resistance follows the dislocation profile, both being W-shaped. When other circumstances prevail during growth, acceptors, in particular the deep level Cr, regulate the SI property. If one postulates that such acceptors are attracted to dislocations via a Cottrell [30] atmosphere, the surrounding region becomes denuded of compensating charges, thus increasing the local conductivity. Extending the argument over an entire slice, one may readily visualize that high dislocation density areas as the size of the denuded sectors increases by overlap, exhibit enhanced conductivity relative to low dislocation density regions. Hence, it is not surprising that the resistivity distribution is M-shaped compared to the W-shaped dislocation profile.

Some supporting evidence for this argument is offered by the investigation of Otsubo *et al.* by SIMS-calibrated photo-current response measurements of Cr concentration [31]. By means of scanning small ($\leqslant 0.03\ cm^2$) surface areas, M-shaped Cr concentrations in different crystallographic orientations across the wafer were determined with values ranging between 10^{16} and $1.5 \times 10^{16}\ cm^{-3}$. The conflicting evidence on resistivity variation challenges the current fundamental understanding of the semi-insulation mechanism. Therefore, in the future a battery of characterization tests must be performed on the same sample coupled with detailed carrier analysis.

Leakage current
A non-destructive determination of small-scale inhomogeneities in SI GaAs is possible by a scanning leakage (I_L) current measurement under illumination. In Figure 11, we show I_L for a SI wafer cut from the top of a LEC boule taken from the work of Matsumoto and Watanabe [32]. Here again an M-shaped distribution is encountered while the corresponding dislocation distribution is W-shaped. These authors have also shown that the resistivity is inversely proportional to I_L. Furthermore, inhomogeneities in I_L can be smoothed out by annealing by 850°C.

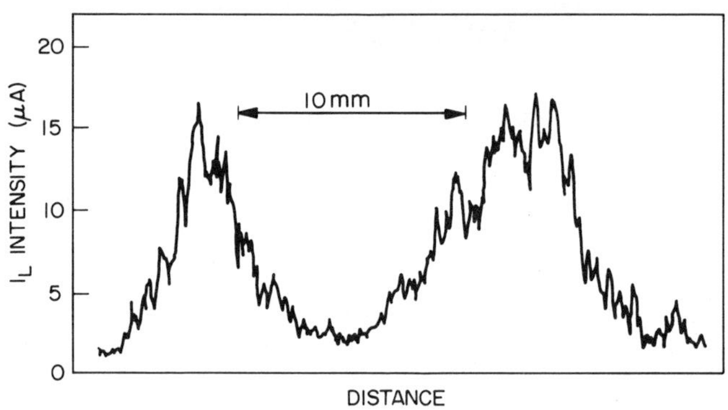

Figure 11 Scanning leakage current versus distance across the wafer.

The correlation between the M-shaped I_L and the dislocation pattern has been confirmed in more detail in both ⟨100⟩ and ⟨110⟩ directions by Miyazawa *et al.* [33]. All the I_L data are consistent with the W-shaped sheet resistance profile of Blunt *et al.* [26].

Effect of dislocations on FETs in device arrays

THRESHOLD VOLTAGE AND DRAIN SOURCE CURRENT VARIATION

Significant studies relating device performance to dislocation density have come from the NTT Laboratories, Japan, where quantitative relationships for the effect of dislocation density of LEC GaAs on source-drain current, I_{ds}, and threshold voltage, V_{th} have been obtained for reasonably large levels of MESFET integration. FET arrays fabricated by direct ion implantation (DII), on (100) SI undoped LEC substrates at 1.5×1.0 mm spacing exhibited significant variation in V_{th} and I_{ds} [34]. Figure 12 shows V_{th} results as a density profile map for 404 devices each having a source-drain spacing of 6 μm, a gate length of 1 μm, and a channel width of 20 μm. Dark areas correspond to low V_{th}. For V_{th} varying between −0.1 V to −0.6 V, a good correlation was obtained with EPD (etch pit density) distribution. The highest threshold voltages appear in areas in which the dislocation EPD is

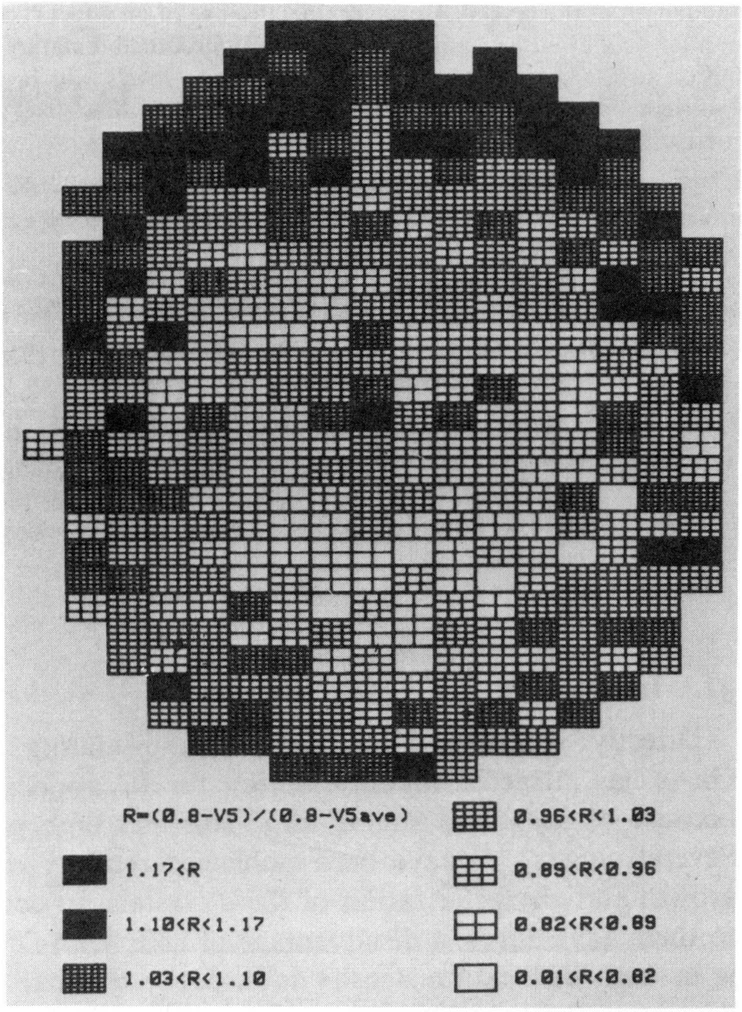

Figure 12 Threshold voltage, V_{th}, density map of 2 inch diameter wafer. Black corresponds to low V_{th}. $R = -(0.8 - V5)/(0.8 - V5_{ave})$, V5 is gate voltage needed to produce 5 μA drain current for 1 V applied source-drain voltage. $V5_{ave} = -0.3$ V.

greatest. Similar results for I_{ds} varying from 0.5 mA to 1.4 mA relate large source-drain currents to wafer areas containing high EPD.

These results were subsequently extended to large array FETs (~20 000 devices) fabricated from low Cr-doped (0.1 ppmw Cr) LEC GaAs on (100) wafers cut normal and parallel to the growth direction [35]. Source-drain current for the (100) wafers measured on devices made along ⟨100⟩ correlated closely with EPD, the centre and periphery regions showing high I_{ds} and large EPD. For a (010) substrate containing 26 000 FETs where the Cr concentration varied from 0.155 ppmw at the top of the substrate to 0.180 ppmw at the bottom, I_{ds} distribution density correlated with EPD, uninfluenced by any effect from Cr variation or other impurity distribution along the length of the crystal. However, the absence of an effect ascribable to Cr cannot be viewed as significant, given the questionable nature of the contribution to the maintenance of SI behaviour at levels which are not greatly different from the concentrations of other shallow background impurities.

FET MICROSCOPY

More recent evidence for the effect of individual dislocations on FETs fabricated by DII relates the position of a dislocation etch pit to the source-drain channel and thus to measured V_{th} for the given FET [36]. These devices, which had a gate length/gate width of 1 μm/5 μm were made on low Cr-doped, ~0.1 ppmw, LEC substrates with 200 μm spacing. By etching with KOH on the fabricated wafer, direct registry of source-drain channel to etch pit location was obtained. The upper region of Figure 13 contains data

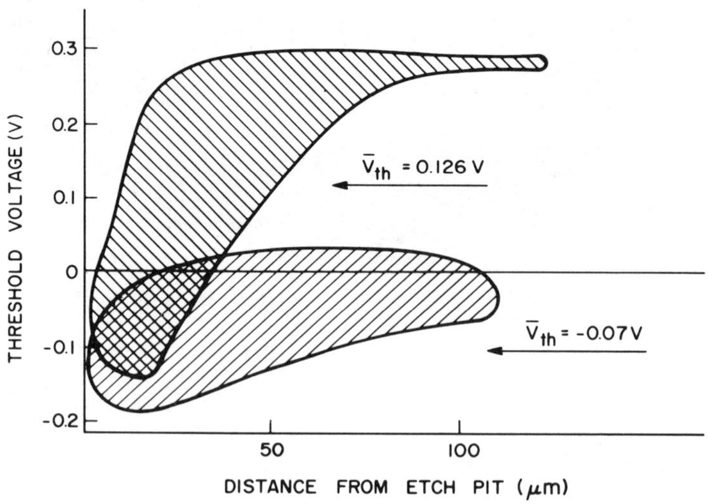

Figure 13 Threshold voltage, V_{th}, versus distance from an etch pit. The upper hatched region envelopes data for 110 FETs measured along ⟨110⟩. In the lower region are similar measurements on a substrate annealed for 14 h at 800°C.

obtained from 110 FETs selected at random along a ⟨110⟩ direction. The result, in concert with previous experiments [37], obtained a $\bar{V}_{th}$ (average) and σV_{th} (standard deviation) of 0.126 V and 0.101 V respectively, for a spread in V_{th} of 0.45 V. The large change in V_{th} occurs at distances <50 μm from the dislocation etch pit. Within this distance SEM CL studies show high image brightness around individual dislocations indicating that the impurity concentration is lowered in this region [38]. Impurities diffuse toward the dislocation core due to strain energy associated with the misfit in the core area. Thus, within the depleted region the Si implant activation efficiency can be expected to be higher, increasing I_{ds} and V_{th}. With annealing, bright CL images become large and less distinct indicating a redistribution of impurity occurs within the depleted zone. The effect of annealing on threshold voltage is shown in Figure 13 by the lower region enveloping measurements along ⟨110⟩ on 110 FETs [36]. An average threshold voltage of −0.069 V was obtained with $V_{th} = 0.05$ V. The maximum variation in V_{th}, 0.025 V is about one-half of that realized when processing unannealed material. This is an excellent result because it relates the change in atmosphere around a dislocation with annealing to attaining improved device parameters. Recent AT&T Bell Laboratories SEM studies on SI boat grown GaAs show no bright CL areas around dislocations [39]. Since boat grown crystals remain at temperatures near the melting point much longer than LEC grown crystals, significant improvements in uniformity may be realized for devices made in such material.

The ways and means of dislocation reduction in GaAs

The correlations between dislocation density distribution and several key physical parameters cited in previous sections have not discouraged the device engineering community in seeking to develop a high speed GaAs IC technology. Consequently, worldwide efforts to prepare more perfect crystals have intensified.

The fundamental understanding of dislocation reduction in III–V compounds is largely based on the quasi-steady state (QSS)heat transfer/thermal stress model that was originated at AT&T Bell Laboratories [27]. Briefly, the temperature distribution in a finite cylindrical crystal during LEC growth has been obtained from a closed-form solution of the QSS partial differential equation for heat conduction. The boundary conditions were formulated by assuming (a) a moving planar solid-liquid interface fixed at the melting point (T_f) and (b) cooling, following Newton's law, by natural convection and radiation from the lateral and top surfaces of the crystal into an ambient at temperature T_a. Subsequently, employing thermoelastic analysis, the radial, tangential and axial stress components caused by the non-uniform temperature gradient were explicitly determined. These stresses permit the evaluation of the 12 resolved shear stress components for the (111), ⟨110⟩ slip system of a boule with either ⟨100⟩ or ⟨111⟩ growth direction. The dislocation density is then postulated to be proportional to the excess stress given by the sum of differences between the absolute value of each resolved shear stress and the critical resolved shear stress (CRSS) if greater than zero [40]. The predicted dislocation distribution contour maps exhibit fourfold and sixfold symmetry [41], respectively, for (100)(see Figure 10) and (111) slices in good accord with etch pit patterns observed on KOH etched GaAs substrates.

Theory suggests two avenues to the reduction and/or elimination of dislocations both confirmed by experiments on GaAs and InP. The first one is a sharp decrease in the ambient temperature gradient [27, 40]. The second alternative is impurity hardening of the lattice and thereby increasing the CRSS [41]. In Figure 14, the dislocation density contour maps for the seed

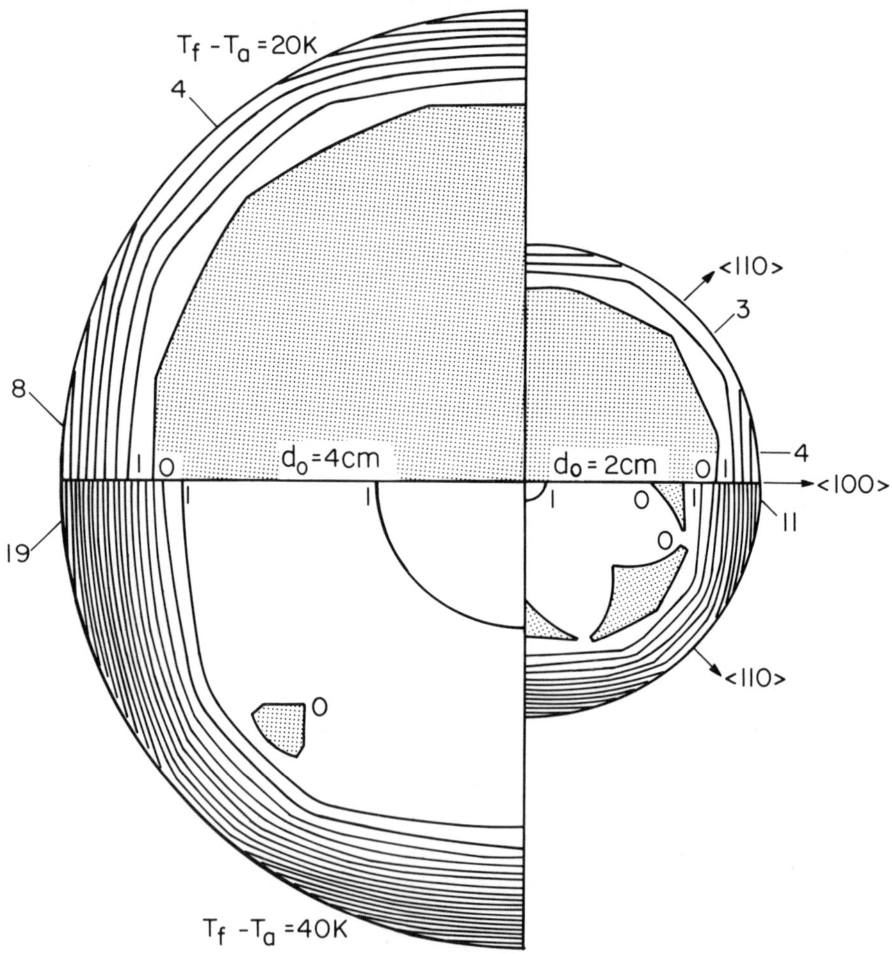

Figure 14 Dislocation distribution in 2 and 4 cm diameter LEC crystals grown at an average ambient temperature of 20 K (top half) and 40 K (bottom half) below the melting point. Shaded areas labelled zero depict dislocation free regions.

end of 2 and 4 cm diameter GaAs boules are shown. The top and bottom halves are for $T_f - T_a = 20$ and 40 K, respectively. At 20 K below T_f the boundary of the dislocation-free region appears approximately at the same relative radius for both 2 and 4 cm diameter crystals [40]. Although at the periphery the dislocation density rises with increasing boule diameter, the larger one includes a substantially greater defect-free area than the smaller one. This clearly illustrates a means to obtain relatively large low-dislocation density regions by increasing the diameter and subsequent coring. As the ambient temperature drops so that $T_f - T_a = 40$ K there is a sharp decline in the dislocation-free segments, especially at 4 cm diameter. Since in a standard LEC puller the actual $T_f - T_a$ may be as large as 200 K, it is not surprising that boules of commercial size are heavily dislocated.

The effect of ambient temperature on dislocation densities has been verified in the case of GaAs and InP at Bell Laboratories [42] and other institutions [43, 44]. We have been able to increase T_a in an experimental low pressure LEC system by raising the height of the B_2O_3 encapsulant and additional measures to modify the thermal environment. Gradients as shallow as 20 K cm^{-1} have been achieved. However, growth under these conditions has led to a highly concave interface accompanied by severe shape instability [42]. By decreasing the temperature gradient in an LEC system an approximately one order of magnitude reduction in defect density occurs. Indeed, the difference between low gradient LEC and HGF is only about a factor of 2. It should be pointed out that a much steeper drop in dislocation density can be realized in Si-doped material [44]. As yet there is no evidence that dislocation densities can be reduced below the 2–4000 cm^{-2} level without impurity hardening or perhaps strict control of stoichiometry by adjusting the vapour pressure of As [45].

Seki *et al.* [46] and Mil'vidskii *et al.* [47] have been the major early proponents of using certain impurities at high concentrations to reduce the dislocation density in GaAs and InP. A recent experimental and theoretical investigation on dislocation formation in Ge-doped InP confirms the view that at elevated temperatures the CRSS is enhanced by increasing impurity additions [41]. Consequently, it is thought that the operative hardening mechanisms are via solid solution and precipitation at low and excessive Ge-doping levels, respectively. The CRSS of heavily Ge-doped (1.3×10^{19} cm^{-3}) InP is sufficiently high to provide dislocation-free material which is in accord with predictions.

Among the highly favoured impurities used to reduce the dislocation density in GaAs are Si and Te in HGF and LEC systems, respectively. Of course, both hardening agents yield n-type boules which are inappropriate for IC use unless an intermediate epitaxial buffer layer is deposited before ion implantation [48].

Presently, a widely used impurity is In, an element isoelectronic with Ga, which leaves the SI character of the boule unchanged [49, 50]. An example from our work of dislocation reduction achievable with an In addition of 0.3 wt.% in the melt is shown in Figure 15. A comparison of standard undoped

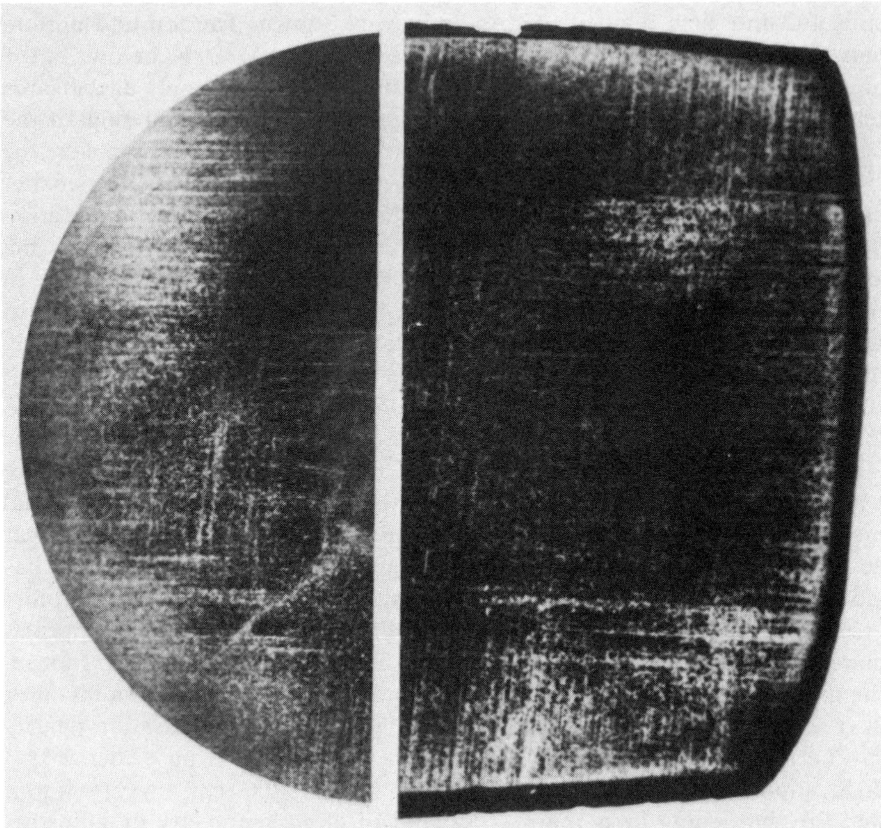

Figure 15 Dislocation structure in an undoped (100) LEC substrate (left) and an In doped (0.2 ppmw) substrate.

and In-doped LEC slices following etching with hot KOH makes the effect of In incorporation immediately obvious. The dislocation density is as low as 500 cm^{-2} in the vicinity of the ⟨110⟩ minima. Still further reductions are obtainable at higher solute concentrations. Among the disadvantages of lattice hardening is the progressive enrichment of the GaAs melt with In (distribution coefficient = 0.08) causing phase separation after 2–3 inches of a single crystal growth. Furthermore, frequently dense slip lines decorated with etch pits appear along the circumference of the substrate.

Some speculations on IC performance limitations caused by dislocations

STATISTICAL ASPECTS

The cited experimental results, correlating the electrical characteristics with dislocation density in currently available GaAs substrates and FET structures bring into question the level of integration ultimately achievable. Here we provide a first order statistical estimate with respect to the deleterious effect of dislocations on IC performance. Two illustrative examples are considered. First, based on the findings of Miyazawa *et al.* [37], it is assumed that a dislocation has a circle of influence, r, on V_{th} and if it intersects the active area of a FET, A, the IC becomes inoperative. Under these conditions it can readily be shown that the probability of failure for a single FET, p, is approximately given by:

$$p = r^2 \pi d_D \tag{1}$$

where d_D is the dislocation density ($\sqrt{A} \ll r$). Consequently, the probability of survival of an IC, P_{IC}, comprising N FETs becomes:

$$P_{IC} = (1 - p)^N \tag{2}$$

Clearly, when a large number of chips is produced P_{IC} can be interpreted as the yield due to dislocation-induced failure.

Alternatively, one may postulate that the direct overlap of an etch pit with the active area of a FET will cause the IC's instant demise. Then, p is of the form:

$$p = Ad_D \tag{3}$$

and the formula for P_{IC} remains as before in equation (2). In Figure 16 we have plotted P_{IC} as a junction of d_D for these disparate cases, taking $r = 20\ \mu m$ and $A = 10\ \mu m^2$. It can be seen that if only the overlap of a dislocation with a FET results in failure then a circuit with 10^4 devices can be fabricated at a 10% yield provided the dislocation density is ~2–3000 cm^{-2}. As mentioned earlier such densities are readily achievable in In-doped LEC or HGF substrates. On the other hand, if V_{th} is a strong function of the distance from a defect as claimed by Miyazawa *et al.* then the same packing at 10% yield would require virtually dislocation-free material [37].

Of course, one cannot exclude the possibility that this first order estimate overstates the difficulty of attaining MSI and LSI in GaAs circuits. For example, we do not know the dependence of r on d_D and FET spacing. Nevertheless, the measurements of Miyazawa *et al.* [37] do not give grounds for optimism in achieving a high degree of integration in heavily dislocated material. This view should be tempered by the announcement of Fujitsu demonstrating a GaAs IC comprising 26 000 FETs. Conceivably, this breakthrough was accomplished in low dislocation density material at a minute yield employing a test circuit that did not exercise *all* the FETs. However, what seems more likely is that the effect of V_{th} variation on device performance is not as profound as initially expected. In the concluding section we shall consider the influence of V_{th} fluctuation on IC behaviour employing a circuit model.

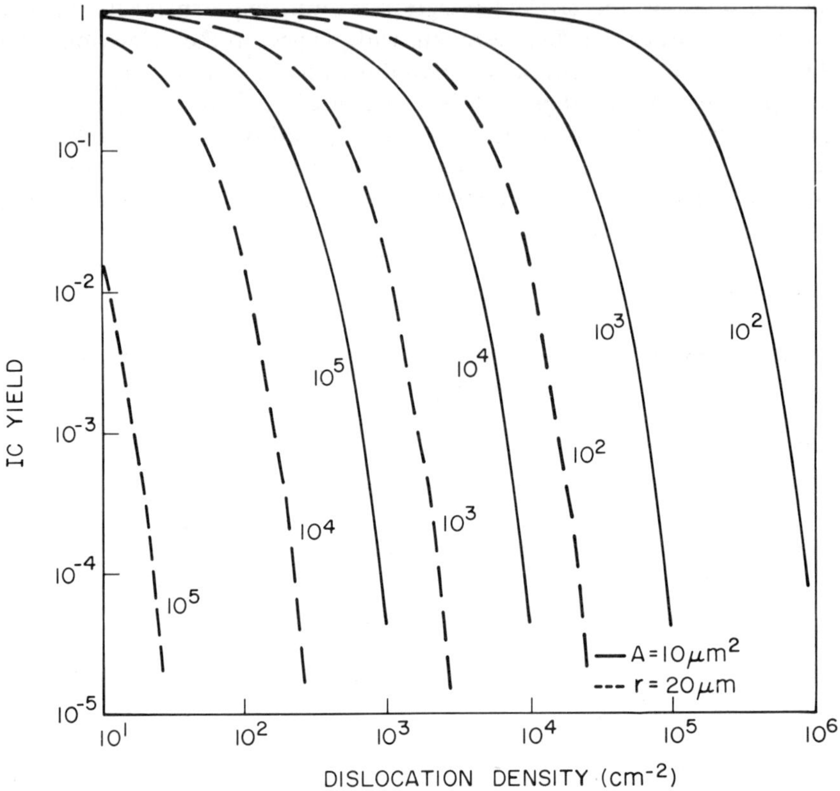

Figure 16 Estimated yield as a function of dislocation density for GaAs ICs. The labels correspond to the FET density within a chip. In constructing the solid and dashed curves, respectively, it is assumed that dislocations overlap with the active area (10 μm^2) and, alternatively, exert their influence over a 20 μm radius.

EFFECTS OF MESFET THRESHOLD VARIATION ON DCFL CIRCUIT PERFORMANCE

The basic DCFL inverter can be characterized by its transfer curve as shown in Figure 17a. The output voltage of an inverter, loaded as it would be in a real circuit by a following inverter stage, is plotted as a function of its input voltage. The stable states of the inverter, shown in Figure 17b are the points where the transfer curve and its reflection about a 45° line intersect and are labelled V_{OH} and V_{OL}. The unity gain points occur where the slope of V_{out} versus V_{in} equals -1. The separation between the stable state and nearby unity gain point is referred to as the noise margin and is a measure of circuit stability.

The following illustration (Figure 18) shows the effects of a variation of enhancement MESFET threshold voltage on the DCFL inverter characteristics. The circuit which was simulated contained a chain of inverters in cascade.

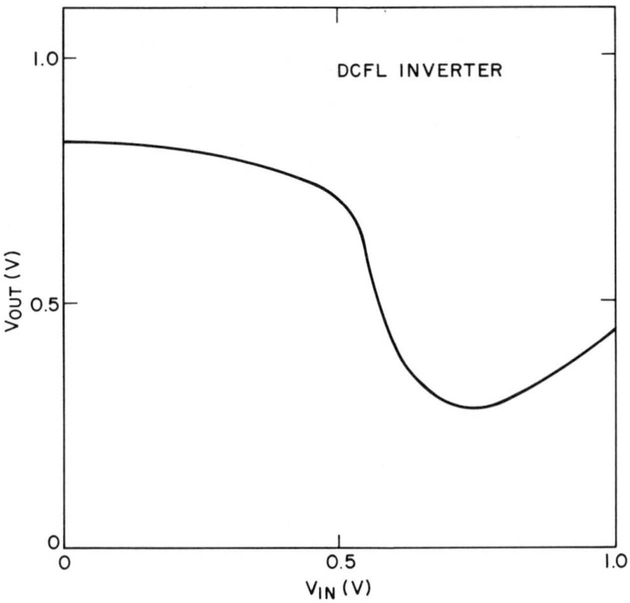

(a)

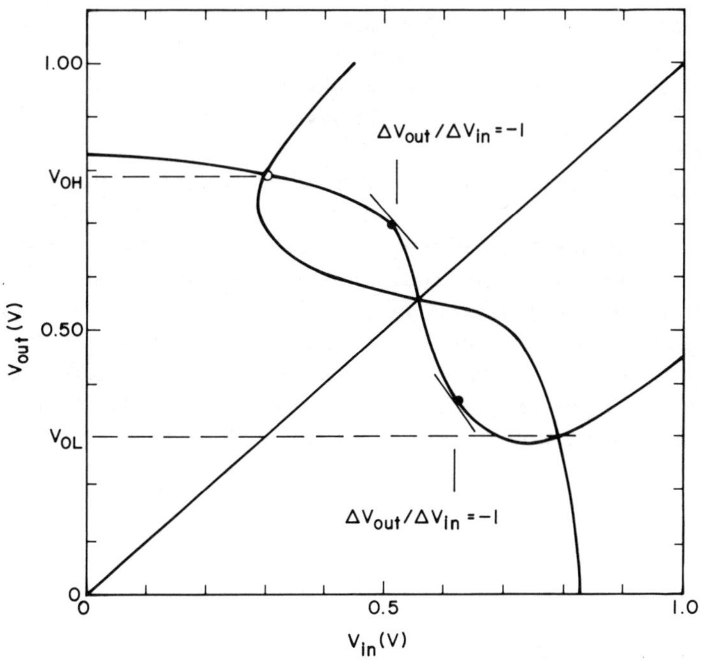

(b)

Figure 17 (a) Transfer curve of a basic DCFL inverter; (b) stable states and the unity gain points of a basic DCFL inverter.

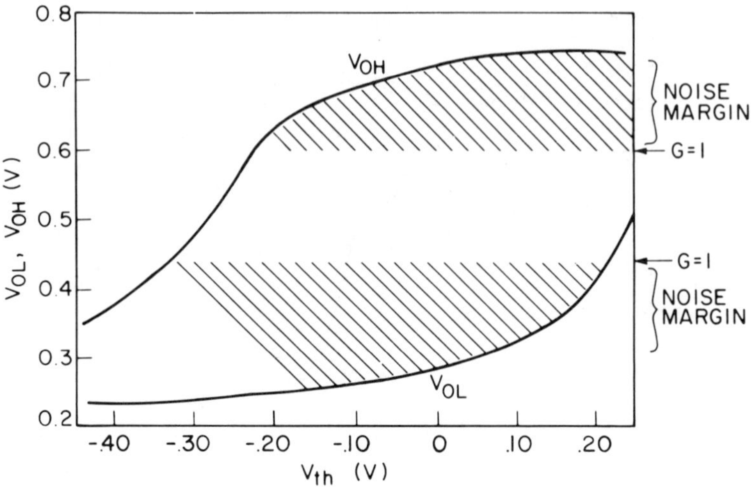

Figure 18 Change in V_{OL} and V_{OH} with driver threshold variation.

The threshold voltage of a single device within this chain was allowed to deviate from its nominal value of $V_{th} = 0$ V and the resulting output voltages were observed. In Figure 18 V_{OL} and V_{OH} are plotted as a function of threshold variation, ΔV_{th}. The noise margins are greater than zero for a range of threshold voltage from roughly −240 mV to +240 mV. Of course, a real circuit requires a certain finite noise margin which restricts the allowable ΔV_{th}. Also, ΔV_{th} affects the current supplied by the inverter during switching transients and hence circuit speed.

In conclusion, if dislocations indeed have as pronounced an effect on V_{th} over a 20 μm or larger radius as reported by Miyazawa *et al.* [37], then the preceding simple statistical argument indicates that VLSI of GaAs ICs is precluded on substrates with dislocations in excess of 100 cm^{-2}. However, the first order circuit simulation of a DCFL inverter chain indicates that noise margins can be maintained over a greater range of V_{th} variations than originally suspected. But this circuit simulation does not include the effects of V_{th} on propagation delay. Thus, the effect of dislocations on circuit yield and performance cannot as yet be forecast with confidence unless more detailed experimental data and sophisticated theoretical circuit models are available.

ACKNOWLEDGEMENTS

The authors are grateful for discussions on this subject with S. H. Wemple, P. G. Flahive, R. Caruso, R. Dingle and C. L. Allyn. The support and interest of J. A. Giordmaine in this work is appreciated.

References

1. Wilson D. and Phillips D. H. (1982). In *GaAs FET Principles and Technology*, pp. 597–619; Massachusetts, Artech House Inc.
2. Eden R. C. and Welch B. M. (1982). In *GaAs FET Principles and Technology*, pp. 672–721; Massachusetts, Artech House Inc.
3. Lehovec K. and Zuleeg, R. (1982). In *GaAs FET Principles and Technology*, pp. 621–669; Massachusetts, Artech House Inc.
4. Jordan A. S., Caruso R. and Von Neida A. R. (1983). *Bell Syst. Tech. J.*, **62,** 477
5. Fukuda T. (1982). *Jap. J. Appl. Phys.*, **22,** 413
6. Akai S., Fujita K., Nishine S., Kito N., Sato Y., Yoshitake S. and Sekinobu H. (1983). *Ext. Abs. Electrochem. Soc.* **83–1,** 508
7. Katsumata T., Nakajima H., Orito F., Terashima K. and Fukuda T. Unpublished
8. Woods J. F. and Ainslie N. G. (1963). *J. Appl. Phys.*, **34,** 1469
9. Grant I., Rumsby D., Ware R. M., Brozel M. R. and Tuck B. (1982). In *Semi-Insulating III–V Materials, Evian*, p. 98 (S. Makram-Ebeid and B. Tuck, eds.); Shiva Publishing, Nantwich, UK
10. Nishine S., Kito N., Fujita K., Sekinobu M., Shikatani O. and Akai S., (1982). Tech. Digest of the IEEE GaAs IC Symposium (New Orleans), p. 58
11. Young M. S. S. Private Communication
12. Martin G. M. (1980). In *Semi-Insulating III–V Materials, Nottingham*, p. 13 (G. J. Rees, ed.); Shiva Publishing, Nantwich, UK
13. Holmes D. E., Elliott K. R., Chen R. T. and Kirkpatrick C. G. (1982). In *Semi-Insulating III–V Materials, Evian*, p. 19 (S. Makram-Ebeid and B. Tuck, eds.); Shiva Publishing, Nantwich, UK
14. Lagowski J., Parsey J. M., Kaminska M., Wada K. and Gatos H. C. (1982). In *Semi-Insulating III–V Materials, Evian*, p. 154 (S. Makram-Ebeid and B. Tuck, eds.); Shiva Publishing, Nantwich, UK
15. Hobgood H. M., Ta L. B., Rohatgi A., Eldridge G. W. and Thomas R. N. (1982). In *Semi-Insulating III-V Materials, Evian*, p. 28 (S. Makram-Ebeid and B. Tuck, eds.); Shiva Publishing, Nantwich, UK
16. Thomas R. N., Hobgood H. M., Barrett D. L. and Eldridge G. W., (1980). In *Semi-Insulating III–V Materials, Nottingham*, p. 76 (G. J. Rees, Ed.); Shiva Publishing, Nantwich, UK
17. Duncan W. M., Westphal G. H. and Sherer J. B. (1983). *IEEE El. Dev. Lett.*, **EDL-4,** 198
18. Vasudev P. K., Wilson R. G. and Evans Jr, C. A. (1980). *Appl. Phys. Lett.*, **36,** 837
19. Klein P. B., Nordquist P. R. and Siebenmann P. G. (1980). *J. Appl. Phys.*, **51,** 4861
20. Jordan A. S. (1982). In *Semi-Insulating III–V Materials, Evian*, p. 253 (S. Makram-Ebeid and B. Tuck, eds.); Shiva Publishing, Nantwich, UK
21. Jordan A. S. and Nikolakopoulou G. A. (1984). *J. Appl. Phys.*, **55**
22. Kanber H., Feng M. and Whelan J. M. (1982). *Appl. Phys. Lett.*, **40,** 960
23. Martin G. M., Jacob G., Poiblaud G., Goltzene A. and Schwab C. (1980). *Inst. Phys. Conf. Ser.*, **59,** 281
24. Brozel M. R., Grant I., Ware R. M. and Stirland D. J. (1983). *Appl. Phys. Lett.*, **42,** 610
25. Blunt R. T. (1982). In *Semi-Insulating III–V Materials, Evian*, p. 107 (S. Makram-Ebeid and B. Tuck, eds.); Shiva Publishing, Nantwich, UK
26. Blunt R. T., Clark S. and Stirland D. J. (1982). *IEEE Trans. Microwave Theory and Tech.*, **MTT-30,** 943
27. Jordan A. S., Caruso R. and Von Neida A. R. (1980). *Bell Syst. Tech. J.*, **59,** 593
28. Bonnet M., Visentin N., Gouteraux B., Lent B. and Duchemin J. P. (1982). Tech. Digest of the IEEE GaAs IC symposium (New Orleans), p. 54
29. Honda T., Yasunobu I., Miyazawa S., Yamazaki H. and Nanishi Y. (1983). *Jap. J. Appl. Phys.*, **22,** L270
30. Cottrell A. H. (1956). In *Dislocations and Plastic Flow*; Oxford University Press, London

31. Otsubo M., Nakatani M. and Shirahata K. (1982). In *Semi-Insulating III–V Materials, Evian*, p. 130 (S. Makram-Ebeid and B. Tuck, eds); Shiva Publishing, Nantwich, UK
32. Matsumoto Y. and Watanabe H. (1982). *Jap. J. Appl. Phys.*, **21,** L515
33. Miyazawa S., Mizutani T. and Yamazaki H. (1982). *J. Appl. Phys.*, **21,** L542
34. Nanishi Y., Ishida S., Honda T., Yamazaki H. and Miyazawa S. (1982). *Jap. J. Appl. Phys.*, **21,** L335
35. Nanishi Y., Ishida S. and Miyazawa S. (1983). *Jap. J. Appl. Phys.*, **22,** L54
36. Miyazawa S., Honda T., Ishii Y. and Ishida S. (1984). *Appl. Phys. Lett.*, **44,** 410
37. Miyazawa S., Ishii Y., Ishida S. and Nanishi Y. (1983). *Appl. Phys. Lett.*, **43,** 853
38. Chin A. K., Von Neida A. R. and Caruso R. (1982). *J. Electrochem. Soc.*, **129,** 2386
39. Chin A. K., Caruso R., Young M. S. S. and Von Neida A. R. *Appl. Phys. Lett.*, in press
40. Jordan A. S., Caruso R., Von Neida A. R. and Nielsen J. W. (1981). *J. Appl. Phys.*, **52,** 3331
41. Jordan A. S., Von Neida A. R. and Caruso R. *J. Crystal Growth*, in press
42. Von Neida A. R., Caruso R. and Jordan A. S. *J. Appl. Phys.*, in press
43. Shinoyama S., Uemura C., Yamamato A. and Tohno S. (1980). *Jap. J. Appl. Phys.*, **19,** 331
44. Elliot, G. Private Communication
45. Parsey J. M., Nanishi Y., Lagowski J. and Gatos H. C. (1982). *J. Electrochem. Soc.*, **129,** 388
46. Seki Y., Watanabe H. and Matsui J. (1978). *J. Appl. Phys.*, **49,** 1978
47. Mil'vidskii M. G., Osvensky V. B. and Shifrin S. S. (1981). *J. Crystal Growth*, **52,** 396
48. Feng M., Eu V. K., Zielinski T., Kanber H. and Henderson W. B. (1984). *IEEE El. Dev. Lett.*, **EDL-5,** 18
49. Jacob G. (1982). In *Semi-Insulating III-V Materials, Evian*, p. 2 (S. Makram-Ebeid and B. Tuck, eds); Shiva Publishing, Nantwich, UK
50. Hobgood H. M., Barrett D. L., Ta L. B., Eldridge G. W. and Thomas R. N. (1983). Presented at A.I.M.E. Electronic Materials Conference, Burlington, VT

Effects of Dislocations on Characteristics of FETs Fabricated on Undoped LEC SI GaAs Substrates

T. Takebe, M. Shimazu, A. Kawasaki, T. Kotani, R. Nakai, K. Kikuchi, S. Murai, K. Tada, S. Akai* and T. Suzuki

R & D Group, Sumitomo Electric Industries Ltd., 1-1-3, Shimaya, Konohana-ku, Osaka, 554 Japan

Abstract

Effects of dislocation distributions in LEC (liquid encapsulated Czochralski) undoped SI GaAs substrates on characteristics of MESFETs (metal-semiconductor field effect transistor) fabricated by Si ion implantation have been investigated. It has been found that the threshold voltage, V_{th}, shifts to more negative values as the dislocation density increases and that the local fluctuation of V_{th} among neighbouring FETs is enhanced by dislocation sub-structures such as cellular structures, lineages and slip-lines, and also that the 'K' value and the Schottky gate characteristics are independent of dislocation distributions. It was thus concluded that to obtain highly uniform FET characteristics, it would be necessary to obtain uniform dislocation distributions without sub-structures or uniform residual impurity distributions by means of improvement of thermal environments in the growth chamber.

Introduction

Recent papers have revealed non-uniform distributions of various properties in Cr-doped and undoped SI GaAs substrates for ICs grown by the LEC method, such as of dislocation density [1–4], Cr concentration [5], resistivity [3], leakage current [6–8], photoluminescence [9–11], EL2 concentration [12, 13] and cathodoluminescence [14–16]. In these papers, it has been pointed out that the observed distributions of the parameters are closely related to the dislocation density distributions. Direct effects of dislocations on electrical properties of the active layer and on the performance of MESFETs formed by ion implantation and annealing were first elucidated by Honda *et al.* [17], Nanishi *et al.* [18, 19] and Miyazawa *et al.* [20]. In particular, it has been reported by Miyazawa *et al.* [20] that the threshold voltage of a FET is closely correlated to the distance between the active layer and the nearest dislocation etch pit. The dislocation density in substrates is, therefore, one of the main parameters which determine the quality and homogeneity of SI GaAs crystals for ICs.

* Semiconductor Division, Sumitomo Electric Industries Ltd., 1-1-1, Koya-Kita, Itami, Hyogo, 664 Japan.

We have developed and supplied Cr-doped 2 inch ϕ and 3 inch ϕ HB (horizontal Bridgman) grown SI crystals [21] for many years, and started developing and producing 2 inch ϕ and 3 inch ϕ undoped and Cr-doped SI LEC crystals some years ago. Our technology for undoped 3 inch ϕ LEC GaAs crystals and a trial to obtain crystals with very low dislocation density by boron doping have been reported elsewhere [22, 23].

In the present work, characteristics of MESFETs fabricated on 2 inch ϕ and 3 inch ϕ LEC undoped SI substrates by Si ion implantation have been investigated in correlation with dislocation density distributions. The gate characteristics, the carrier concentration profile in the active layer and cathodoluminescence from the substrate and the ion implanted and annealed layer have been measured and used to explain the non-uniform distribution of the threshold voltage.

Substrates

The (001) substrates investigated were cut from the seed end of a 3 inch ϕ LEC undoped crystal grown from a directly synthesized melt (As/Ga = 0.992) in a PBN crucible with a B_2O_3 encapsulant containing 600 ppmw H_2O under N_2 ambient pressure of 6.0 kg cm^{-2}. The resistivity, Hall carrier concentration and Hall mobility at 300 K were 8.0×10^7 Ω cm, 1.4×10^7 cm^{-3} and 4600 cm^2 V^{-1} sec^{-1}, respectively.

Etch pit density distribution

Figure 1a shows the distribution of etch pits revealed by KOH etching (400°C, 20 min) of a quarter section of the wafer, including the centre. As is typical in LEC crystals, marked groupings of dislocations (dislocation sub-structures), such as cellular structures and lineages in the centre and the intermediate regions and slip-lines at the periphery, are present. The radial distribution of the EPD (etch pit density) is so-called W-shaped (Figure 1b), ranging from 6000 (locally 3000) to 80 000 cm^{-2}.

FET fabrication and characterization

A FET matrix consisting of identical FETs with the structure specified in Table 1 was formed on a mirror-polished and etched substrate by selective Si ion implantation and metallization by lift-off. The FETs and active layers were characterized as follows: the drain saturation current, I_{ds}, is expressed in the following form as a function of the gate voltage, V_g, at a fixed drain voltage, V_{ds}:

$$I_{ds} = K (V_g - V_{th})^2 \quad (1)$$

where V_{th}, the threshold voltage and K, the so-called 'K value', are the most important parameters characterizing FET performance. The values of V_{th}

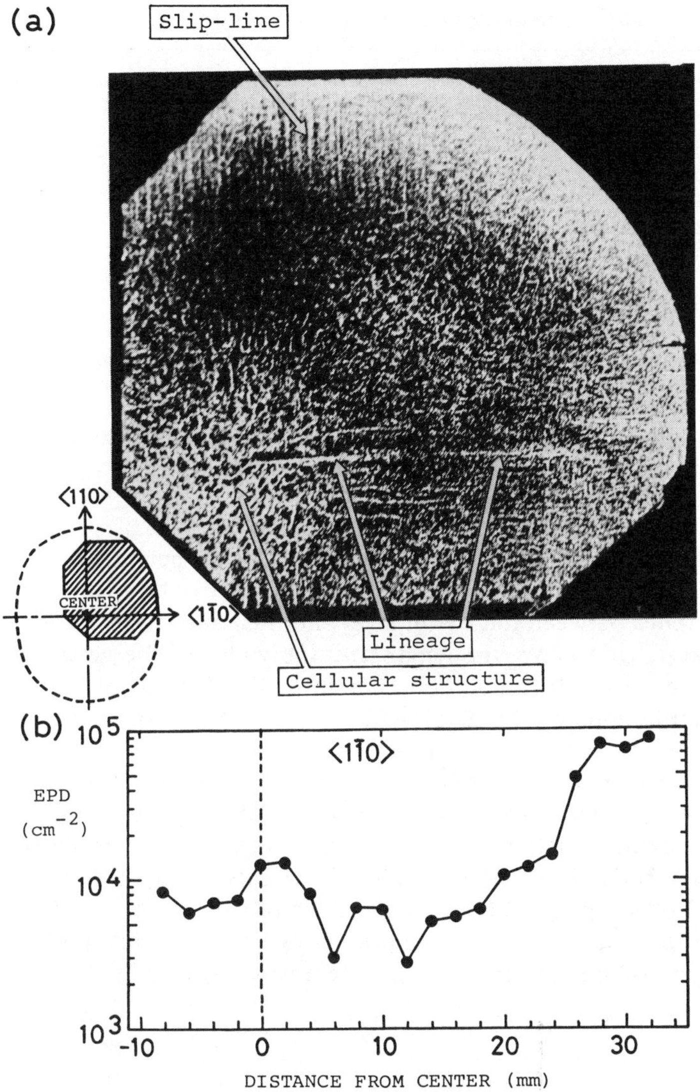

Figure 1 (a) Distribution of etch pits revealed by KOH etching, showing various dislocation sub-structures. (b) Radial distribution of etch pit density in ⟨1$\bar{1}$0⟩ direction.

and K were determined by least-squares fitting of the experimentally obtained $\sqrt{I_{ds}}$ versus V_g relation at $V_{ds} = 1.0\,V$ to the above equation. These two parameters are expressed by the following equations:

$$V_{th} = \phi_B - (E_c - E_f) - \frac{q}{2\varepsilon}(N_D - N_A)R^2 \tag{2}$$

$$K = \frac{\varepsilon}{2} \times \frac{W_g}{L_g} \times \frac{\mu_d}{R} \tag{3}$$

Table 1 *MESFET structure used for present study*

Parameter	Value
Gate length	1.0 μm
Gate width	5.0 μm
Gate direction	⟨110⟩
Source-drain distance	5.0 μm
Si ion implantation for active layer	180 keV 1.5×10^{12} cm^{-2}
Si ion implantation for contact layer	50 keV 2.0×10^{13} cm^{-2}
Activation annealing	820°C for 20 min in N_2 flow
Ohmic contact	Au–Ge–Ni alloyed at 400°C for 5 min
Schottky gate	Ti/Au double layer
Interval between adjacent FETs	200 μm

In the above equations, ϕ_B is the Schottky barrier height, $N_D - N_A$ the net carrier concentration in the active layer, R the active layer thickness, $E_c - E_f$ the difference between the Fermi level and the conduction band edge in the active layer, L_g and W_g the length and the width of the gate, respectively, and μ_d the carrier drift mobility. Therefore, distributions of V_{th} and K come from distributions of the gate characteristics (ϕ_B) and/or the active layer parameters ($N_D - N_A$, R, and μ_d).

The value of ϕ_B was estimated, together with the ideality factor, n, from the current-voltage (I–V) characteristics of the gate on the basis of the thermionic emission theory [24]. The carrier concentration profile in the active layer was obtained from the depletion layer capacitance-voltage (C–V) relation [24] of a 'fat' FET (gate area = 200×140 μm^2) fabricated in the same way on the adjacent wafer. In the present experiment, a computer-controlled automatic probing and measuring system was employed.

Results

DISTRIBUTION OF V_{th}

A plot of the V_{th} distribution in the wafer is shown in Figure 2. The variation of V_{th} is expressed in a 9-level gradation with 100 mV step, in which V_{th} is more negative as the shading becomes darker. A close correspondence is obvious between the EPD distribution in Figure 1a and the V_{th} distribution in Figure 2. That is: (a) macroscopically, V_{th} shifts to more negative values as EPD increases; and (b) microscopically, V_{th} of FETs located on and near dislocation sub-structures shifts to more negative values, e.g. FETs in the centre region where the cellular structure is evident, those on the lineages in the intermediate region, and those on the slip-lines running at the periphery.

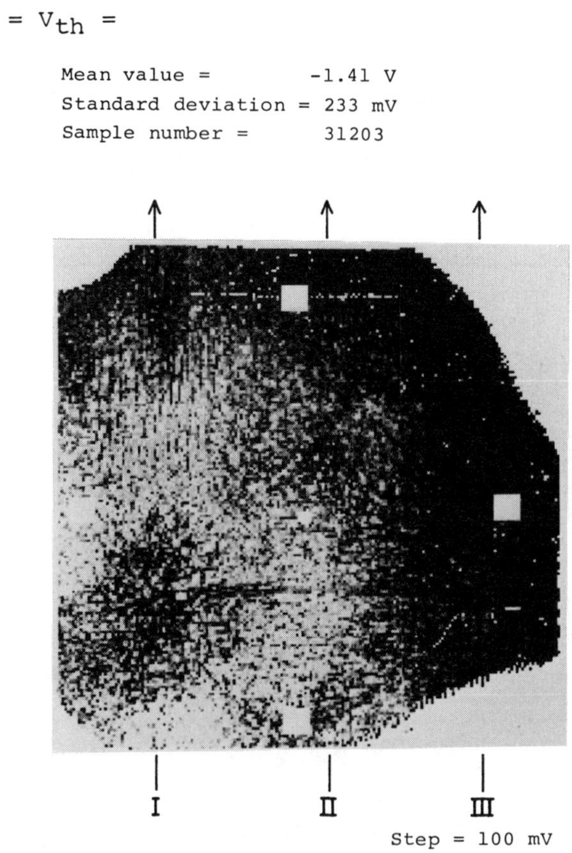

Figure 2 Distribution of threshold voltage in wafer.

These points are expressed graphically in Figure 3, where the V_{th} distributions along the lines indicated in Figure 2 are shown. The macroscopic M-shaped V_{th} distribution in line I passing through the centre (Figure 3) clearly corresponds to the W-shaped EPD distribution (Figure 1b) and the local fluctuations of V_{th} between the neighbouring FETs located in the centre and the intermediate regions are caused by the presence of dislocation sub-structures. A point to be noted is that the local fluctuations of V_{th} are reduced at the periphery where dislocation sub-structures become finer as EPD increases, and in low-EPD regions without appreciable sub-structures (indicated by arrow). In order to treat this point in a statistical manner, the whole region was divided into $2 \times 2\,mm^2$ regions, each containing 100 FETs. The mean value $\bar{V}_{th}$ and the standard deviation σV_{th} for each region are correlated with EPD, as shown in Figures 4a and 4b. $\bar{V}_{th}$ shifts monotonically from -1.1 to -1.8 V as EPD increases from 6000 to 80 000 cm^{-2}. σV_{th} seems to have a peak around 8000 cm^{-2} ($\sigma V_{th} = 150$ to 200 mV). The reduction of σV_{th} at higher EPDs (the minimum of σV_{th} is 33 mV at $\bar{V}_{th} = -1.75$ V) and at lower

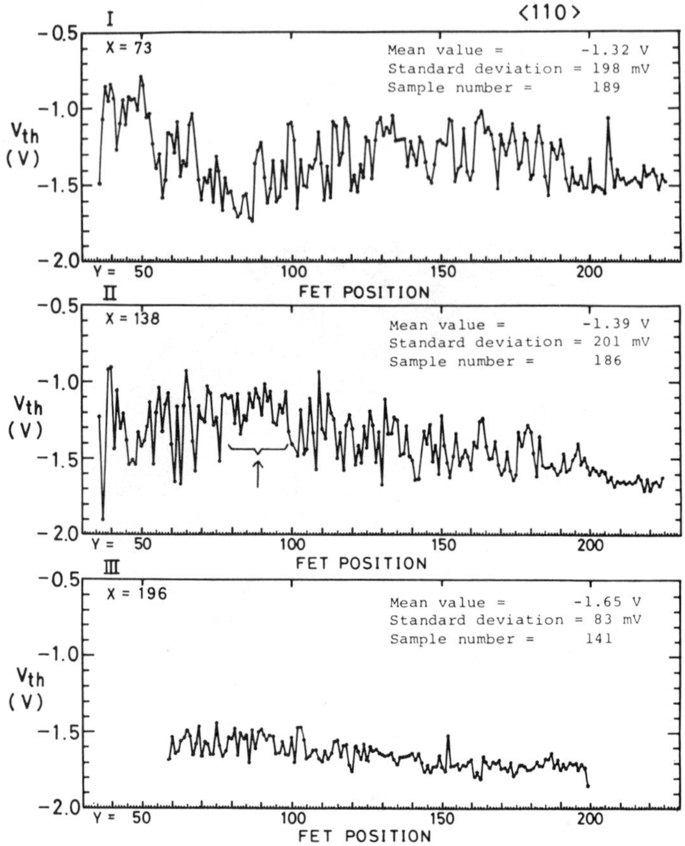

Figure 3 Variation of threshold voltage along lines indicated in Figure 2.

EPDs (the minimum of σV_{th} is 106 mV at $\bar{V}_{th} = -1.16$ V) is due to the disappearance of dislocation sub-structures. σV_{th} is in the range of 150–200 mV at EPD around 8000 cm^{-2} due to dislocation sub-structures. It reduces to around 50 mV (with a minimum of 33 mV) at higher EPDs with finer sub-structures and tends to reduce (with a minimum of 106 mV) at lower EPDs without appreciable sub-structures.

It is then of much interest to investigate whether the observed EPD-dependent V_{th} distribution is attributed to a distribution of ϕ_B or to that of $N_D - N_A$ and R. The answer is given in Figure 5b, where two FETs showing different V_{th} have almost the same ϕ_B (about 0.65 eV) and n (about 1.4). This result was confirmed for all the FETs on the line, which indicates that the surface or the interface properties are not influenced by dislocations. On the other hand, Figure 6 shows that in high-EPD regions the peak carrier concentration, n_p, is higher and the distance from the surface at which the carrier concentration drops to 10^{16} cm^{-3}, d_{16}, is longer than in low-EPD regions. Figure 5 compares the radial distributions of EPD, V_{th}, n_p and d_{16}

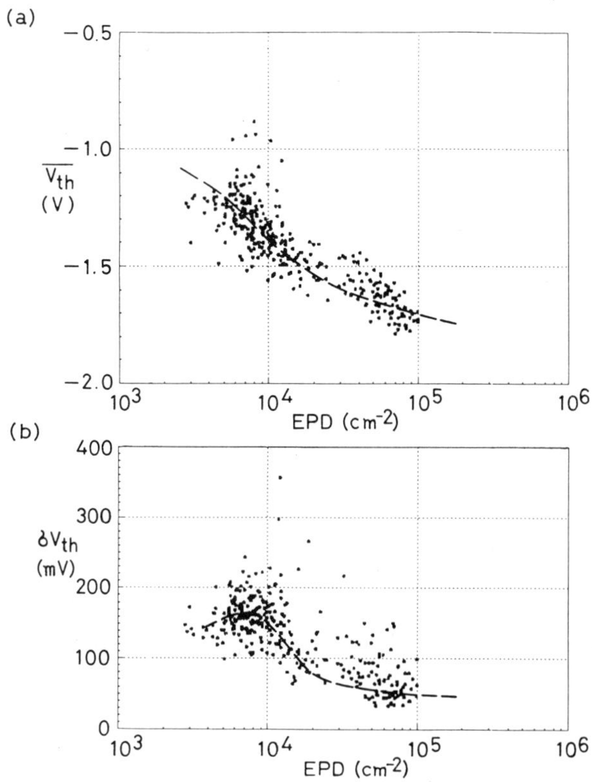

Figure 4 Correlation of (a) mean value and (b) standard deviation of threshold voltage with etch pit density for groups of 100 FETs each.

and demonstrates a close correspondence between them, leading to the conclusion that the 'on-shift' of V_{th} for FETs in high-EPD regions is caused by enhancement of the activation of the implanted impurity. It should be noted that the EPD-dependent impurity activation tends to level off for EPD > 20 000 cm^{-2}. This contributes to the reduction of σV_{th} in the high-EPD region. Substituting the measured values of ϕ_B (about 0.65 eV), n_p for $N_D - N_A$, and 0.9 d_{16} for R into equation (1), we can reproduce the macroscopic V_{th} distribution as shown with a dashed line in Figure 5b. Hence, the 0.4 V difference of V_{th} corresponds to the 6×10^{15} cm^{-3} difference of n_p. The observed microscopic V_{th} difference, 0.7 V, between low-EPD and high-EPD areas amounts to an approximately 1×10^{16} cm^{-3} difference of n_p.

DISTRIBUTION OF K

Figure 7 shows a plot of K expressed in the same manner as in Figure 2. The value of K is shown in a gradation of 5×10^{-3} mA V^{-2} steps. The K value

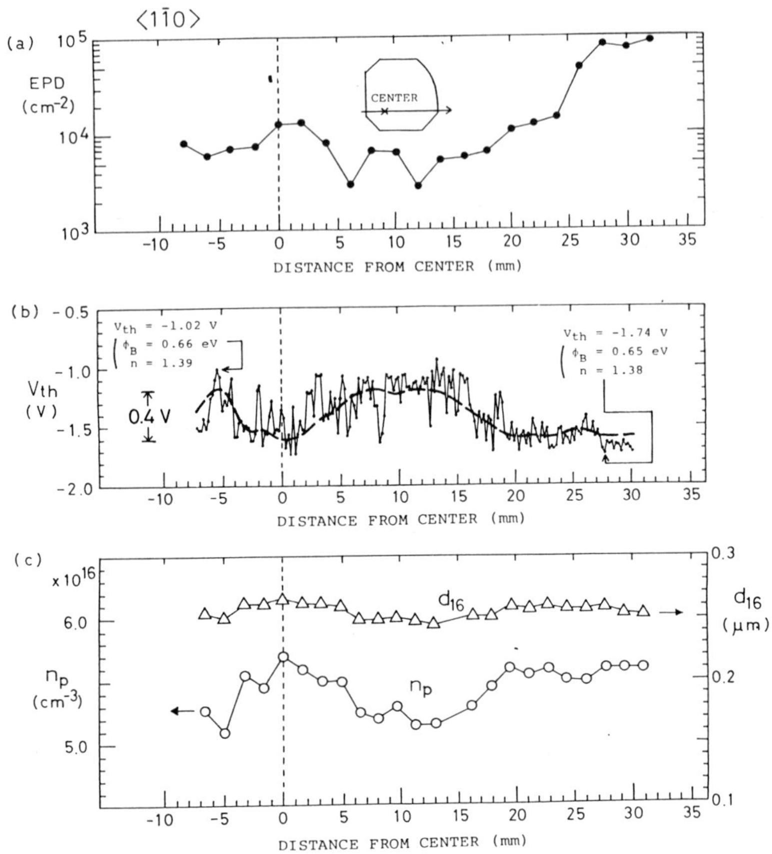

Figure 5 Radial distributions of (a) EPD, (b) threshold voltage and (c) active layer parameters. Values of Schottky barrier height and ideality factor for two FETs having different threshold voltage values are also shown in (b).

distribution in the wafer is independent of the EPD distribution, a striking contrast to V_{th}. It is due to: (a) a uniform distribution of the Hall mobility in the active layer, 4200 $cm^2 V^{-1} sec^{-1}$, across the wafer; and (b) a small variation of d_{16} as shown in Figure 6.

Similar results were obtained on distributions of V_{th} and K value for 2 inch ϕ undoped substrates.

Discussion and conclusion

It is interesting to investigate whether the n_p distribution is caused by: (a) a difference in the activation efficiency of the Si implants; or (b) a difference in compensating impurity concentrations in the substrate between low-EPD and high-EPD areas. Here, the obtained results, especially the dependence

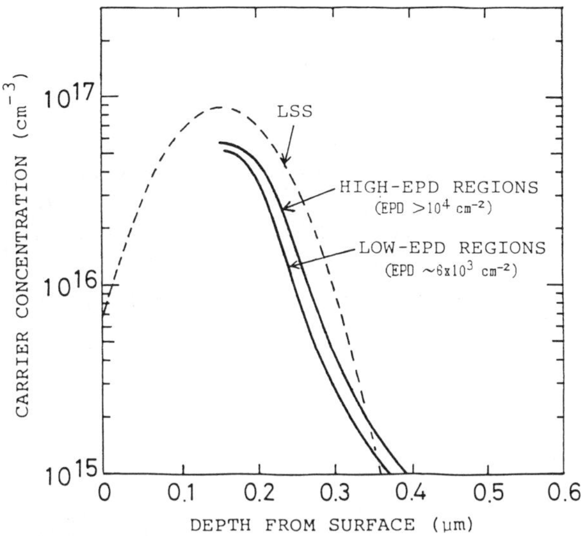

Figure 6 Comparison of carrier concentration profile in active layer between high-EPD and low-EPD regions.

= K =

Mean value = 1.07×10^{-1} mA/V^2
Standard deviation = 1.17×10^{-2} mA/V^2
Sample number = 31418

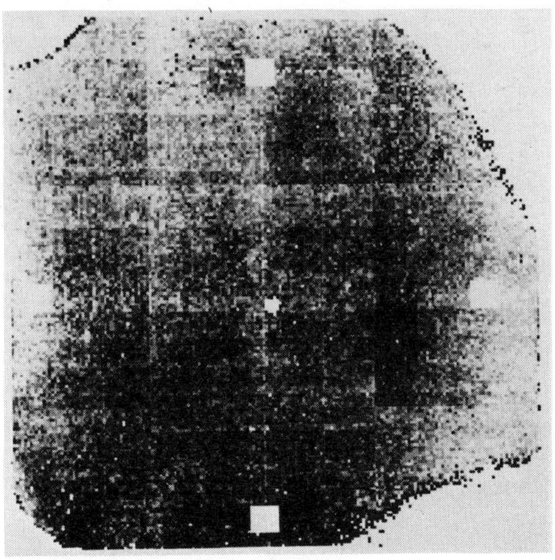

Step = 5×10^{-3} mA/V^2

Figure 7 Map of distribution of K value in wafer.

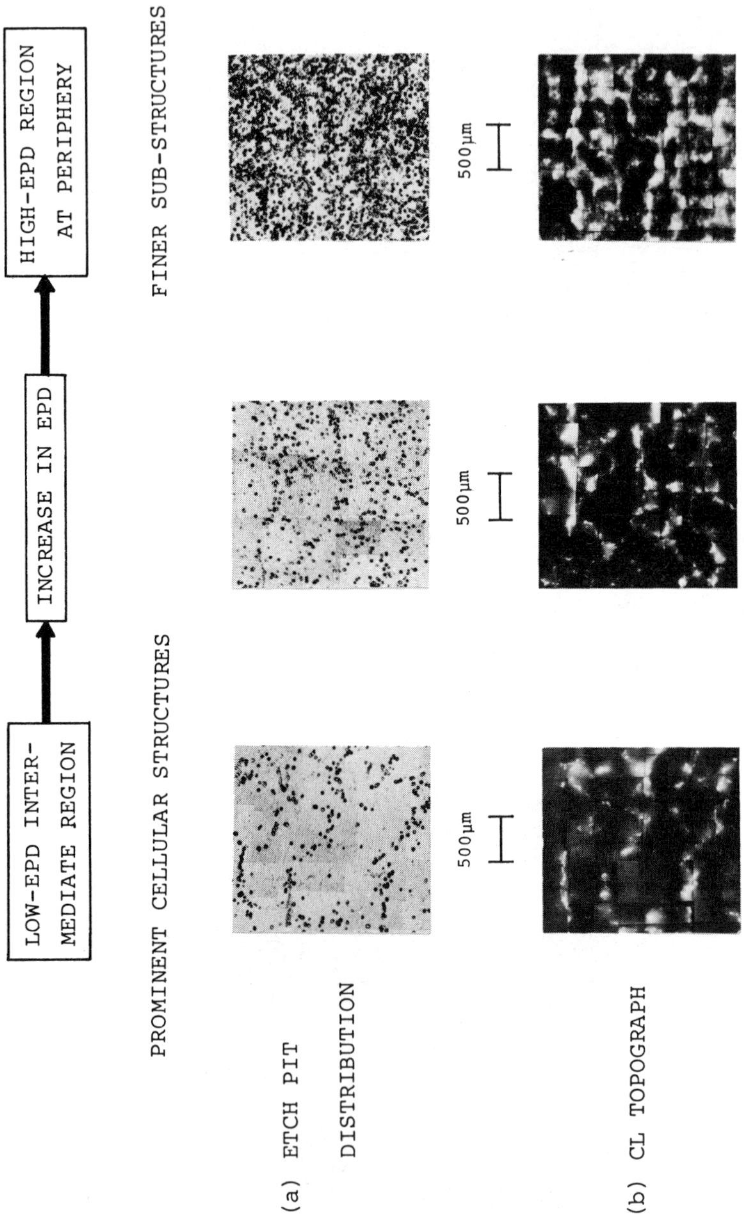

Figure 8 Etch pit distributions and cathodoluminescence topographs at selected areas of as-grown substrate.

of the V_{th} distribution on the EPD distribution, are discussed on the basis of cathodoluminescence (CL) studies for the substrate. In Figure 8, room temperature cathodoluminescence topographs taken at selected areas of the as-grown substrate are shown together with etch pit distributions at the corresponding areas. The bright regions (of several tens of microns width) correlate well with the dislocation sub-structures. The CL intensity increases as the EPD increases, and the bright regions increasingly overlap with each other as the sub-structures become finer. The CL distribution is considered to reflect a distribution of residual impurities or defects, which can be electrically active, depending on the dislocation sub-structures. The CL distribution for an Si-implanted and annealed wafer was basically the same as for the as-grown substrate except that the CL intensity was enhanced and distributed more uniformly. From the above observation, it is evident that the distributions of $\bar{V}_{th}$ and σV_{th} shown in the previous section basically originate from a partial distribution of residual impurities around dislocation sub-structures. In particular, the levelling off of V_{th} in the high-EPD region can be explained by the overlap of the CL bright regions. Based on the data mentioned above, a CL bright region is depleted of compensating impurities by $1 \times 10^{16}\ cm^{-3}$ compared with a dark region surrounded by a bright region.

The mechanism of CL intensity distribution around dislocations is considered to be impurity gettering at dislocations [14–16] during or after solidification of crystals from melts. As a result, to realize a uniform distribution of FET characteristics throughout the wafer, it is necessary to develop substrates which have: (a) uniform EPD distributions without sub-structures; or (b) uniform residual impurity distributions by suppression of impurity gettering at dislocations or by freeing of gettered impurities through, for example, improvement of thermal environments in the growth chamber.

ACKNOWLEDGEMENTS

The authors would like to express their deep gratitude to Messrs Ohta and Tanaka of the Physical and Analytical Office of Sumitomo Electric for the CL measurements.

References

1. Blunt R. T., Clark S. and Stirland D. J. (1982). *IEEE Trans. El. Dev.*, **ED-29,** 1039
2. Bonnet M., Lent B., Visentin N., Duchemin J. P. and Oberti J. D. (1982). In *Semi-Insulating III–V Materials, Evian,* p. 92 (S. Makram-Ebeid and B. Tuck, eds.); Shiva Publishing, Nantwich, UK
3. Grant I., Rumsby D., Ware R. M., Brozel M. R. and Tuck B. (1982). In *Semi-Insulating III–V Materials, Evian,* p. 98 (S. Makram-Ebeid and B. Tuck, eds.); Shiva Publishing, Nantwich, UK
4. Chen R. T. and Holmes D. E. (1983). *J. Cryst. Growth,* **61,** 111

5. Otsubo M., Nakatani M. and Shirahata K. (1982). In *Semi-Insulating III–V Materials, Evian*, p. 130 (S. Makram-Ebeid and B. Tuck, eds.); Shiva Publishing, Nantwich, UK
6. Matsumoto Y. and Watanabe H. (1982). *Jap. J. Appl. Phys.*, **21,** L515
7. Miyazawa S., Mizutani T. and Yamazaki H. (1982). *Jap. J. Appl. Phys.*, **21,** L542
8. Matsumura T., Emori H., Terashima K. and Fukuda T. (1983). *Jap. J. Appl. Phys.*, **22,** L154
9. Tajima M. (1982). *Jap. J. Appl. Phys.*, **21,** L227
10. Kitahara K., Nakai K. and Shibatomi A. (1982). *J. Electrochem. Soc.*, **129,** 880
11. Kitahara K., Ozeki M. and Shibatomi A. (1983). *Appl. Phys. Lett.*, **42,** 188
12. Brozel M. R., Grant I., Ware R. M. and Stirland D. J. (1983). *Appl. Phys. Lett.*, **42,** 610
13. Holmes D. E., Chen R. T., Elliott K. R. and Kirkpatrick C. G. (1983). *Appl. Phys. Lett.*, **43,** 305
14. Chin A. K., Von Neida A. R. and Caruso R. (1982). *J. Electrochem. Soc.*, **129,** 2386
15. Kamejima T., Shimura F., Matsumoto Y., Watanabe H. and Matsui J. (1982). *Jap. J. Appl. Phys.*, **21,** L721
16. Franzosi P. and Salviati G. (1983). *J. Cryst. Growth*, **63,** 419
17. Honda T., Ishii Y., Miyazawa S., Yamazaki H. and Nanishi Y. (1983). *Jap. J. Appl. Phys.*, **22,** L270
18. Nanishi Y., Ishida S., Honda T., Yamazaki H. and Miyazawa S. (1982). *Jap. J. Appl. Phys.*, **21,** L335
19. Nanishi Y., Ishida S. and Miyazawa S. (1983). *Jap. J. Appl. Phys.*, **22,** L54
20. Miyazawa S., Ishii Y., Ishida S. and Nanishi Y. (1983). *Appl. Phys. Lett.*, **43,** 853
21. Nishine S., Kito N., Fujita K., Sekinobu M., Shikatani O. and Akai S. (1982). In IEEE GaAs IC Symposium Technical Digest, New Orleans, p. 58
22. Nakai R., Kotani T., Kawasaki A., Takebe T., Shimazu M., Takenaka K., Kishi M., Mizuuchi K., Murai S., Tada K. and Yamaguchi T. (1984). Sumitomo Electric Technical Review, No. 23, 159
23. Tada K., Kawasaki A., Kotani T., Nakai R., Takebe T., Akai S. and Yamaguchi T. (1982). In *Semi-Insulating III–V Materials, Evian*, p. 36 (S. Makram-Ebeid and B. Tuck, eds.); Shiva Publishing, Nantwich, UK
24. Sze S. M. (1981). *Physics of Semiconductor Devices* (2nd edn.), Ch. 5; John Wiley & Sons, New York

Orientation Dependence of Device Uniformity in GaAs Integrated Circuits

C. P. Lee, M. F. Chang, P. M. Asbeck, L. D. Hou, R. P. Vahrenkamp and C. G. Kirkpatrick

Rockwell International, Microelectronics Research and Development Center, Thousand Oaks, California 91360, USA

Abstract

Device uniformity is one of the most important factors in limiting the yield of GaAs integrated circuits. It is reported in this paper that the non-uniformity of GaAs FET (field effect transistor) characteristics can be caused by non-uniformly distributed stresses across the wafer through the piezoelectric effect. This non-uniformity depends on the orientation of the devices on the wafer. On (100) oriented substrates, the characteristics of FETs oriented in the ⟨011⟩ and $\langle 01\bar{1} \rangle$ directions have a strong dependence on their radial positions on the wafer. However, FETs oriented in the ⟨001⟩ and ⟨010⟩ directions do not display such dependence and, therefore, have better device uniformity.

Introduction

As GaAs ICs are moving rapidly toward LSI complexity and large wafer processing, increased demands are put on the control of wafer processing in order to achieve good device uniformity and reasonable circuit yield. However, the problems encountered during wafer fabrication and their impact on the GaAs device performance are often different from those in Si devices. Some of these problems are unique to GaAs devices and are related to the GaAs crystal structure. Understanding the impact and the physical mechanisms of these problems, and controlling them are critical to the successful fabrication of GaAs FET devices and ICs.

The most pronounced effect on GaAs devices due to the nature of the GaAs crystal, a zincblende crystal structure, is the orientation dependence of the electrical characteristics of GaAs MESFETs (metal-semiconductor FET) [1]. The device orientation on a GaAs crystal has a strong influence on device performance, uniformity and reproducibility. The orientation dependence and the preferred device orientation can be varied with either device structure or processing techniques [1, 2].

In this paper, the orientation dependence of the device uniformity across 3 inch GaAs wafers is described. It was found that in certain crystal orientations, device characteristics can be extremely uniform and, therefore, are preferred for circuit fabrication.

Orientation effect

On (100) oriented GaAs substrates, the characteristics of MESFETs oriented in the ⟨011⟩ direction are different from those of MESFETs in the $\langle 01\bar{1}\rangle$ direction, while the characteristics of FETs in the ⟨001⟩ and ⟨010⟩ directions are identical [1]. For selectively ion-implanted planar GaAs MESFETs [3], the threshold voltages of devices oriented in the ⟨011⟩ direction are always lower, and the threshold voltages of devices in the $\langle 01\bar{1}\rangle$ direction are always higher than those of devices oriented in other directions. Figure 1 shows the drain current versus gate voltage characteristics for five identical 1 μm × 50 μm FETs oriented in five different directions. It is clearly shown that the device characteristics are functions of the orientation, θ. The current and the threshold voltage increase continuously with angle θ, where $\theta = 0$ corresponds to the ⟨011⟩ direction.

The physical mechanism underlying the orientation effect has been proposed to be a stress-induced piezoelectric effect [4]. Elastic stresses are often unintentionally introduced into the GaAs substrate during device fabrication. The stresses can be generated during thermal processing, metal contact alloying and, more importantly, from dielectric overlayers [5]. To illustrate

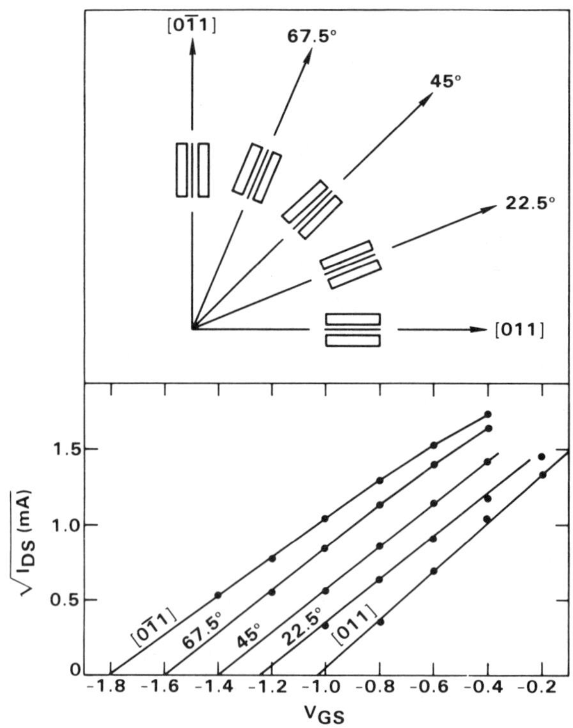

Figure 1 Orientation dependence of FET characteristics. Transfer curves ($\sqrt{I_{DS}}$ versus V_{GS}) are shown for FETs oriented at angles of 0°, 22.5°, 45°, 67.5° and 90° with respect to the ⟨011⟩ crystal direction.

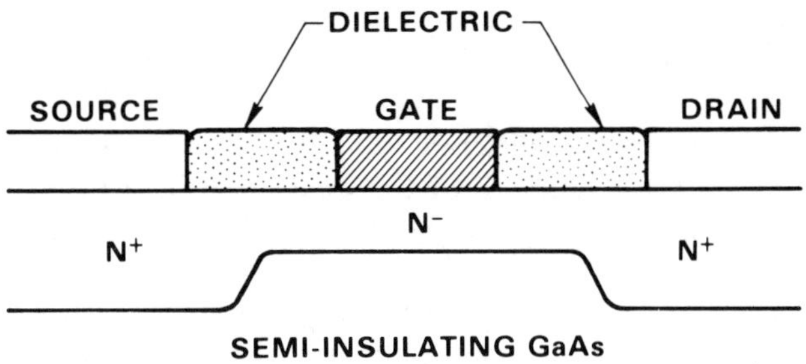

Figure 2 Structure of a planar selective ion implanted GaAs MESFET.

this phenomenon, the structure of a planar GaAs MESFET is shown in Figure 2. The surface of the FET channel is covered by a Si_3N_4 layer, which is used as an annealing cap and a protective film for the device. The gate metal and ohmic contacts are deposited in the etched windows in the Si_3N_4 dielectric openings where the Schottky gate is seated could be as high as can impart stresses to the underlying semiconductor substrate, and these stresses can build up to a substantial high value near the openings in the films [5]. Depending on the deposition technique, stresses in the vicinity of dielectric openings where the Schottky gate is seated could be as high as 5×10^4 N cm^{-2} [6]. Because GaAs, unlike Si, is a piezoelectric material, such high stresses can induce piezoelectric charges in the FET channels causing significant shifts in threshold voltage, V_{th}. Due to the nature of the piezoelectric tensor of GaAs, the sign and the amount of the threshold voltage shifts are dependent on the orientations of the devices on the wafer.

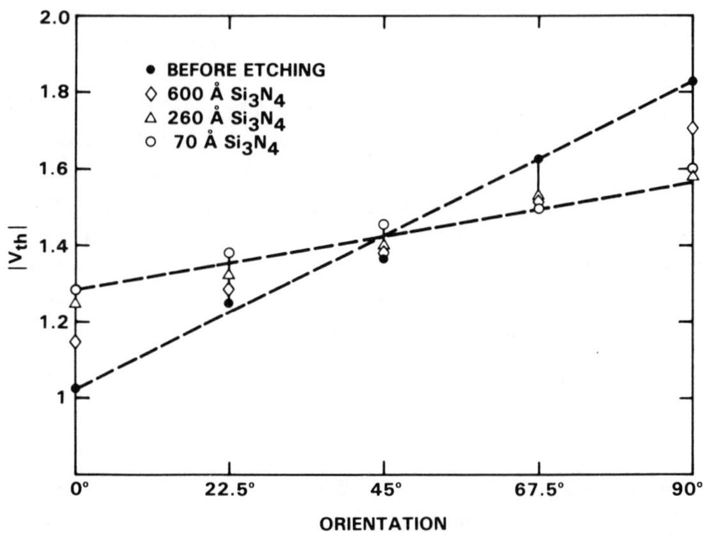

Figure 3 Threshold voltages of FETs oriented at angles of 0°, 22.5°, 45°, 67.5° and 90° with respect to the ⟨011⟩ direction as the surface Si_3N_4 film was thinned.

Since the amount of stress created by the surface dielectric film is a function of its thickness, the orientation effect is sensitive to the film thickness change. Evidence of such an effect is shown in Figure 3. Here, the threshold voltages of five FETs oriented in five different directions (same as those shown in Figure 1) are plotted as functions of the angle θ with respect to the ⟨011⟩ direction. The threshold voltages were measured as the surface Si_3N_4 film was gradually thinned by plasma etching. Before the Si_3N_4 was removed, the threshold voltage had a very strong angular dependence. The 90° FET had a threshold voltage of 1.8 V, which was nearly two times that of the 0° FET. As the Si_3N_4 layer was gradually removed, the threshold voltages of the FETs oriented with angles smaller than 45° increased, while the threshold voltages of the FETs oriented with angles larger than 45° decreased. The largest changes occurred for 0° and 90° FETs, and the 45° FETs had nearly no change in threshold voltage. This result clearly indicates the stress dependence of the orientation effect. The characteristics of the 45° FETs or the FETs oriented in the ⟨010⟩ or ⟨001⟩ direction are obviously not as process dependent as those of the ⟨011⟩ and $\langle 01\bar{1}\rangle$ FETs.

Device uniformity

As shown above, FETs with certain orientations are sensitive to the process parameters such as surface dielectric layer thickness, but there are also FETs with some other orientations that do not show such dependence. In order to study the orientation dependence of device uniformity, we have fabricated FETs along different crystal directions on 3 inch GaAs substrates. The substrates used were round, undoped, SI LEC GaAs with (100) surfaces. Low threshold voltage depletion mode FETs were fabricated using selective ion implantation. The Schottky gate was 1 μm long and 50 μm wide. The source to drain spacing was 2.5 μm.

Saturation currents (I_{dss}) and threshold voltages (V_{th}) of FETs in different orientations were measured as functions of their radial locations on the wafer. It was found that device characteristics have a strong radial dependence for FETs oriented along the ⟨011⟩ and $\langle 01\bar{1}\rangle$ directions. An example of this radial dependence on a 3 inch wafer is shown in Figure 4. Each data point in the figure represents the average threshold voltage of several FETs with similar radial positions, and the error bars indicate the spreading of the threshold voltages of the FETs. The FETs along the $\langle 01\bar{1}\rangle$ direction always have higher threshold voltages than their ⟨011⟩ counterparts. The radial dependence of device characteristics is opposite in these two directions. V_{th} increases with radius for ⟨011⟩ FETs, but decreases for $\langle 01\bar{1}\rangle$ FETs. The difference in threshold voltage between ⟨011⟩ and $\langle 01\bar{1}\rangle$ FETs can be as high as 0.7 V at the centre of the wafer and as low as 0.2 V at the edge. But, for the ⟨010⟩ and ⟨001⟩ FETs which are at 45° to the ⟨011⟩ and $\langle 01\bar{1}\rangle$ FETs, no radial dependence was observed. Measurement of the saturation currents also reveals the same dependence.

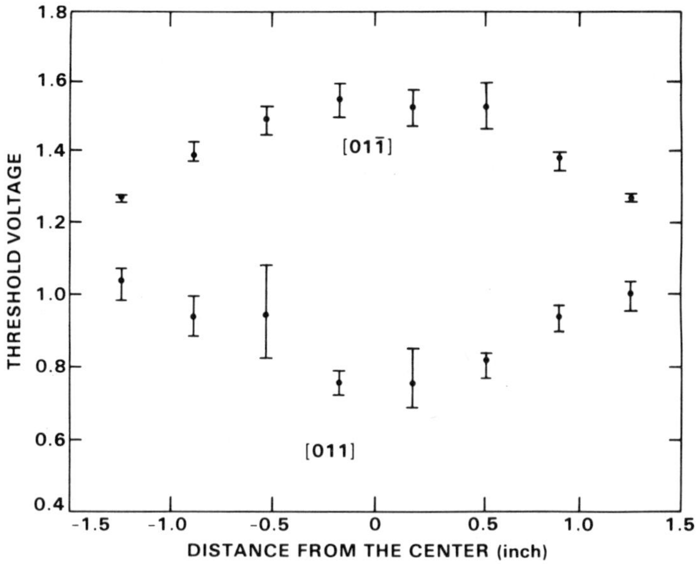

Figure 4 Radial dependence of the threshold voltages of ⟨011⟩ and ⟨01$\bar{1}$⟩ FETs on a 3 inch GaAs wafer.

The orientation dependence of device uniformity described above can be satisfactorily explained by the piezoelectric effect. Because of the anisotropy of GaAs crystals, the density of the piezoelectric charges induced by the surface dielectric layer in the channels of ⟨011⟩ FETs is of opposite sign with respect to that in the ⟨01$\bar{1}$⟩ FETs, and it diminishes for the FETs in the ⟨001⟩ and ⟨010⟩ directions. This is why the orientation dependences of the FETs characteristics are observed. In the channels of ⟨011⟩ and ⟨01$\bar{1}$⟩ FETs, the magnitude of the piezoelectric charge density, ρ_{pz}, is proportional to the thickness of the dielectric overlayers and inversely proportional to the square of the length of film openings in the gate area [4]. Since the piezoelectric charges are induced by the stresses from the dielectric film on the surface, ρ_{pz} is strongly dependent on the processing parameters. The variations in the thickness of the dielectric film and in the size of the opening in the film for gate deposition have a very strong influence on the value of ρ_{pz}. For large diameter GaAs wafers, the radial dependence of these processing parameters is sometimes difficult to avoid. The deposited dielectric films are generally thinner around the wafer edge and thicker in the centre of the substrate. In addition, the amount of undercut of dielectric openings by plasma etching often appears to be less near the centre than at the edge of the wafer. The non-uniformly distributed stresses, stronger at the centre and weaker at the edge, therefore cause non-uniform ρ_{pz} across the wafer and the radial dependence of the FET characteristics described above.

The three-dimensional plots of the threshold voltage distribution across a 3 inch wafer are shown in Figure 5. The ⟨011⟩ FETs have a convex type

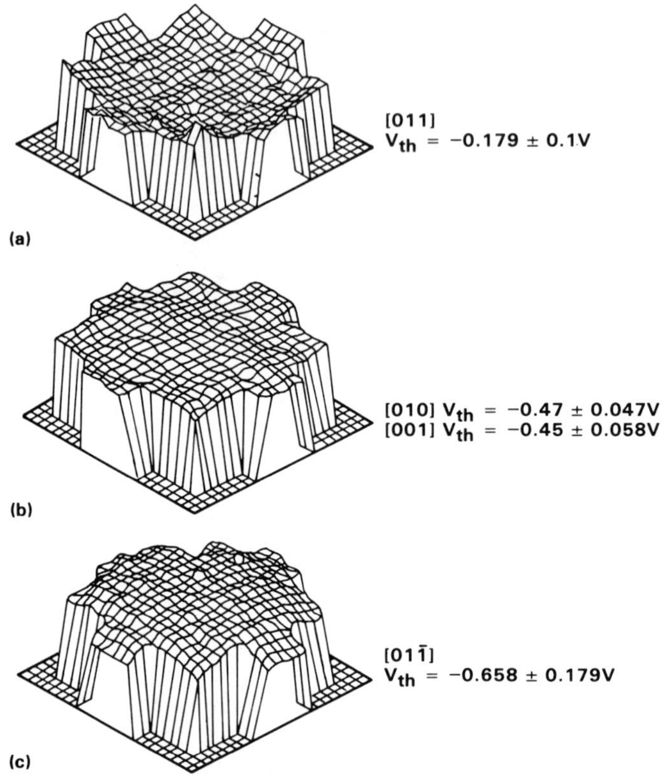

Figure 5 Two-dimensional distribution of FET threshold voltage on a 3 inch GaAs wafer: (a) for ⟨011⟩ FETs; (b) ⟨010⟩ FETs; and (c) for ⟨01$\bar{1}$⟩ FETs.

distribution and the ⟨01$\bar{1}$⟩ have a concave type distribution, while the ⟨010⟩ FETs have a flat distribution. The average threshold voltage of the ⟨011⟩ FETs is −0.179 V ± 100 mV; for the perpendicular ⟨01$\bar{1}$⟩ FETs, V_{th} is −0.658 V ± 179 mV. However, for the ⟨010⟩ FETs, V_{th} is −0.47 V ± 47 mV, and for the ⟨001⟩ FETs, it is −0.45 V ± 58 mV. Because of the lack of radial dependence, both ⟨001⟩ and ⟨010⟩ FETs display less than half the standard deviation of their ⟨01$\bar{1}$⟩ and ⟨011⟩ counterparts.

Based on the results presented above, unless the radial dependence of the process non-uniformity can be totally eliminated, it is desirable to choose a FET orientation in which the process effect on the device characteristics is minimized. Our experimental results clearly show ⟨010⟩ or ⟨001⟩ are the preferred orientations of FET fabrication because of their insensitivity to the process variations. Another advantage of using ⟨010⟩ and ⟨001⟩ FETs is the increased packing density. Because the device characteristics of these two kinds of FETs are identical, it is not necessary to have all the FETs in a circuit oriented in the same direction to minimize variations.

In conclusion, we have found that FETs oriented in the ⟨010⟩ and ⟨001⟩ directions, which were at 45° angles from the ⟨011⟩ cleavage planes, do not display stress-related threshold voltage shifts and have better uniformity than their ⟨011⟩ and $\langle 01\bar{1}\rangle$ counterparts. The radial dependences are explained by the anisotropically and non-uniformly distributed piezoelectric charge densities in the GaAs substrate induced by dielectric and etching inhomogeneities.

ACKNOWLEDGEMENTS

This work was supported, in part, by the Defense Advanced Research Projects Agency of the US Department of Defense and monitored by the Air Force Office of Scientific Research under Contract No. F49620-83-C-0042.

References

1. Lee C. P., Zucca R. and Welch B. M. (1980). *Appl. Phys. Lett.*, **37,** 311
2. Yokoyama N., Onodera H., Ohnishi T. and Shibatomi A. (1983). *Appl. Phys. Lett.*, **42,** 270
3. Welch B. M., Shen Y. D., Zucca R., Eden R. and Long S. I. (1980). *IEEE Trans. El. Dev.*, **ED-27**, 1116
4. Asbeck P. M., Lee C. P. and Chang M. F. *IEEE Trans. El. Dev.*, in press
5. Blech I. A. and Meieran E. S. (1967). *J. Appl. Phys.*, **38(7),** 2913
6. Kirby P. A., Selway P. R. and Westbrook L. D. (1979). *J. Appl. Phys.*, **50(7),** 4567

Effects of Impurity Compensation on I–V Characteristics of n–i–n Structures and Backgating in GaAs Integrated Circuits

K. Horio, T. Ikoma* and H. Yanai

Shibaura Institute of Technology, 3-9-14, Shibaura, Minato-ku, Tokyo 108, Japan

Abstract

Current-voltage characteristics of an n–i–n structure have been computed. It is found that an effective resistivity in the low voltage region depends on acceptor and trap densities and the length of the i-layer. To achieve good isolation between two devices in GaAs ICs, it is suggested that the shallow acceptor density as well as the trap density must be larger than a critical value. An analytical model has been presented to estimate the effective resistivity and the onset voltage of current rise. The backgating effect has also been analysed in terms of a separation distance and an acceptor density.

Introduction

Because of its high resistivity ($>10^7$ Ω cm), SI GaAs is used as a substrate for MESFETs and high speed ICs to provide isolation between devices. A SI material is achieved by impurity compensation [1] and hence device characteristics strongly depend on junction properties between an (n-type) active region and a SI substrate [2–5]. In GaAs MESFETs, drain current decreases through a modulation of interface space charge when negative voltage is applied to the substrate. This is the so-called backgating effect and is a serious problem in complex GaAs ICs [6, 7].

In order to understand isolation effects with a heavily compensated i-layer, we have investigated I–V characteristics and space charge distributions in an n–i–n structure by computer aided analysis. Emphasis is placed on influences of impurity compensation on these characteristics.

Physical model

A one-dimensional n–i–n structure is considered. We assume undoped SI LEC (liquid encapsulated Czochralski) GaAs, where deep donors 'EL2' compensate shallow acceptors. The donor density in the n-layer, N_D, is 2×10^{16} cm^{-3}, and the length of the n-layer is 2 μm. The length of the i-layer,

* Institute of Industrial Science, University of Tokyo, 7-22-1, Roppongi, Minato-ku, Tokyo 108, Japan.

Table 1 *Physical parameters used in the calculation: trap parameters for EL2*

Energy level (eV) $E_C - E_T$	Capture cross-section (cm^2)	
	Electrons	Holes
0.8	1×10^{-13}	2×10^{-18}

d, is typically 20 μm. The Poisson equation and current continuity equations are solved for electrons and holes in the steady state. Recombination and generation via EL2 are taken into account by using Shockley–Read–Hall statistics. Trap parameters for EL2 are given in Table 1. Both diffusion and drift currents are included. The Einstein relation is assumed between mobility and diffusivity. Field dependent mobilities are also taken into account both for electrons and holes as given in Table 2. The basic equations are put into discrete form by the finite difference technique using the Scharfetter–Gummel formulation [8], and are solved by the usual Newton–Raphson method. Equilibrium carrier concentrations are assumed at the edges of the n-layers as boundary conditions, but no restriction is placed at the n–i interfaces.

Table 2 *Physical parameters used in the calculation: mobilities*

	Equations	Parameters
Electrons	$\mu_n = \mu_{no} \dfrac{1 + (v_{ns}/\mu_{no}E)(E/E_0)^4}{1 + (E/E_0)^4}$	$\mu_{no} = 8000\ cm^2\ V^{-1}\ s^{-1}$ $v_{ns} = 8.5 \times 10^6\ cm\ s^{-1}$ $E_0 = 4000\ V\ cm^{-1}$
Holes	$\mu_p = \mu_{po} \dfrac{1}{1 + \mu_{po}E/v_{ps}}$	$\mu_{po} = 400\ cm^2\ V^{-1}\ s^{-1}$ $v_{ps} = 1 \times 10^7\ cm\ s^{-1}$

I–V characteristics

NUMERICAL RESULTS

Calculated I–V characteristics at room temperature are shown by solid lines in Figure 1. In Figure 1, the ratio of the shallow acceptor density N_A to the deep donor density N_T in an i-layer is 0.09, so that the equilibrium hole and electron concentrations are equal to $2.9 \times 10^7\ cm^{-3}$ and $1.6 \times 10^5\ cm^{-3}$, respectively, corresponding to an equilibrium resistivity ρ_0 of $4.9 \times 10^8\ \Omega$ cm. The calculated curves consist of three regions: (a) linear relationship (Ohmic current) (region 1); (b) steep rise of current (trap filling current) (region 2);

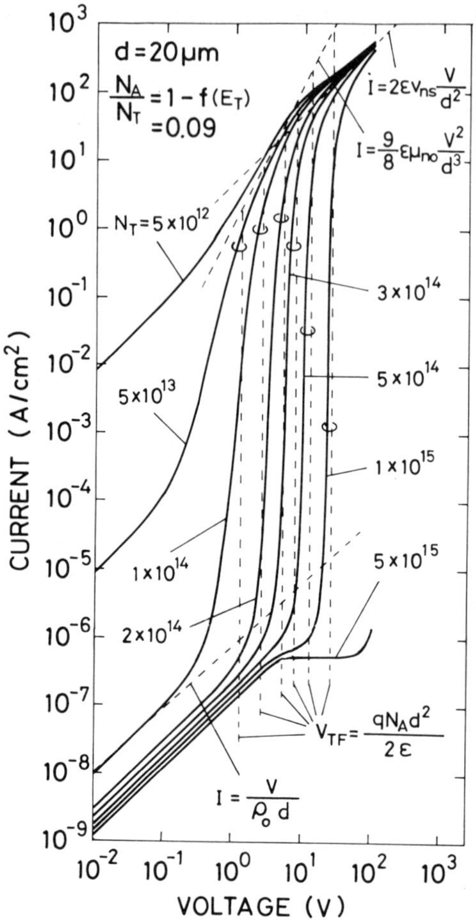

Figure 1 Calculated I–V characteristics (solid lines) of a GaAs n–i–n structure with a heavily compensated i-layer. Broken lines show analytical approximations (see text).

and (c) space-charge-limited current (region 3). These curves are qualitatively consistent with the prediction by Lampert [9]. However, important differences are observed in regions 1 and 2.

In region 1, the effective resistivity ρ_{eff}, which is defined as $\rho_{eff} = (V/I)/d$, depends on N_T, N_A and d. It becomes smaller as $N_T(N_A)$ becomes smaller, though the equilibrium resistivity ρ_0 is constant. When $N_T = 5\times10^{12}\ cm^{-3}$ and $N_A = 4.5\times10^{11}\ cm^{-3}$, ρ_{eff} is $5\times10^2\ \Omega$ cm, six orders of magnitude smaller than ρ_0, while, when $N_T > 1\times10^{14}\ cm^{-3}$ and $N_A > 9\times10^{12}\ cm^{-3}$, ρ_{eff} becomes larger than ρ_0. This is because holes do not contribute to current in an n–i–n structure.

In region 2, the trap filling voltage V_{TF}, defined as $V_{TF} = q(1-f_T)N_Td^2/2\varepsilon = qN_Ad^2/2\varepsilon$ by Lampert [9], and shown in Figure 1 by dotted lines, does not correspond to the onset voltage of current rise. Here q, ε and f_T are electron charge, permittivity of GaAs and the occupation function of the trap, respectively. V_{TF} is much larger than the onset voltage and the difference becomes a factor of ten in some cases. When $N_T = 5\times10^{15}\ cm^{-3}$ and $N_A = 4.5\times10^{14}\ cm^{-3}$, a current saturation is observed. This is due to the negative differential mobility and a charge accumulation is observed (Gunn effect).

ANALYTICAL EXPRESSIONS

We present here analytical approximations for the effective resistivity ρ_{eff}, the onset voltage of current rise V_{Th} and the space-charge-limited voltage V_{SCL}. V_{Th} is defined as the voltage at an intersection of extrapolated Ohmic and trap-filling currents, and V_{SCL} is the voltage at the intersection of the extrapolated trap-filling and space-charge-limited currents in the log I–log V plot. In region 3, the analytical expressions of space-charge-limited current are shown by dotted lines for two cases: $I = 2\varepsilon v_{ns} V/d^2$ (constant velocity: v_{ns}) and $I = (9/8)\varepsilon\mu_{no} V^2/d^3$ (constant mobility: μ_{no}). Because of the high electric field, the constant velocity is applied to most of the characteristics.

Figure 2 shows ρ_{eff}, V_{Th} and V_{SCL} as a function of N_T for the cases shown in Figure 1. The analytical expression for ρ_{eff} in the range $>10^7\ \Omega$ cm is given by:

$$\rho_{eff} = \rho_{no}(d - 2w_0)/d \tag{1}$$

which is plotted by the solid line in Figure 2. Here ρ_{no} and w_0 are given by:

$$\rho_{no} = (q\mu_{no}n_0)^{-1} = (C_n/e_n)/(q\mu_{no}(N_T/N_A - 1)) \tag{2}$$

and

$$w_0 = (2\varepsilon V_B/qN_A)^{1/2} \tag{3}$$

where, n_0 is the equilibrium electron concentration in the i-layer, C_n and e_n are the electron capture coefficient and the electron emission rate of the deep donor, respectively, and V_B is the built-in potential at the n-i junction, given by $(kT/q)\log(N_D/n_0)$, where k is Boltzmann's constant and T is the absolute temperature. Equation (1) is valid only when $d > 2w_0$.

Two remarkable features may be deduced from equation (1):

1. ρ_{eff} depends on ρ_{no} and not ρ_0. This means that the effective resistivity is determined by electron current. Therefore, ρ_{eff} can be larger than the equilibrium resistivity ρ_0, as shown in Figure 2. The linear relationship between voltage and current in the low voltage region is not a normal Ohmic current. In this region, holes are depleted because the deep trap ('EL2') has a small capture cross-section for holes as compared with that for electrons.
2. A dramatic feature arises from the term $d - 2w_0$ in equation (1). It is understood that the resistance in the space charge region at an n–i junction

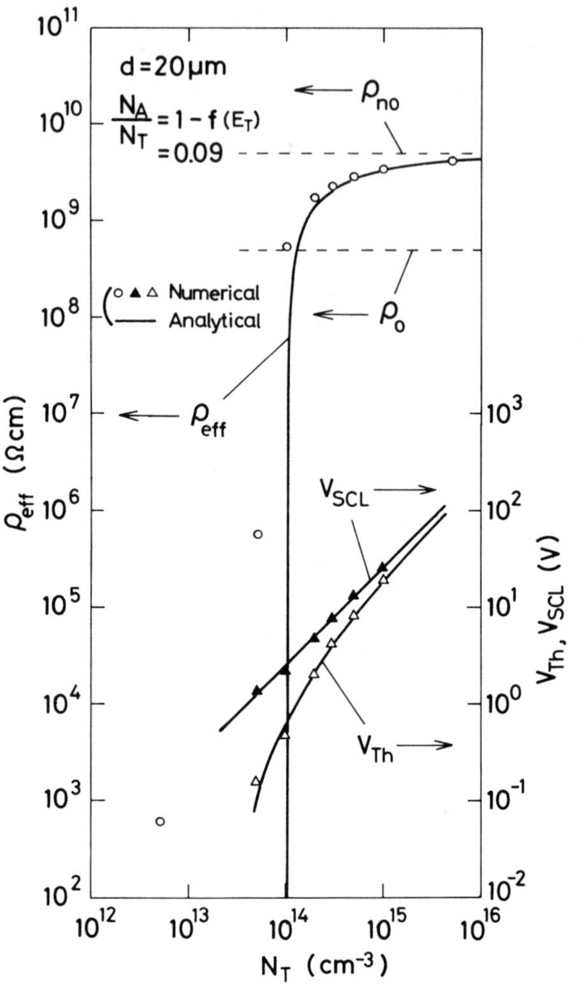

Figure 2 Effective resistivity ρ_{eff}, onset voltage of current rise V_{Th} and space-charge-limited voltage V_{SCL}, as a function of N_T. Analytical expressions (solid line) are compared with numerical values (○, △, ▲) obtained from calculated I–V curves in Figure 1.

is small, and the electrically neutral region $(d-2w_0)$ mainly determines the resistivity. Hence, if $w_0 \simeq d/2$, the effective resistivity becomes very small, as was shown in Figure 1. The small effective resistivity stems from the fact that electrons are injected from the one n-layer and drift to the other n-layer without being trapped when the density of empty traps $(1-f_T)N_T$ or N_A is small. The critical behaviour depends mainly on d and N_A, because w_0 is determined essentially by N_A as is shown in equation (3).

We next discuss V_{Th} and V_{SCL}. The analytical expression of V_{SCL} is given by:

$$V_{SCL} = qN_A d^2/2\varepsilon \tag{4}$$

as shown by the solid line in Figure 2. This is exactly equal to the trap-filling voltage V_{TF} as introduced by Lampert [9]. Note that in this case, N_A is equal to the density of empty traps under equilibrium. The voltage V_{Th} at the onset of the current rise is also fitted well by:

$$V_{Th} = qN_A(d - w_0)^2/2\varepsilon \quad (5)$$

as shown by the solid line in Figure 2. V_{Th} becomes very small when d becomes comparable to w_0. Therefore, both V_{Th} and V_{SCL} are determined by N_A and d.

Effects of doping level on backgating

ISOLATION EFFECT

A good substrate for GaAs ICs must meet several conditions. First, it must remain highly resistive between closely placed devices. Limiting factors for isolation can be estimated from the results in the previous section. ρ_{eff} and V_{Th} must be larger than certain values. From equations (1) and (5) these conditions are expressed as:

$$d > 2w_0/(1 - \rho_{eff}/\rho_{no}) \quad (6)$$

and

$$d > w_0 + (2\varepsilon V_{Th}/qN_A)^{1/2} \quad (7)$$

These two equations give necessary conditions for the minimum distance d between two devices and the minimum acceptor density N_A. In addition to equations (6) and (7), the following equations should be satisfied:

$$\rho_{eff} < \rho_{no} \quad \text{and} \quad \rho_{eff} < \rho_0 \quad (8)$$

ρ_{no} and ρ_0 are plotted as a function of N_A/N_T in Figure 3a. From this figure, the range in which N_A/N_T must fall can be determined for a given ρ_{eff}.

Figures 3b and 3c show the relations between d and N_A (equations (6) and (7)) for $\rho_{eff} > 10^7$ Ω cm and $V_{Th} > 5$ V, respectively. From these figures, it is apparent that for a critical separation distance d, N_A must be larger than a certain value. For an example, to attain $\rho_{eff} > 10^7$ Ω cm, N_A must be larger than 3.5×10^{13} cm^{-3} for d = 10 μm, and 3.5×10^{15} cm^{-3} for d = 1 μm. On the other hand, to attain $V_{Th} > 5$ V, N_A must be larger than 1.2×10^{14} cm^{-3} for d = 10 μm and 1.2×10^{16} cm^{-3} for d = 1 μm. From equation (8), it is seen that N_T must range from N_A to 5×10^3 N_A. It is concluded that to permit a small separation of active devices in GaAs ICs, both shallow acceptor and deep donor concentrations in a SI substrate must be large.

BACKGATING EFFECT

We next show another substrate-related effect, the so-called backgating effect where a drain current is modulated by a substrate bias. This effect is related

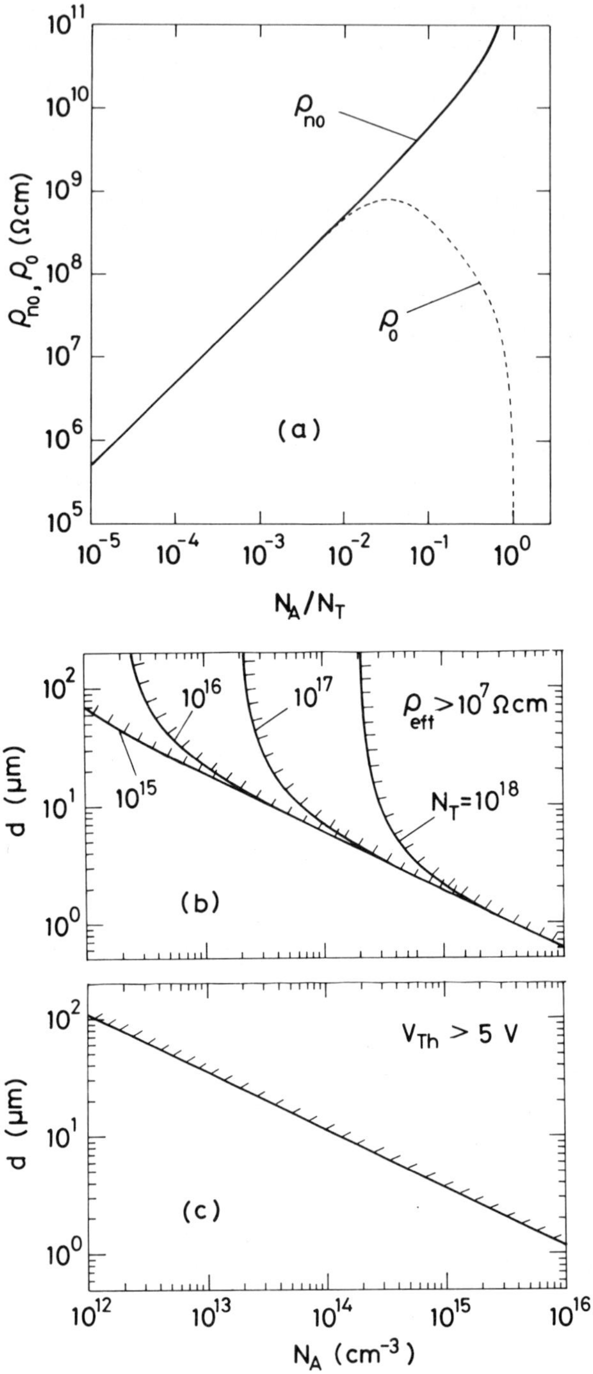

Figure 3 (a) ρ_{no} and ρ_0 as a function of N_A/N_T. (b) Minimum separation distance d as a function of N_A for $\rho_{eff} > 10^7$ Ω cm, with N_T as a parameter. (c) Minimum separation distance d as a function of N_A for $V_{Th} > 5$ V.

to how deep the space charge region is broadened into the n-layer at the n–i junction when bias voltage is applied to an n–i–n structure.

Figure 4a shows examples of variations of the potential drop ψ_n within an n-layer as a function of reverse voltage, with length of an i-layer as a parameter. In this figure ψ_{no} is the value of ψ_n at equilibrium and w_n is the

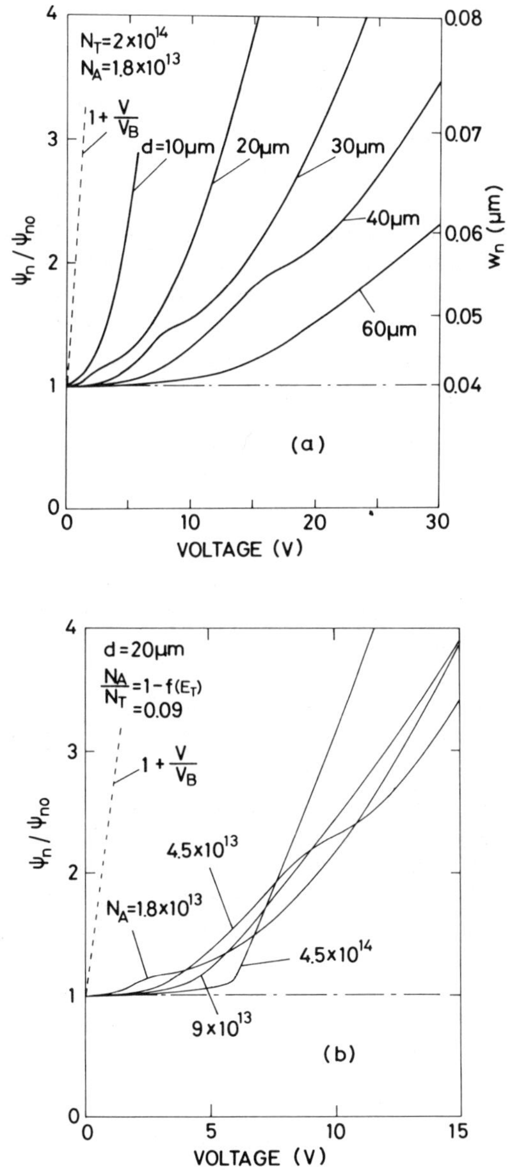

Figure 4 Variations of relative potential drop ψ_n/ψ_{no} within an n-layer as a function of reverse voltage. (a) $N_A = 1.8 \times 10^{13}$ cm^{-3} and $N_T = 2 \times 10^{14}$ cm^{-3}; w_n is an effective channel modulation width, given by $(2\varepsilon\psi_n/qN_D)^{1/2}$, where N_D is 2×10^{16} cm^{-3}. (b) d = 20 μm, $N_A/N_T = 0.09$.

width of the space charge layer in the n-layer, defined by $(2\varepsilon\psi_n/qN_D)^{1/2}$, where N_D is 2×10^{16} cm^{-3} in this case. w_n is an effective channel modulation width due to substrate bias. The dashed line represents a case where the normal abrupt depletion approximation would hold. In the case of a thicker i-layer, the variation of ψ_n is smaller at a given voltage, hence backgating can be smaller, as expected.

Figure 4b shows ψ_n/ψ_{no} as a function of reverse voltage for $d=20$ μm, where N_A and N_T are varied but still keeping $N_A/N_T=0.09$. It is recognized from Figure 4b and Figure 1 that the voltage where ψ_n starts to increase corresponds to the voltage where the I–V curve starts to deviate from the Ohmic behaviour. This voltage is smaller than V_{Th}. In the low voltage region, ψ_n/ψ_{no} is smaller for larger $N_A(N_T)$. However, ψ_n/ψ_{no} increases more rapidly for larger N_A. This is understood as follows: at low voltages, injected electrons are captured by unoccupied traps near the cathode. Therefore, no increase in ψ_n is observed and the I–V characteristics are linear. When injected electrons reach the other n–i junction, I–V characteristics deviate from Ohmic and unoccupied traps near the anodic junction capture electrons to create space charges; hence ψ_n starts to increase. To minimize this space charge effect, the Ohmic region must be made wide. This can be realized by increasing N_A.

Conclusions

The I–V characteristics of an n–i–n structure consist of three regions: (a) linear relationship; (b) steep rise of current; and (c) space-charge-limited current. These regions are qualitatively consistent with Lampert's theory. However, important differences have been found in regions 1 and 2. The effective resistivity can be different from that at thermal equilibrium, because it is determined by electron current and by the width of the electrically neutral i-region, which depends on the shallow acceptor density, the trap density and the length of the i-layer. The onset voltage for current rise is smaller than the trap-filling voltage by a factor of ten in some cases, and again is determined by the acceptor density and the length of the i-layer. The presented analytical model suggests that to achieve a good isolation between two devices in GaAs ICs, the shallow acceptor density as well as the trap density must be larger than a critical value.

References

1. Martin G. M., Farges J. P., Jacob G., Hallais J. P. and Poiblaud G. (1980). *J. Appl. Phys.*, **51,** 2840
2. Houng Y. M. and Pearson G. L. (1978). *J. Appl. Phys.*, **49,** 3348
3. Itoh T. and Yanai H. (1980). *IEEE Trans. El. Dev.*, **ED-27,** 1037
4. Kocot C. and Stolte C. A. (1982). *IEEE Trans. El. Dev.*, **ED-29,** 1059

5. Lee C. P., Lee S. J. and Welch B. M. (1982). *IEEE Trans. El. Dev. Lett.*, **EDL-3,** 97
6. Birrittella M. S., Seelbach W. C. and Goronkin H. (1982). *IEEE Trans. El. Dev.*, **ED-29,** 1135
7. D'Avanzo D. (1982). *IEEE Trans. El. Dev.*, **ED-29,** 1051
8. Scharfetter D. L. and Gummel H. K. (1969). *IEEE Trans. El. Dev.*, **ED-16,** 64
9. Lampert M. A. and Mark P. (1970). *Current Injection in Solids*; Academic Press, New York

MBE-Grown GaAs–AlGaAs Heterostructures on SI GaAs Substrates: Applications to Optoelectronic Integrated Circuits and HEMT LSIs

S. Hiyamizu, K. Kondo, T. Fujii, J. Saito, T. Ishikawa, S. Sasa and S. Tatsuta

Fujitsu Limited, 1677 Ono, Atsugi 243-01, Japan

Abstract

This paper describes MBE-grown GaAs–AlGaAs heterostructures on SI GaAs substrates and their device applications, that is, applications to: (a) a GaAs–AlGaAs quantum well (QW) laser monolithically integrated with GaAs MESFETs (metal-semiconductor field effect transistor) on a Cr-doped HB GaAs substrate (EPD of 3000–3500 cm^{-2}); and (b) a HEMT 1Kb static RAM which was fabricated with a selectively doped GaAs/N–AlGaAs heterostructure grown on an undoped LEC GaAs substrate (EPD of 30 000 cm^{-2}).

Introduction

Molecular beam epitaxy (MBE) [1–3] is known as the best crystal growth technology for very fine multi-layer structures such as GaAs–AlGaAs quantum well structures or selectively doped GaAs/N–AlGaAs heterostructures since it provides ultimately precise control over layer thickness (in units of monoatomic layer), doping profile, and composition profile in ternary or quaternary compound semiconductors. In addition, it provides very high quality GaAs–AlGaAs heterostructures, and high uniformity of epitaxial layers over a considerably large area. In recent years, with the use of this growth technology, many novel devices with GaAs–AlGaAs multi-layer structures have been grown on SI GaAs substrates to achieve high-speed performance.

GaAs–AlGaAs QW lasers were grown on SI GaAs substrates and were monolithically integrated with GaAs MESFETs for application to optoelectronic integrated circuits (OEIC) [4, 5]. OEICs have been investigated to construct optical transmission and data-processing systems because of their high-speed and low-noise performance, high reliability, and compactness of circuits.

Selectively doped GaAs/N–AlGaAs heterostructures grown by MBE on SI GaAs substrates exhibited extremely high electron mobility, especially at low temperatures because of spatial separation between electrons and their parent ionized donors [6–14]. High electron mobility transistors (HEMT), which were first developed with the use of GaAs/N–AlGaAs heterostructures in 1980 [15], showed the highest speed performance with low power dissipation for semiconductor devices [16, 17]. HEMT ICs fabricated on SI GaAs substrates have been investigated intensively for applications to high-speed logics and memories [16–23].

This paper describes GaAs–AlGaAs heterostructures grown by MBE on SI GaAs substrates and their applications to OEICs and HEMT LSIs.

MBE growth

Varian's MBE/GEN-II system and home-made MBE system were used to grow GaAs–AlGaAs heterostructures. By using rotating substrate technology, we obtained an epitaxial layer with high uniformity over a large area.

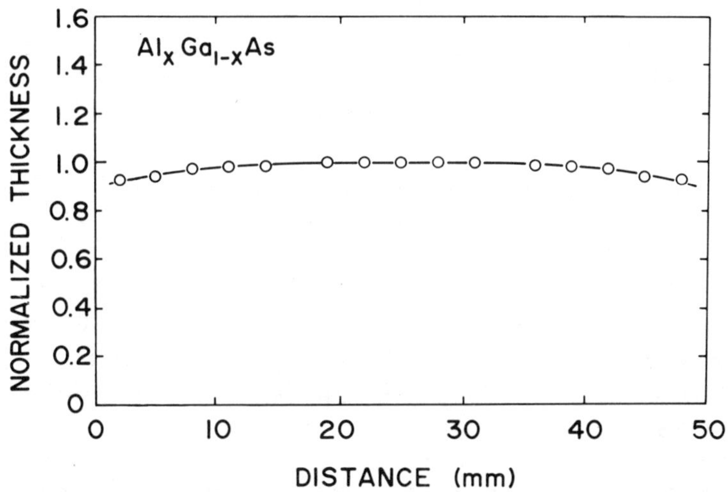

Figure 1 Normalized thickness of MBE-grown $Al_xGa_{1-x}As$ (x = 0.3, absolute thickness of about 1 μm) along a diameter of a 2 inch wafer. In the central region with an approximate diameter of 30 mm, variation of thickness was only ±1%.

Figure 1 shows a thickness profile of a Si-doped $Al_xGa_{1-x}As$ layer (x = 0.3, thickness of about 1 μm, Si-doping concentration of 5×10^{17} cm^{-3}), which was grown on a rotating, 2 inch diameter GaAs substrate with a stripe shadow masking. The layer thicknesses were measured by Talystep. Normalized thicknesses were plotted along a diameter. In the central region with an approximate diameter of 30 mm, variation of the layer thickness was within ±1%. Variation of other parameters such as AlAs mole fraction and a doping concentration of Si was also found to be within ±1.5% in this central region. Utilizing this high uniformity, we prepared GaAs–AlGaAs heterostructures for OEICs and HEMT LSIs on SI GaAs substrates.

The major problem of MBE growth of GaAs and AlGaAs has been surface defects [24, 25]. These defects are three-dimensional GaAs or AlGaAs crystals anomalously grown and existing on an MBE layer with a typical density of 500 cm^{-2} and with size and height of 0.5–1.0 μm for a 1 μm thick layer. There are several kind of surface defects. A typical one is an 'oval defect'. A high density of crystal defects, such as stacking faults, exist in a region surrounding an 'oval defect' [26]. The surface defect density increases with the thickness of the MBE layer. The causes of these defects, which have not yet been investigated sufficiently, include local contamination of the GaAs substrate prior to MBE growth, small Ga particles spit from a boiled Ga source in a heated crucible (1000–1100°C), and GaO from the Ga source. No direct correlation between the surface defect density and EPD of the GaAs substrate has been observed.

Optoelectronic integrated circuits

Si-doped or Cr-doped HB GaAs substrates were used for MBE growth of GaAs–AlGaAs QW lasers. Specifications of these substrates are listed in Table 1. The EPD of the Cr-doped GaAs substrate is 3000–3500 cm^{-2}, which

Table 1 *Specification of HB GaAs substrates*

	Si-doped HB GaAs	Cr-doped HB GaAs
Doping concentration	$1–2\times10^{18}$ cm^{-3}	0.2–0.5 ppmw
Resistivity	1×10^{-3} Ω cm	1×10^{7} Ω cm
EPD	$1–2\times10^{3}$ cm^{-2}	$3–3.5\times10^{3}$ cm^{-2}

is the lowest value for commercially available GaAs:Cr substrates. Multi-quantum well (MQW) lasers and graded-index waveguide separate confinement heterostructure (GRIN-SCH) lasers were grown at a temperature of 710°C by MBE. Figure 2 shows the schematic diagram of a GaAs–AlGaAs GRIN-SCH laser grown on the Si-doped HB GaAs substrate. The total thickness was about 7 μm. Si and Be were used for n-type and p-type dopants, respectively. The AlAs mole fraction, x, in a graded composition N-$Al_xGa_{1-x}As$ layer (0.2 μm) varied from 0.45 (the same as the x value of a 1.5 μm thick cladding layer) to 0.18. The active layer was an undoped GaAs QW layer (60 Å thick) sandwiched between AlGaAs graded composition layers with $x = 0.18$ at the GaAs–AlGaAs interface.

The spatial distribution of photoluminescence (PL) intensity from the QW active layer of the GRIN-SCH laser structure [5] is shown in Figure 3. The origin is the centre of the rotation, and PL intensity is plotted along a radius. Very high uniformity was achieved, that is, in a central region with an approximate diameter of 30 mm the variation of PL intensity was within ±3%. In the peripheral region, PL intensity decreased significantly, possibly

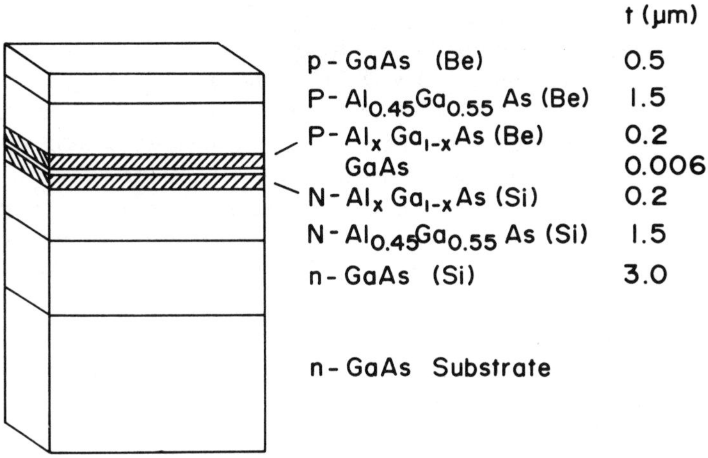

Figure 2 Schematic cross-sectional diagram of a GaAs–AlGaAs GRIN-SCH laser grown by MBE.

due to an unintentional decrease of substrate temperature and contamination by gases from a substrate heating system during MBE growth. Almost the same uniformity of PL intensity was also obtained in a Si-doped GaAs–AlGaAs layer ($n = 5 \times 10^{17}$ cm^{-3}) grown on the Cr-doped HB GaAs substrate.

The peak energy of a PL spectrum from the active layer of the laser structure (1.575 eV at 77 K) showed excellent agreement with the calculated value (the difference between the ground energy levels of electrons and holes in the corresponding potential well), indicating the realization of an almost ideal

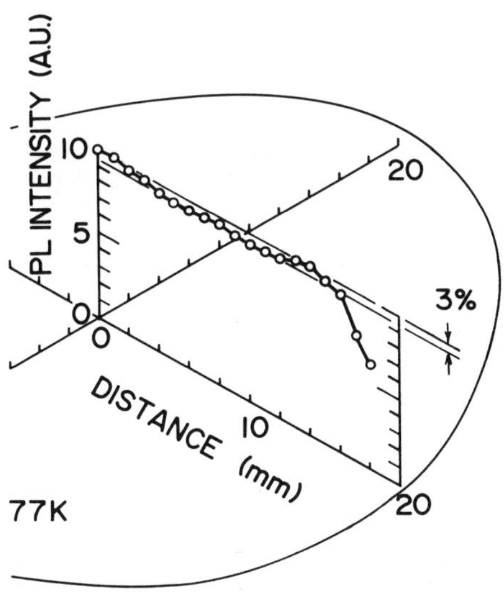

Figure 3 Distribution of PL intensity at 77 K from the GaAs quantum well layer (60 Å) of the GRIN-SCH laser structure grown on a Si-doped HB GaAs substrate.

QW structure [5]. The variation of the peak energy was only ±1 meV over the wafer, which corresponds to the change of barrier composition, Δx, of ±0.007 from x = 0.18 when the QW width is supposed to be rigidly constant.

Broad-area contact GRIN-SCH lasers grown on the Si-doped HB GaAs substrates (cavity length of 400 μm) achieved threshold current density as low as 230 to 260 A cm^{-2} at room temperature. Ridge-waveguide stripe GRIN-SCH lasers (cavity length of 300 μm and stripe width of 3 μm) exhibited a typical threshold current of 20 mA. In our experiments, lower threshold current was obtained for GRIN-SCH lasers than for MQW lasers, contrary to previously reported work in which almost the same threshold current densities were observed for both lasers [27, 28]. These results indicate that GRIN-SCH lasers are very promising for optoelectronic integrated circuits because of their low power dissipation and low current operation.

A schematic diagram of a ridge-waveguide stripe GRIN-SCH laser monolithically integrated with two GaAs MESFETs on the Cr-doped SI HB GaAs substrate is illustrated in Figure 4. The integrated lasers also achieved a typical threshold current of about 20 mA [5], which is almost the same as that of lasers fabricated on the Si-doped GaAs substrates, in spite of the p-side up configuration for the integration.

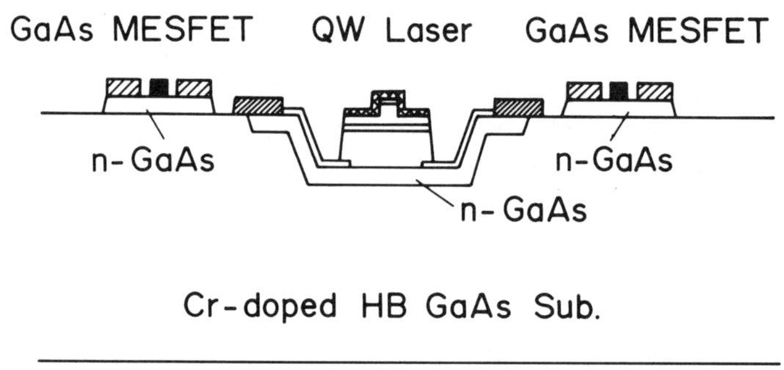

Figure 4 Schematic diagram of a ridge-waveguide stripe GaAs–AlGaAs GRIN-SCH laser monolithically integrated with two GaAs MESFETs on a Cr-doped HB GaAs substrate.

We are doing an ageing test of the MQW lasers (100 Å thick GaAs and 30 Å thick $Al_{0.18}Ga_{0.82}As$, five wells) grown on the Si-doped HB GaAs substrate with EPD of about 2000 cm^{-2}. In this test, eight lasers with uncoated facets are driven to produce a constant output-power of 3 mW from one facet at a temperature of 25°C in N_2 gas atmosphere. Half of the lasers are still operating stably, and operating time reached about 4000 hours. The other four lasers showed rapid degradation by growing dark line defects in the active region within 200 hours. There are two dominant causes of the rapid degradation. One is the surface defect (which was mentioned above) in the stripe contact region of the lasers. In the case of GaAs–AlGaAs QW laser structures, surface defect density reached 10^3–10^4 cm^{-2} because of the large thickness of the MBE layer (or the long growth time). This problem of MBE technology should be resolved in the near future. The other cause is, as is well known, dislocations in the active region of the lasers, which originate from dislocations in the GaAs substrate. We are now planning an ageing test of integrated GRIN-SCH lasers on the Cr-doped HB GaAs substrate. It is supposed that the Cr-doped HB substrate would deteriorate lasers more and the yield of long-life lasers would be reduced because of its higher EPD. Hence, it is important to further reduce EPD of the Cr-doped HB GaAs substrate for practical application of the lasers to optoelectronic integrated circuits.

HEMT LSIs

HEMTs were fabricated with the use of selectively Si-doped GaAs/N–AlGaAs heterostructures grown on SI GaAs substrates by MBE. Figure 5 shows a schematic diagram of a selectively doped GaAs/N–AlGaAs heterostructure grown at temperature about 680°C by MBE. It consists of a 0.8 μm thick undoped GaAs layer and a 0.1 μm Si-doped N–$Al_xGa_{1-x}As$ ($x = 0.3$) layer. The Si doping concentration, N_d, in the N–AlGaAs layer was $1–2 \times 10^{18}$ cm^{-3}. Usually an undoped AlGaAs spacer-layer (typical thickness, $\Delta t =$ 60 Å) was introduced between the Si-doped AlGaAs and GaAs layers. Some electrons originating from donor impurities in the AlGaAs layer transfer to the GaAs region because of the larger electron affinity of GaAs and form a two-dimensional electron gas (2DEG) accumulating at the heterojunction interface. This 2DEG exhibits extremely high mobility, especially at low temperatures because it is spatially separated from its parent ionized Si impurities.

Selectively doped GaAs/N–AlGaAs heterostructures were grown on both an undoped LEC GaAs substrate and a Cr-doped HB GaAs substrate, both

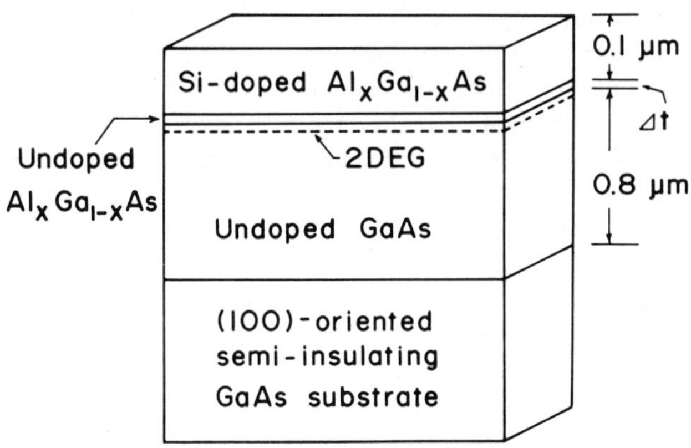

Figure 5 Schematic diagram of a selectively Si-doped GaAs/N-AlGaAs heterostructure. A 2DEG with extremely high mobility is formed at the GaAs–AlGaAs heterojunction interface.

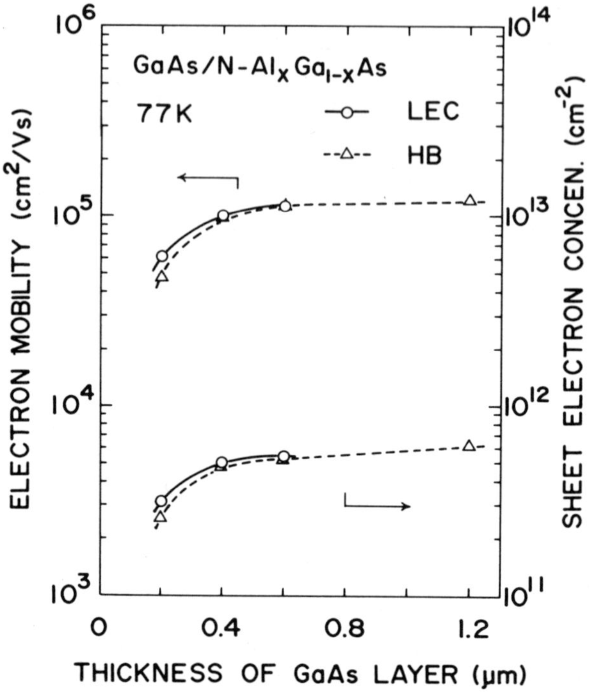

Figure 6 The 2DEG mobility along the heterojunction interface and sheet electron concentration at 77 K in GaAs/N–$Al_xGa_{1-x}As$ heterostructure ($x = 0.3$, $\Delta t = 60$ Å, $N_d = 2 \times 10^{18}$ cm^{-3}), grown on undoped LEC GaAs substrates and on Cr-doped HB GaAs substrates, as a function of thickness of the undoped GaAs layer, i.e. the distance between the 2DEG and the substrate interface.

of which had EPDs of about 3×10^4 cm^{-2}, and electrical properties of these epitaxial layers were examined. Figure 6 shows 2DEG mobility and sheet electron concentration, N_s, of the GaAs/N–AlGaAs heterostructures ($x = 0.3$, $\Delta t = 60$ Å, $N_d = 2 \times 10^{18}$ cm^{-3}) on the two kinds of substrates as a function of the thickness of the undoped GaAs layer, that is, the distance between the 2DEG and the substrate. In the region of GaAs layer thickness larger than 0.6 μm, 2DEG mobility and N_s were almost constant (2DEG mobility of about 1.15×10^5 cm^2 V^{-1} s^{-1} and $N_s = 5.5 \times 10^{11}$ cm^{-2}, typical values of GaAs/N–AlGaAs heterostructures). As the layer thickness decreased from 0.6 μm, both 2DEG mobility and N_s decreased significantly. The 2DEG mobility and N_s for the Cr-doped HB substrate decreased more than those for the undoped LEC GaAs substrate. At first, this was thought to be due to diffusion of Cr into the GaAs channel region from the Cr-doped HB substrate during MBE growth (growth temperature was about 680°C). The Cr concentration in the GaAs layer, however, was found to be below the SIMS detection limit (10^{14}–10^{15} cm^{-3}). Hence, larger decreases of 2DEG mobility and N_s for the HB substrate are possibly due to a higher density of electron traps at the interface between the GaAs layer and the GaAs substrate. The electron trap density at the interface between a MBE-grown GaAs layer and an n-GaAs substrate was found to be about 10^{12} cm^{-2}. Then a slight increase of the trap density for the HB GaAs substrate might decrease the 2DEG concentration and 2DEG mobility apparently. (In the region of $0 < N_s < 6 \times 10^{11}$ cm^{-2}, lower N_s results in lower 2DEG mobility.) Furthermore, in the case of the HB GaAs substrate, variation of 2DEG mobility and N_s over a wafer was also large compared with that of the LEC GaAs substrate. These results indicate that an undoped LEC GaAs substrate is preferable for MBE growth of GaAs/N–AlGaAs heterostructures.

Recently, extremely high 2DEG mobility was obtained in a GaAs/N–AlGaAs heterostructure ($x = 0.3$, $N_d = 1 \times 10^{18}$ cm^{-3}, $\Delta t = 200$ Å) [14]. As temperature decreased, electron mobility increased steeply and reached a maximum value of 1.25×10^6 cm^2 V^{-1} s^{-1} with $N_s = 3 \times 10^{11}$ cm^{-2} at 5 K as shown in Figure 7. After the sheet electron concentration was increased up to 5.04×10^{11} cm^{-2} by exposure to light at 5 K, 2DEG mobility rose to

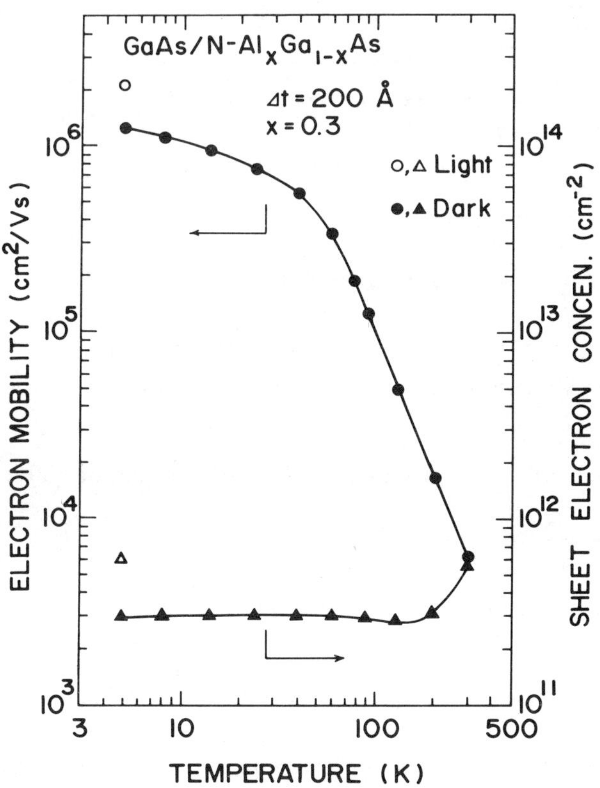

Figure 7 The 2DEG mobility along the heterojunction interface and sheet electron concentration, N_s, in a GaAs/N–Al$_x$Ga$_{1-x}$As heterostructure ($x = 0.3$, $\Delta t = 200$ Å, $N_d = 1 \times 10^{18}$ cm^{-3}) as a function of temperature. A 2DEG mobility as high as 2.12×10^6 cm^2 V^{-1} s^{-1} with $N_s = 5.04 \times 10^{11}$ cm^{-2} was achieved at 5 K by photo-enhancement.

2.12×10^6 cm^2 V^{-1} s^{-1}. This is, to our knowledge, the highest electron mobility ever observed in a semiconductor material. A concentration of background impurities in the GaAs channel region of this heterostructure, the region within about 100 Å from the GaAs–AlGaAs interface is assumed to be as low as 10^{14} cm^{-3} or less. This implies that very high quality GaAs/N-AlGaAs heterostructures can be grown on undoped LEC GaAs substrates by MBE.

Taking advantage of the high mobility of the GaAs/N–AlGaAs heterostructure, HEMT has been developed [15–23]. A cross-sectional diagram of an enhancement-mode (normally off) HEMT (E-HEMT) is shown in Figure 8. In the gate region, the n-GaAs top layer was selectively etched off by reactive ion etching technology [29]. When a bias higher than the threshold voltage, V_{th} (>0), is applied to the gate, a 2DEG is induced in the region under the gate and we have source-drain current due to 2DEG flow.

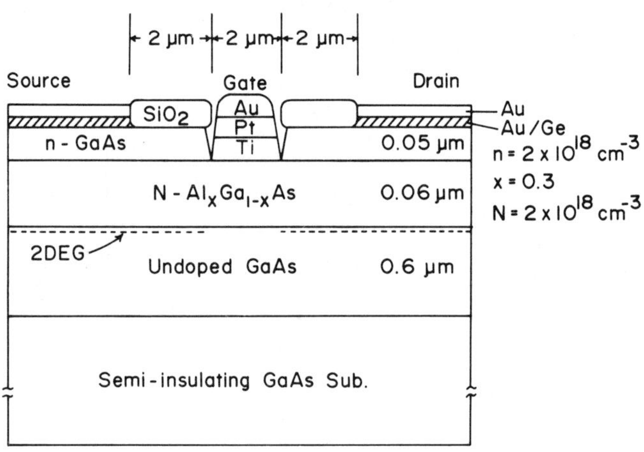

Figure 8 Schematic cross-sectional structure of an enhancement-mode HEMT.

V_{th} of the E-HEMT is given by the following equation [30]:

$$V_{th} = \psi - \frac{qN_d}{2\varepsilon} d^2 - \Delta E_c$$

where ψ is the Schottky barrier potential, q is the electron charge, ε and d are the dielectric constant and the layer thickness of the N–AlGaAs layer, respectively. ΔE_c is the discontinuity of the conduction band edge at the GaAs–AlGaAs interface. When $\psi = 1.2$ V, $N_d = 2 \times 10^{18}$ cm^{-3}, $\varepsilon = 12.2\varepsilon_0$, and $\varepsilon_0 = 8.85 \times 10^{-14}$ F cm^{-1}, d becomes 246 Å for $V_{th} = 0$. (The value of d in Figure 8 is much larger than the calculated one, probably due to the less effective donor concentration in an early stage of the development.) The partial derivative of V_{th} with respect to d, $(\partial V_{th}/\partial d)$ is $-qN_d d/\varepsilon$. This indicates that an increase of d by 2.8 Å (thickness of the monoatomic layer of AlGaAs) results in 20 mV decrease of V_{th} from $V_{th} = 0$. A Si-doped AlGaAs layer in

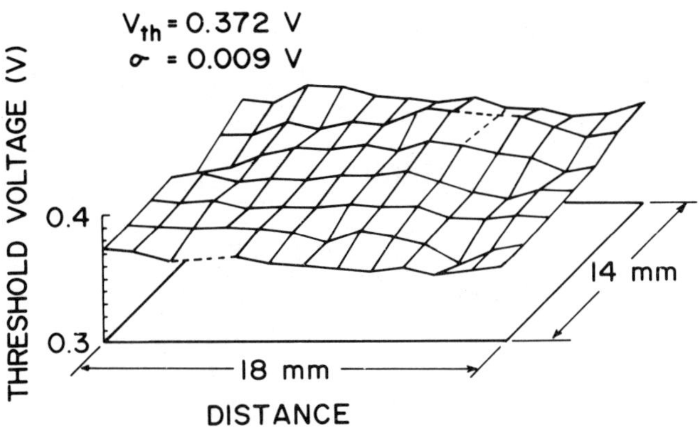

Figure 9 Variation of the threshold voltage, V_{th}, of an E-HEMT over an area of 18 mm × 14 mm in the high uniformity region of an MBE-grown GaAs/N–AlGaAs wafer. The standard deviation of V_{th} is as small as 9 mV.

the HEMT structure, designed to provide a certain V_{th} with respect to d, x, and N_d, has been grown successfully with the precise control of MBE.

Figure 9 shows the variation in E-HEMT V_{th} over an area of 18 mm × 14 mm [20]. The number of E-HEMTs was 89. These devices were fabricated in the central high uniformity area of an MBE-grown wafer. The average V_{th} was 0.372 V, and the standard deviation of V_{th} was as small as 9 mV. This uniformity of V_{th} is two or three times higher than that of conventional GaAs MESFETs fabricated by ion implantation. This small variation in V_{th} suggests an ultimately high uniformity of a GaAs/N–AlGaAs heterostructure grown by MBE. In the corresponding area, the variation of d is estimated to be within one-monolayer thickness.

A HEMT 1Kb static RAM has been fabricated with the use of an MBE-grown GaAs/N–AlGaAs wafer [23]. A total of 7244 HEMTs was integrated on a 2.9 mm × 3 mm chip. The gate length of a driver HEMT was 2 μm. Address access time of 0.9 ns and operating power of 360 mW have been achieved at 77 K.

Next, we characterize an undoped, 1 μm thick GaAs layer (almost the same thickness as a whole GaAs/N–AlGaAs heterostructure) grown on a SI GaAs substrate. The growth temperature was 680°C. Figure 10 shows photoluminescence (PL) spectra at 4.2 K for an undoped GaAs layer grown on an undoped LEC GaAs substrate and for an undoped GaAs layer grown on a Cr-doped HB GaAs substrate. A large free-exciton peak (1.515 eV) was observed in both samples. PL peaks were not observed for any impurities except carbon. Furthermore, we did not observe any EL2 or Cr peaks in the wavelength range from 1.05 μm to 2.6 μm. Carbon is a major background impurity of MBE-grown GaAs and is usually considered to be incorporated into GaAs layers from residual, background gases during MBE growth. This result indicates that a very high purity GaAs layer was obtained on an undoped LEC GaAs substrate as well as on a Cr-doped HB GaAs substrate.

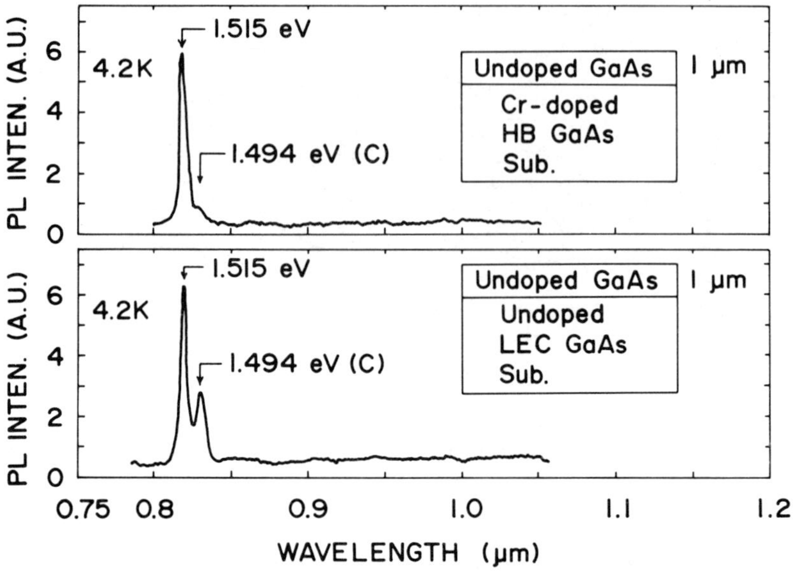

Figure 10 Photoluminescence spectra at 4.2 K for an undoped GaAs layer (1 μm thick) grown on a Cr-doped HB GaAs substrate (upper) and for an undoped GaAs layer (1 μm thick) grown on a undoped LEC GaAs substrate (lower). Both undoped GaAs layers were grown at 680°C by MBE.

The PL intensity profile at 4.2 K of an undoped LEC GaAs substrate and an undoped GaAs layer (1 μm thick) on a similar LEC GaAs substrate are shown in Figure 11. PL at 1.494 eV (C) was used to obtain these PL profiles. The diameter of the focussed Ar laser beam was about 100 μm and power was 30 mW. The excitation beam was scanned along the (100) axis of the wafer. In the edge region, strong PL was obtained in both samples, reflecting a typical 'U' shape distribution of dislocation density in the LEC GaAs substrate. The PL profiles of the MBE layer, which are much different from the flat PL profiles of a VPE-grown GaAs layer on an LEC GaAs substrate [31], indicate that a considerable concentration of non-radiative centres exists in the MBE-grown GaAs layer. Gettering of these non-radiative centres by dislocations results in a strong PL intensity in the high dislocation-density region. The PL profile in Figure 11 (upper) is considered to indicate a spatial distribution not of C impurities in the MBE-layer but of PL efficiency of the layer since a similar PL profile was also obtained with the use of the free-exciton peak (1.515 eV).

We did not characterize HEMT fabricated in such high dislocation density regions because the uniformity of the MBE-layer is not sufficiently good in the corresponding area as can be seen in Figure 1. It is assumed that HEMTs, fabricated in the peripheral region after the improvement of uniformity in MBE layers, may exhibit some uncontrollable device characteristic variations such as threshold voltage shift. Hence, it is desirable to improve uniformity

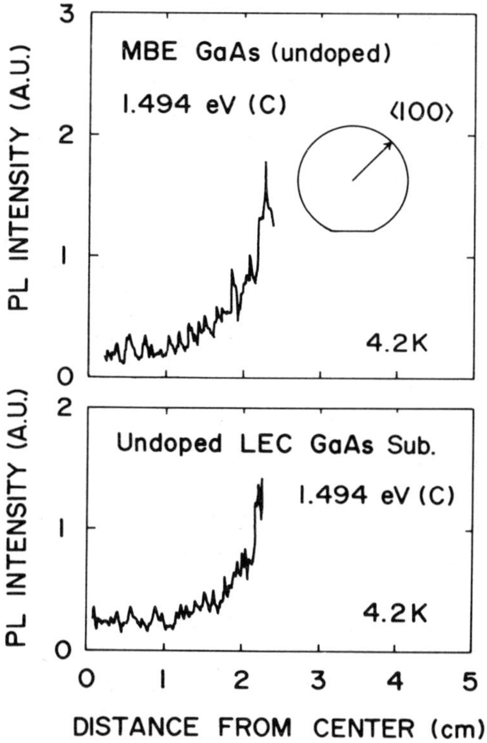

Figure 11 Photoluminescence intensity profiles (4.2 K) of an undoped GaAs layer grown on an undoped LEC GaAs substrate (upper) and of an undoped LEC GaAs substrate itself (lower). The MBE-grown GaAs layer showed a 'U' shape PL profile similar to that of the undoped LEC GaAs substrate.

of the dislocation density as well as to reduce the absolute value of the dislocation density in LEC GaAs substrates for application to HEMT ICs.

The LEC GaAs substrate hardly degraded the uniformity of GaAs/N–AlGaAs MBE-layers in the central region of a wafer as mentioned above. Meanwhile, surface defects in this central region are also a serious problem for HEMT LSI fabrication. When a surface defect lay under the gate, the threshold voltage was observed to shift significantly or that HEMT did not even work. Hence, this phenomenon reduced the yield of HEMT LSI chips with the scale of larger than 1K gate, when the surface defect density was 500 cm^{-2}. At present, this surface defect problem seems to be more serious than that of non-uniformity of undoped LEC GaAs substrates for MBE growth of HEMT wafers.

Summary

High quality GaAs–AlGaAs GRIN-SCH lasers and HEMT structures (selectively doped GaAs/N–AlGaAs heterostructures) were grown by MBE on

Cr-doped GaAs substrates and on undoped LEC GaAs substrates, respectively. A GRIN-SCH laser achieved a threshold current as low as 20 mA at room temperature. No Cr diffusion into the epitaxial layer during MBE growth was observed. Excellent uniformity of threshold voltage, V_{th}, for E-HEMTs was obtained in the central region of the MBE-wafer (about 30 mm diameter). The standard deviation of V_{th} was as small as 9 mV in an area of 18 mm × 14 mm.

The existence of surface defects on MBE-grown GaAs–AlGaAs heterostructures is a serious obstacle for device applications. These defects should be eliminated in the near future. Uniform distribution of dislocations and reduction of dislocation density of the SI HB and LEC GaAs substrates are very important for applications to OEICs as well as HEMT LSIs.

ACKNOWLEDGEMENTS

The authors would like to thank K. Kitahara for PL measurements and valuable discussions on SI GaAs materials, S. Yamakoshi for useful suggestions on OEICs, and K. Nishiuchi, T. Mimura and M. Abe for discussions on HEMT LSIs. They also thank T. Kotani, O. Ryuzan and T. Misugi for encouragement in this work.

This work is supported by the Agency of Industrial Science and Technology, MITI of Japan in the frame of the National Research and Development Projects 'Scientific Computing System' and 'Optical Measurement and Control Systems'.

References

1. Cho A. Y. and Arthur J. R. (1975). *Prog. Solid State Chem.*, **10,** 157
2. Ploog K. (1980). *Crystal Growth, Properties and Applications* (Freyhardt, ed.), **3,** 73
3. Wood C. E. C. (1980). *Physics of Thin Films*, p. 35 (C. Hoff and M. Francombe, eds.); Academic Press, New York
4. Sanada T., Yamakoshi S., Wada O., Fujii T., Sakurai T. and Sasaki M. (1984). *Appl. Phys. Lett.*, **44,** 325
5. Fujii T., Yamakoshi S., Nanbu K., Wada O. and Hiyamizu S. *J. Vac. Sci. Technol.*, in press
6. Stormer H. L., Dingle R., Gossard A. C., Wiegmann W. and Sturge M. D. (1979). *Solid-State Commun.*, **29,** 705
7. Hiyamizu S., Mimura T., Fujii T. and Nanbu K. (1980). *Appl. Phys. Lett.*, **37,** 805
8. Witkowski L. S., Drummond T. J., Stanchak M. C. and Morkoc H. (1980). *Appl. Phys. Lett.*, **37,** 1033
9. Hiyamizu S., Mimura T., Fujii T., Nanbu K. and Hashimoto H. (1981). *Jap. J. Appl. Phys.*, **20,** L245
10. Drummond T. J., Morkoc H., Hess K. and Cho A. Y. (1981). *J. Appl. Phys.*, **52,** 5231
11. Stormer H. L., Gossard A. C., Wiegmann W. and Baldwin K. (1981). *Appl. Phys. Lett.*, **39,** 912
12. Hiyamizu S. and Mimura T. (1982). *J. Cryst. Growth*, **56,** 455
13. Saito J., Nanbu K., Ishikawa T. and Hiyamizu S. (1983). *Jap. J. Appl. Phys.* **22,** L79
14. Hiyamizu S., Saito J., Nanbu K. and Ishikawa T. (1983). *Jap. J. Appl. Phys.*, **22,** L609

15. Mimura T., Hiyamizu S., Fujii T. and Nanbu K. (1980). *Jap. J. Appl. Phys.*, **19,** L255
16. Abe M., Mimura T., Yokoyama N. and Ishikawa H. (1982). *IEEE Trans. Microwave Theory and Techniques*, **MTT-30,** 992
17. Lee C. P., Miller D. L., Hou D. and Anderson R. J. (1983). Abstracts 41st Device Res. Conf., IIA-7
18. Mimura T., Joshin K., Hiyamizu S., Hikosaka K. and Abe M. (1981). *Jap. J. Appl. Phys.*, **20,** L598
19. Tung P. N., Delescluse P., Delagebeaudeuf D., Laviron M., Chaplart J. and Linh N. T. (1982). *Electron. Lett.*, **18,** 517
20. Abe M., Mimura T., Nishiuchi K., Shibatomi A. and Kobayashi M. (1983). Technical Digest GaAs IC Symp., p. 158
21. Lee C., Hou D., Lee S., Miller D. and Anderson R. (1983). Technical Digest GaAs IC Symp., 162
22. Mimura T., Nishiuchi K., Abe M., Shibatomi A. and Kobayashi M. (1983). Technical Digest IEDM, 99
23. Nishiuchi K., Kobayashi N., Kuroda S., Notomi S., Mimura T., Abe M. and Kobayashi M. (1984). Digest of Technical Papers ISSCC, 48
24. Chai Y. G. and Chow R. (1981). *Appl. Phys. Lett.*, **38,** 796
25. Bachrach R. Z. and Krusor B. S. (1981). *J. Vac. Sci. Technol.*, **16,** 756
26. Suzuki Y., Seki M., Horikoshi K. and Okamoto H. Unpublished
27. Tsang W. T. (1981). *Appl. Phys. Lett.*, **39,** 786
28. Tsang W. T. (1982). *Appl. Phys. Lett.*, **40,** 217
29. Hikosaka K., Mimura T. and Joshin K. (1981). *Jap. J. Appl. Phys.*, **20,** L847
30. Kuroda S., Mimura T., Nishiuchi K., Kobayashi N., Notomi S., Saito J. and Abe M. Unpublished
31. Kitahara K., Itoh H., Komeno J. and Ozeki M. (1984). This volume, p. 141

Material Parameters Affecting Surface Leakage in GaAs Integrated Circuits

M. F. Chang, C. P. Lee, R. P. Vahrenkamp, L. D. Hou, D. E. Holmes and C. G. Kirkpatrick

Rockwell International, Microelectronics Research and Development Center, Thousand Oaks, California 91360, USA

Abstract

A pseudo two-dimensional multi-conduction path model has been developed to explain substrate current leakage in GaAs integrated circuits based on trap-fill-limited carrier injection. Injection occurs at the n^+–p–n^+ region near the substrate surface as a result of the outdiffusion of deep traps. The amount of current leakage and the threshold voltage for sudden current increase are correlated with the EL2 and carbon distribution in the substrate. Good agreement between the experimental results and the theoretical predictions is achieved.

Introduction

The substrate leakage current in processed undoped SI GaAs wafers follows nonlinear current injection behaviour [1]. At low voltages, the current conduction is ohmic with very high resistance, but as the voltage reaches a certain threshold (the trap-fill-limited voltage), the current increases suddenly and the I–V characteristic becomes nonlinear. Based on the carrier injection model, the trap-fill-limited voltage, V_{TFL}, is related to the trap concentration in the GaAs substrate. However, the observed V_{TFL} is lower than that expected from the measured trap concentration in the bulk of the substrate [2], indicating the possible role of redistribution effects at the surface. In undoped SI GaAs, the major electron traps are EL2 deep donors compensated by carbon acceptors [3]. In case the EL2 outdiffuses during the processing, the concentration of empty EL2 (electron traps) in the vicinity of the surface can be much lower than that in the bulk, and therefore causes a low trap-fill-limited voltage for substrate conduction. Recent experiments have suggested a correlation between the V_{TFL} and the EL2 distribution [4, 5]. However, the depletion of EL2 leads to a p-type surface layer [3]. To explain the low threshold voltage carrier injection through a p-type conversion surface within two n^+ electrodes (Figure 1), a multiconduction path structure of the GaAs substrate is presented in this paper. The leakage current can be decomposed into three components: the space-charge limited (I_{SCL}) injection through the surface depletion region, the reverse saturation current through the n^+–p–n^+ transistor sitting below the surface depletion zone (I_r) and the ohmic leakage

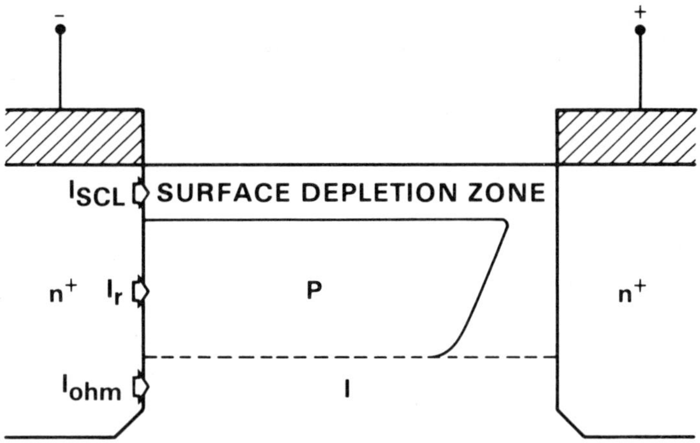

Figure 1 Cross-sectional view of isolation test device. I_{SCL} is the space-charge-limited injection through the surface depletion zone, I_r is the reverse saturation current of the n^+–p–n^+ transistor, and I_{ohm} is the bulk leakage current.

current through the bulk of the substrate (I_{ohm}). An analytical pseudo two-dimensional model was therefore developed to correlate the injection level and the threshold voltage with EL2 and carbon distributions. Quantitative agreement was achieved, and the results of an isolation experiment agree well with the model presented.

Pseudo two-dimensional approximation

For the sake of simplicity, a structure with two n^+ electrodes separated by a gap much narrower than the widths of the electrodes is used in the following analysis. The current flowing through the surface depletion region from the cathode to the anode in this case can be viewed as a two-dimensional boundary value problem. The boundary conditions for this structure, shown in Figure 2, are: the upper surface of the depletion zone is biased by the surface potential V_s, the cathode is grounded, the anode is externally biased

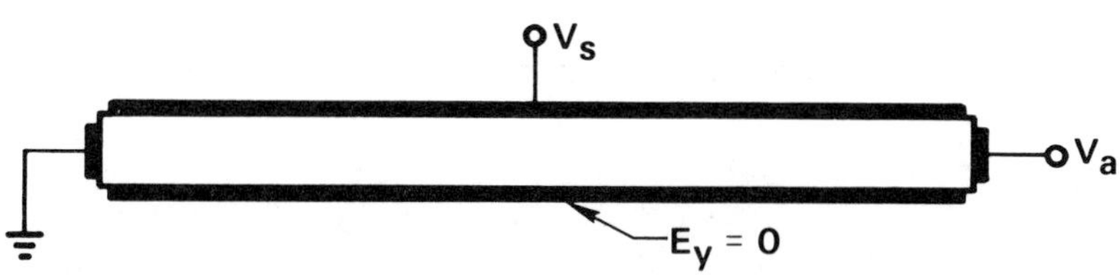

Figure 2 Boundary conditions for surface space-charge-limited injection.

by V_a, and the vertical component of the electric field is zero at the lower boundary of the depletion zone because of the neutral p-type region underneath.

Since the major traps are ionized EL2 donors, the equations which determine the injection behaviour are:

$$\nabla \cdot E(x, y) = -\frac{q}{\varepsilon}(n + n_t + (N_A - N_t)) \tag{1}$$

$$\nabla \cdot J_{SCL}(x, y) = 0 \tag{2}$$

$$J^x_{SCL} = q\mu n E_x + qD\frac{\partial n}{\partial x} \tag{3}$$

$$J^y_{SCL} = q\mu n E_y - qD\frac{\partial n}{\partial y} \tag{4}$$

$$n_t = \frac{N_t}{1 + (N/gn)} \tag{5}$$

$$N = N_c \exp\left(\frac{E_t - E_c}{kT}\right) \tag{6}$$

where n and n_t are the electron concentrations in the conduction band and at the trap level, respectively. N_A is the carbon concentration; J^x_{SCL} and J^y_{SCL} are the components of current flow in x (horizontal) and y (vertical) directions, respectively; N_t stands for the concentration of ionized EL2 traps, N_c for the density of states in the conduction band, E_t and E_c for the energy levels of the trap state and the conduction band edge, respectively; g is a degeneracy factor for the electron population; and N is introduced as a matter of convenience.

To solve the equations analytically, the following assumptions are made: (a) the surface depletion is caused by the surface potential; (b) there is no net current in the y direction and the diffusion current in the x direction can be neglected; (c) N_t, n and n_t are only functions of y and N_A is a constant which is much higher than $N_t(y)$ within the depletion zone. All these assumptions are made to reduce the equations into a pseudo two-dimensional form.

Assuming:

$$\frac{\partial E_y}{\partial y} = -\frac{q}{\varepsilon}(N_A - N_t(y)) \tag{7}$$

then, for a certain depth y down from the surface, equation (1) becomes:

$$\frac{\partial E_x}{\partial x} = -\frac{q}{\varepsilon}(n(y) + n_t(y)) \tag{8}$$

and the equation of continuity is:

$$J^x_{SCL} = q\mu n E_x = \text{constant} \tag{9}$$

with a constraint:

$$n(y) = n(0)\exp\left(-\frac{1}{kT}\left(\frac{q^2}{\varepsilon}\right)N_A\left(d_s y - \frac{y^2}{2}\right)\right) \tag{10}$$

where n(0) is the electron concentration at the upper surface and d_s is the depth of the depletion zone. Equations (8), (9) and (10) are then used to calculate the SCL injection through the surface depletion region with EL2 traps.

CASE OF SHALLOW TRAPPING

As long as E_f lies below E_t by more than kT, $1 + N/gn \sim N/gn$, the traps are said to be shallow, and the ratio of free to trapped electron densities is a constant θ, which is independent of n and can be expressed as:

$$\theta = \frac{n(y)}{n_t(y)} = \frac{N}{gN_t} \tag{11}$$

In the condition of $\theta \ll 1$, most of the injected electrons are immobilized in such traps. For the case of EL2 traps, $E_t = E_c - 0.83$ eV so that:

$$\theta = \frac{1.88 \times 10^3}{N_t} \tag{12}$$

Assuming the trap concentration of EL2 ranges from 10^{14} cm^{-3} to 10^{16} cm^{-3} in the substrate [6], θ is extremely small, as is the SCL injection in this case.

CASE OF DEEP TRAPPING

If E_f lies above E_t by more than kT, $1 + N/gn \sim 1$, the traps are said to be deep, and they are filled with electrons. In this case $n_t(y)$ is almost equal to $N_t(y)$. Manipulating equations (8) and (9), the current density of SCL injection in the deep trap case can be found as:

$$J^x_{SCL}(y) \sim \frac{9}{8}\frac{\mu\varepsilon V_a^2}{W^3} - q\mu N_t(y)\left(\frac{V_a}{W}\right) \tag{13}$$

After the deep traps are filled, all carriers will be injected into the conduction band leading to trap-fill-limited SCL current. The total injection current is:

$$I_{SCL} = \int_0^d J^x_{SCL}(y)\,dy \tag{14}$$

The upper limit of the integration is determined by the constraint coming from equation (10), in which the surface potential confines all electron carriers into a surface conduction layer with a thickness of:

$$d = d_s - \sqrt{d_s^2 - \frac{2D\varepsilon}{\mu q N_A}} \tag{15}$$

where d_s is determined by the surface potential V_s and the carbon concentration N_A. The magnitude of d is around several hundred angstroms.

CASE OF SHALLOW TO DEEP TRANSITION

During the transition, since the injection level changes so abruptly, there is no simple form which could be applied for current injection. In the present case, when the injection increases, the Fermi level moves across the trap

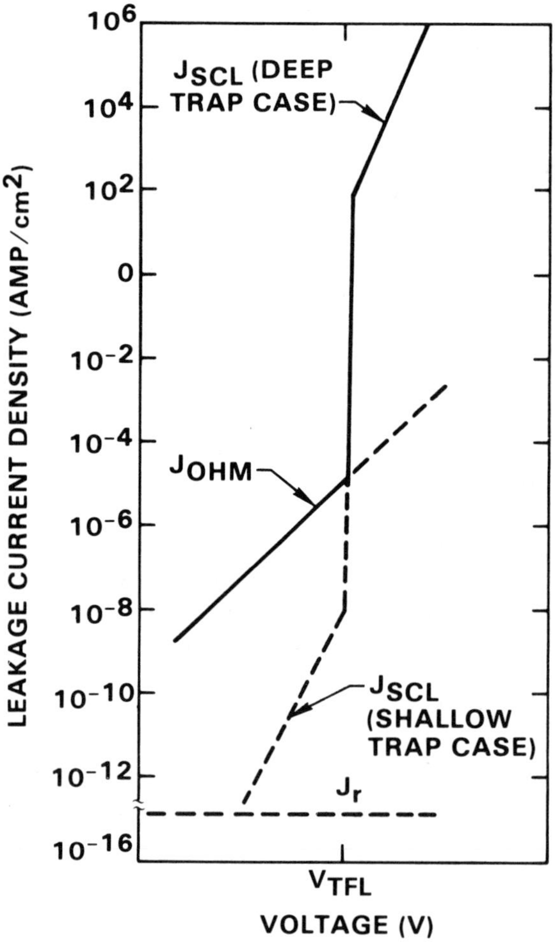

Figure 3 Calculated current densities for I_{SCL}, I_r and I_{ohm}.

level. From the classical trap-fill-limited law of Lampert [7], the transition takes place at V_{TFL}, which is:

$$V_{TFL} = \frac{qN_t^s}{C} \tag{16}$$

where C is the capacitance per unit area seen between cathode and anode, and N_t^s is the empty EL2 concentration on the surface. As illustrated in Figure 3, when the applied voltage reaches V_{TFL}, the SCL flow switches suddenly from one limiting case (shallow trap) to another limiting case (deep trap).

Reverse saturation current through n^+–p–n^+

For a nearly floating base transistor, the major current is the collector reverse saturation current, I_r, because most of the bias voltage drops across the collector space charge region. In the case of $V_a \gg 25$ mV, the current density, I_r, is approximately equal to 1×10^{-15} A cm^{-2}. As shown in Figure 3, the reverse saturation current is too small to be responsible for the substrate leakage.

Bulk leakage current

Although EL2 outdiffusion causes surface leakage, the concentration of empty EL2 normally remains high in the bulk. Since the trap-fill-limited voltage (V_{TFL}) is linearly proportional to the empty trap density, the bulk V_{TFL} should be much higher than its surface counterpart. In principle, we can write:

$$V_{TFL}^{(bulk)} = \left(\frac{N_t^B}{N_t^s}\right) \times V_{TFL}^{(surface)} \tag{17}$$

Previous work has demonstrated $N_t^B/N_t^s \sim 10^2$ after thermal annealing [6]. Under such circumstances, the bulk leakage current does not initiate trap-fill-limited injection for substrate leakage. For a typical undoped SI GaAs wafer (10^7 Ω cm), the electron concentration is about 2×10^7 cm^{-3}. Based on this number, the calculated current densities are illustrated in Figure 3. It is interesting to find that for V lower than V_{TFL}, ohmic current through the bulk dominates, while for voltage higher than V_{TFL}, the injection through the surface depletion zone dominates.

Experiments

An experiment was performed to evaluate the dependence of V_{TFL}, after thermal annealing, on the concentration of EL2 and carbon to test the multi-conduction path model. Four undoped SI LEC wafers were processed,

Table 1 *Dependence of V_{TFL} on concentration of carbon and EL2*

Wafer no.	[Carbon]	[EL2]	V_{TFL}
1	4.6×10^{16}	1.6×10^{16}	4 ~ 6 V
2	5.4×10^{15}	2.3×10^{16}	6.5 ~ 7.5 V
3*	5.4×10^{15}	2.3×10^{16}	6.5 ~ 7 V
4	7×10^{15}	4.8×10^{15}	1.6 ~ 3 V

*Surface treated with hot H_3PO_4 before Si_3N_4 cap

in which the EL2 concentrations were varied from $4.8 \times 10^{15}\ cm^{-3}$ to $2.3 \times 10^{16}\ cm^{-3}$, and the carbon concentrations were varied from $7 \times 10^{15}\ cm^{-3}$ to $4.6 \times 10^{16}\ cm^{-3}$, as shown in Table 1. No correlation was found between V_{TFL} and carbon concentration. But V_{TFL} increased with the bulk EL2 concentration, which is in agreement with the model. These results suggest that the surface EL2 concentration is lower than that of the residual carbon.

The second experiment was performed to check quantitatively the injection level with 3 μm spaced isolation devices after thermal annealing. The experimental data are shown as a solid line in Figure 4.

For the calculation of I_{SCL}, the following distribution of EL2 close to the surface was assumed [6]:

$$N_t(y) = N_t^S + (N_t^B - N_t^S)\left(\frac{y}{y_0}\right)^2 \tag{18}$$

where N_t^S and N_t^B are the EL2 concentrations on the surface and in the bulk, respectively. y_0 is the depth of the outdiffusion zone. From the measured V_{TFL} of 0.55 V, an N_t^S of $4.55 \times 10^{13}\ cm^{-3}$ was calculated from equation (16). A y_0 of 0.5 μm and an EL2 concentration of $1.0 \times 10^{16}\ cm^{-3}$ were obtained by etching experiments [5]. Assuming $\mu = 5000\ cm^2\ V^{-1}\ s^{-1}$, $N_A = 5 \times 10^{15}\ cm^{-3}$ and $D = 130\ cm^2\ s^{-1}$, we found $d = 100$ Å from equation (15). The total injection current is:

$$I_{SCL} = V_a \times d\left(1.49\ V_a - 1.7 \times 10^{-14}\ N_t^S - 57.7\left(\frac{d}{y_0}\right)^2\right) \tag{19}$$

which is plotted in Figure 4. Again, very close agreement is achieved; for $V < V_{TFL}$ bulk ohmic current dominates, while for $V > V_{TFL}$ surface SCL dominates.

Summary and conclusion

A multi-conduction path model has been used to model the substrate leakage current in undoped annealed LEC substrates. The consistency between the theory and experiment suggests that ionized EL2 concentration near the

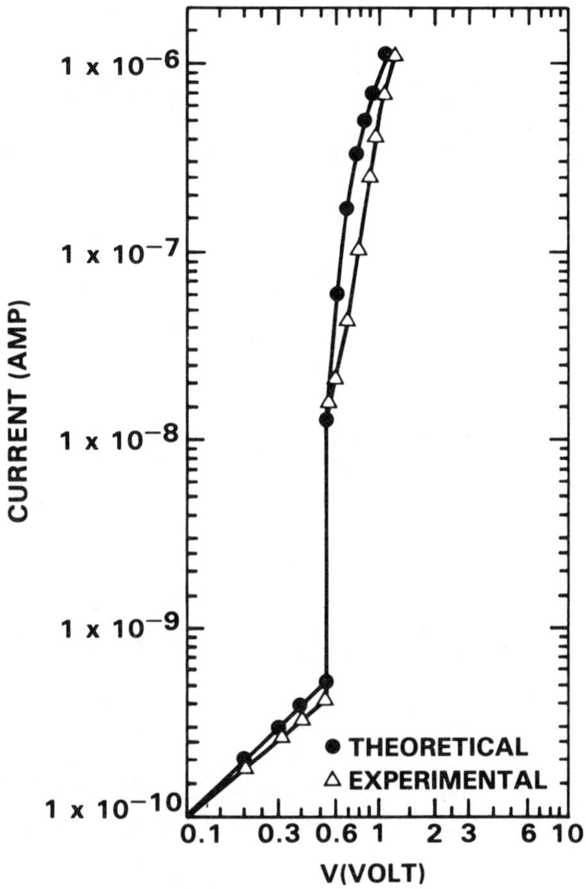

Figure 4 Measured and calculated I–V characteristics of GaAs surface conduction.

surface determines the V_{TFL} for substrate conduction. Calculations also suggest that the substrate leakage current includes three components: the SCL injection through a surface depletion zone, the reverse saturation current through the undoped p region, and the ohmic current through the bulk of the substrate.

ACKNOWLEDGEMENTS

The authors would like to thank Dr Ken Elliott for helpful discussions and reading the manuscript. This work was supported, in part, by the Defense Advanced Research Projects Agency of the US Department of Defense under Contract No. MDA903-83-C-0067.

References

1. Lee C. P., Lee S. J. and Welch B. M. (1982). *IEEE Electron. Dev. Lett.*, **EDL-3(4),** 97
2. Lee C. P., Vahrenkamp R. P., Lee S. J., Shen Y. D. and Welch B. M. (1982). Proc. of 1982 IEEE GaAs IC Symp., 169
3. Holmes D. E., Chen R. T., Elliott K. R., Kirkpatrick C. G. and Yu P. W. (1982). *IEEE Trans. MTT*, **30(7),** 949
4. Hughes B. and Li C. (1982). *Inst. Phys. Conf. Ser.*, **65**, 57
5. Chang M. F., Lee C. P., Hou D. L., Vahrenkamp R. P. and Kirkpatrick C. G. (1984). *Appl. Phys. Lett.*, **44**, 869
6. Makram-Ebeid S., Gautard D., Devillard P. and Martin G. M. (1982). *Appl. Phys. Lett.*, **40(2),** 161
7. Lampert M. A. (1962). *Proc. IRE*, **50,** 1781

Improved GaAs MESFETs Transconductance by Gettering

Faa-Ching Wang and Marina Bujatti

Santa Rosa Technology Center, Hewlett-Packard Company, 1412 Fountaingrove Parkway, Santa Rosa, California 95401, USA

Abstract

The relationship between the characteristics of undoped SI GaAs substrates and the performance of MESFET (metal-semiconductor field effect transistor) devices fabricated by direct ion implantation into the substrates was investigated. A positive correlation was observed between dislocation EPD (etch pit density), as-grown mobility and device transconductance. Localized substrate inhomogeneities appear to affect the device performance via the mobility. Gettering experiments were carried out in order to improve the substrate quality. The best results were obtained by introducing mechanical damage on the backside of the wafers, followed by heat treatment. A transconductance improvement of up to 50% was achieved on MESFETs fabricated on gettered substrates as compared to as-grown substrates for otherwise identical processing conditions.

Introduction

There has recently been an increasing interest in investigating the relationship between substrate properties and MESFET performance [1–3]. In particular, correlations between dislocation density and device parameters such as pinch-off voltage and saturation currents have been observed. In this paper we report the results of an investigation aimed at identifying material properties which affect the device performance most directly.

Experimental

Five undoped SI wafers taken at 2 cm intervals from a typical ⟨100⟩ GaAs ingot grown by the low pressure LEC (liquid encapsulated Czochralski) method were used for this investigation. The ingot was approximately 2.5 inches in diameter and had passed the standard tests for SI quality and thermal conversion. One half of each wafer was used for material characterization including optically-assisted resistance profile (OARP) [4–6], dislocation EPD, and van der Pauw (VDP) measurements. The OARP was measured on SI wafers by using ultrasonic soldering to provide good ohmic contact [5–6]. Resistivity readings were taken over 2 mm square areas at 3 mm intervals. The EPD was obtained by counting etch pits on photographs covering 1.1 mm square areas after molten KOH etching. Finally, VDP measurements were carried out on 4 mm square samples to determine the as-grown Hall mobility and free carrier concentration.

The other half of the wafer was processed to produce discrete MESFETs. Each wafer was implanted with Si at 200 and 60 keV with a 5×10^{12} cm^{-2} dose for each energy in order to produce an active and an n+ layer respectively. A conventional recessed gate process was used with 1 μm length, 350 μm width, and 3.5 μm source-drain spacing. A Ti/Pt/Au metallization was deposited for the gates. Every six devices (1.8 mm), a Schottky barrier test pattern covering a 70×70 μm square area enabled the capacitance to be measured. Transconductance (g_m), threshold voltage (V_{th}) and saturation current (I_{ds}) were measured on the FETs. Hall mobility and capacitance were measured on the test patterns.

Results and discussion

The results of material and device measurements taken on adjacent areas of a wafer were compared. As summarized in Figures 1 and 2, the transconductance appears to be positively correlated to the mobility of the as-grown substrate both for the ingot as a whole and on individual wafers. For the first two thirds of the ingot, the transconductance and as-grown mobility also appear to be positively correlated to the EPD, such as for the wafer shown in Figure 1. No other clear correlation was observed between other material and device parameters. In particular, no clear correlation was observed between the free carrier concentration and other material parameters such as EPD. The positive experimental correlation between EPD, mobility and transconductance leads us to speculate that dislocations act as gettering centres for imperfections, therefore forming purer bulk regions with higher mobility in between [6]. Similar experimental evidence has been reported elsewhere [1–3, 7–9]. It is difficult to conceive that a single defect or impurity could have such a strong effect on mobility without affecting free carrier concentration. Therefore, although at present we do not know the mechanism for gettering, we feel that more than a single imperfection is likely to be involved in the dislocation gettering process. It seems therefore that no significant quality improvement of GaAs substrates for directly ion implanted majority carrier devices can be expected by the reduction of dislocations alone. Preliminary results of an investigation aimed at clarifying the nature of the correlation between microdefects and device properties are reported elsewhere [10].

Prompted by these facts, gettering experiments were carried out aimed at reducing the overall imperfection density in the active region of the substrates. The gettering technique has not been widely applied to GaAs. A reduction of impurities in GaAs was reported using a glass surface layer gettering [11]. Backside damage followed by heat treatment at 800°C was applied to getter Cr and Au in GaAs substrates [12, 13]. More recently, the effects of heat

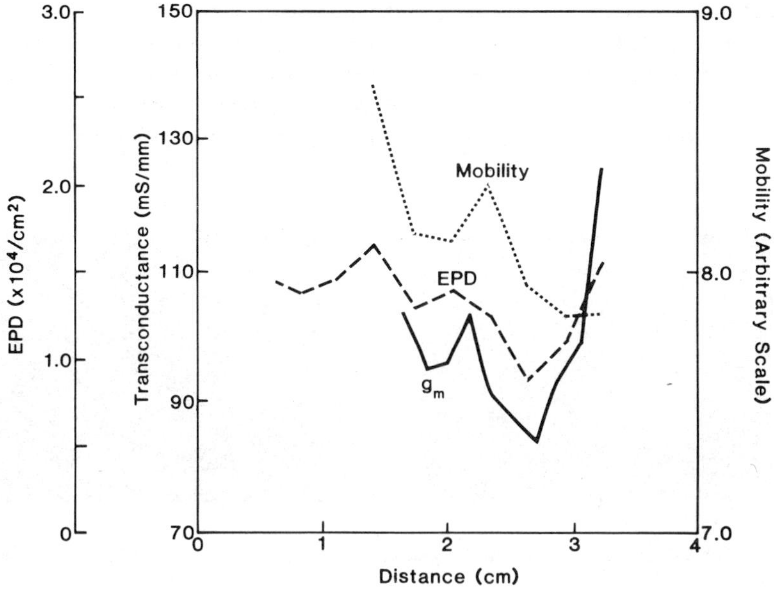

Figure 1 Radial position dependence of dislocation EPD, substrate mobility (Ref. 6) and MESFET transconductance on the wafer.

treating both wafer and ingot forms of GaAs, at relatively high temperatures (~800°C), were reported [14–16]. In all cases, a quality improvement as measured by different material parameters was achieved.

In our investigation, mechanical damage was introduced by bead blasting the backside of the wafer followed by a low temperature (400–600°C) heat

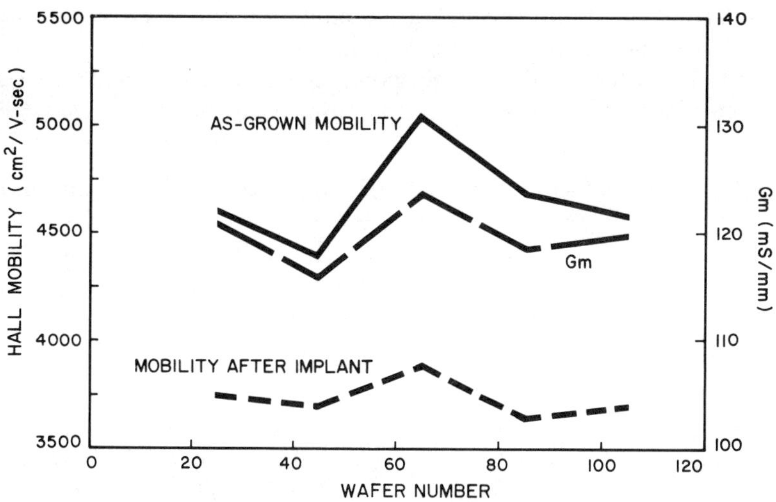

Figure 2 The position dependence along the ingot for averaged values of mobility before and after ion implantation, and MESFET transconductance on the wafer.

treatment in forming gas for half a day up to five days. The wafers were capped with SiO_2 before bead blasting for protection purposes. Subsequent to the gettering treatment, MESFETs were fabricated and measured in exactly the same way as that described earlier. A substantial improvement in MESFET transconductance was obtained by using gettered substrates and otherwise identical processing and doping.

(a) (b)

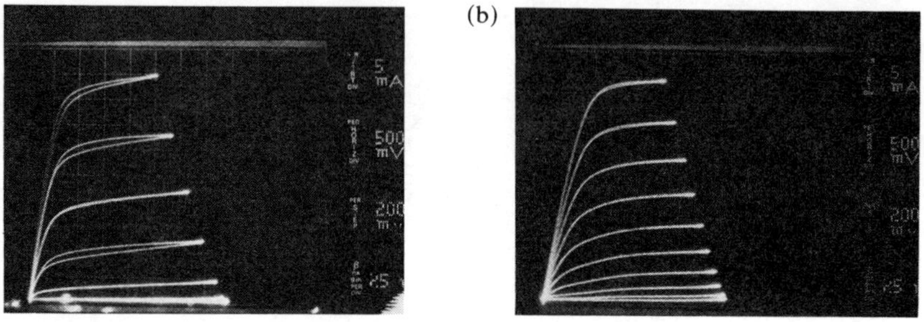

Figure 3 The I–V curves of direct ion implanted MESFETs on (a) gettered and (b) as-grown substrates.

In Figure 3, the current-voltage characteristics of two MESFETs fabricated on gettered and as-grown substrates, respectively, are compared. Transconductance values as high as 190 mS mm^{-1} were measured for devices fabricated on gettered substrates as compared to 130 mS mm^{-1} on as-grown substrates, while the capacitance was unaffected. However, a larger output conductance (a factor of 1.5 to 2) was also observed for MESFETs fabricated on gettered substrates as indicated by the higher slope of the saturated current region in Figure 3. On the other hand, no significant improvement was obtained with other gettering techniques (heavy ion implantation and interface mismatch) or with heat treatment without previous mechanical damage.

Table 1 shows the results of transconductance measurements on wafers from three different vendors. An average improvement of about 40–50% is

Table 1 *Transconductance of MESFETs fabricated on gettered substrates*

Ingot No.	Transconductance (mS mm^{-1})		Improvement (%)
	Gettered wafer	Control wafer	
Vendor 1	170	120	42
Vendor 1	190	130	46
Vendor 1	160	120	33
Vendor 1	170	110	55
Vendor 1	170	130	31
Vendor 2	170	120	42
Vendor 3	150	110	36

clearly demonstrated. Although data from VDP measurements were more scattered, at least a 25% increase in mobility before and after ion implantation was observed in the gettered substrates as compared to the as-grown substrates. In addition, KOH etch was used after gettering, with no significant change observed in pattern or magnitude.

Conclusions

An experimental correlation was established between dislocations, mobility and transconductance in directly ion implanted GaAs MESFETs. Dislocations are believed to be indirectly related to the device performance through mobility improvement due to dislocation gettering of imperfections. The substrate quality can be improved by an extrinsic gettering technique consisting of wafer backside damage followed by low temperature heat treatment.

ACKNOWLEDGEMENTS

We wish to acknowledge the help of Linda Tewksbury and Donna Spitzer in processing the wafers.

References

1. Nanishi Y., Ishida S., Honda T., Yamazaki H. and Miyazawa S. (1982). *Jap. J. Appl. Phys.*, **21,** L335
2. Nanishi Y., Ishida S. and Miyazawa S. (1983). *Jap. J. Appl. Phys.*, **22,** L54
3. Miyazawa S., Ishii Y., Ishida S. and Nanishi Y. (1983). *Appl. Phys. Lett.*, **43,** 853
4. Blunt R. T., Clark S. and Stirland D. J. (1982). *IEEE Trans. El. Dev.*, **ED-29,** 1038
5. Wang F-C. and Li C. (1983). Seventh Conf. on Crystal Growth, Fallen Leaf Lake, California, Section V-2
6. Wang F-C. and Bujatti M. (1984). *IEEE El. Dev. Lett.*, in press
7. Kitahara K., Nakai K. and Shibatomi S. (1982). *J. Electrochem. Soc.*, **129,** 880
8. Kamojima T., Shimura F., Matsumoto Y., Watanabe H. and Matsui J. (1982). *Jap. J. Appl. Phys.*, **21,** L721
9. Chin A. K., Von Neida A. R. and Caruso R. (1982). *J. Electrochem. Soc.*, **129,** 2386
10. Ponce F. A., Wang F-C. and Hiskes R. This volume, p. 68
11. Dastidar R., Kavindran K. and Pank H. (1969). *Electron. Lett.*, **5,** 553
12. Magee T., Peng J., Hong J., Evans C. and Deline V. (1979). *Phys. Stat. Solidi*, **A55,** 159
13. Magee T., Peng J., Hong J., Katz W. and Evans C. (1979). *Phys. Stat. Solidi*, **A55,** 161
14. Matsumoto Y. and Watanabe H. (1982). *Jap. J. Appl. Phys.*, **21,** L515
15. Palmateer S. C. and Eastman L. F. (1983). Fifth Molecular Beam Epitaxy Workshop, Atlanta, Georgia, section TP3
16. Rumsby D., Ware R., Smith B., Tyjberg M., Brozel M. and Foulkes E. (1983). GaAs IC Symposium, Phoenix, Arizona, p. 34

Low Frequency Noise in GaAs FETs

L. Forbes, P. Canfield, R. Gleason* and A. McCamant*

Electrical and Computer Engineering, Oregon State University, Corvallis, Oregon 97331, USA

Abstract

A survey has been made of low frequency noise in different types of GaAs FET (field effect transistor) structures. A major component of the noise in GaAs JFETs (junction FET) is shown to be generation-recombination which can be correlated to conductance transients measured on the same devices. Measurements at high temperatures, however, show that all devices have a residual 1/f noise component which can not be attributed to generation-recombination and/or trapping.

Low frequency noise in GaAs FETs has been variously attributed to surface phenomena or surface states outside the gate region; surface states in the Schottky barrier gates; deep level traps in the gate region [1], in the back-gate region, or channel itself; inter-valley transfer of hot carriers; temperature fluctuations in the channel; or, even more recently, diffusion of ions along dislocations [2]. Similar reports have been made on HEMT (high electron mobility transistors) like structures attributing all of the low frequency noise to generation-recombination and/or trapping [3].

In the process of investigating these we have characterized HEMTs (still noisy), junction gate versus MESFETs (metal-semiconductor FET) (no significant difference), implanted versus VPE (vapour phase epitaxial) grown active layers, FETs versus FET-like gaps with no gate, different substrate materials, Be-implanted p-type regions behind the channel (to eliminate backgating), different source-gate spacings and n-channel versus p-channel devices (still noisy). Also, we have studied different gate lengths, widths and surface passivations. These measurements have been done over a wide range of bias conditions and temperatures from 13 K to 500 K. In addition, conductance DLTS (deep level transient spectroscopy) scans have been made on the same devices under the same operating bias voltages [4].

Figure 1 shows a summary of conductance DLTS results and shows the different types of conductance transients which have been observed previously [5, 6] as well as those more recently on Tektronix JFETs. Also shown are the conductance transients observed on JFETs from another source, McDonnell–Douglas (M–D). All GaAs FETs seem to demonstrate an electron trap which has a time constant of about 1 s at 300 K. This has usually been attributed to the so-called EL2 level in GaAs. As will be shown later it is this imperfection

* Applied Research, Tektronix Inc., Beaverton, Oregon 97077, USA.

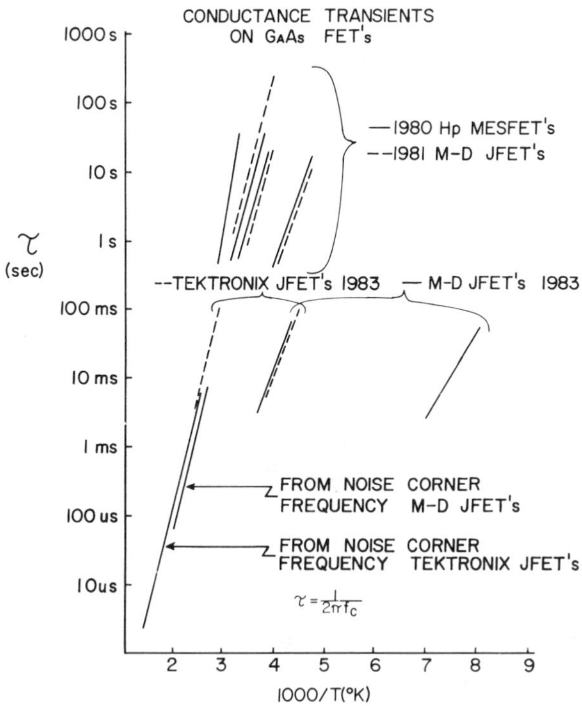

Figure 1 Conductance transients on Hewlett-Packard, McDonnell-Douglas and Tektronix GaAs FETs showing the most recent results on JFETs in the centre. Also shown are the corner frequencies which can be observed in the JFET noise spectra.

centre which is responsible for, or is the primary source of, generation-recombination noise. A variety of other conductance transients are observed on JFETs, and although these may have large amplitudes they do not seem to contribute any significant trapping noise. These other centres correspond to those in Figure 1 with millisecond time constants at values of 1000/T of four or larger; in other words, very fast at room temperature.

Figure 2 shows the noise spectra observed on a Tektronix JFET at high temperatures, all above room temperature. In this particular example the generation-recombination component is very obvious, and is characterized by the corner frequency which changes with temperature. The same type of measurements have also been done on JFETs at low temperatures, in the range 13 K to 300 K. In this case, however, no such structure is evident, indicating that only one centre is contributing any significant generation-recombination or trapping noise. When the corner frequency or time constant associated with the generation-recombination noise is plotted on Figure 1, it corresponds closely to the conductance transients observed on the same device—specifically the electron trap with a time constant of around 1 s at room temperature. As shown in Figure 1, a similar generation-recombination

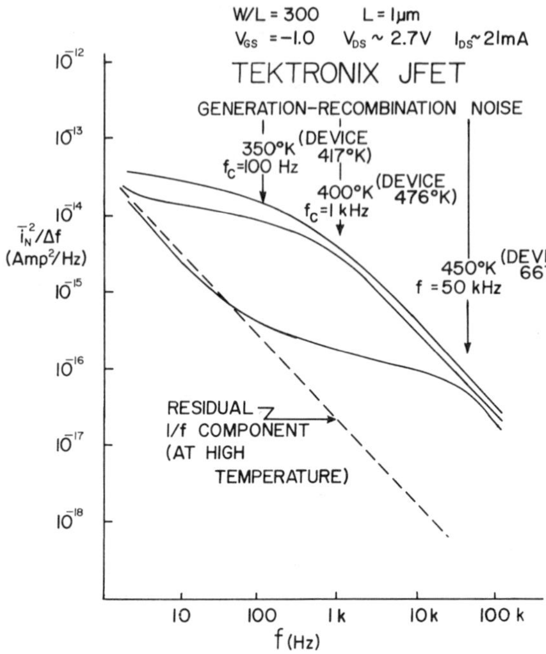

Figure 2 Noise spectra observed at low frequencies on Tektronix JFET.

noise component is observed on McDonnell-Douglas JFETs. In general, generation-recombination noise is observed easily only on JFETs, not MESFETs; and then, because of the long time constant, only at temperatures above room temperature.

Figure 3 shows, by contrast, the noise spectra at low frequencies of a Tektronix MESFET. In the case of MESFETs it is difficult to observe the corner frequencies associated with any generation-recombination and/or trapping component. MESFETs, in general, tend to show a 1/f noise spectra which is peculiarly independent of temperature. The MESFET shown in Figure 3 is unusual in that it does show some generation-recombination component, and then only because it has been biased closer to pinch-off to accentuate this effect. The noise spectra of GaAs MESFETs are very similar to those of silicon MESFETs, primarily 1/f but much higher on the GaAs MESFETs.

In conclusion, then, we attribute the low frequency noise in GaAs JFETs to:

1. Generation-recombination or trapping at an electron trap which has a time constant of about 1 s at room temperature, usually attributed to the EL2 level in GaAs; there may be, and usually are, other centres which result in much larger conductance transients but they do not seem to contribute significantly to the low frequency noise.

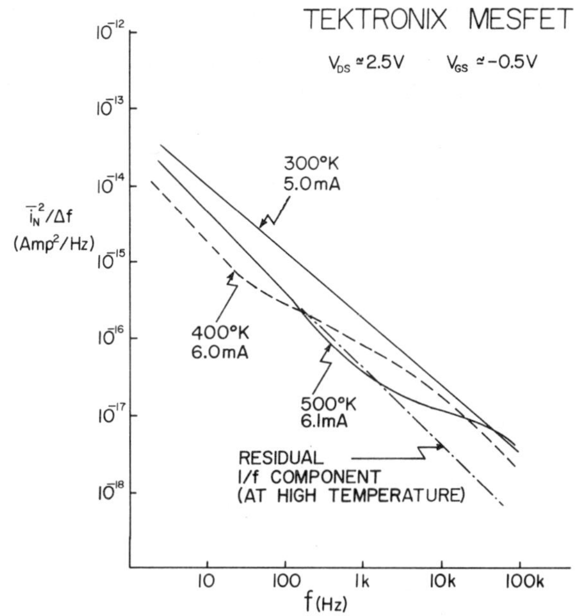

Figure 3 Noise spectra on a Tektronix MESFET.

2. A 1/f component, probably associated with surface states outside the gate region, and which is most easily identified by measurements at high temperatures where all the generation-recombination or trapping time constants become very fast.

The low frequency noise on GaAs MESFETs we attribute to:

1. Primarily a 1/f component, probably associated with surface states both outside and within the gate region, and one which is peculiarly independent of temperature.
2. Secondarily, generation-recombination noise or trapping noise which is not easily identifiable in most measurements on GaAs MESFETs.

The conclusions presented here are significantly different from those presented in the literature [1, 2]; however, we believe this is the result of having used JFETs and not just MESFETs, having done the measurements at high temperatures, and having done conductance DLTS scans on the same devices.

References

1. Su C. Y., Rohdin H. and Stolte C. (1983). *Proc. IEDM,* 601
2. Peczalski A., van der Ziel A. and Zuleeg R. (1983). *Solid St. Electron.,* **26,** 861

3. Loreck L., Dambkes H., Heime K., Ploog K. and Weimann G. (1981). *IEEE El. Dev. Lett.*, **EDL-5,** 9
4. Forbes L. and Chang C. D. (1979). *Proc. 1st Int. Circuits Symp.*, 16
5. Chang C. D. and Forbes L. (1980). In *Semi-Insulating III–V Materials, Nottingham*, p. 329 (G. J. Rees, ed.); Shiva Publishing, Nantwich, UK
6. Forbes L., Chang C. D. and Zuleeg R. (1981). *Inst. Phys. Conf. Ser.*, **63,** 179

The Effect of Mobility Profiles on Implanted GaAs MESFET Performance

D. C. Look, S. Chaudhuri, A. Ezis* and M. Moloney†

University Research Center, Wright State University, Dayton, Ohio 45435, USA

Abstract

Mobility (μ) profiles, carrier concentration (n) profiles, and I–V characteristics have been obtained for seven, identically-processed, SI GaAs substrates, each implanted with $4 \times 10^{12}\,\mathrm{cm}^{-2}$, 100 keV Si ions, then annealed and fabricated with a test pattern that contains gated Hall-bar, van der Pauw, and various FET (2, 4, 6, 10 and 50 μm gate length) devices. An improved mobility-profiling technique has been employed. The I–V characteristics have been analysed by a numerical model, which allows the μ and n profiles to be explicitly entered. In general, for the samples studied here, we find that n profiles are much more important than μ profiles in determining FET (field effect transistor) behaviour. The μ profiles are typically rather flat from 0.05 to 3.0 μm, varying from about 3500–4500 $\mathrm{cm}^2\,\mathrm{V}^{-1}\,\mathrm{s}^{-1}$.

It has been conjectured that mobility (μ) profiles may play an important role in GaAs MESFET (metal-semiconductor FET) behaviour. Recently a convenient method, involving the geometrical magnetoresistance (GMR) effect has been proposed and implemented in actual FET structures [1–4]. We have improved this technique by including finite gate current in the analysis, which allows measurements to be taken at forward gate bias [5]. Thus, the near-surface region is more accessible, with the minimum depth typically dropping by about a factor of two. The working equation, if parasitic resistances can be neglected (usually the case) is:

$$\mu_{GMR}(z) = \frac{1}{B}\left[1 - \frac{\Delta I_{SDG}(B)}{\Delta I_{SDG}(0)}\right]^{1/2} \qquad (1)$$

where $I_{SDG} \equiv (I_S + I_D)/2 - (I_G/2) f(\alpha)$, $f(\alpha) = \coth(\alpha) - \alpha^{-1}$, $\alpha = eV_{SD}/2kT$. Here I_S, I_D and I_G are the source, drain and gate currents, respectively, and $\Delta I_{SDG} = I_{SDG}(V_{SG} + \Delta V_{SG}) - I_{SDG}(V_{SG})$. The various approximations used in the derivation of equation (1), as well as a treatment of parasitic resistances, may be found in Ref. 5. The depletion depth z is determined from the usual relationship, $z = \varepsilon A/C$, but if the capacitance results are invalid, due to series resistance or abrupt-profile effects, then μ_{GMR} can still be measured as a function of V_{SG}, and the magnitudes will be correct.

* Universal Energy Systems Inc., 4401 Dayton-Xenia Rd., Dayton, Ohio 45432, USA.
† Rose-Hulman Institute of Technology, Physics Department, Terre Haute, Indiana 47803, USA.

Table 1 *I–V, C–V and mobility data for implanted fat FETs (50 × 400 μm) on different wafers*

Sample	Growth method	Specific doping	[a] Activation (%)	[b] g_m(mS mm^{-1})	[c] I_{dss}(mA)	[d] V_p(V)	[e] $\mu(\times 10^3\ cm^2\ V^{-1}\ s^{-1})$
A	LEC	none	26 ± 2	8 ± 1	4 ± 1	2.6 ± 0.1	4.2 ± 0.1
B	LEC	none	29 ± 2	10 ± 1	5.5 ± 0.5	2.0 ± 0.3	4.2 ± 0.2
C	LEC	none	23 ± 1	4.0 ± 0.5	1.5 ± 0.3	1.4 ± 0.1	3.5 ± 0.2
D	HB	Cr	19 ± 2	4.0 ± 0.2	1.5 ± 0.1	1.4 ± 0.1	4.1 ± 0.2
E	VPE	Cr (light)	22 ± 3	5.0 ± 0.5	1.8 ± 0.2	1.4 ± 0.1	4.3 ± 0.2
F	LEC	none	37 ± 8	14.0 ± 0.2	15 ± 2	4.7 ± 0.3	3.9 ± 0.3
G	LEC	Cr (light)	25 ± 15	8 ± 3	5 ± 3	2.7 ± 0.8	3 ± 1

[a] Defined, for comparative purposes, as the integrated carrier concentration from 0.1 to 1.0 μm, divided by the implant dose.
[b] Taken at $V_{SD} = 4$ V, between $V_{SG} = 0$ and $V_{SG} = -0.5$ V.
[c] Saturated current at $V_{SG} = 0$.
[d] Pinch-off voltage, as defined by the intercept of an I_D versus V_{SG} plot, at $V_{SD} = 0.2$ V.
[e] The maximum mobility in the implanted layer.

We have measured μ-profiles (by GMR) and n-profiles (by C–V), in seven, identically-processed, SI GaAs substrates, each implanted with 4×10^{12} cm^{-2}, 100 keV Si ions, and then capped with Si_3N_4 and annealed at 850°C for 15 min. The C–V, I–V and GMR results are summarized in Table 1. A 'fat' FET (50 × 400 μm) was chosen as the test structure to avoid the more likely processing problems associated with small devices. The error bars give an indication of the variation between different devices on the same wafer. Samples A and B are two wafers from the same boule, and give an indication of intra-boule variations, as well as processing variations between wafers. It is clear that the variations between boules are larger than the variations within a single wafer, or the variations within a single boule, especially if sample G is ignored. Samples F and G, from the same manufacturer, showed higher activation variations than the other crystals, although they also showed, on the average, higher conductances. Of all seven samples, only G could clearly be labelled a 'bad' performer, and it also was the only sample with a relatively low and variable mobility. Furthermore, it failed an 850°C conversion test, which probably accounts for its poor performance. Thus, from the meagre statistics of this study, it could be concluded that 'good' substrates give an FET implanted layer with a fairly flat mobility profile in the 3500–4500 cm^2 V^{-1} s^{-1} range. The actual profiles are shown in Figure 1.

In contrast to the μ-profiles, the n-profiles differ considerably from wafer to wafer, possibly due to dopant, impurity, and defect diffusion problems. Representative results are presented in Figure 1, although it should be noted that the accuracy of the C–V n-profiles may suffer from the well-known

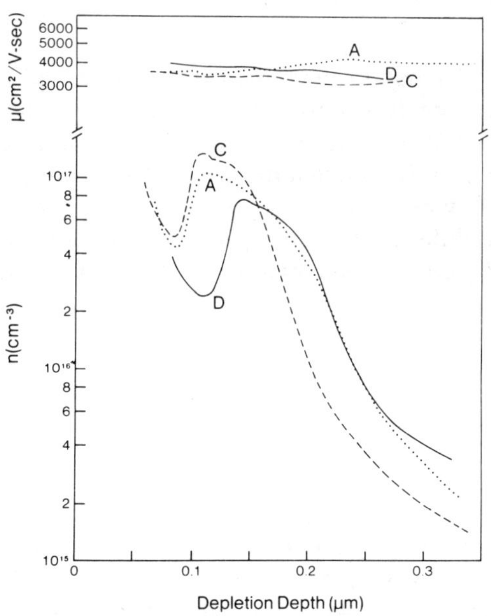

Figure 1 Mobility and carrier-concentration profiles for samples A, C and D.

series-resistance, abrupt profile and gate-current effects. By comparing the profile data in Figure 1 with the FET data in Table 1, it may be seen that n-profile variations can affect the FET parameters considerably. For example, note that samples A and C have similar activation energies, but much different pinch-off voltages, evidently due to profile differences in their respective tail regions. Similar results have been observed before, but below we present a systematic way to study such variations.

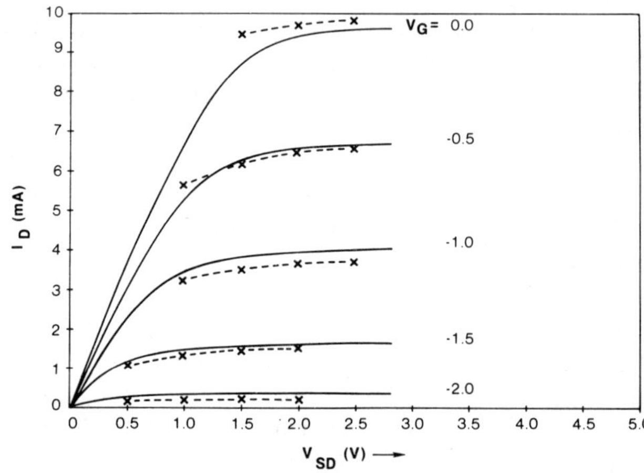

Figure 2 Experimental (solid line) and theoretical (dashed line) I–V curves for a 2 μm gate MESFET on sample A.

A theoretical model has been implemented which allows n-profiles, μ-profiles, velocity-field relationships and other pertinent data to be explicitly entered. Results for a 2×100 μm gate MESFET, fabricated on substrate A, are shown in Figure 2. It is seen that the data in the various saturation regions are fitted quite well, with no arbitrary parameters in the theory. Such good agreement gives confidence that this model can be used to test sensitivities of the various FET parameters to changes in the material parameters. Work along these lines will be continued.

ACKNOWLEDGEMENTS

We would like to sincerely thank T. A. Cooper, J. Ehret, C. Geesner, D. E. Johnson and L. Horn for help with the various aspects of this project. The work was performed at the Avionics Laboratory, Wright–Patterson AFB, Ohio under Contract F33615-81-C-1406 and F33615-82-C-1716.

References

1. Sites J. R. and Wieder H. H. (1980). *IEEE Trans. El. Dev.*, **ED-27,** 2277
2. Jay P. R. and Wallis R. H. (1981). *IEEE El. Dev. Lett.*, **EDL-2,** 265
3. Wallis R. H. and Jay P. R. (1982). In *Semi-Insulating III–V Materials, Evian,* p. 344 (S. Makram-Ebeid and B. Tuck, eds.); Shiva Publishing, Nantwich, UK
4. Ford W. and Barrera J. (1982). In *Semi-Insulating III–V Materials, Evian,* p. 352 (S. Makram-Ebeid and B. Tuck, eds.); Shiva Publishing, Nantwich, UK
5. Look D. C. and Cooper T. A. *Solid St. Electron,* in press

FET Arrays on In-Alloyed GaAs Substrates

H. V. Winston, A. T. Hunter, H. M. Olsen, R. P. Bryan and R. E. Lee

Hughes Research Laboratories, Malibu, California 90265, USA

Abstract

The threshold voltages of a FET (field effect transistor) array fabricated by direct ion implantation on an LEC (liquid encapsulated Czochralski) $In_{0.003}Ga_{0.997}As$ wafer are more uniform than for corresponding conventional LEC GaAs substrates. The values of V_{th} (the threshold voltage) are functions primarily of position on the wafer. The In-alloyed GaAs substrate contains sizeable regions of low or zero dislocation density but the dislocations do not influence V_{th} directly, either through their proximity to a gate or through local average dislocation density.

Uniformity of electrical behaviour of FET devices across a substrate wafer is a major issue for the practical realization of ICs (integrated circuits). This paper is a study of the non-uniform electrical activation of Si donors implanted to form an active layer in wafers of semi-insulating In-alloyed GaAs grown by the high-pressure LEC process from pyrolytic BN crucibles without intentional electrical doping (undoped, UD). We find that lightly In-alloyed GaAs substrates (very similar electrically to conventional LEC UD GaAs but differing in having significant regions of low or zero dislocation density [1, 2]) lead to more uniform activation and electrical behaviour than do similarly processed conventional unalloyed substrates.

We fabricated FET arrays on SI substrates by a planar process that included direct implantation of 60 keV Si ions followed by an 830°C/20 min furnace anneal with a Silox cap. The FETs, shown in Figure 1, had source-drain spacings of 5 μm and 1 μm × 5 μm gold-based Schottky barrier gates. They

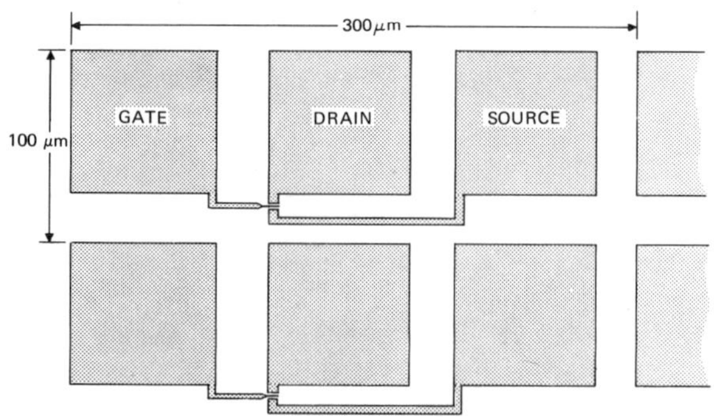

Figure 1 Two FETs of the array.

were arranged on a 100 μm × 300 μm grid, which allowed space for 75 μm square contact pads, and were electrically isolated by an oxygen damage implantation. Values of V_{th}, the threshold or pinch-off voltage, were obtained with an Accutest Model 3000 automatic d.c. parameter test system. Then we used molten KOH to reveal etch pits corresponding to dislocations; the KOH also removed the metallization but the locations of gates and pads are still clearly visible, as shown in Figure 2.

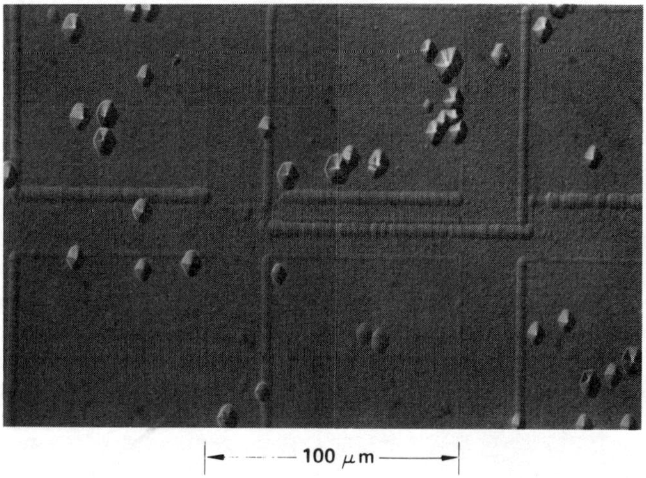

Figure 2 Etched FETs, showing pits and location of gates and pads.

Figure 3 displays V_{th} for four columns of 450 FETs in a 1.2 mm × 45 mm stripe along a ⟨110⟩ diameter of a 2 inch In-alloyed GaAs wafer. Along the diameter, dislocation density (DD) is greater than 3×10^4 cm^{-2} at the centre of the wafer (2.25 cm). Leaving the centre, DD falls below 2500 cm^{-2} at 1.5 and 3.0 cm, and to zero at 0.75 and 3.75 cm, and rises again to about 10^4 cm^{-2} close to the edge. In Figure 3 the values of V_{th} in the regions of highest (2.25 cm) and lowest (0.75 and 3.75 cm) DD are nearly equal; there is no correlation of V_{th} with DD. The four parallel columns of FETs (spaced 0.3 mm apart) exhibit remarkably similar trends of V_{th} with position. This behaviour is hard to reconcile with the report [3] that V_{th} is controlled by the distance from the FET to the nearest dislocation. We checked the effect of dislocation proximity by comparing V_{th} of FETs further from and closer to an etch pit than 100 μm. The sample of FETs was the set of 450 in one of the columns in Figure 3. As Figure 4 shows, the distribution of V_{th} is essentially the same for FETs near to, or far from, a dislocation. Setting the boundary between near and far at 50 μm or 25 μm does not change the shape of the histograms or the average value of V_{th}. We conclude that the distance to a dislocation does not influence V_{th} of a FET.

Even though dislocation distance and dislocation density seem to have no influence on V_{th} in FETs on In-alloyed GaAs, the total excursion of V_{th} values

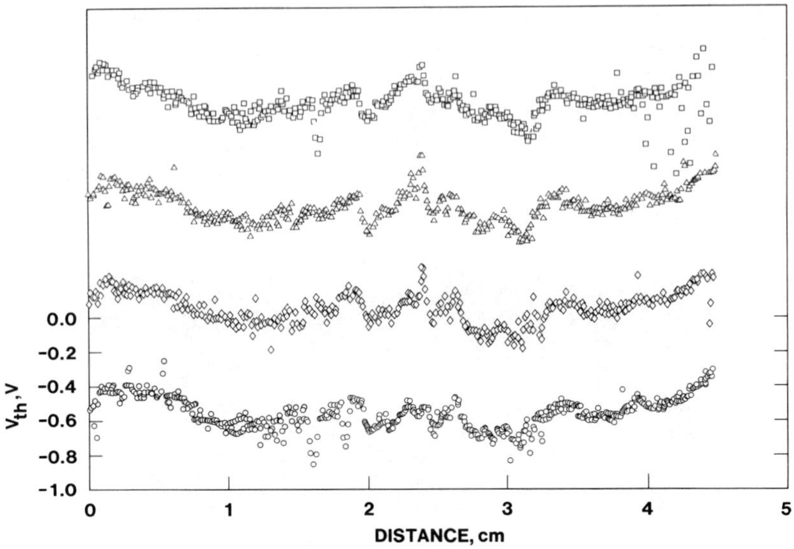

Figure 3 V_{th} versus distance along ⟨110⟩ diameter for 4 adjacent columns of FETs on an $In_{0.003}Ga_{0.997}As$ substrate implanted with 2.4×10^{12} Si^+ cm^{-2} at 60 keV. The columns are spaced 300 μm apart and the FETs in each column are 100 μm apart. Each column contains 450 FETs. The scale of V_{th} is shifted up by 0.6 V for each successive column.

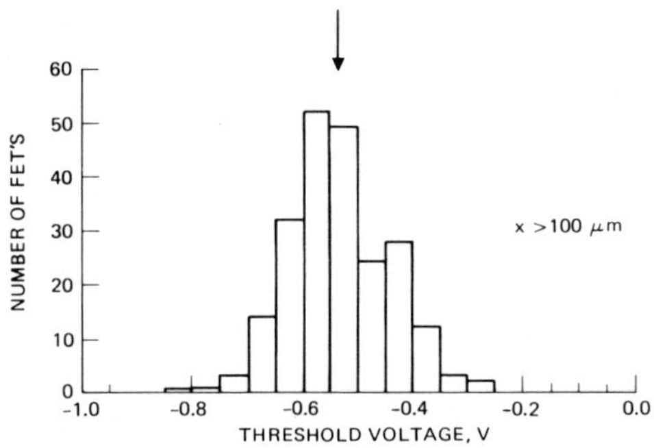

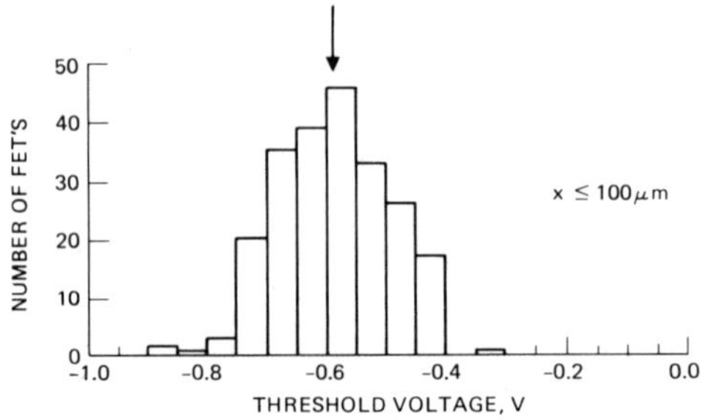

Figure 4 Histograms of V_{th} for FETs in $In_{0.003}Ga_{0.997}As$ for nearest gate-pit distance greater than and less than 100 μm. The arrows indicate the average value for each set.

across a wafer is less than half that for similar arrays on conventional LEC UD GaAs substrates with the usual range of dislocation densities across the wafer of 10^4 to 10^5 cm^{-2}. The vast majority of the FETs in Figure 3 have V_{th} values within a 400 mV range. For a correspondingly processed conventional wafer, the V_{th} values span 1 V [4]. We suggest that the improvement in uniformity may result from the fact that our In-alloyed GaAs was pulled at a slower rate than conventional unalloyed material to avoid crystal growth problems inherent in the composition difference between melt and crystal caused by segregation of In [5]. The growing crystal therefore remained at high temperature near the melting point for longer times than a conventional crystal as it was slowly withdrawn from the melt. This difference in thermal history may produce uniformity improvements like those reported for post-growth heat treatments [6].

ACKNOWLEDGEMENTS

We wish to express our appreciation for the growth of crystals by H. Kimura and C. B. Afable, wafer polishing by J. Johnston, the design of the mask set for the FET array by P. T. Greiling and R. E. Lohr, and the fabrication of FET arrays by the personnel of the GaAs Device Fabrication Laboratory, with special thanks to V. J. Miller. The work was supported in part by the Defense Advanced Research Projects Agency under Naval Research Laboratory Contract N00014-83-C-2087.

References

1. Jacob G., Duseaux M., Farges J. P., van den Boom M. M. B. and Roksnoer P. J. (1983). *J. Cryst. Growth*, **61**, 417
2. Hobgood H. M., Barrett D. L., Ta L. B., Eldridge G. W. and Thomas R. N. *1983 Electronic Materials Conference, Burlington, VT*
3. Miyazawa S., Ishii Y., Ishida S. and Nanishi Y. (1983). *Appl. Phys. Lett.*, **43**, 853
4. Winston H. V., Hunter A. T., Olsen H. M., Bryan R. P. and Lee R. E. Unpublished
5. Kimura H., Afable C. B., Olsen H. M., Hunter A. T. and Winston H. V. Unpublished
6. Miyazawa S., Honda T., Ishii Y. and Ishida S. (1984). *Appl. Phys. Lett.*, **44**, 853

GaAs MESFET Anomalies Related to Minority Carrier Injection into SI Substrate

S. Mottet and C. Le Mouellic

CNET Route de Trégastel, 22301 Lannion, France

Abstract

Side gating effects in implanted GaAs MESFETs have been characterized. From experimental results it has been found that impact ionization could occur in the channel, giving rise to non-negligible hole densities. The purpose of this paper is to describe the active layer-substrate interface behaviour and to show the influence of the hole current on the equilibrium. A physical model of the interface behaviour is proposed, which makes possible a description of the measured side gating effects.

The GaAs MESFETs (metal-semiconductor field effect transistor) we studied are fabricated on layers obtained by direct implantation of Se into Cr-doped SI substrates. In order to study the active layer-substrate interface behaviour, a side gate electrode composed of an ohmic contact on the same implanted layer is located at about 200 μm from the FET. Isolation between the FET active layer and the side gate is achieved by a boron implantation as shown in Figure 1. The influence of a side gate potential on the static characteristics of the FET is measured. When applying a negative bias V_{sub} (typically 5 V) on the side gate, a decrease in the saturation current is observed, as shown in Figure 2. This well known phenomenon is explained by an extension of the interface space charge region into the active layer when it is increased under reverse bias. But it has to be noted that when V_{ds} increases over a

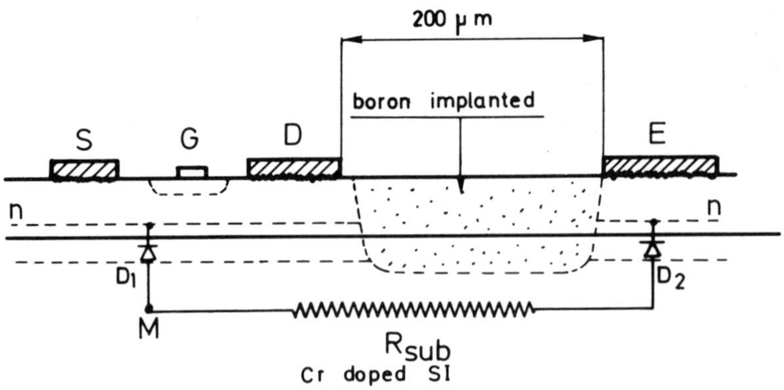

Figure 1 Schematic representation of the sample and equivalent circuit of the substrate conduction.

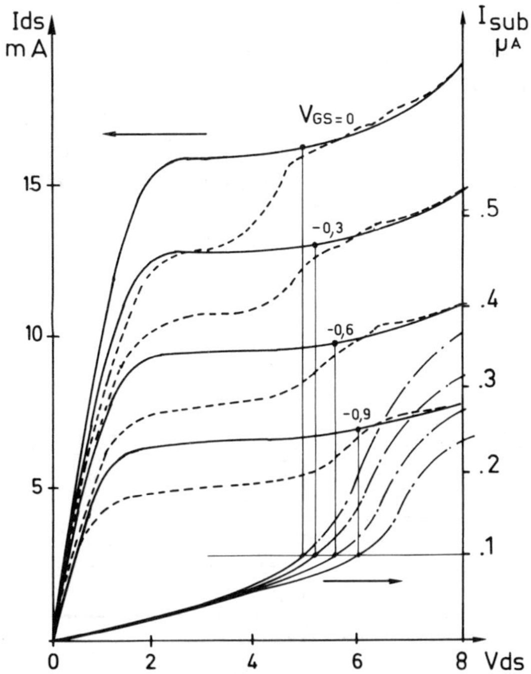

Figure 2 Static characteristics of the transistor for $V_{sub} = 0$ (solid line) and $V_{sub} = -5V$ (dashed line) and I_{sub} as a function of V_{ds} (dot-dash line).

critical value V_c, which depends on V_{gs}, the saturation current recovers the value it has for $V_{sub} = 0$; this phenomenon occurs for any V_{sub} value. The I_{sub} current, plotted in Figure 2, increases rapidly when $V_{ds} > V_c$. We find that $V_{ds} = V_c$ corresponds to $I_{sub} \simeq 100$ nA which is independent of V_{gs}. In measuring I_g (V_{ds}) in FETs, which have low surface leakage current (about 0.5 μA), we find that when increasing V_{ds} over V_c the I_g current rises to some μA. So, when $V_{ds} > V_c$ an excess current appears at the gate as well as at the side gate. Considering the transverse electric fields (under the gate and at the interface) it can only be a hole current coming from the channel. In fact, such a minority carrier current had already been found in silicon JFETs (junction FET) and MOSFETs (metal-oxide-semiconductor FET), the origin of which is attributed to impact ionization in the channel [1, 2]. In a previous paper we have shown that this generation of holes could modify the surface space charge region between source and gate when surface charge densities are present, due to poor surface passivation [3]. This current may also modify the active layer substrate interface equilibrium. As the substrate is doped by a deep acceptor it can be established that the interface acts like a diode junction. The I(V) characteristic is:

$$I(V) = I_d(\exp(qV/kT) - 1) + I_r(\exp(qV/2kT) - 1)$$

If hole generation occurs in the active region the characteristic becomes $I'(V) = I(V) - I_p$ where I_p is the hole current flowing in the surface diode. On this basis, conduction between the side gate and the active region can be described by the equivalent circuit composed of the interface diode under the side gate D_2, the SI resistance R_{sub} and the interface diode D_1 under the channel (Figure 1). In Figure 3 are plotted the I(V) characteristics of the D_1

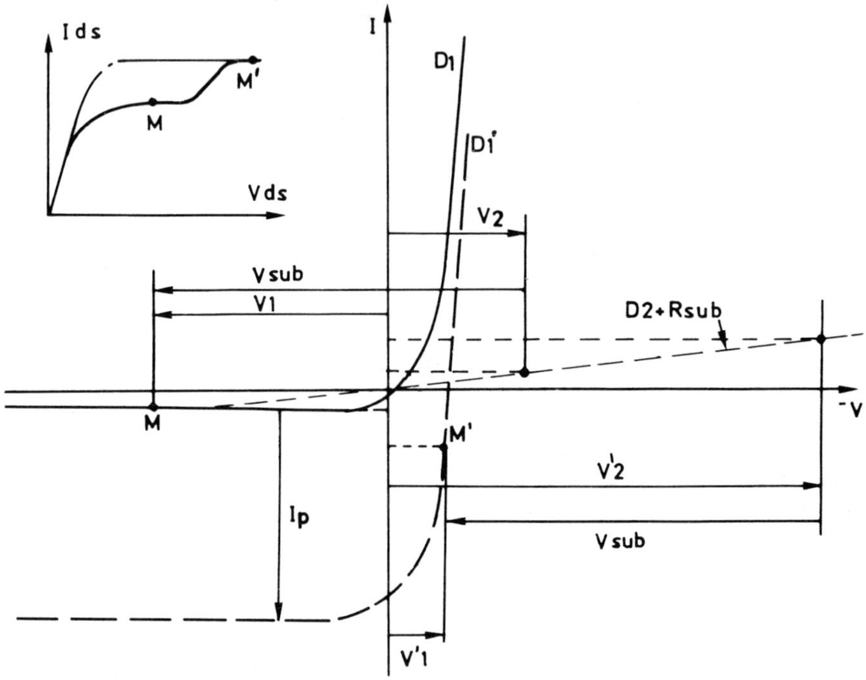

Figure 3 Equilibrium of the interface diode as a function of the hole excess density.

diode and the D_2 diode, including the series resistance R_{sub}. The equilibrium (M point) is given by $I_1 = -I_2$ and $V_{sub} = V_1 - V_2$. For a given V_{sub}, without hole generation ($V_{ds} < V_c$), V_1 is negative: the interface junction is reverse biased which causes a decrease in the channel thickness and, consequently, a decrease in the saturation current. If hole generation occurs ($V_{ds} > V_c$), the I(V) characteristic of D_1, due to the excess hole concentration becomes D'_1. The new equilibrium gives the M′ point for which V'_1 is positive: the interface diode is forward biased. The interface space charge region is minimum, which explains a maximum of I_{ds} saturation current, which becomes independent of V_{sub}.

The interface diode bias depends on V_{ds}. Therefore, when a V_{ds} pulse, higher than V_c, is applied, a forward bias pulse is applied to the interface diode which permits electron capture at deep centres in the interface region. After the pulse, emission of electrons from the centres (for example, EL2) occurs and changes the space charge density. Evidence of these phenomena has been published by Makram-Ebeid *et al.* who have performed DLTS, (deep level transient spectroscopy) measurements on I_{ds} as a function of the V_{ds} pulse height [4]. The amplitude of the induced hysteresis phenomena, when V_{ds} varies, depends strongly on the Cr and deep centre profiles, that is to say there is a large influence of the annealing conditions [5].

In conclusion, impact ionization in the channel can be large enough to generate non-negligible densities of holes. For most of the usual applications of FETs, the hole currents can be neglected, but they must be taken into account to describe the behaviour of the space charge region of the active layer substrate interface. This leads to side gate effects which are critical for integrated logic circuits and depend strongly on the impurity and defect redistribution after implantation annealing.

References

1. Fowler E. P. (1968). *Electron. Lett.*, **4,** 216
2. Mo D. L. (1970). *Proc. IEEE* (*Lett.*), **58,** 1161
3. Le Mouellic C., Mottet S., Dumas J. M. and LeCrosnier D. (1984). *Rev. Phys. Appl.*, **19,** 149
4. Makram-Ebeid S., Mitonneau A. and Laurence G. (1982). In *Semi-Insulating III-V Materials, Evian*, p. 336 (S. Makram-Ebeid and B. Tuck, eds.); Shiva Publishing, Nantwich, UK
5. Mottet S. and Le Mouellic C. (1984). *European GaAs IC, Workshop, Nice*

Mobility Profiling of Implanted GaAs Layers for MESFET Devices

C. P. Stewart, J. D. Medland and D. K. Wickenden

GEC Research Laboratories, Hirst Research Centre, Wembley, UK

Abstract

A method is outlined for calculating the low-field, low-frequency transconductance of a MESFET (metal-semiconductor field effect transistor) from the data obtained from a chemically etched van der Pauw sample. Results are presented for two implanted layers, one directly into an undoped LEC (liquid encapsulated Czochralski) semi-insulating substrate, the other into a high resistivity buffer layer grown on a HB (horizontal Bridgman) Cr–O doped substrate. Good agreement was obtained between the measured and predicted characteristics for the two layers. The higher transconductance in the buffer layer device is attributed to a higher mobility and a lower device parasitic resistance.

Conductivity and Hall voltage measurements made on clover-leaf geometry samples from which thin layers are removed by a controlled surface stripping technique allow profiles of the sheet conductivity $\sigma_s(t)$, carrier concentration n(t) and Hall mobility $\mu_H(t)$ to be obtained as a function of etch depth, t [1]. Using these data it is possible to predict the variation in the low-field, low-frequency transconductance, g_m of a MESFET made on such material provided that the full geometry of the device and the total contact resistance at the source and drain are known.

The total resistance between drain and source, R_{ds}, is equal to the sum of the channel resistance and the parasitic resistance, R_p, which is independent of the gate bias. At small V_{ds}:

$$R_{ds} = R_{con} + R_A + \frac{L}{Z\sigma_s(y)} \tag{1}$$

where the parasitic resistance is composed of two parts, the total contact resistance at source and drain, R_{con} and the access resistance from the source and drain to the channel region of the device, R_A; L is the gate length, Z the gate width and $\sigma_s(y)$ the sheet conductivity of the undepleted part of the channel which varies with the width of the depletion region y.

The transconductance at small V_{ds} is given by:

$$g_m = \frac{dI_{ds}}{dV_g} = V_{ds}\frac{d(1/R_{ds})}{d\sigma_s(y)} \times \frac{dy}{dV_g} \times \frac{d\sigma_s(y)}{dy} \tag{2}$$

The first term on the right-hand side of equation (2) may be evaluated from equation (1) and the second term by use of Poisson's equation and the differential form of the Petritz equation [1, 2], so that equation (2) normalized with respect to drain-source voltage becomes:

$$\frac{g_m}{V_{ds}} = \frac{LZ}{(ZR_p\sigma_s(y)+L)^2} \times \frac{\varepsilon_0\varepsilon_r}{y}\frac{\mu_H(y)}{r} \tag{3}$$

where r is the Hall scattering factor as discussed in Ref. 2.

In an actual device, a thickness T is removed from the surface of the material before the gate metallization (i.e. for gate recessing). If the electrical profiles are obtained from a sequentially etched sample, as in the present study, then equation (1) may be approximated using the data from the ith and (i+1)th step by:

$$R_{ds_i} = R_p + \frac{L}{Z(\sigma_{si} + \sigma_{s(i+1)})/2} \tag{4}$$

and equation (3) by:

$$\frac{g_m}{V_{ds}} = \frac{LZ}{(ZR_p((\sigma_{si} + \sigma_{s(i+1)})/2) + L)^2} \times \frac{\varepsilon_0\varepsilon_r\mu_{Hi}}{((t_i + t_{(i+1)})/2) + d_i - T} \tag{5}$$

for all $(t_i + t_{(i+1)})/2 > T$, where t_i corresponds to the thickness of material etched from the clover-leaf sample for the ith point and the term $d_i = \sqrt{(2V_B\varepsilon_0\varepsilon_r/qn_i)}$ approximates to the depletion thickness at the surface. The scattering factor r is assumed to be equal to unity. Values for μ_{Hi} and n_i are obtained from the differential form of the Petritz equations as in Ref. 1.

MESFET layers were produced by implanting (100) GaAs wafers at room temperature with $^{29}Si^+$ 4×10^{12} ions cm^{-2} at 200 keV and $^{29}Si^+$ 8×10^{12} ions cm^{-2} at 25 keV, followed by annealing under an arsine atmosphere at 850°C for 20 minutes [3]. Clover-leaf structures were prepared on pieces of material cleaved from the implanted wafers before device processing. The electrical measurements were carried out in the dark at room temperature and low magnetic fields (<0.4 T) using a fully automated apparatus controlled by a desk-top computer. Throughout the measurement the sample was positioned in an etching vessel fixed in the poles of an electromagnet, the etching, cleaning and drying of the sample between sheet measurements being carried out by the opening and closing of a series of computer-operated valves. The etch used was 2.5 g NaOH/2.5 ml H_2O_2/1000 ml H_2O, which has been used previously for GaAs Hall samples [4]. The total thickness removed from each sample was measured at the end of the run using a Dektak surface profiler. Nominally 1 μm gate length FET structures were made on the remaining portion of the slice, ohmic contacts being put down on the top surface of the material with the gate metallization being carried out after a source-drain etch [5]. The variation of g_m as a function of the drain-source conductance ($1/R_{ds}$) was measured at low V_{ds} (50 mV) using a lock-in technique at 4 kHz.

The carrier concentration n(y) and mobility $\mu_H(y)$ profiles for the two layers are shown in Figure 1. For the implant into a 3 μm high resistivity buffer layer the mobility (A) was low at the surface in the highly doped

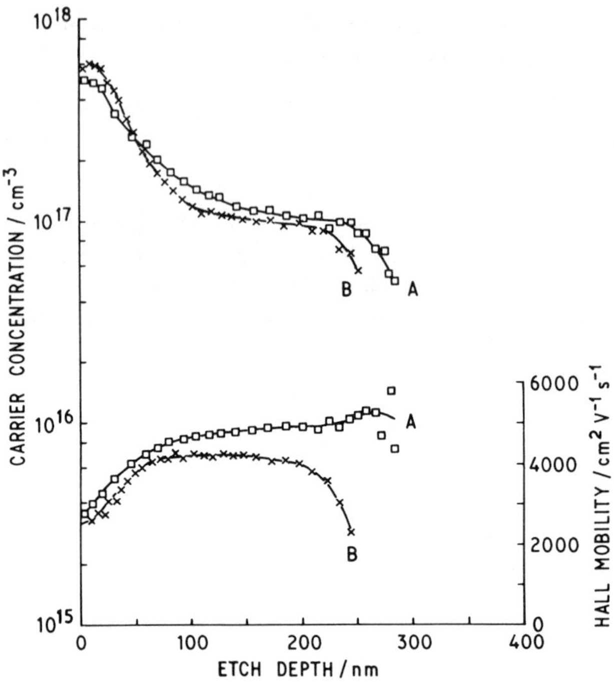

Figure 1 Graphs of carrier concentration and Hall mobility versus etch depth for a double Si implant into buffer layer (A) and into an undoped substrate (B). For implantation and anneal conditions see text.

contact region but rose steadily to a high value ($\sim 5000\ cm^2\ V^{-1}\ s^{-1}$) in the tail of the implant. The second sample produced by directly implanting into an undoped SI substrate showed a mobility profile (B) with a lower peak value ($\sim 4300\ cm^2\ V^{-1}\ s^{-1}$) and a marked fall in mobility towards the tail of the implant.

The values of g_m/V_{ds} calculated from equation (5) as a function of the drain source conductance for the two layers are shown in Figure 2, the smooth curves corresponding to the results measured on devices made on the same material. The variation in the calculated g_m/V_{ds} was found to be sensitive to the contact resistance R_{con} and gate length L. These parameters were varied to obtain the best fit to the measured data. The values used for the two devices together with the derived values of R_p are given in the caption of Figure 2. The values of R_{con} are well within experimental error. The error bars in Figure 2 correspond to the range in the calculated g_m obtained by varying the gate length by $\pm 0.1\ \mu m$ and the contact resistance by $\pm 1\ \Omega$. Measurements on these and other layers suggest that the best fit can usually be obtained by increasing the gate length to $\sim 1.2\ \mu m$. The increased value of gate length agrees well with that determined using the method of Fukui [6] for the present FET design [7].

The higher g_m obtained for the implant into the buffer layer is due to two factors: first the lower parasitic resistance R_p which is especially important

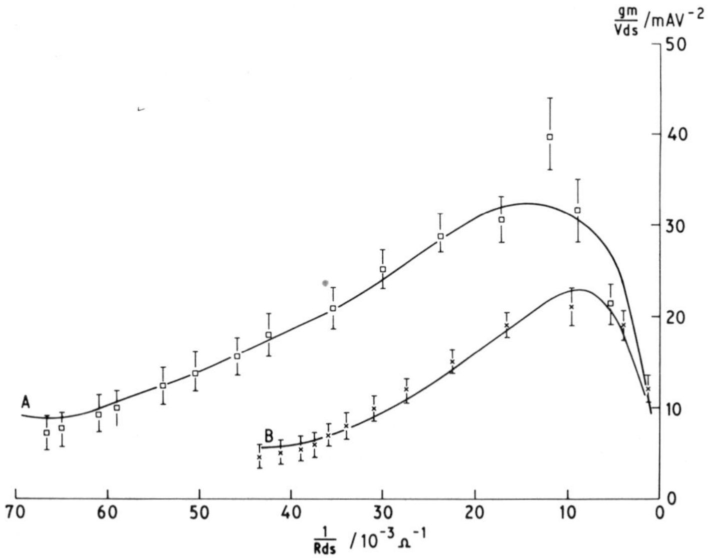

Figure 2 Graphs of calculated and measured g_m/V_{ds} versus drain-source conductance, $1/R_{ds}$, for a buffer layer FET (A) and a directly implanted FET (B). Nominal device dimensions were $L = 1$ μm, $Z = 300$ μm, source-drain spacing 5 μm. Best fit to measured data was obtained using $L = 1.2$ μm, $R_{con} = 3$ Ω, $R_p = 11.5$ Ω for device A and $L = 1.3$ μm, $R_{con} = 6$ Ω and $R_p = 18.5$ Ω for device B.

at near open-channel conditions (i.e. large $1/R_{ds}$); and secondly the high mobility in the tail of the implant, which dominates the g_m characteristic near pinch off when $1/R_{ds}$ is small.

ACKNOWLEDGEMENTS

The authors would like to thank D. C. Bartle for preparing the implanted layers, R. Tayrani for the device processing and R. H. Wallis for his encouragement and advice, particularly with regard to the device measurements.

References

1. Baron R., Shifrin G. A. and Marsh O. J. (1969). *J. Appl. Phys.*, **40,** 3702
2. Petritz R. L. (1958). *Phys. Rev.*, **110,** 1254
3. Grange J. D. and Wickenden D. K. (1983). *Solid St. Electron.*, **26,** 313
4. Esteve J., Ponse F. and Bachem K-H. (1981). *Thin Solid Films*, **82,** 287
5. Tayrani R. (1982). *Inst. Phys. Conf. Ser.*, **65,** 347
6. Fukui H. (1979). *Bell Syst. Tech. J.*, **58,** 771
7. Hughes E. Private Communication

Section 6:
Theory and impurity/defect state characterization

Recent Advances in the Theory of Deep Level Defects

G. A. Baraff

AT&T Bell Laboratories, Murray Hill, New Jersey, USA

Abstract

The importance of deep level defects to the production of semi-insulating III–V materials has led to intensive study of their properties. Concomitant with the development of experimental techniques for their study has been the development of increasingly effective theoretical techniques for calculating their properties. Much of the recent work has been reviewed by Blakemore and Rahimi [1]. In the present paper, we shall mention progress made since the completion of their review, but the main focus will primarily be on the strengths and limitations of the various approaches, with particular emphasis on the potential for contributing to the central problem of defect identification.

Deep level point defects are interesting both from a technological and a fundamental point of view. Of particular interest to this conference is their importance in compensating shallow donors and acceptors thus producing SI (semi-insulating) semiconductor substrates. Additionally, they are important as recombination centres, affecting carrier lifetimes and optical efficiencies. The wave functions of these deep level states are compact spatially, being strongly concentrated in the defect region. In some cases, their occupancy, whether they are full or empty, has remarkably little effect on the charge distribution at the centre [2–4]. In other cases, their occupancy strongly affects the strength of the bonds in their neighbourhood [5, 6] and as a result there are often large rearrangements of the atomic positions at the defect when there is a change in the charge state. This can lead to various metastable situations [7–9] and may contribute to device degradation as well. For these reasons, there has been intense interest in the characterization, identification and understanding of deep levels in semiconductors. The detailed experimental investigations have been paralleled by a marked increase in the amount of calculational and theoretical effort devoted to these systems.

The present paper is a survey of some of the recent effort in the theoretical and calculational areas. The intended audience is the experimentalist who is presumably concerned with two broad classes of phenomena. The first class consists of those phenomena which provide clues to the identification of an observed defect with a microscopic model of its structure. Included here are those properties which, in their totality, Jaros [10] has called the 'signature' of the defect. These include the ionization energy for electrons and holes, the symmetry (both spatial and spin) of the state, the internal electronic excitation energies within a defect, the various thermal and optical electronic cross-sections and the effects of external probes, such as pressure, stress, temperature and magnetic field. The second broad class is dynamic, including the defect reactions which lead to interconversion of defects, to annealing, and to a defect chemistry. The fundamental questions here involve the energy changes which, from their endothermic or exothermic nature, govern which defects should be abundant and which rare. Equally fundamental are the reaction rates. Specializing further in this direction, one needs to know how these properties are influenced by the presence or absence of free electrons or holes in the defect environment.

None of these properties can be calculated with absolute confidence, and for that reason, one wants to know which of these can be calculated at all, how reliably, and what are the prospects of doing better. In an attempt to answer these questions, I shall present the results of many calculations, side by side, so that they can be compared and conclusions reached thereby. One can assess their accuracy by seeing how well they compare with the fully self-consistent Green's function calculations carried out by one of the following four groups, namely, the IBM group of Pantelides, Bernholc, Lipari and collaborators, the Bell group of Baraff, Schluter and collaborators, the SERI group of Zunger and collaborators, and recently, a European group composed of Matthias Scheffler, who worked with the IBM group, and Giovanni Bachelet, who worked with the Bell Labs group. Although I use these calculations as a standard for reasons which will be clearer below, there is a strong limitation to what is meant by 'standard'. It certainly does not mean 'ultimate' or 'best'. The situation will now be described.

At the most fundamental level, there exists a description of the defect as a many body system of fully correlated electrons and nuclei which is too difficult to use. Therefore, approximations are essential. At each level of approximation, we throw away some aspect of reality we may later regret having had to discard. In Table 1, I have listed a series of approximations that lead to the one-electron picture, which is where much of the recent progress has been made, and the associated losses of information, which are cumulative as one reads down the chart.

Eventually one reaches the third level of approximation, the one-electron self-consistent potential, with local density and pseudopotential. This can be implemented (but very laboriously, it is true) for an isolated point defect in an otherwise perfect infinite crystal. The four groups mentioned above have implemented this approximation in three widely different approaches and, in those cases where comparison is possible, they agree well amongst themselves. For this reason, calculations of this sort do serve as a standard. When further approximations are made, i.e. beneath that row, calculated results do not agree with each other nearly as well.

Calling these calculations a standard is done with reservations for two reasons. Firstly, looking at the loss column in Table 1, we see that a number of things have been discarded. Any one of these may contain the interesting physics of the defect. In this regard, the necessity for dealing realistically

Table 1 *Approximations commonly made to reduce the defect problem to a tractable computational form*

Description	Approximation	Result	Loss
Many body, correlated motion of all the electrons and nuclei	Born-Oppenheimer, which separates the nuclear motion and electronic motion	An electronic many body hamiltonian with fixed nuclei plus a potential for the nuclear motion in each electronic state	Carrier capture cross-section
Electronic many body hamiltonian with fixed nuclei	Pseudopotential	Valence electronic many body hamiltonian with fixed nuclei	Core state information. Valence state information at nucleus, e.g. hyperfine parameters
Valence electron many body hamiltonian	Local Density	One electron self-consistent hamiltonian	Correct band gap? All correlation information, multiplet structure
One electron self-consistent hamiltonian	Empirical Potential	Tight binding, with or without self consistency	Confidence

with exchange and correlation effects is particularly important. Secondly, we are forced to make further approximations because of the laborious nature of these calculations. At present, it is not economical for us to calculate a full sequence of defects or a sequence of hosts so as to establish general trends at the full one-electron level. It is even very difficult, though it is being done, to calculate enough geometries, i.e. atomic positions, near a single defect to learn how it wants to relax in the host crystal. Therefore, there is still value in approximating further so as to get greater flexibility.

The number of calculations done at the 'standard' level is miniscule compared to the total. There are three main directions in which simplifications have been made, and virtually every other theoretical result incorporates one or more of these additional approximations. First of all, one can still retain the defect in an infinite crystal, but can opt for an empirical tight-binding description. Second, that description can be made more realistic by allowing the empirical parameters of the tight binding method to incorporate some self-consistency, but necessarily at a rudimentary level. Thirdly, one can give up on the infinite crystal and use instead a small cluster of atoms to simulate the defect. This latter approximation is especially easy (but certainly not trivial) to implement when it is combined with the further approximation of replacing the true self-consistent potential near each atom by its spherical average, and replacing the true self-consistent potential between atoms by a constant. This set of approximations is usually called the X-alpha scattered wave method.

In so far as these methods are approximations to the best of what can be done within one-electron theory, it is useful to compare them. To keep the presentation balanced, one should also compare the best of the one-electron work and reality. Unfortunately, there is no means to make a systematic comparison because, in many cases, we don't know what the reality is to compare with.

For the purpose of comparing the various one-electron calculations to the standard, let us start with a model defect which doesn't really exist, namely the ideal neutral unrelaxed vacancy in silicon. This is not a frivolous choice because, in terms of the input information needed, this is absolutely the simplest system one can have, and therefore, differences in results come about more because of differences in method than because of differences in input information. There is only one type of atom in the system, its atomic levels are known if one wants to think about the problem that way, its perfect crystal band structure is known if one wants to think about the problem that way. Each atom except the missing one is in its perfect crystal position, so the question of parameters near the missing atom being altered by distortion never arises.

Figure 1 contains the results of every calculation I could find which located the level of the neutral vacancy relative to the band gap. These are plotted according to the submission date on the publication because sometimes it is interesting to compare a calculation with what is the generally accepted truth at the time it is performed. The undistorted vacancy gives rise to localized states of two symmetry types, namely A_1, which is s-like and indicated by a circle, and T_2, which is p-like, and is indicated by a square. An open circle or square indicates that the level was calculated self-consistently, and a solid circle or square indicates that the calculation was not self-consistent.

Prior to 1975, Watkins' experimental work [11] had already indicated that the vacancy had a T_2 level which was slightly affected by a measured Jahn–Teller distortion of the defect, that that level should lie somewhere in the lower part of the gap, and that there should exist an A_1 level, unobserved, somewhere in the valence band below. Figure 1 shows that every calculation predicts a T_2 state in the gap, and that most of them also give an A_1 state, actually a resonance, not a fully localized state, within 1 eV of the top of the valence band.

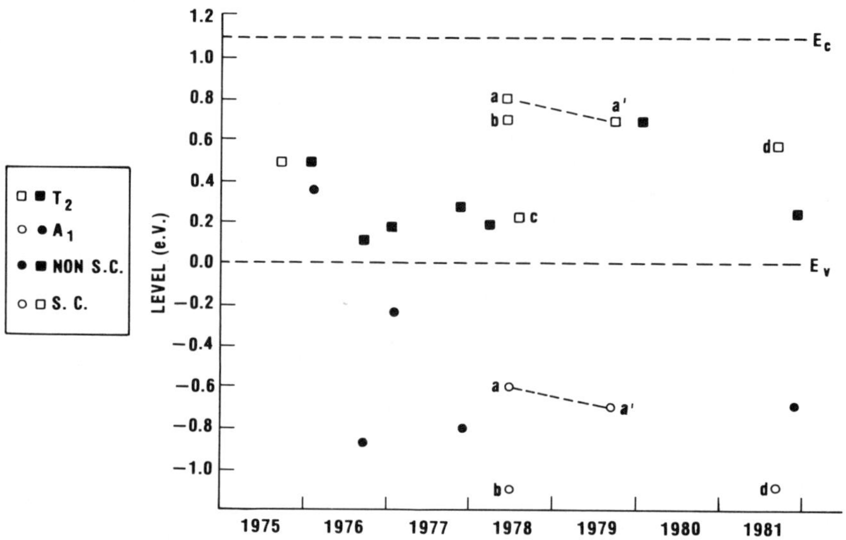

Figure 1 Electronic levels calculated for the neutral ideal vacancy in silicon versus the time at which the calculation was performed (i.e. the submission date on the published article).

There is an enormous variation in where the various calculations put that unobserved A_1 state. The very first calculation, a self consistent one, found the anticipated T_2 level to lie at 0.5 eV, but did not correctly identify the A_1 level at all. Three months later, the first tight binding empirical calculation was carried out. The parameters were adjusted to fit the perfect crystal band structure and the vacancy was produced by removing the central atom, that is, deleting all its orbitals and the matrix elements involving them. Only first nearest neighbour interactions were retained. That neglect of second nearest neighbour interactions kills the T_2-A_1 splitting, and so the two levels should really be coincident. However, the authors guessed that there should be 0.25 eV separating the A_1 and T_2 levels, and on the basis of that guess, which put their T_2 level at exactly the same place as the self-consistent result, they arrived at the values listed.

The next non self-consistent calculation, in late 1976, was carried out using the same physics as the one earlier in the year, but now second nearest neighbour interactions were retained, and now a T_2–A_1 splitting emerges. The splitting is rather smaller than the result given by the standard calculations. This is a persistent characteristic of almost every calculation of vacancies in Si and in the III–Vs. The T_2 level now appears low in the gap, and the A_1 resonance is about a volt below it. Most of the other non self-consistent calculations confirm this result.

Results obtained by the self-consistent Green's function first appear in mid-1978. Three sets of calculations appear at that time, but I think that the result marked c should be ignored because the computing power available to that group severely restricted their accuracy. One should concentrate on the results marked a, (which were altered only very slightly in the follow-up publication marked a′) and b, and on the self-consistent calculation in late 1981. These are the three calculations of the same defect by the three competing groups whose agreement leads to some confidence in the calculational technique. Notice two things. First, the self-consistent calculations place the T_2 level about 0.5 eV higher in the gap than do the empirical tight binding calculations, and secondly, the A_1–T_2 splitting is 1.4 or 1.8 eV, which is 50 to 90% larger than the tight binding result. Further calculations show that the T_2 level drops in energy when Jahn–Teller distortion is included, so the midgap T_2 level drops to its observed position, while a low T_2 level would, if allowed to Jahn–Teller distort, drop out of the gap. Agreement with experiment is especially impressive here, leading in fact to the prediction that the Si vacancy would be a negative U system, and predicting parameters which are fairly close to what was later measured [12–14].

Why does the tight binding method misplace the T_2 level by so much? The authors of a′ suggested that it is because the empirical tight binding parameters do a poor job of generating the conduction band structure. They suggested that a better set of tight binding parameters which did as good a job on the conduction bands as on the valence bands would raise the tight binding T_2 level up to where the self-consistent calculations had placed it. Their suggestion was taken up immediately [15]. The authors of the tight binding 1980 paper chose a set of parameters which also give good conduction bands, and the T_2 gap level comes out just right. This result was widely disseminated and from this time on, all tight binding calculations of other defects have been parametrized with emphasis on also getting the conduction bands right.

During this same period however, there developed a growing suspicion that things are not really as good as they appear here, and in late 1981 Talwar and Ting [16] purposely went back once again to the silicon vacancy with a set of parameters which parametrize the conduction and valance bands as well as the 1980 calculation did [15]. They found results which are very close to all the other empirical tight binding results, namely a low T_2 state and a small T_2–A_1 splitting. This shows that there is still uncertainty associated with the choice of tight binding parameters, a situation that Das Sarma and Madukhar [17] had explicitly emphasized from the tight binding point of view, and which Krieger and Lauffer [18] had been stressing from a more abstract point of view.

Figure 2 contains an example which is closer to the main subject of this volume. It is the neutral undistorted arsenic vacancy (another non-existent situation) in GaAs. The first tight binding calculation in 1975 ignored second nearest neighbour interactions and therefore obtained no T_2–A_1 splitting. All calculations put the T_2 level somewhere in the upper part of the gap. The T_2–A_1 splittings are about 1.0 eV for the non self-consistent calculations while the self-consistent calculations in 1981 and 1982 gave splittings of 1.8 to 2.0 eV. The 1979 self-consistent calculation was carried out using the scattered wave method, which does a spherical average of the potential. Zunger has pointed out at the 1982 Gordon conference on point defects that that method also underestimates the T_2–E splittings for transition metal defects in silicon. I suspect that this is the cause of the discrepancy, and this calculation should not be taken too seriously.

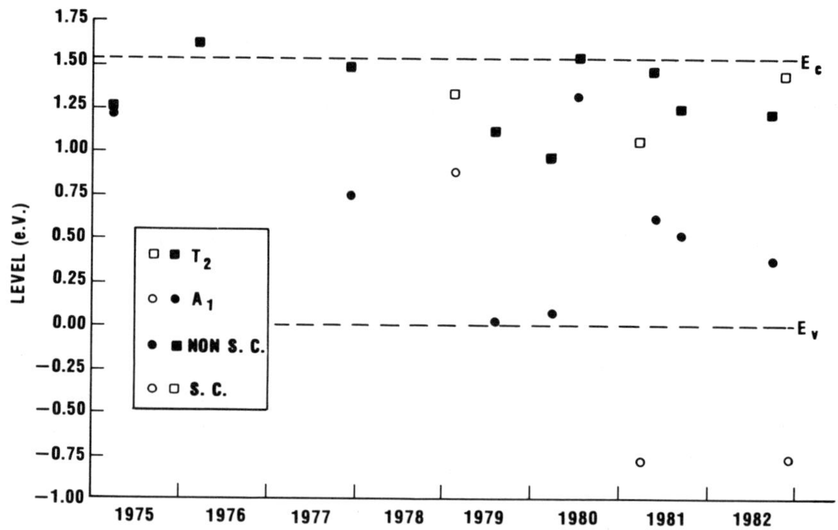

Figure 2 As Figure 1, but for the arsenic vacancy in GaAs.

Notice the large A_1–T_2 splitting in the 1982 self-consistent calculation. Unlike all the other self-consistent calculations mentioned thus far, this was accomplished with a very modest computation. It is in fact, the self-consistent tight binding calculation reported at the Amsterdam defects meeting by Van der Rest and Pecheur [19]. They understood that the small T_2–A_1 splitting of the other empiric calculations implied that the second nearest neighbour interactions at the vacancy site were too small, even though they had been chosen to give a good band structure for the perfect crystal. They argued that removing the atom to form the vacancy allowed the remaining dangling bonds on the second nearest neighbours (first nearest neighbours to the vacancy) to overlap more, increasing the T_2–A_1 splitting, and they gave a chemical method for estimating the expected enhancement. This tight binding calculation indicates that there is danger in taking tight binding parameters from one situation to another quite different one. Differences in bonding situations can be significant on the scale that defect energies must be known.

In Figure 3, calculations of another defect of particular interest are summarized, the arsenic antisite in GaAs, which has a suspected relationship with EL2. The first calculation of its neutral level, in early 1981, is by Lin-Chung

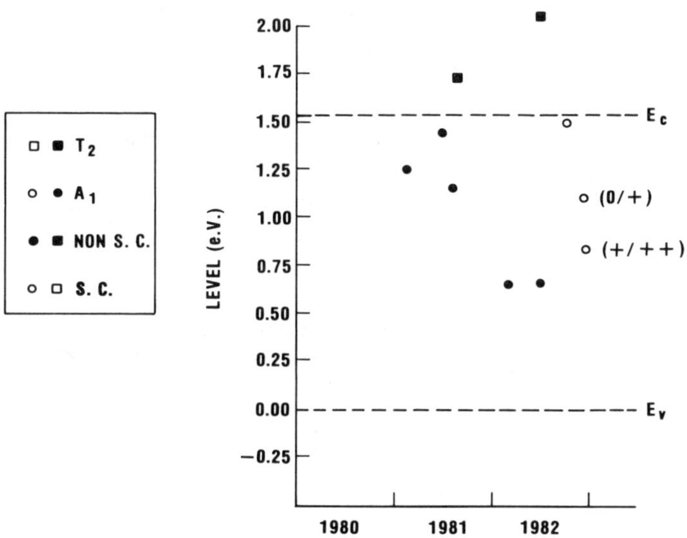

Figure 3 As Figure 1, but for the arsenic antisite in GaAs.

and Reinecke [20]. By mid-1982, when the level had been measured at 0.75 eV [21], they comment that their calculated value of 1.25 for the antisite rules out the identification of the antisite with EL2. Two other non self-consistent calculations in mid-1981 also place the level rather high. Two non SCF calculations in 1982 give a level low enough to be consistent with EL2. The one self-consistent Green's function calculation of the antisite, done by Bachelet, Schluter and Baraff, being fully self-consistent, is able to calculate both the single and double donor levels. The spacing between them is in exact accord with experiment, but both of them lie too high by 0.3 eV. This is a bit high, and perhaps allowing the adjacent atoms to relax might bring the calculated level into better agreement. I have some doubts that this will be enough.

Let us now consider another example where even the best of the self-consistent Green's function calculations was seemingly in disagreement with experiment, because the ultimate resolution of this paradox does provide an informative example of the utility of a strong interplay between theory and experiment. The example is the identification of the gallium vacancy in GaP

[22–24]. First, however, recall some history in the deep level defects field which took place before the present era of computation had begun. The early theorists were very concerned with the effects of electron correlation and with multiplet structure [25, 26]. They were thinking of the defects as being composed of atoms, and on the scale of energy of interest, the atomic correlation energies they had in mind would indeed be large. For this reason, the early calculations were done on the smallest possible of systems, the so called defect molecule, composed of the four dangling orbitals nearest the vacancy in diamond, which was the example they were studying.

This scheme allowed them to use some of the techniques that theoretical chemists use to study correlation in small molecules, but it made it completely impossible to calculate levels relative to band edges. The hope was that this scheme would at least give the right ground state symmetry and perhaps the symmetry and ordering of excited states. The results were inconclusive to poor, and this line of approach came to a sudden end with George Watkins' experiments and the theoretical framework he proposed [27].

The framework he proposed for the s–p defects ignored spin and correlation altogether, except for the rule that electrons go into the lowest energy states available which, in accord with the Pauli principle, means with paired spins, two to a state, when there are enough electrons. If, under this scheme, a T_2 state, which is 6-fold degenerate, was only partly occupied, then a Jahn–Teller distortion would occur to remove the degeneracy, see Figure 4. This scheme served to explain a vast amount of data and laid the framework for the modern period of calculations, based on the one electron theory, where spin and correlations are neglected.

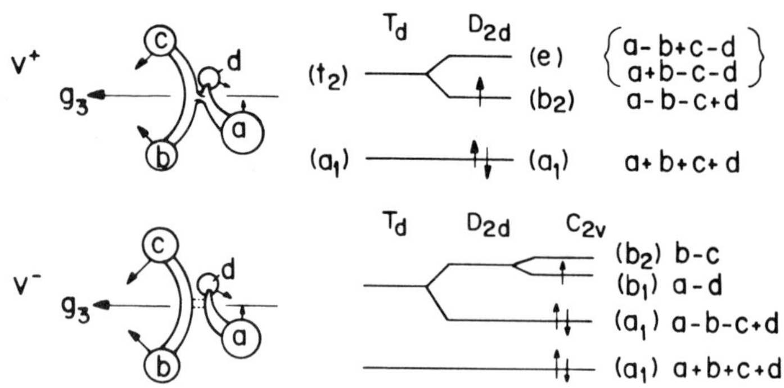

Figure 4 Watkins' description of Jahn–Teller splitting for an s–p vacancy in a tetrahedral host. The particular example here is for the silicon vacancy.

When Kennedy and Wilsey [22] first observed the ESR (electron spin resonance) signal of a defect they identified as the gallium vacancy in GaP, the signal they found indicated that there was no Jahn–Teller distortion associated with the defect. This indicated, according to the one electron theory, that the T_2 state which should be associated with the vacancy had to be either completely full, containing 6 electrons, or completely empty. The only way that this situation could occur was either that the defect was in an exceedingly unlikely state of triple negative charge, or that the ordering of the levels in the gap was such that the A_1 level was above the T_2 level, a situation that most theorists felt was highly unlikely. Finally, the IBM group [23] calculated the electronic structure of the defect and confirmed what most theorists thought, namely that the T_2–A_1 ordering was normal, with the T_2 in the gap and the A_1 completely filled beneath it.

On the basis of this discrepancy, the IBM group concluded that the alleged vacancy really had to be something else, and they suggested some defects which would give an A_1 state in the gap instead, and thus be compatible with the fully symmetric situation as observed. Kennedy and Wilsey were unwilling, on chemical identity grounds, to accept any of the IBM suggestions for alternative defects, and this situation remained unresolved for about three years. Finally Kennedy *et al.* [24] were able to do stress experiments which showed that the spin of the state was $\frac{3}{2}$, which meant that three electrons had gone in with all three spins lined up, obeying Hund's rule, with one electron in each of the px, py, and pz orbitals, a situation which should indeed be symmetric as observed. Thus, the calculations of the IBM group were perfectly correct, the identification by Kennedy and Wilsey of their defect as the gallium vacancy was correct, and the three years dispute was caused by the assumption on everybody's part that the independent one-electron picture with electron spins pairing whenever possible applied in this case. It is correlation which causes the spins to line up, and neglecting it in the one-electron theory caused the trouble.

The question of whether the spins line up is the question of the relative strength of the Jahn–Teller splittings and the correlation effects, or Hund's rule terms. There are no general answers to this, but Michel Lannoo has, in a series of papers, set up a framework to think about this question and has applied that framework to several situations [28–30].

It was a surprise that, for the gallium vacancy, the exchange energies, which favour the high spin state, were stronger than the Jahn–Teller energies, which favour the low spin state. The other extreme, where Hund's rule effects are known to be important, is found in the compact d states of the transition metal ions when they are incorporated into the semiconductor as substitutional or interstitial impurities. The basic theoretical framework for dealing with these systems was laid down five years ago by Hemstreet and Dimmock [3, 31] who, in two early papers, already revealed many of the important trends. Their approach was to calculate the electronic structure of the system using one-electron theory, and then, using those few states which were clearly associated with the d states of the transition metal impurity, they obtained matrix elements for the electron-electron interaction and, in perturbation theory, included their effects on the one-electron energies. These electron-electron terms were not calculated *ab initio*, but instead were scaled down from matrix elements that had been fit to experimental free atom spectra, the amount of scaling being determined by how strongly the free atom d-wave function contributed to the spread out d-like states they had calculated for the transition metal atom defect in the host crystal.

As far as I know, there have been no conceptual improvements in this basic idea, although there have been improvements in its implementation. DeLeo *et al.* [32] have used the same one-electron calculational scheme for silicon as Hemstreet and Dimmock had used for GaAs, namely, the X-alpha method, where the crystal is reduced to a small cluster of atoms, and the potential near the impurity is replaced by its spherical average, and they used the Hemstreet–Dimmock scheme to include many electron effects. They carried it one step further, using the Slater transition state method to obtain electron levels, in addition to the inter-defect excitation energies which the method gives [33].

Zunger has criticized the scattered wave method because the spherical averaging of the potential gives an artificially small T_2–E splitting (these are the symmetries the atomic d states take when the transition metal atom is placed in the semiconductor). This small splitting artificially favours the Hund's rule high-spin situation which is observed. His own one-electron calculations were carried out within the self-consistent Green's function method. They gave a larger T_2–E splitting and, in contrast to experiment, predict low spin for some defects which are known to give high spin [34].

I think that this failure of his prediction is a positive step forward, for the following reason: Zunger's calculational apparatus was especially designed to deal with the compact transition metal states we are now discussing, and I believe that for this purpose, it may be the most accurate method for solving the one-electron equations that exists. Therefore, (discounting the possibility of relaxation of the nearest neighbour atoms), the difference between reality and Zunger's one-electron calculations must be a measure of the specific electron-electron correlations. At the present time, there is no really good understanding of electron correlations in localized states immersed in extended systems, other than the sort of rough and ready scaling that Hemstreet and Dimmock had proposed. This suggests that Zunger's one-electron calculations might be used reliably to subtract off the one-electron terms from the complicated reality, and arrive thereby at what a correct theory of correlations should produce. Fazzio *et al.* have already reported a start in this direction [35]. They find, quite convincingly, that although defect levels do not change monotonically with smooth changes in host ionicity or defect valence, the parameters which describe the underlying many-electron effects do exhibit the smooth variation. An advance in understanding the many-electron effects in general would have far-reaching implications for the understanding of transition metal impurities in GaAs, where they may again become interesting for their value in producing semi-insulating material.

One of the other significant defect signatures which has been studied recently is the photoionization cross-section. I have listed some of the papers which I found interesting in this field [36–43]. None of them purport to be realistic calculations. They are all model calculations which, if read carefully, tell something about the various approximations which one could use to make a realistic calculation tractable. I read the results rather pessimistically because, taken as a whole, they invalidate many of the simplifying approximations that one would like to make. I really feel that, at present, the experimenter has as much theoretical information from the knowledge of the calculated gap levels and sharp resonances as he would have from the calculated cross-section, even if it could be calculated well.

There is one final area in deep level theory which has emerged recently and promises to be important in adding to our ability to help with defect identification. Up till now, I have not stressed the fact that all of the calculations of electronic structure I mentioned have been carried out with the atoms which constitute the defect and its neighbourhood residing at preassigned locations. In most cases, these locations are either the perfect crystal locations, or distortions whose symmetry may have been suggested by experiment but whose amplitudes are usually unknown. Occasionally, a series of calculations will be made with a variety of atomic trial locations just to see how sensitively the calculated levels depend on the atomic positions. The recent advance which, potentially can remove the uncertainty associated with this unwanted degree of freedom is the development of force and total energy programs [44–48].

With their use, it becomes possible to adjust the atoms to the position of lowest energy, or what is the same thing, to the position where each atom is acted upon by zero force [48]. The levels calculated from these starting points are, in many cases, the physically relevant ones. Calculations in which the total energy is computed have the additional power of being able to add to our knowledge of defect reactions and barriers, how these might be altered by the presence of minority carriers, such as occurs under injection or optical illumination [49, 50]. Moreover, they will make it possible to say something about the chemical heats of defect reaction, a subject which has not been updated since Van Vechten's thermochemical work [51, 52].

Clearly, a paper of this length cannot be a review of all the significant recent work in deep level theory. Many of the topics alluded to as being of interest to the experimentalist have not been mentioned here simply because, in my judgement, reasonable calculational techniques have yet to be demonstrated. Equally obviously, there has been progress in some narrowly defined areas. I have described these as best I could. If I seemed to dwell overly on the shortcomings of the theory, it is because I feel that, while there have been some really useful and original contributions which originated with the theorists, I also feel that the theoretical work has not been subject to the scrutiny that hindsight can give. I felt that, after a decade of real progress in the deep level theory, the time was ripe for such a backward look, provided that it was directed at techniques, and not at the practitioners.

References

1. Blakemore J. S. and Rahimi S. (1984). In *Semiconductors and Semimetals Volume* 20 (R. K. Willardson and A. C. Beer, eds.); Academic Press, New York, p. 233

2. Haldane F. D. W. and Anderson P. W. (1976). *Phys. Rev.*, **B13,** 2553
3. Hemstreet L. A. and Dimmock J. O. (1978). *Phys. Rev.*, **B20,** 1527
4. Zunger A. and Lindefelt U. (1982). *Phys. Rev.*, **B26,** 5789
5. Baraff G. A., Kane E. O. and Schluter M. (1981). *Phys. Rev. Lett.*, **47,** 601
6. Hauenstein R. J., McGill T. C. and Feenstra R. M. (1984). *Phys. Rev.*, **B29,** 1858
7. Lang D. V. and Logan R. A. (1977). *Phys. Rev. Lett.*, **39,** 635
8. Vincent G. and Bois D. (1978). *Solid-State Commun.*, **27,** 431
9. Levinson M., Stavola M., Benton J. L. and Kimerling L. C. (1983). *Phys. Rev.*, **B28,** 5848
10. Jaros M. (1982). *Deep Levels in Semiconductors*; Adam Hilger Ltd., Bristol
11. Watkins G. D. (1973). *Radiation Damage and Defects in Semiconductors*, p. 228 (J. E. Whitehouse, ed.); Institute of Physics, London
12. Baraff G. A., Kane E. O. and Schluter M. (1979). *Phys. Rev. Lett.*, **43,** 956
13. Baraff G. A., Kane E. O. and Schluter M. (1980). *Phys. Rev.*, **B21,** 5662
14. Newton J. L., Chatterjee A. P., Harris R. D. and Watkins G. D. (1983). *Physica*, **116B,** 219
15. Popaconstantopoulos D. and Economu E. N. (1980). *Phys. Rev.*, **B26,** 5689
16. Talwar D. N. and Ting C. S. (1982). *J. Phys. C.*, **15,** 6573
17. Das Sarma S. and Madukhar A. (1980). *J. Vac. Sci. Technol.*, **17,** 1120; (1981). *Phys. Rev.*, **B24,** 2051
18. Krieger J. B. and Lauffer P. M. (1981). *Phys. Rev.*, **B23,** 4063
19. Van der Rest J. and Pecheur P. (1983). *J. Physique*, **44,** 1297
20. Lin-Chung P. J. and Reinecke T. L. (1981). *J. Vac. Sci. Technol.*, **19,** 443
21. Weber E. R., Ennen H., Kaufmann U., Windscheif J., Schneider J. and Wosinski T. (1982). *J. Appl. Phys.*, **53,** 6140
22. Kennedy T. A. and Wilsey N. D. (1978). *Phys. Rev. Lett.*, **41,** 977
23. Scheffler M., Pantelides S. T., Lipari N. O. and Bernholc J. (1981). *Phys. Rev. Lett.*, **47,** 413
24. Kennedy T. A., Wilsey N. D., Krebs J. J. and Stauss G. A. (1983). *Phys. Rev. Lett.*, **50,** 1281
25. Coulson C. A. and Kearsley M. J. (1957). *Proc. R. Soc.*, **241A,** 433
26. Yamaguchi T. (1962). *J. Phys. Soc. Jap.*, **17,** 1359
27. Watkins G. D. (1975). *Inst. Phys. Conf. Ser.*, **23,** 1
28. Lannoo M. (1979). *Inst. Phys. Conf. Ser.*, **44,** 1
29. Lannoo M., Baraff G. A. and Schluter M. (1981). *Phys. Rev.*, **B24,** 913
30. Lannoo M., Baraff G. A. and Schluter M. (1981). *Phys. Rev.*, **B24,** 943
31. Hemstreet L. A. and Dimmock J. O. (1979). *Solid-State Commun.*, **31,** 461
32. DeLeo G., Watkins G. D. and Fowler W. B. (1982). *Phys. Rev.*, **B25,** 4962
33. DeLeo G., Watkins G. D. and Fowler W. B. (1982). *Phys. Rev.*, **B25,** 4972
34. Zunger A. and Lindefelt U. (1982). *Phys. Rev.*, **B26,** 5989
35. Fazzio A., Caldas M. and Zunger A. (1984). *Bull. Am. Phys. Soc.*, **29,** 250
36. Blow K. J. and Inkson J. C. (1980). *J. Phys. C*, **13,** 359
37. Blow K. J. and Inkson J. C. (1982). *J. Phys. C*, **15,** 3711
38. Blow K. J. and Inkson J. C. (1983). *J. Phys. C*, **16,** 5267
39. Burt M. G. (1980). *J. Phys. C*, **13,** 1825
40. Dzwig P., Crum V. and Chantre A. (1982). *J. Phys. C*, **15,** L1177
41. Dzwig P., Burt M. G., Inkson J. C. and Crum V. (1982). *J. Phys. C*, **15,** 1187
42. Banks P. W., Brand S. and Jaros M. (1980). *J. Phys. C*, **13,** 6167
43. Jaros M. and Banks P. W. (1982). *J. Phys. C*, **15,** 5965
44. Baraff G. A. and Schluter M. (1983). *Phys. Rev.*, **B28,** 2296
45. Bar Yam Y. and Joannopoulos J. (1984). *Phys. Rev. Lett.*, **52,** 1129
46. DeLeo G., Fowler W. B. and Watkins G. D. (1984). *Phys. Rev.*, **B29,** 3193
47. Car R., Kelly P. J., Pantelides S. T. and Oshijawa A. (1984). *Bull. Am. Phys. Soc.*, **29,** 250
48. Scheffler M., Vigneron J. P. and Bachelet G. B. (1982). *Phys. Rev. Lett.*, **49,** 1765
49. Troxell J. R., Chatterjee A. P., Watkins G. D. and Kimerling L. C. (1979). *Phys. Rev.*, **B19,** 5336
50. Baraff G. A., Schluter M. and Allan G. (1983). *Phys. Rev. Lett.*, **50,** 739
51. Van Vechten J. A. (1975). *J. Electrochem. Soc.*, **122,** 419
52. Van Vechten J. A. (1975). *J. Electrochem. Soc.*, **122,** 423

Investigation of Donors Shallower than EL2 in LEC GaAs

A. T. Hunter, J. P. Baukus, O. J. Marsh, H. V. Winston,
M. H. Young and R. Hart*

Hughes Research Laboratories, Malibu, California 90265, USA

Abstract

Donors shallower than EL2 are important in controlling the compensation in undoped, semi-insulating GaAs grown by the LEC (liquid encapsulated Czochralski) method. The concentration of such donors is typically in the low $10^{15}\,cm^{-3}$ range. Temperature dependent Hall measurements and DLTS measurements on initially semi-insulating material doped n-type by neutron transmutation doping reveal the presence of two such levels, one at 0.2–0.3 eV (Hall and DLTS), the other at ~0.6 eV (DLTS). The combined concentration of these donors is $>10^{15}\,cm^{-3}$. However, there are insufficient donors shallower than EL2 to correspond to the first ionization of a hypothetical double donor the second ionization of which would correspond to EL2. The high room temperature mobility of our semi-insulating material (6000–7000 $cm^2\,V^{-1}\,s^{-1}$) is consistent with a relatively small concentration of ionized impurities or defects.

Introduction

A great deal of progress has been made in the last few years in developing a model for the electronic levels in semi-insulating LEC-grown GaAs [1, 2]. High resistivity is maintained by the presence of a partially occupied state near midgap. In undoped material, the partially occupied level is EL2 [1], a donor level located 0.77 eV from the conduction-band edge. Partial occupation of the level requires that acceptors in the lower half of the gap outnumber donors shallower than EL2. The acceptor that compensates those shallower donors in undoped material grown from PBN crucibles is apparently carbon [2, 3].

The concentration and distribution of the shallower donors may have important effects on the electrical properties of SI material. (In this paper we use ‘shallow donor’ to refer to any donor shallower than EL2. The term ‘effective-mass donor’ is used to identify the very shallow donors 6 meV [4] from the band edge.) This study is designed to investigate the influence of these donors on properties such as the mobility and the uniformity of the carrier concentration.

* Texas A & M University, College Station, Texas 77843, USA.

Experimental

All material used in this study was grown in our laboratory by the LEC technique in PBN crucibles. The melt was nearly stoichiometric, and was undoped. The water content of the B_2O_3 encapsulant was varied to provide a range of carbon concentrations [3] from <0.5 up to 8×10^{15} cm^{-3}.

We irradiated several samples from one crystal with thermal neutrons to dope the material with Ge and Se [5]. Various irradiation times were employed to provide a range of added Ge and Se. The concentration of atoms transmuted in each sample was calculated by measuring the activity of Fe standards irradiated along with the GaAs. Known values of nuclear cross-sections were then employed to derive the concentration of Ga transmuted to Ge, and of As transmuted to Se. Samples were subsequently annealed at 800°C for 1 h to remove damage created by the irradiations.

Measurements of the strength of the local vibrational mode absorption [6] to obtain the carbon concentration were made at 77 K, unless otherwise noted. We used our own calibration [3] to convert absorption to concentration: for an $\alpha\Delta$ product of 1 cm^{-2} (where α is the maximum absorption, and Δ the width at half maximum, both in cm^{-1}) we would calculate 1.1×10^{16} cm^{-3} carbon.

Results

Values for the compensation were derived from an analysis [7] of temperature dependent Hall effect measurements on SI samples. Figure 1 shows the

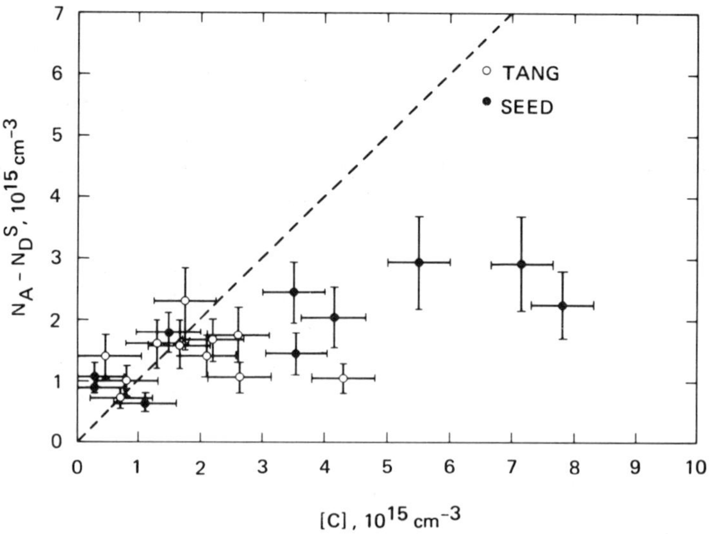

Figure 1 $N_A - N_D^S$, determined from temperature dependent Hall measurements, plotted against carbon concentration, determined from infrared absorption. The dashed line would be the value of $N_A - N_D^S$ if carbon were the only acceptor, and no shallow donors were present.

compensation, for those samples showing an 0.77 eV activation energy, plotted against the carbon concentration determined from infrared absorption. Compensation agrees approximately with carbon content for carbon below 2×10^{15} cm^{-3}, but is consistently lower for higher carbon values.

As shown in Table 1, n and $N_A - N_D^S$, the compensation, vary across a wafer. Most of the difference in n is caused by a change in compensation (see Samples 1 and 5 in Table 1). The difference in compensation is not quite large enough by itself to account for the increase in n from edge to centre; an EL2 increase from edge to centre by a factor of 1.4 is implied. Carbon content is nearly constant across an adjacent wafer at 4.3×10^{15} cm^{-3}, suggesting that variation of N_D^S causes the change in compensation. Absorption measurements at 1.0 μm confirm an increase of EL2 from edge to centre, though only by about 20%.

Table 1 *Carrier concentration, mobility and compensation measured for samples cut from a diameter of a ~60 mm wafer of undoped GaAs*

Sample	n (cm^{-3})	μ (cm^2 V^{-1} s^{-1})	$N_A - N_D^S$
1 (Edge)	2.14×10^7	6572	1.65×10^{15} cm^{-3}
2	2.98×10^7	6759	
3	3.62×10^7	7066	
4	3.33×10^7	6798	
5 (Centre)	4.83×10^7	6866	1.06×10^{15} cm^{-3}
6	Not a Hall sample		
7	3.33×10^7	6751	
8	3.52×10^7	6699	
9	3.22×10^7	6876	

Another means of investigating donors shallower than EL2 is the analysis of initially SI samples doped by neutron transmutation, which form Ge and Se. An analysis [5] of high dose data (up to 4.9×10^{17} cm^{-3} added Ge plus Se) shows that only a small fraction of the Ge is incorporated as an acceptor. For these samples, the electron concentration after annealing is within 5% of the concentration of Ge and Se added by transmutation. Therefore, we can use transmutation doping to add a known uniform concentration of effective-mass donors to our initially SI samples. At lower doses the resulting carrier concentration also depends significantly on the number of ionized deeper donors that must be filled before added electrons go into the conduction band.

Figure 2 shows temperature dependent Hall results for a sample that became n-type conducting after neutron irradiation and annealing. We used a computer fitting routine to vary the parameters in the charge balance equation to reproduce the observed data. The data below ~80 K were not used for this analysis, since the charge balance equation we used does not

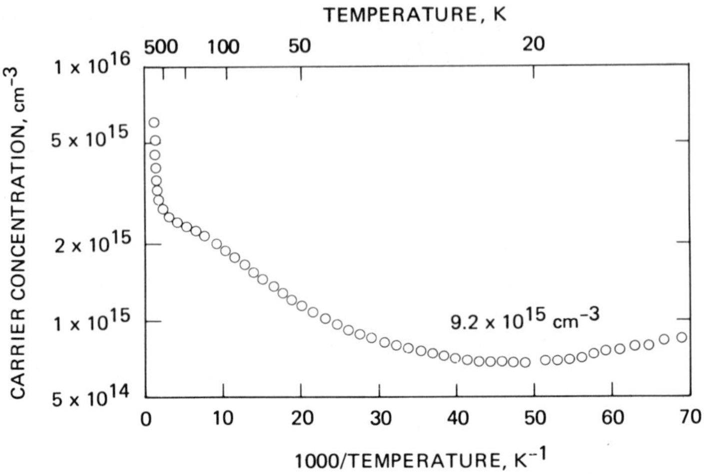

Figure 2 Carrier concentration versus inverse temperature for a sample of initially SI GaAs transmutation doped with 9.2×10^{15} cm^{-3} Se+Ge. For the range in 1000/T from ~2 to 7 K^{-1}, only about one quarter of the measured data were plotted, to avoid overlap of points.

take into account the impurity-band conductivity responsible for the apparent rise in n at lower temperatures. Successful fitting required three levels. The first was a very deep level, needed to explain the steep increase in carrier concentration above 700 K. We set its activation energy at the EL2 value of 0.77 eV. An effective-mass donor level (with an activation energy of 0.006 eV) was needed to explain the slight decrease in carrier concentration below room temperature. A third level was required to reproduce the carrier concentration between 500 and 700 K.

The best fit to the data under the assumptions listed above was achieved with parameters shown in Table 2. The root mean square error for the fitting procedure was 0.8%, and the maximum error was 2%, indicating a very good fit to the data. The concentration of the effective-mass donors, 8.1×10^{15}, agrees fairly well with the concentration of Ge plus Se added to the sample.

Table 2 *Activation energies and concentrations of three different donor levels*

Level	E (eV)	N (cm^{-3})
1	0.006	8.1×10^{15}
2	0.22	8.4×10^{14}
3	0.77	5.9×10^{16}
A		5.5×10^{15}

Total acceptor (A) concentration determined from fitting data measured on sample with 9.2×10^{15} cm^{-3} Se plus Ge added by neutron transmutation doping.

The EL2 concentration is very close to the value of 5.6×10^{16} cm^{-3} determined by temperature dependent Hall analysis in an unirradiated SI sample from the same wafer.

We also ran DLTS measurements on the same sample. With a rate window of 30 ms, we observed a peak at 139 K which we tentatively correlate with the 0.22 eV level observed in the Hall effect measurement. A second peak appears at 250 K with a concentration of $\sim 10^{15}$ cm^{-3}. Its activation energy ($\sim$0.6 eV) is probably close enough to that of EL2 so that it could not be observed as a separate level in the Hall effect analysis at such a low concentration.

Since donors shallower than EL2 are ionized in SI material, the presence of these donors should affect the mobility in our samples. The third column in Table 1 shows the room temperature mobility in the eight samples used for the uniformity study discussed earlier. All of the mobility values are above 6500 cm^2 V^{-1} s^{-1}. These are typical for samples cut from the tail end of our SI crystals. Seed-end mobilities, however, tend to be lower and are often in the 5000–6000 cm^2 V^{-1} s^{-1} range.

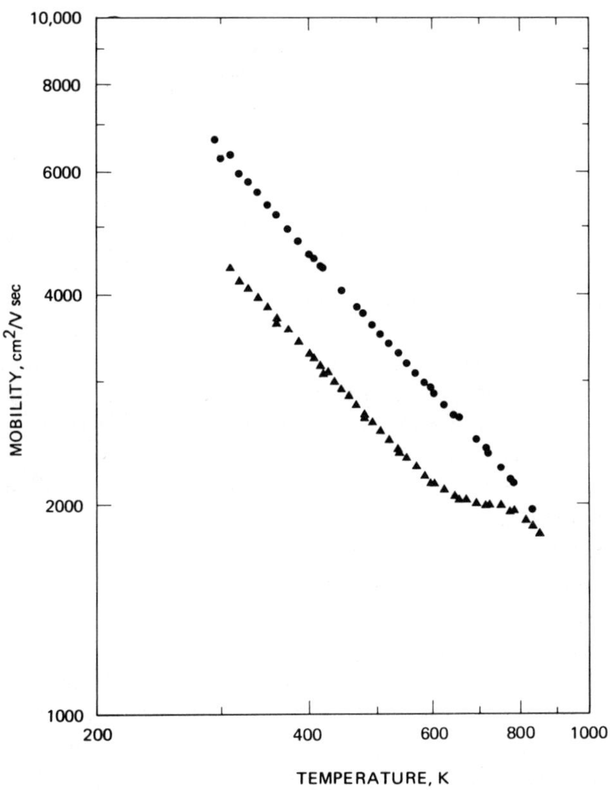

Figure 3 Mobility versus temperature (both on log scales) for two different SI samples. The lower mobilities were measured on a sample from the seed end of a crystal, the higher mobilities on a sample from the bottom of the same crystal.

To gain further insight into possible reasons for this seed-to-tail difference in mobility, we plotted the mobility as a function of temperature in Figure 3 for samples from the seed and tail of a SI crystal. The mobility at the tail decreases smoothly from 6700 $cm^2\ V^{-1}\ s^{-1}$ at room temperature to 2000 $cm^2\ V^{-1}\ s^{-1}$ at 850 K. The data for the seed sample, with a much lower room temperature mobility of 4800 $cm^2\ V^{-1}\ s^{-1}$, are very different. Rather than asymptotically approaching the higher mobility curve, the lower curve parallels the first up to ~600 K, then shows almost no temperature dependence until it reaches 800 K. Above 800 K, the two curves approach the same values. While details vary, similar trends are observed in samples from other crystals.

Discussion and conclusions

The comparison of compensation with carbon concentration made in Figure 1 implies the presence of a few times $10^{15}\ cm^{-3}$ shallow donors in many of our SI samples. So long as the activation energy of the samples is the same (all are 0.77 eV), the room temperature carrier concentration is proportional to:

$$\frac{N_{EL2}-(N_A-N_D^S)}{N_A-N_D^S}.$$

Variations in N_D^S can therefore lead to non-uniformities in the electron concentration. The data in Table 1 provide a specific example. While the average carbon concentration was determined to be $4.3\times10^{15}\ cm^{-3}$, the compensation $(N_A-N_D^S)$ is much less and varies: $1.65\times10^{15}\ cm^{-3}$ at the edge and $1.05\times10^{15}\ cm^{-3}$ at the centre. This suggests that shallower donors not only affect the carrier concentration, but non-uniformities in the shallow-donor concentration cause non-uniformities in carrier concentration. The infrared absorption measurements at 1.0 μm [8] on the optical sample are consistent with the temperature dependent Hall results in showing that EL2 variations are less important than compensation variations in controlling the electron-concentration uniformity.

The transmutation doping experiment places some restrictive upper limits on the total concentration of shallow donors. In the sample of Figure 2, the room temperature electron concentration is $2.8\times10^{15}\ cm^{-3}$. The low temperature Hall data show that the shallowest occupied level is an effective-mass donor level. This allows us to place an upper limit of $6.4\times10^{15}\ cm^{-3}$ (the difference between the added donor concentration and the room-temperature electron concentration) on the concentration of donors deeper than 6 meV that were filled with electrons by the transmutation doping and annealing. This concentration must include the ionized EL2, which was measured to be $1.5\times10^{15}\ cm^{-3}$ in the starting material. Therefore, the upper limit of the concentration of donors shallower than EL2 and deeper than the effective mass level is 4.9×10^{15} cm^{-3}. The EL2 concentration in this sample determined by temperature dependent Hall measurements is $>5\times10^{16}$ cm^{-3}. Therefore, our data indicate that the EL2 defect does not have an ionization anywhere between 0.77 eV and 6 meV, in contrast to a recent proposal [9]. In fact, if such a level were present at the concentration of EL2, its activation energy would control the carrier concentration at low temperature. This is obviously not observed. The good fit to our temperature dependent Hall data with approximately the concentration of effective-mass donors added by transmutation is evidence against an effective mass donor level at 6 meV corresponding to a first ionization of the EL2 defect. We conclude that the EL2 level at 0.77 eV is the first ionization of the corresponding defect.

Finally the mobility data are consistent with a fairly low concentration of ionized impurities in SI material. It is the samples with room-temperature mobilities between 6000 and 7000 cm^2 V^{-1} s^{-1} that show 'normal' high-temperature mobilities which agree closely with the shape and magnitude observed for high-purity low resistivity samples [10]. The kinks around 700 K in μ versus T for samples with room-temperature μ less than 6000 cm^2 V^{-1} s^{-1} suggest that some mechanism other than additional ionized impurity scattering, possibly electrical non-uniformity [11], causes the lower mobility. Therefore, we can use the higher room temperature mobility values as characteristic of the ionized impurity content in our material. These high values for the room temperature mobility in SI GaAs are not consistent [12] with a fully-ionized shallow level at the concentration of EL2.

To summarize, we have shown that donors shallower than EL2 contribute to the compensation in our SI samples. These shallow donors may contribute to the electron concentration variations we observe. Finally, the concentration of such donors is in the low 10^{15} cm^{-3} range, too low to allow for the presence of a shallow first ionization of the EL2 defect.

ACKNOWLEDGEMENTS

We are grateful to H. Kimura, C. Afable and H. Olsen for providing the SI material. We also wish to thank C. Haeussler and H. Rudolph for sample preparation and Hall effect measurements, and R. Wong Quen for the infrared absorption measurements.

References

1. Martin G. M., Farges J. P., Jacob G. and Hallais J. P. (1980). *J. Appl. Phys.*, **51,** 2840.
2. Holmes D. E., Chen R. T., Elliott K. R. and Kirkpatrick C. G. (1982). *Appl. Phys. Lett.*, **40,** 46
3. Hunter A. T., Kimura H., Baukus J. P., Winston H. V. and Marsh O. J. (1984). *Appl. Phys. Lett.*, **44,** 74
4. See, for example, Sze M. S. (1981). *Physics of Semiconductor Devices*, p. 21; Wiley-Interscience, New York

5. Hunter A. T., Winston H. V., Marsh O. J. and Hart R. R. To be published in *Proceedings of the Fifth Int. Conf. on Neutron Transmutation Doping, STP 850* (D. C. Gupta, ed.); ASTM
6. Brozel M. R., Clegg J. B. and Newman R. C. (1978). *J. Phys. D*, **11,** 1331
7. Hunter A. T. and Baron R. Unpublished
8. Martin G. M. (1981). *Appl. Phys. Lett.*, **39,** 747
9. Walukiewicz W., Lagowski J. and Gatos H. C. (1983). *Appl. Phys. Lett.*, **43,** 112
10. Blood P. (1972). *Phys. Rev. B*, **6,** 2257
11. Herring C. (1960). *J. Appl. Phys.*, **31,** 1939
12. Walukiewicz W., Pawlowicz L., Lagowski J. and Gatos H. C. (1982). In *Semi-Insulating III–V Materials, Evian*, p. 121 (S. Makram-Ebeid and B. Tuck, eds.); Shiva Publishing, Nantwich, UK

High-Frequency Transient-Resistance Spectroscopy of Deep Levels in SI GaAs

A. C. Seabaugh, M. I. Bell, R. D. Larrabee and J. D. Oliver, Jr*

Semiconductor Materials and Processes Division, National Bureau of Standards, Washington DC 20234, USA

Abstract

A new photoinduced transient-resistance technique is used to characterize deep levels in SI GaAs. In this technique, termed photoresistance deep-level transient spectroscopy (PR-DLTS), an optical pulse is used to generate excess carriers which are trapped by deep levels in the material. The a.c. resistance of the specimen is monitored, and the resistance transient which occurs afer the illumination ends is signal processed in the same way as the capacitance transient in conventional DLTS. Comparison of this technique with the d.c. current-transient measurement, photoinduced transient spectroscopy (PITS), shows that it is sensitive to the same trapping/detrapping phenomena. PR-DLTS data for the Cr-related deep level is consistent with published DLTS results. Results are reported for materials grown by the horizontal Bridgman (HB) method and by the liquid encapsulated Czochralski (LEC) technique, both with and without chromium doping. Nineteen specimens from ten different manufacturers are compared.

Introduction

It is not yet possible to specify fully the electrical properties of SI GaAs which lead to high yield and performance in devices and circuits. This is due in part to the difficulty in making and interpreting transport measurements on high-resistivity semiconductors. In particular, the study of deep levels by conventional (capacitance) DLTS [1] is not possible since an electrical bias cannot be used to alter the occupancy of the deep levels. On the other hand, optical excitation of excess carriers can be used to change the occupation of deep-level traps, and the transient response which results after the light is extinguished (and the photogenerated excess carrier recombine) is a function of the deep-level detrapping kinetics.

Hurtes *et al.* [2] and Martin and Bois [3] have measured the temperature-dependent d.c. photocurrent transient of a SI GaAs photoconductor, using the rate-window method [1] employed in DLTS. They refer to this technique as optical transient-current spectroscopy (OTCS); it has also been termed

* Electrical Engineering Department, Thornton Hall, University of Virginia, Charlottesville, Virginia 22901, USA.

photoinduced transient spectroscopy (PITS) [4], which is the designation which will be used here. These workers and others [5–11] have interpreted PITS data in terms of the emission of carriers trapped by deep levels in the material. The characteristic parameters (emission rate, capture cross-section and activation energy) of deep levels determined from PITS results are in general agreement with those obtained from traps seen by DLTS in conducting specimens. Nevertheless, great care must be used in the interpretation of these quantities. As noted by Look [12], the apparent emission rate determined by PITS may not be directly related to the emission process, either because more than one level emits at a significant rate in the temperature range of interest or because the temperature dependence of such quantities as the optical absorption, carrier lifetime or deep-level capture cross-section are not taken into account in the analysis. In addition, the interpretation may depend upon recapture of emitted carriers by other centres. Similar considerations may apply in the case of DLTS, and Martin *et al.* [13] and Milnes [14] state that the measured emission rate should be regarded as a trap signature which may not have a simple physical interpretation.

Despite their limitations, DLTS and PITS have provided much of the available information on deep levels in GaAs. In the present work we describe a variation of the PITS technique which offers several advantages. The method, which we call photoresistance deep-level transient spectroscopy (PR-DLTS), differs from PITS in that instead of measuring the d.c. photocurrent transient we measure the a.c. photoresistance transient using an r.f. bridge circuit. Since the contacts need not pass d.c. current, ohmic contacts are not necessary. This fact eliminates any concern that the thermal processing required to form ohmic contacts may have altered the material properties of interest. In addition, the technique is essentially non-destructive since the unalloyed contacts can be removed following the measurement. We have found good agreement between PR-DLTS and PITS results on the same specimen, and between PR-DLTS and published DLTS data for the chromium-related deep level.

Apparatus

The test specimen geometry used in the PR-DLTS measurement was a parallel-plate capacitor structure consisting of a transparent top metal electrode (an evaporated metal grid) and a back-surface contact. Light was directed through the transparent top-surface contact on to the sample surface. Specimens were held on a grounded wafer chuck by vacuum and contacted on the transparent top surface electrode by a probe. The wafer chuck system [15] was mounted on a wafer prober housed in a dry nitrogen box. The chuck can be cooled with liquid nitrogen to −196°C and resistively heated to around 200°C.

A 50 mW HeNe laser operating at 632.8 nm was used as an illumination source. The laser was chopped at frequencies of typically 10–100 Hz. To obtain fast transitions, a 10× microscope objective was used to focus the laser beam to roughly 2 μm at the chopping point. The light was then recollimated by a second 10× objective. Transitions of roughly 1 μs at 50 Hz were achieved using this arrangement (see Figure 1).

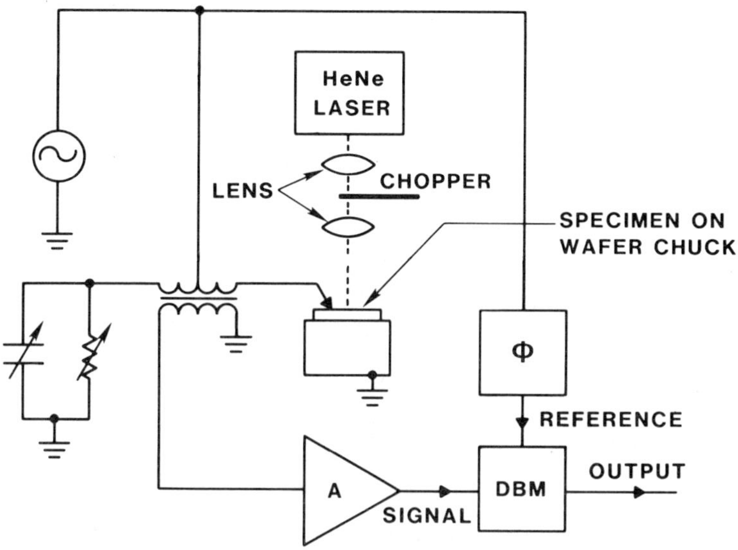

Figure 1 Schematic diagram of the PR-DLTS measurement where A is a wideband amplifier, ϕ is a variable delay line and DBM is a double-balanced mixer.

The resistance transient was measured in a bridge circuit previously used for capacitance DLTS measurements [16]. The circuit is shown schematically in Figure 1. The specimen was connected to one arm of a centre-tapped, 1 : 1, wideband transformer. The opposite arm of the transformer was terminated in a variable resistor and capacitor connected in parallel. An r.f. signal, 10–20 MHz, was applied through the centre tap of the transformer. To measure the resistance transient, the circuit is balanced such that the difference signal at the transformer output is minimized. Under this condition, illumination of the specimen upsets the bridge balance and induces an output signal in the transformer secondary. The double-balanced mixer (DBM) was used as a phase detector and its output is proportional to the change in resistance of the specimen. This DBM output signal was then processed in a conventional boxcar DLTS system [16].

Experiment

In order to compare the PR-DLTS and PITS techniques, a different specimen geometry was used. Two Au : Ge (88 : 12) top-surface electrodes 1.2 mm wide

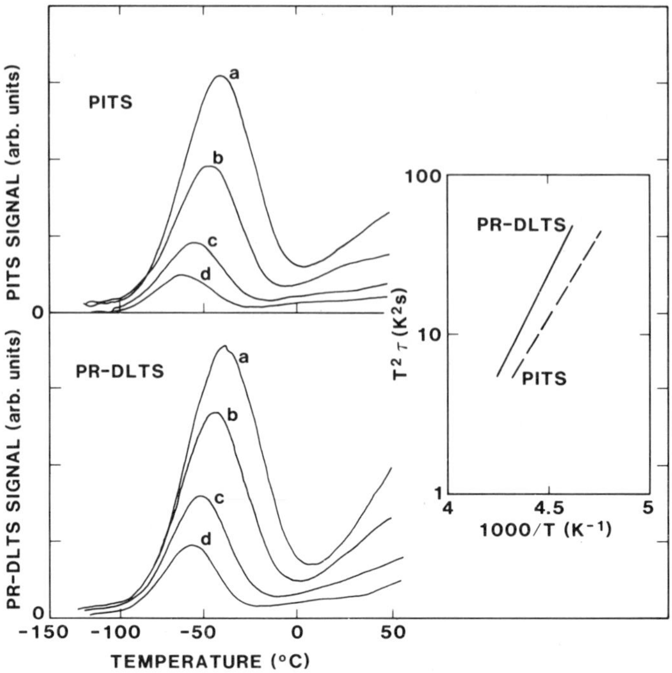

Figure 2 PITS and PR-DLTS measurements on the same specimen of Cr-doped HB-grown GaAs at (a) $t_1 = 100\ \mu s$, (b) $t_1 = 200\ \mu s$, (c) $t_1 = 500\ \mu s$ and (d) $t_1 = 990\ \mu s$ and $t_2/t_1 = 10$. Inset: Arrhenius-type plot comparison of the specimens.

separated by a 125 μm gap were deposited on a Cr-doped HB-grown specimen. The same structure was measured by both techniques and the results are shown in Figure 2. The spectra have similar general appearance in terms of number of peaks, but differ slightly in lineshape and peak position. While some of the differences might be attributed to the influence of the contacts on the PITS measurement or to differences in the temperature dependence of the signal magnitude for the two techniques, these effects have not yet been isolated. An Arrhenius-type plot of the PITS and PR-DLTS spectra in Figure 2 is shown in the inset, where we analyse both spectra assuming a transient decay of the form $\lambda\ e^{-\lambda t}$ [2], where λ is the apparent emission rate equal to the reciprocal of the transient time constant, $1/\tau$. (These plots are described as 'Arrhenius-type', to emphasize the fact that their slopes do not necessarily represent the activation energies of well-defined levels.) This form is obtained if all the emitted carriers are swept out of the material and collected at the contacts. (An alternative assumption, that all excess carriers recombine before reaching the contacts, leads to a transient of the form $e^{-\lambda t}$. This possibility will be considered later in analysing the temperature dependence of the PR-DLTS signal.) Although the slopes of the lines in the inset of Figure 2 are different, the apparent emission rate of the level as measured by the two techniques is very nearly the same. Since the emission rate is a strong function of temperature, it is not surprising that temperature spectroscopy of the same level (even by different techniques) yields a similar signature.

Analysis of a model consisting of a single trapping centre predicts that the photoinduced transient is proportional to $\lambda\, e^{-\lambda t}$ [2], where the temperature dependence is contained in the apparent emission rate, λ. Given this form for the transient, the temperature dependence of the emission rate can be obtained from a single spectrum in a way analogous to lineshape analysis in DLTS [17]. The PITS signal ΔS is proportional to $\lambda(e^{-\lambda t_1}-e^{-\lambda t_2})$ where t_1 and t_2 are the sampling times and $t_1<t_2$. Itoh and Yanai [7] have shown that for $t_2 \gg t_1$ the peak, of magnitude ΔS_{max}, occurs when $\tau=1/\lambda \cong t_1$. The time constants associated with temperatures other than that at the peak of the spectrum can be related to the magnitude of the signal by the transcendental expression $(\Delta S/\Delta S_{max})=(t_1/\tau)\exp(1-t_1/\tau)$. The temperature dependence of τ can also be obtained by varying t_1 and t_2 (as in Figure 2). If the two methods fail to agree, we have an indication that multiple emitting centres are present or that the temperature dependence of the signal is not adequately described by the assumed model.

Using this analysis on the PITS spectra in Figure 2, we find that these signals are not simply related to the transient form, $\lambda\, e^{-\lambda t}$. Examination of the literature on transient photocurrent measurements [2–11] reveals no reported observation of a single isolated level which agrees with the shape predicted by this model. Deveaud and Toulouse [6] showed that the spectral shape could be predicted by assuming the presence of multiple levels contributing to the transient signal, but even then there was no guarantee that the fit was unique. As in the case of the PITS results, lineshape analysis of the PR-DLTS spectra of Figure 2 shows that the signals are not simply described by the above trap model [2] nor by a model where the transient is proportional to $e^{-\lambda t}$. This prevents quantitative analysis of the PITS and PR-DLTS transients. However, as with DLTS and PITS, the PR-DLTS signal will be regarded here as a signature of the trapping centre and used to compare specimens.

Table 1 lists and labels nineteen specimens from ten different manufacturers which were measured by PR-DLTS. In addition to HB, specimens grown by both high and low pressure (HP and LP) LEC and gradient-freeze (GF) growth techniques were measured. Figure 3 is a composite Arrhenius-type plot summarizing the results for comparative purposes. As noted above, uncertainty in the form of the PR-DLTS transient precludes a quantitative interpretation of the data. It can be concluded, however, that data widely separated in temperature in the plot of Figure 3 result from different centres and that closely spaced data could represent the same centre. There are several trends worth noting.

The signatures from $1000/T=5$ and above, corresponding to 'shallow' centres (approximately 0.1–0.3 eV [14]), are all from LEC-grown specimens. All of these low temperature peaks are multiple and appear to result from several deep levels, which may be one reason for their large range of slopes.

Table 1 *Specimen identification*

Producer	Specimen	Added dopant	Growth technique
1	a, b, c, d, e, f	None	HP-LEC
2	g	Cr	HB
3	h, i	None	HP-LEC
4	j, k	None	HP-LEC
5	l	Cr	HP-LEC
6	m	Cr	HB
	n	Cr	LP-LEC
7	o	Cr	GF
8	p	Cr	LP-LEC
9	q	Cr	HB
10	r	Cr	HB
	s	None	LP-LEC

All of the undoped LEC specimens measured have significantly more shallow centres than the HB-grown specimens; specimens b and c also show low temperature signals, but unambiguous peaks were not discernible. The Cr-doped LP-LEC specimens do not show any of the shallow centres observed in the undoped LEC specimens, while in the Cr-doped HP-LEC specimen a shallow centre was observed. This centre, labelled l, yields an unusually broad peak and an equally unusual signature. As with the other low temperature centres, this signature is likely to be associated with several deep traps.

A comparison of LP-LEC and HB material from producer number 6 shows two impurities common to both growth techniques (see levels m1 and n1, and m2 and n3) pointing to the possibility that these impurities were present in the source material itself. The cluster of levels in the range $1000/T = 4$–5 occurs in all of the HB and GF-grown specimens observed in this study. The high temperature levels from specimens g, m, n and p as well as the previously mentioned level cluster are all found in Cr-doped specimens. The high temperature centre is likely to be the same as the hole trap level HB1 seen in Cr-doped LPE [18] attributed to Cr. These and the centres which correspond well with other published DLTS and PITS results are shown in Figure 4. Several other correlations are suggested by these data and are compiled in Table 2 and compared in Figure 4.

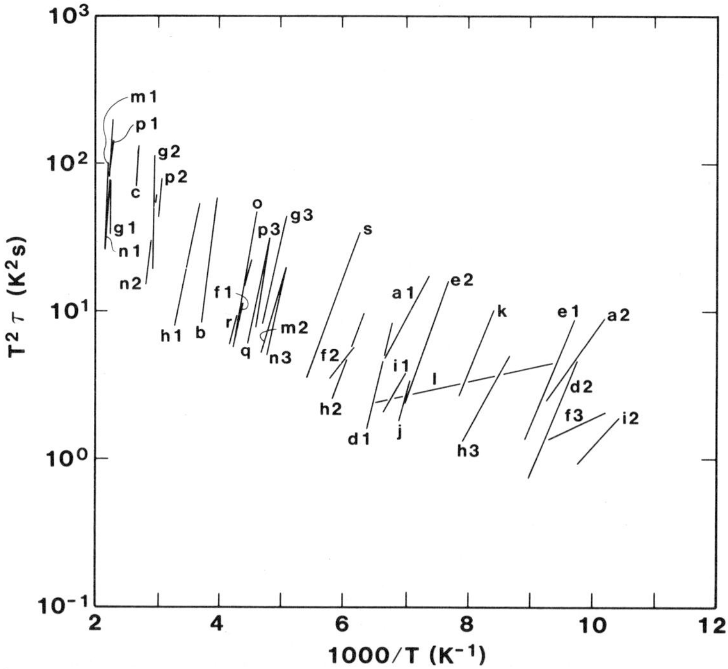

Figure 3 Arrhenius-type plot comparison of nineteen SI GaAs specimens from ten different manufacturers. The specimen identification letters correspond to Table 1; the numbers label the signatures of specimens from which more than one discernible peak was observed. The numbering is taken from left to right, high temperature to low temperature, on the plot.

Summary

We have examined a non-destructive a.c. method of measuring photoinduced resistance transients in SI GaAs. The technique does not require ohmic

Table 2 *Possible correlation between PR-DLTS signatures and published DLTS and PITS results*

Specimen	Correlation	Measurement	Observation	Ref.
g1, m1, n1	HB1	DLTS	Cr-doped LPE	18
and p1	OA	DLTS	VPE on Cr-doped or Te-doped substrates	20
n2 and p2	D	Drain-current-DLTS	Ion-implant into undoped substrate	19
o	XP1	PITS		7
p3 and q	HB5	DLTS	as-grown LPE	18
g3	EC6	DLTS		13
s	EC8	DLTS		13
a2	EC12	DLTS		13

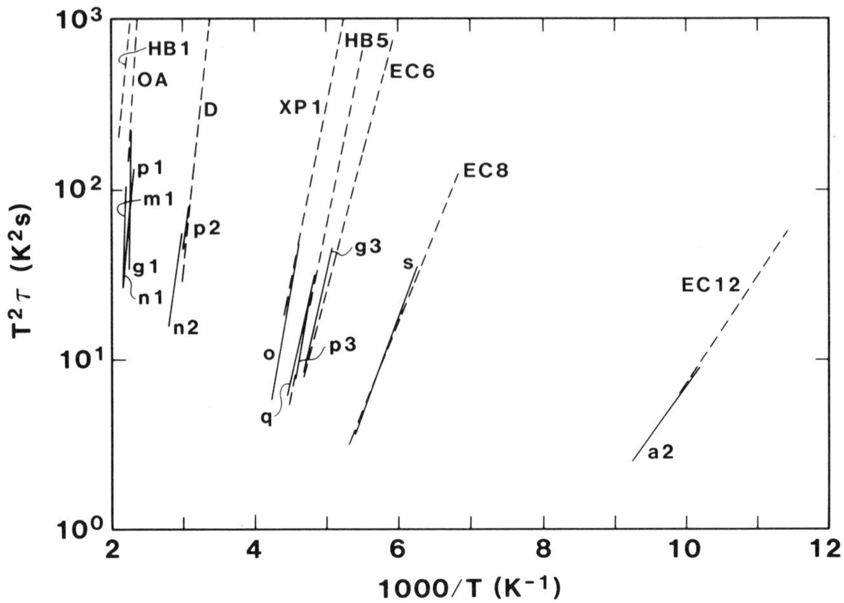

Figure 4 An Arrhenius-type plot of selected centres and their correlation with published DLTS and PITS data. The specimen identification and correlating level designation correspond to Table 2.

contacts which eliminates any concern that thermal processing may have altered the material properties of interest. Comparing the results with a photoinduced transient current measurement under d.c. bias showed that the technique is sensitive to the same trapping/detrapping processes. A comparison of our Cr-doped specimens with published DLTS results on Cr-doped GaAs reveals a similar signature in our specimens. A comparison of LEC and HB-grown specimens from 10 manufacturers indicates that undoped LP and HP-LEC wafers have significantly more shallow centres than boat-grown specimens.

References

1. Lang D. V. (1974). *J. Appl. Phys.*, **45**, 3023
2. Hurtes Ch., Boulou M., Mitonneau A. and Bois D. (1978). *Appl. Phys. Lett.*, **32,** 821
3. Martin G. M. and Bois D. (1978). *Electrochem. Soc. Proc.*, **78–3,** 32
4. Fairman R. D., Morin F. J. and Oliver J. R. (1979). *Inst. Phys. Conf. Ser.*, **45,** 134
5. Hurtes Ch., Hollan L. and Boulou M. (1980). *Inst. Phys. Conf. Ser.*, **45,** 342
6. Deveaud B. and Toulouse B. (1980). In *Semi-Insulating III–V Materials, Nottingham*, p. 241 (S. Makram-Ebeid and B. Tuck, eds.); Shiva Publishing, Nantwich, UK
7. Itoh T. and Yanai H. (1980). *Inst. Phys. Conf. Ser.*, **56,** 537
8. Yuba Y., Gamo K. and Namba S. (1981). *Inst. Phys. Conf. Ser.*, **63,** 221
9. Fairman R. D., Chen R. T., Oliver J. R. and Ch'en D. R. (1981). *IEEE Trans. El. Dev.*, **28,** 135

10. Rhee J. K., Bhattacharya P. K. and Koyama R. Y. (1982). *J. Appl. Phys.*, **53,** 3311
11. Rhee J. K. (1982). *PhD Thesis, Oregon State University*
12. Look D. C. (1983). In *Semiconductors and Semimetals*, vol. 19, p. 75 (R. K. Willardson and A. C. Beer, eds.) Academic Press, New York
13. Martin G. M., Mitonneau A. and Mircea A. (1977). *Electron. Lett.*, **13,** 191
14. Milnes A. G. (1983). *Adv. Electronics Electron. Phys.*, **61,** 63
15. Koyama R. Y. and Buehler M. G. (1979). *Rev. Sci. Instrum.*, **50,** 983
16. Oettinger F. F. and Larrabee R. D. (1980). National Bureau of Standards Interagency Report 80-2061, 9. Unpublished
17. Vasil'ev A. V., Smagulova S. A. and Shalmeev S. S. (1983). *Sov. Phys. Semicond.*, **17,** 103
18. Mitonneay A., Martin G. M. and Mircea A. (1977). *Electron. Lett.*, **13**, 666
19. Sriran S. and Das M. B. (1983). *IEEE Trans. El. Devices*, **ED-30**, 586
20. Ozeki M., Komeno J., Shibatomi A. and Ohkawa S. (1979). *J. Appl. Phys.*, **50**, 4808

Origin of the Non-Uniformities in the Infrared Transmission of LEC-Grown GaAs Crystals

M. S. Skolnick, D. A. O. Hope and B. Cockayne

Royal Signals and Radar Establishment, Malvern, Worcestershire, UK

Abstract

Photoquenching experiments at low temperatures are employed to demonstrate in a clear way that the inhomogeneities in IR (infrared) transmission of LEC (liquid encapsulated Czochralski) crystals of GaAs arise from fluctuations in the concentration of the deep donor EL2 in its neutral charge state. Results obtained in the transition region between dislocation-free and dislocated regions in small diameter crystals are shown to provide new information on the formation mechanisms of EL2 centres.

Undoped SI GaAs grown by the LEC technique is of great importance at the present time as a substrate material for integrated circuit fabrication. One of the most important properties of the substrate, which is required for the successful production of large scale digital integrated circuits, is that it should be uniform over large areas.

In order to investigate the uniformity of LEC GaAs crystals new IR imaging techniques have been developed recently [1, 2]. The observed IR inhomogeneities have been attributed to fluctuations in the concentration of the deep donor $(EL2)^{\circ}$ (in its neutral charge state) which controls the resistivity of the SI material [1, 2]. Variations in the concentration of $(EL2)^{\circ}$ ($[EL2]^{\circ}$) of up to 90% across lineage features, corresponding to regions of high dislocation density at low angle tilt boundaries [3], were found in slices cut from 7.5 cm diameter crystals [2]. The attribution of the IR non-uniformities to $[EL2]^{\circ}$ fluctuations was based primarily on the similarity of the IR absorption spectrum at regions of high and low absorption [2] to that already reported for $(EL2)^{\circ}$ [4]. However, the $(EL2)^{\circ}$ absorption spectrum is broad and featureless, particularly at 300 K, extending from 0.7 to 1.4 eV, and so it is difficult to be sure of the origin of the IR non-uniformities on this basis alone.

In the present paper clear evidence that the IR non-uniformities do indeed arise from $[EL2]^{\circ}$ fluctuations is presented. These conclusions are reached on the basis of low temperature (~100 K) photoquenching experiments which demonstrate in a direct way that the IR non-uniformities show exactly the same 'bleaching' behaviour as that previously reported for EL2 [5, 6]. Preliminary details of the quenching experiments have been given previously [7], but much clearer IR images of the structure observed before and after quenching are presented here. Quantitative spectroscopic information on the magnitude of the [EL2]° fluctuations before quenching, and of independent near band-edge absorption fluctuations after quenching is given, and the two absorption processes discussed. In addition, new results on the forms of the fine structure in IR images of small diameter (2.0 cm) LEC crystals which are dislocation free for a length of 3.5 cm from the seed end, are presented.

In Figure 1a an IR image taken with light of energy 1.38 eV (resolution 30 meV), with the sample held at 30 K in an optical crysostat is shown. An image of the same sample at 300 K has exactly the same form. The sample is of 1.2 cm diameter and 3 mm thick. It is cut from the centre of a 5.6 cm diameter, (001) slice of a 3 kg SI undoped crystal [2] (A238) grown at Cambridge Instruments. The images are obtained with a silicon vidicon television camera. The sample is illuminated with light from a tungsten-halogen projector lamp dispersed by a 0.25 m focal length spectrometer. Dark regions in the figure correspond to high IR absorption. The forms of fine structure in the IR image are very clearly visible and are typical of those observed in large diameter LEC crystals grown in a high pressure puller. The non-uniformities in Figure 1a can be divided into two classes (for more details see Ref. 2), namely 'sheets' of high IR absorption running along ⟨110⟩ diameters and cellular structure (ce) corresponding to polygonalized dislocation networks.

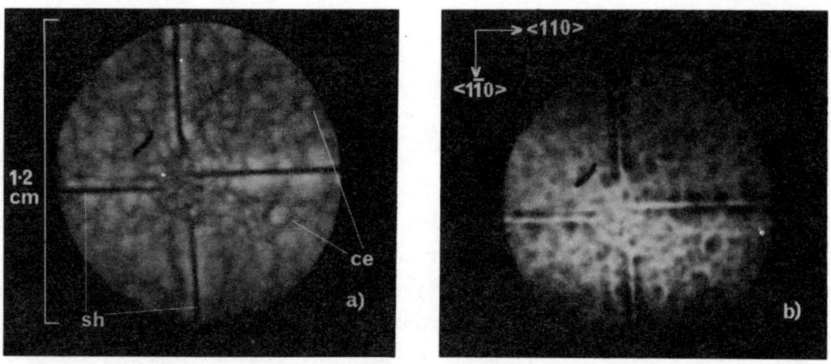

Figure 1 Infrared images at 30 K taken before (a) and after (b) low temperature quenching, at photon energies of 1.38 and 1.48 eV (30 meV resolution) respectively. The views are parallel to the ⟨001⟩ growth axis of the crystal. Sheet (sh) and cellular structure (ce) regions of high (EL2)° absorption are labelled in (a). Dark features correspond to high IR absorption.

The sample at 30 K was then exposed to a more intense ($\sim 10\ \mathrm{W\,cm^{-2}}$) broad band, second light source for 30 seconds. After removal of the second light source, the sample was viewed again with zero order light from the spectrometer. All the 'black on white' fine structure of Figure 1a was found to have been quenched-out but was replaced by just discernible 'white on black' structure. The 'white on black' structure is found to be much more easily visible when the spectrometer is tuned close to the band gap of GaAs. The result is shown in Figure 1b for 1.48 eV light. Comparison of Figures 1a and 1b shows that all the 'black on white' non-uniformities of Figure 1a are replaced by reversed contrast 'white on black' features in Figure 1b. This result is the same as that presented previously [7], but the IR images of Figure 1 are of considerably better quality, and the one-to-one correspondence of the non-uniformities of Figures 1a and 1b can be very easily seen. The spectral dependence of the bleaching of the absorption of Figure 1a, having a maximum rate for quenching light at 1.1 eV, and the temperature dependence of the regeneration rate above 110 K are found to be very similar to those reported from photocapacitance measurements for the deep donor EL2 [5–7]. EL2 is the only centre in GaAs which has been reported to show this quenching behaviour. Thus, these results provide convincing evidence that the non-uniformities in the IR images of LEC GaAs crystals arise from fluctuations in $(EL2)^o$ concentration.

Further information on the magnitudes of the changes in IR absorption across the 'sheet' features in Figures 1a and 1b and their spectral dependence can be obtained by direct analysis of the video signal corresponding to one line of the TV image. The intensity variation across one line of the image was displayed directly on an oscilloscope using the delayed trigger facility. This signal was then analysed using a PAR 162 boxcar by scanning a 10 nsec gate across the line of interest. The boxcar output was then recorded on an X–Y plotter. The results are shown in Figure 2 for line scans across the 'sheet' features in the before quenching (a) and after quenching (b) states. (a) is taken at 1.38 eV and (b) at 1.48 eV with spectral resolution of 30 meV in both cases. The reversal of the sense of variation of the transmitted intensity across the 'sheet' region is clearly visible between (a) and (b).

Absorption coefficient changes corresponding to the variation in transmitted light intensity across the 'sheet' features were then calculated from the line scans of Figure 2. The results are presented as a function of photon energy in Figure 3. The vertical bars in Figure 2 indicate the changes of transmitted intensity across the sheets used in the calculation of the data points of Figure 3. The light intensity away from the sheet in the quenched state was taken as the reference intensity, corresponding to zero absorption but including reflection losses, in the absorption coefficient calculations. This method should be reliable except in the estimation of the EL2 absorption coefficient in the near band-edge region due to the influence of the residual absorption processes in this spectral region in the quenched state. For this reason no data points (circles) are included in Figure 3 above 1.41 eV for the before quenching state.

The circles on Figure 3 represent the change in absorption coefficient across the 'sheet' before quenching. Their spectral variation is of the general form expected for $[EL2]^o$. Using the calibration curves of Martin [4], the change in absorption coefficient at 1.24 eV corresponds to a change in $[EL2]^o$ of $6.5 \times 10^{15}\ cm^{-3}$ and is equivalent to a 70% increase in $[EL2]^o$ compared to its value away from the sheet.

The absorption fluctuations of Figure 1b after quenching extend ~60 meV (15 meV resolution) below the low temperature band edge E_g of 1.52 eV (the

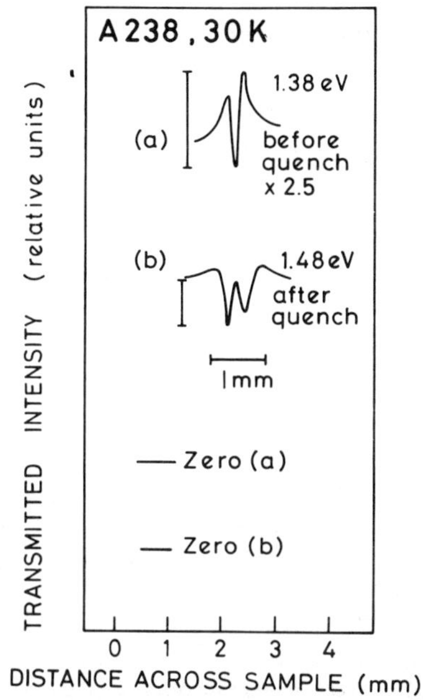

Figure 2 Scans of transmitted light intensity at 30 K across sheet features before (a) and after (b) quenching. The results are obtained by scanning a boxcar gate across one line of the video signal corresponding to a position ~2 mm from the top of the image in Figure 1. The scale for the plot in the unquenched state in (a) is multiplied by 2.5 relative to (b). The vertical bars to the left of the scans represent the magnitudes of the variations across the sheets used to calculate the absorption coefficient changes of Figure 3.

crosses in Figure 3). The origin of the quench-resistant structure is not yet clear. It was suggested earlier [7] that it might arise from modulation of the band gap in the strain fields of dislocations, since it has been shown that the regions of high $[EL2]^{o}$ correspond to regions of high dislocation density [1, 2, 8]. However, it should be noted that the absorption spectrum represented by the crosses is of the spectral form expected for band tailing effects below the band gap, induced by random charge variations on a microscopic scale [9, 10]. Maximum band tail absorption would be required in the regions adjacent to the high dislocation density sheet and cellular structure features to explain the observed results. A third possibility which should be considered is that the near band edge features could arise from absorption processes which raise an electron from ionized carbon shallow acceptors to the conduction band [11]. Carbon is known to be the dominant acceptor in these GaAs crystals from photoluminescence measurements [12]. Furthermore, the hole binding energy to the carbon acceptor is 27.0 meV [13] and so absorption from carbon acceptors which have been compensated by shallow and deep (EL2) donors in these SI crystals will have an onset at $E_g - 27.0$ meV. This is reasonably close to the observed onset of the quench-resistant spectrum of Figure 3 of $E_g - 60$ meV, particularly given the spectrometer resolution of 15 meV and the possible influence of band-tail effects. This raises the intrigu-

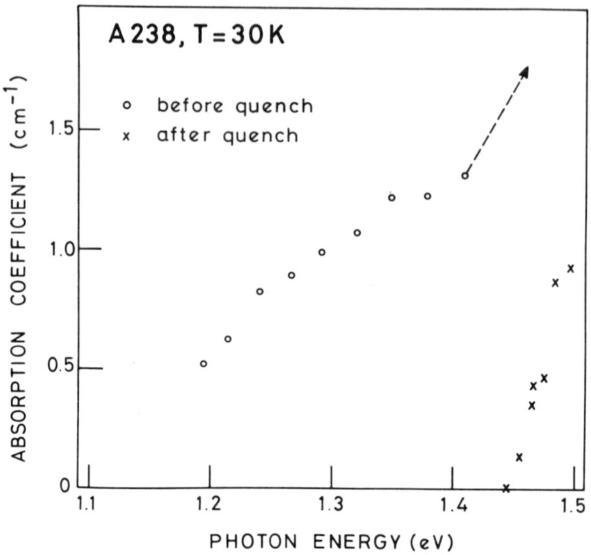

Figure 3 Spectral variation of absorption coefficient changes across sheet before (O) and after (×) quenching, calculated from the magnitude of the vertical bars of Figure 2. The light intensity away from the sheet in Figure 2 in the quenched state is taken as the reference intensity corresponding to zero absorption in the calculations.

ing possibility that the carbon acceptor is non-uniformly distributed and is gettered to the regions surrounding the cell walls and sheets. This would provide a natural explanation for the anti-correlation in the spatial distribution of the before and after quenching IR images, since in regions of high ionized carbon concentration, low $[EL2]^o$ concentration would be expected due to compensation effects, if the total EL2 concentration is assumed uniform. This suggestion is only speculative but experiments are being carried out to test its validity. However, it should be mentioned that microscopic fluctuations in carbon concentrations have been reported by Kamejima *et al.* in scanning SIMS (secondary ion mass spectrometry) measurements [14].

In the final part of this paper IR images obtained for a 2.2 cm maximum diameter, SI, undoped crystal (D57) grown in the ⟨100⟩ direction by the low pressure LEC technique are presented. The first 3.5 cm of the crystal are dislocation free as demonstrated by molten KOH and A–B etching. The room temperature resistivity of the dislocation free part of the crystal is 7.3×10^6 Ω cm, with $n_e = 1.7 \times 10^8$ cm^{-3} and $\mu_{300} = 5100$ cm^2 V^{-1} sec^{-1}. The IR image of the dislocation free region is uniform, as expected from studies of larger diameter crystals where IR non-uniformities correlate with non-uniformities in dislocation density. Dislocations begin to nucleate at a critical diameter of ~1.8 cm in the low pressure LEC crystal. Their arrangement is consistent with slip in the (111) planes. Towards the tail of the crystal a more random, partially polygonized dislocation configuration, indicative of substantial climb, develops.

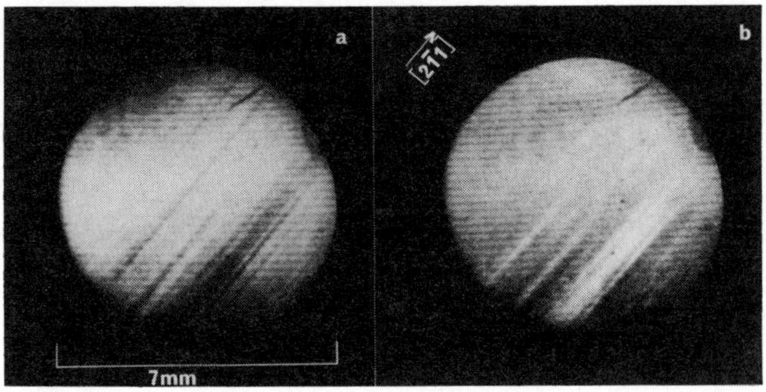

Figure 4 Infrared images at 20 K of the 7 mm disk cut from the region of crystal D57 where dislocations in slip planes first start to nucleate. The crystal is viewed in the ⟨011⟩ direction. (a) is before quenching and (b) after quenching. The top left hand portion of the crystal is dislocation free.

A low temperature (20 K) IR image for crystal D57 taken in the region where the slip bands appear is shown in Figure 4a. The crystal face is a (011) plane and the regions of high absorption lie in the $\langle 2\bar{1}1\rangle$ direction, the direction of intersection between the $\langle\bar{1}\bar{1}1\rangle$ slip plane and the (001) face. The IR absorption shows the same quenching behaviour as that reported here for the large diameter crystals, as shown in Figure 4b. Once again the regions of higher than average absorption before quenching have lower than average absorption after quenching. This shows that the reversal of contrast in the IR images in the regions of dislocations is a general phenomenon in SI crystals. Furthermore, the separate observation of IR quenching in dislocation free regions demonstrates the presence of EL2 in this crystal. This shows that dislocations are not necessary for EL2 production, but that the concentration of $(EL2)^{o}$ is perturbed in the vicinity of dislocations. Further details of the analyses of the dislocation-free crystals will be given elsewhere [15].

ACKNOWLEDGEMENTS

We thank I. Grant and R. M. Ware for the provision of the large diameter crystal and L. J. Reed and A. D. Pitt for invaluable experimental assistance.

References

1. Brozel M. R., Grant I., Ware R. M. and Stirland D. J. (1983). *Appl. Phys. Lett.*, **42,** 610
2. Skolnick M. S., Brozel M. R., Reed L. J., Grant I., Stirland D. J. and Ware R. M. (1984). *J. Electron. Mat.*, **13,** 107
3. Brown G. T. Private communication
4. Martin G. M. (1981). *Appl. Phys. Lett.*, **39,** 747
5. Mitonneau A. and Mircea A. (1979). *Solid-State Commun.*, **30,** 157

6. Vincent G., Bois D. and Chantre A. (1982). *J. Appl. Phys.*, **53,** 3643
7. Skolnick M. S., Reed L. J. and Pitt A. D. (1984). *Appl. Phys. Lett.*, **44,** 447
8. Stirland D. J., Grant I., Brozel M. R. and Ware R. M. (1983). *Inst. Phys. Conf. Ser.*, **67,** 285
9. Redfield D. and Afromowitz M. A. (1967). *Appl. Phys. Lett.*, **11,** 138
10. Dow J. D. and Redfield D. (1972). *Phys. Rev.*, *B***5,** 594
11. Johnson E. J. and Fan H. Y. (1965). *Phys. Rev.*, **A1991,** 139
12. Skolnick M. S. Unpublished
13. Ashen D. J., Dean P. J., Hurle D. T. J., Mullin J. B. and White A. M. (1975). *J. Phys. Chem. Sol.*, **36,** 1041
14. Kamejima T., Shimura F., Matsumoto Y., Watanabe H. and Matsui J. (1982). *Jap. J. Appl. Phys.*, **21,** L721
15. Hope D. A. O., Skolnick M. S. and Cockayne B. Unpublished

Far-Infrared Absorption Measurements of the 78 meV Acceptor in GaAs

W. J. Moore, B. V. Shanabrook and T. A. Kennedy

Naval Research Laboratory, Washington DC 20375, USA

Abstract

Samples of GaAs with conductivity dominated by the 78 meV acceptor have been studied in infrared transmission as a function of 2 MeV electron irradiation. A reduction in the strength of the absorptions of the shallow acceptors and the 78 meV acceptor is observed with irradiation. An increase in absorption due to the negative charge state of a double acceptor at 203 meV is observed as the Fermi level shifts through the 78 meV acceptor level. The results provide strong evidence for the identification of the 78 meV and the 203 meV levels with the neutral and negatively charged states of a double acceptor, respectively.

Introduction

Undoped LEC (liquid encapsulated Czochralski) GaAs crystals pulled from Ga-rich melts often exhibit an acceptor level with a binding energy of 78 meV [1–6]. The properties of this defect have been elucidated by infrared intra-centre absorption [4], photoluminescence [1, 4, 5, 6] and Hall effect [2, 3, 5] measurements. These studies have suggested that the 78 meV level arises from the neutral charge state of a double acceptor. However, there is a disagreement on the binding energy characteristic of the negative charge state of this centre. Specifically, Yu *et al.* [5] and Elliott [4] have suggested that this level exhibits a binding energy of 236 meV and 200 meV, respectively. Thus, there is uncertainty in the binding energy of the negative charge state of the 78 meV centre and in the origin of the 200 meV and 236 meV levels.

In order to address some of these questions, we have performed a series of infrared transmission measurements of p-type GaAs crystals. Neutral shallow acceptors in GaAs exhibit a characteristic pair of intra-centre electronic transitions in the infrared spectral region which are denoted as D ($1s_{3/2}$ to $2p_{5/2}(\Gamma_8)$) and C ($1s_{3/2}$ to $2p_{5/2}(\Gamma_7)$) lines [7]. The separation between the D and C lines is a constant (16 cm^{-1}) for all uncharged, shallow acceptors in GaAs with effective mass-like excited states. This includes both single and neutral double acceptors. In GaAs, the binding energy of neutral acceptors may be determined by adding the 7.7 meV binding energy of the excited state to the D absorption line [7] observed in the infrared measurements. For double acceptors in the negative charge state, the separation between the D and C lines will be approximately four times larger than that observed for the neutral charge state. In this case, the activation energy may be determined by adding 30.8 meV to the energy of the D transition. If one had a systematic method for shifting the Fermi level from a position below the 78 meV level to a location between the neutral and negative charge state of this double acceptor, one should observe the growth of D and C transitions with a separation characteristic of a negatively charged double acceptor (i.e. about 64 cm^{-1}). One method for systematically shifting the Fermi level of p-type GaAs crystals towards the centre of the energy gap, which we have employed, is the irradiation of the crystals with 2 MeV electrons [8]. The irradiation produces hole traps that move the Fermi level of the sample toward mid-gap by compensating shallow acceptors. If the 78 meV acceptor does not form complexes with irradiation-induced defects and if the Fermi level of the sample is situated within the thermal energy of the neutral charge state of this double acceptor, then we should observe intra-centre transitions which are characteristic of both charge states of this double acceptor.

Experimental

All of the infrared transmission measurements were conducted on samples mounted immediately in front of a gallium-doped germanium detector on the cold finger of a liquid helium container. Both the sample and detector were cooled to near 4.2 K. Spectra were acquired with a Bomem FTIR spectrometer with mylar and KBr beam-splitters to cover the spectral range of 80–3000 cm^{-1}. Sample thicknesses were approximately 0.5 mm resulting in approximate minimum detectable concentrations of 1×10^{13} cm^{-3} and 1×10^{15} cm^{-3} for shallow acceptors (carbon, magnesium, zinc and the 78 meV level) and deeper acceptors near 200 meV, respectively. Under our experimental conditions, we would be able to observe the local vibrational modes of boron on the Ga site at a concentration of 5×10^{15} cm^{-3}. Irradiations were performed at the NRL electron van de Graaff with the samples on a water-cooled block.

We discuss infrared measurements performed on three LEC-grown GaAs samples. The first two, 120-W and 41-R, were pulled from Ga-rich melts and exhibited the 78 meV acceptor level. Sample 41-R was intentionally doped with Si donors. The third sample, 10-35-N, exhibited significant contamination by carbon acceptors but did not exhibit the 78 meV acceptor.

Results and discussion

Shown in Figure 1a is the infrared transmission spectrum obtained from the as-grown GaAs sample 120-W. Between 130 cm^{-1} and 210 cm^{-1} intra-centre transitions characteristic of typical shallow acceptors found in GaAs (predominantly carbon, magnesium and zinc) are observed. This observation implies that the Fermi level in this sample is at about 30 meV from the valence

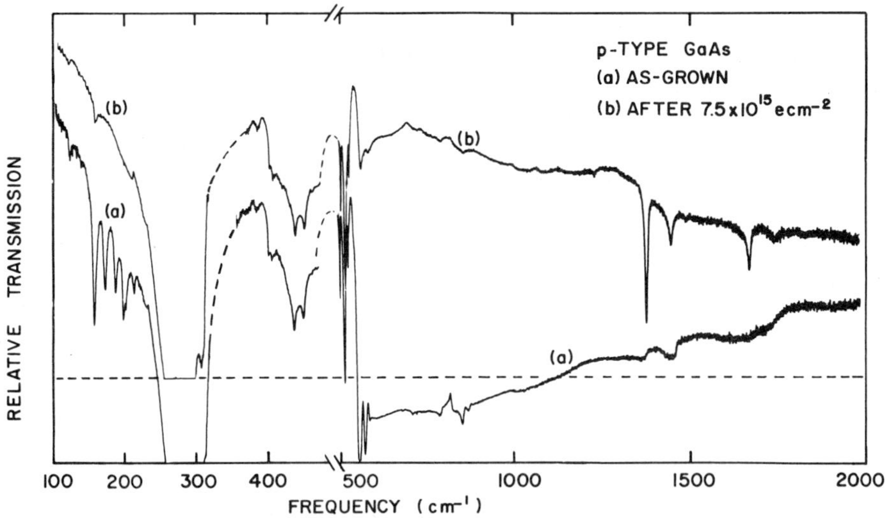

Figure 1 Infrared transmission spectra of p-type GaAs sample 120-W before and after electron irradiation. Shallow acceptors are seen in the region 140 cm^{-1}–210 cm^{-1}, the 78 meV level is observed near 600 cm^{-1} and the absorption characteristic of the negative charge state of a double acceptor is observed near 1400 cm^{-1}.

band edge and indicates that the 78 meV level, whose intra-centre absorption occurs near 600 cm^{-1}, is totally uncompensated. The transmission spectrum of this sample after an electron irradiation of 7.5×10^{15} electrons cm^{-2} is shown in Figure 1b. In this case, the intra-centre transitions due to the shallowest acceptors (D and C lines between 140 cm^{-1} and 210 cm^{-1}) are no longer observed and the absorption due to the 78 meV level is significantly reduced. In addition, we observe the growth of new absorption lines at 1388.2 cm^{-1} and 1457.6 cm^{-1} (with phonon replicas at 1681.3 cm^{-1} and 1750.7 cm^{-1}). These lines, which exhibit a separation of 69 cm^{-1}, are characteristic of the negative charge state of a double acceptor with a binding energy of 203 meV. However, we have not observed these new lines in either the carbon doped sample (10-35-N) after a similar fluence of electron irradiation or in sample 120-W after a smaller fluence of 4×10^{14} electrons cm^{-2}, which left the Fermi level still near the shallowest acceptors. The lines around 1400 cm^{-1}, which were observed previously by Elliott [4] in an as-grown sample, occur in sample 41-R as-grown. The only other intra-centre absorption lines seen in sample 41-R arise from the 78 meV acceptor. This observation implies that the Fermi level in the as-grown sample 41-R is pinned at the 78 meV level.

Quantitative results are derived from infrared spectra by determining either the integrated absorption strength of the D and C lines or the magnitude of the 1s continuum absorption. Both of these procedures yield absorption strengths which are proportional to the concentration of an acceptor level in a particular charge state. When the D and C transitions are not over-absorbed, the integrated absorption from these transitions is a more accurate measure of the concentration of a particular charge state of an acceptor than is the 1s continuum. However, because the large concentration of the 78 meV acceptors in the samples 120-W and 41-R result in an over-absorption of the D and C transitions, the magnitude of the 1s continuum absorption is a better measure of the concentration of the 78 meV acceptor in these samples. The 1s continuum absorption can clearly be seen for the 78 meV acceptor in Figure 1 above 600 cm^{-1}.

As shown in Figure 2, the 1s continuum absorption of the 78 meV acceptor level in both samples is observed to decrease as a function of increasing

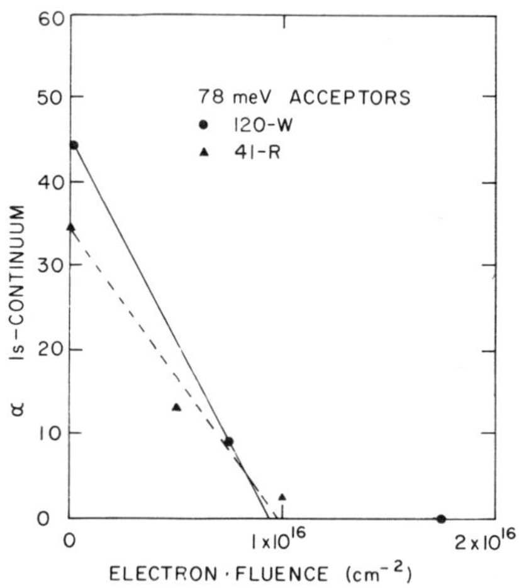

Figure 2 Absorption due to the 78 meV acceptor as a function of 2 MeV electron irradiation. The lines shown indicate compensation which is linear with fluence.

electron fluence. This behaviour can be compared to that shown in Figure 3 for the integrated absorption strengths of the D and C lines of the 203 meV acceptor level as a function of electron fluence. The integrated absorption of the D and C lines of the 203 meV level observed in sample 41-R increases from a finite concentration at zero fluence to a maximum value at about 1×10^{16} cm^{-2} and falls practically to zero by a fluence of 4×10^{16} cm^{-2}. The behaviour of the concentration of the 203 meV level as a function of electron fluence is not as well studied for sample 120-W, but this sample also exhibits an increase in the concentration of the 203 meV level for small fluences (i.e. $<2 \times 10^{16}$ electrons cm^{-2}).

The data which have been presented in Figures 1–3 can be explained by two models. According to model (a) the main effect of the electron irradiation on each sample is to produce deep hole traps at a constant introduction rate. Furthermore, these hole traps have binding energies which are greater than 203 meV and do not form complexes with the 78 meV acceptor, to a significant degree. In other words, the main effect of the irradiation is to produce hole traps which compensate the existing 78 meV centres. Once the 78 meV level

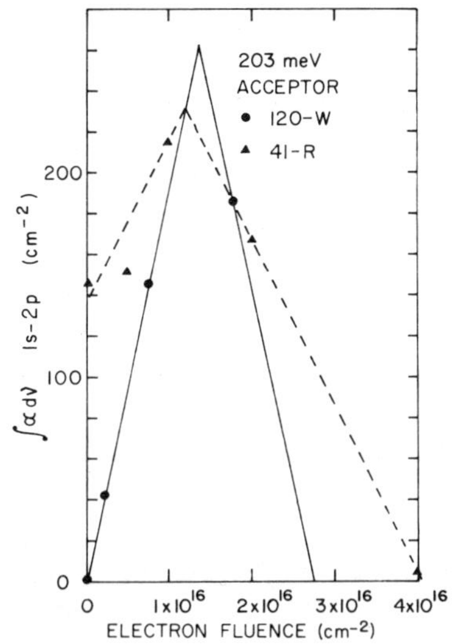

Figure 3 Integrated absorption coefficient for the 203 meV acceptor as a function of 2 MeV electron irradiation. The lines shown indicate compensation which is linear with fluence.

has been partially compensated, we observe the growth of D and C intra-centre absorption lines characteristic of the negative charge state (203 meV level) of the double acceptor. From this model we conclude that the 78 meV and 203 meV levels *are* the neutral and negatively charged states of a double acceptor. According to model (b) the 203 meV level, which we know exhibits D and C intra-centre absorption lines characteristic of the negative charge state of a double acceptor, can be created by the irradiation and is not associated with the negative charge state of the 78 meV acceptor. In the discussion which follows, we will indicate the relative strengths and weaknesses of these two models in explaining the data which have been presented above.

Model (a) suggests that the integrated absorption of the 78 meV level should exhibit a gradual reduction in intensity as a function of increasing fluence. Furthermore, the maximum concentration of the 203 meV level would occur at the fluence where the D and C lines of the 78 meV level just disappear. At higher fluences, the 203 meV level should begin to be compensated and therefore exhibit a decreasing intensity. This behaviour is represented in Figures 2 and 3 by straight lines and compared with the experimental points. We conclude that the agreement between experiment and the predictions of model (a) is rather good. However, since the slopes of the lines shown in Figures 2 and 3 are different for samples 120-W and 41-R, we conclude that the rates of compensation caused by the electron irradiation must be different for these two samples. Specifically, this rate must be less for sample 41-R. In support of model (a), it is important to note that the carbon doped GaAs sample, which did not exhibit the 78 meV level, did not exhibit a 203 meV level at comparable fluences of electron irradiation.

As stated earlier, model (b) indicates that the 203 meV level, which exhibits D and C lines characteristic of the negative charge state of a double acceptor, may be created by the electron irradiation and furthermore that the neutral charge state of this level is not the 78 meV acceptor. It is important to note that for model (b) to be correct, it is necessary that the 203 meV level can occur in as-grown samples (for example 41-R) or be created by electron irradiation. Furthermore, this model would require that either the 78 meV level is a single acceptor, or that the 78 meV double acceptor forms complexes with irradiation produced defects before we ever observe its second level. In addition, the energy level of the neutral charge state for this double acceptor remains undetermined because we have not, for any of the samples examined or for any of the electron fluences employed, observed the *growth* of an intra-centre transition which is characteristic of the *neutral* charge state of a double acceptor.

In view of the arguments stated in the preceding paragraphs, we conclude that the 78 meV and the 203 meV levels which sometimes occur in LEC-grown GaAs crystals are related to the neutral and negatively charged states of a double acceptor.

Conclusion

Samples of GaAs exhibiting the 78 meV acceptor have been studied in infrared transmission as a function of 2 MeV electron irradiation. A reduction in the strength of the intra-centre electronic transitions (1s-2p) of shallow acceptors and the 78 meV acceptor is observed with irradiation. As the Fermi level moves through the 78 meV acceptor level, we observe the growth of a new absorption. The characteristics of this absorption imply that it arises from the negative charge state of a double acceptor with a binding energy of 203 meV. We conclude that this observation provides rather strong evidence for associating the 78 meV and 203 meV level with the neutral and negative charge states, respectively, of a double acceptor. The chemical identity of the defect which gives rise to this double acceptor cannot be determined from these measurements. In contrast to earlier suggestions [4, 5], we believe that additional studies, such as electron spin resonance, are necessary before the chemical identity of the double acceptor can be established firmly.

ACKNOWLEDGEMENTS

This work was supported in part by the Office of Naval Research. The authors wish to thank E. M. Swiggard of NRL, K. R. Elliott of Rockwell International and H. M. Hobgood of Westinghouse for generously supplying the samples examined in this study.

References

1. Yu P. W., Holmes D. E. and Chen R. T. (1982). *Inst. Phys. Conf. Ser.*, **63,** 209
2. Elliott K. R., Holmes D. E., Chen R. T. and Kirkpatrick C. G. (1982). *Appl. Phys. Lett.*, **40,** 898
3. Ta L. B., Hobgood H. M. and Thomas R. N. (1982). *Appl. Phys. Lett.*, **41,** 1091
4. Elliott K. R. (1983). *Appl. Phys. Lett.*, **42,** 274
5. Yu P. W., Mitchel W. C., Mier M. G., Li S.S. and Wang W. L. (1982). *Appl. Phys. Lett.*, **41,** 532
6. Bishop S. G. and Shanabrook B. V. This volume, p. 302; Bishop S. G., Shanabrook B. V. and Moore W. J. Unpublished
7. Kirkman R. F., Stradling R. A. and Lin-Chung P. J. (1978). *J. Phys. C: Solid State Phys.*, **11,** 419
8. Woodhead J., Newman R. C., Grant I., Rumsby D. and Ware R. M. (1983). *J. Phys. C: Solid State Phys.*, **16,** 5523

Photothermal Spectroscopy and Imaging of GaAs

N. M. Amer, E. E. Haller and E. R. Weber

Center for Advanced Materials, Lawrence Berkeley Laboratory, University of California, Berkeley, California 94720, USA

Abstract

Using photothermal spectroscopy, we have performed spatially-resolved imaging of LEC (liquid encapsulated Czochralski) undoped GaAs wafers. We report the observation of sub-surface and surface inhomogeneities a few microns in dimensions. Upon heating the wafers to 250°C in a nitrogen atmosphere these inhomogeneities disappear. We conclude from analysis of the photothermal signal that the inhomogeneities are arsenic-rich inclusions. This conclusion is verified by scanned Auger spectroscopy. We have also used photothermal spectroscopy to investigate gap-state absorption (0.5–1.5 eV) of undoped, Cr, and Si-doped crystals and to study the effect of plastic deformation on the absorption spectrum. In addition to reproducing the known absorption features, we observe in some of the samples a peak at 0.5–0.6 eV which we tentatively attribute to Fe impurities.

The small rise in temperature associated with the absorption of electromagnetic radiation has provided the basis for a new spectroscopic tool generally known as photothermal spectroscopy. A characteristic of this type of spectroscopy is its sensitivity and the relative ease with which it can be implemented. Until recently, the more familiar version of this technique has been photoacoustic spectroscopy where the optical heating is converted into sound and is detected with a suitable transducer, e.g. a microphone or a piezoelectric transducer. Although this technique is useful, the ultimate achieved sensitivity is limited by scattering of light on the transducer. Furthermore, for experiments requiring a wide range of temperatures and pressures, or those involving a hostile environment, photoacoustic detection cannot be employed.

To overcome these limitations, the optical heating can be exploited in a different manner. The heating causes a change in the index of refraction of the medium surrounding the sample. This change is probed with a weak laser beam which is deflected as the index of refraction changes. Using phase-sensitive methods, the amplitude and the phase of the deflection can be readily measured with a position-sensitive detector (Figure 1). Thus by varying the wavelength of the optically-exciting (pump) beam, the deflection of the probe beam is a measure of the optical absorption. This type of spectroscopy is known as photothermal deflection spectroscopy [1].

For thermally-thin materials, the photothermal signal S is quantitatively related to the optical absorption coefficient in the following manner [1]:

$$S = A(1 - \exp(-\alpha l)) \qquad (1)$$

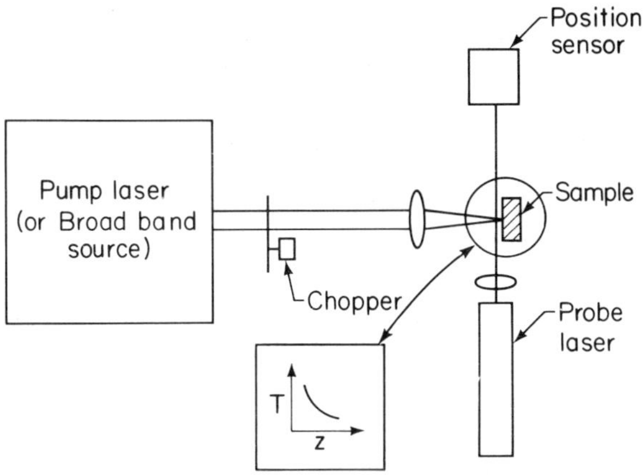

Figure 1 Schematics of the experimental setup for photothermal deflection spectroscopy.

where A is a constant which can be determined empirically, α is the optical absorption coefficient and l is the sample thickness.

Sensitivities of $\alpha l \simeq 10^{-7}$ have been achieved and the superiority of this technique in terms of sensitivity and flexibility has been demonstrated in a recent study of the optical properties of defects in amorphous semiconductors [2].

There exists a class of experimental conditions for which both photoacoustic and photothermal deflection tehniques are not suited for studying the optical and thermal properties of matter. Examples of such experiments are those which require vacuum and/or cryogenic temperatures. Such are the conditions encountered in the study of surfaces and interfaces. Similar requirements exist for the task of *in situ* and real-time characterization of thin-film semiconductors. These conditions motivated the use of optical heating in a yet another manner. Since heating results in the expansion and buckling of the illuminated surface, a measurement of the surface displacement is a means of determining the optical and thermal properties of the illuminated solid. This small displacement can be detected in a variety of ways, the simplest of which is by reflecting a weak probe laser beam off the illuminated surface. As the surface is displaced upon being optically heated, the probe beam will be deflected from its original reflection direction. This deflection is then detected with a position-sensitive detector (Figure 2). Typically, αl of 10^{-6} are readily achievable. This version of photothermal spectroscopy, which is called photothermal displacement spectroscopy [3], has been employed recently to determine unambiguously the nature of the surface reconstruction of Si(111) 2×1 [4].

An important and characteristic parameter in photothermal spectroscopy is the thermal diffusion length L_{th} which defines the depth from which the photothermal signal originates. L_{th}, which is also called the thermal

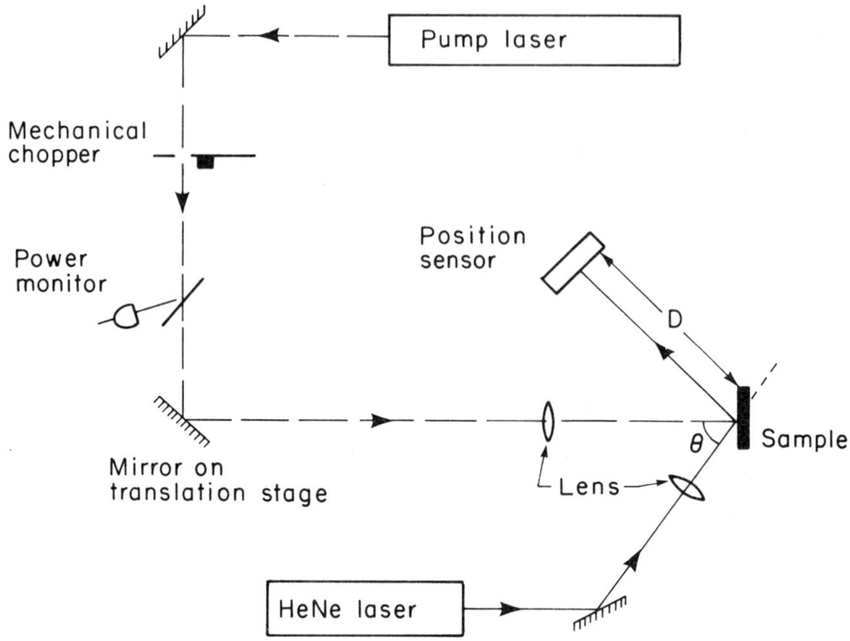

Figure 2 Schematics of the experimental arrangement for photothermal displacement spectroscopy.

wavelength, is given by:

$$L_{th} = (K_{th}/\omega\rho C)^{1/2} \quad (2)$$

where K_{th} is the thermal conductivity of the sample, ω is the modulation frequency of the light, ρ is the density and C is the heat capacity.

The modulation dependence of the thermal diffusion length is what is exploited to perform non-destructive depth profiling and imaging of solids. The higher the modulation frequency, the shorter the thermal length, and the closer to the surface the information is derived from. On the other hand, for low modulation frequencies, the photothermal signal is a probe of the bulk. Furthermore, by focussing the pump and probe beams, scanned microscopy can be performed. The achieved spatial resolution is a function of the size of the focal spot of the pump beam as well as the thermal diffusion length. Resolution of 1 μm has been attained.

We have performed scanned photothermal microscopy of SI LEC undoped GaAs wafers. The pump beam was an Ar^+ ion laser focussed to 1 μm diameter on the wafer. The probe beam was a 0.5 mW HeNe laser. Both the deflection and the displacement schemes were used and yielded identical results. As can be seen (Figure 3), both the phase and the amplitude of the photothermal signal exhibit small features which are due to the scattering of the thermal wave by thermal inhomogeneities in the wafer. By varying the modulation

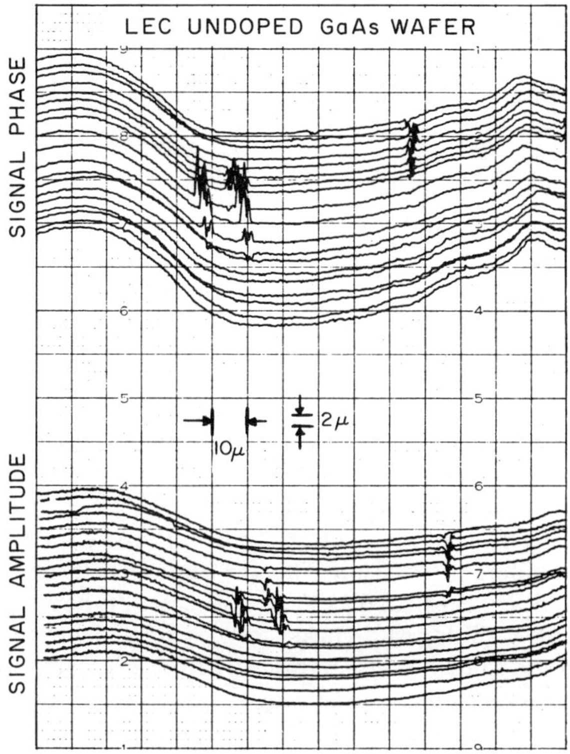

Figure 3 Spatially-resolved photothermal scans of LEC undoped GaAs wafers.

frequency, we were able to determine that these features lie on and near the surface. By analysing the photothermal signal using the theory described in Ref. [1], we were able to deduce that the inhomogeneities are arsenic-rich inclusions. Our preliminary estimate of the degree of arsenic enhancement in these inclusions is approximately 14% averaged over the 1 μm pump-beam diameter. To test the validity of our conclusion, scanned Auger spectroscopy (2 μm resolution) was performed on these inclusions. The Auger spectra verified our conclusion and gave an arsenic enhancement of 12%. Upon heating the wafers in a nitrogen atmosphere at 250°C, the arsenic-rich inclusions were no longer observed in the photothermal scans. This could be due to the arsenic diffusing into the matrix or evaporating out of the wafer. It should be noted that anion inclusions have been observed in cleaved III–V surfaces using scanned Auger spectroscopy [5].

We have also investigated the optical absorption spectra of undoped, Cr, and Si-doped GaAs crystals in the 0.5–1.5 eV range. In Figure 4 we compare the photothermal deflection spectra of Cr and Si-doped materials. As can be seen, the SI Cr-doped sample reproduces the main features obtained using conventional techniques [6]. In addition, we observe a small peak at 0.5–0.6 eV

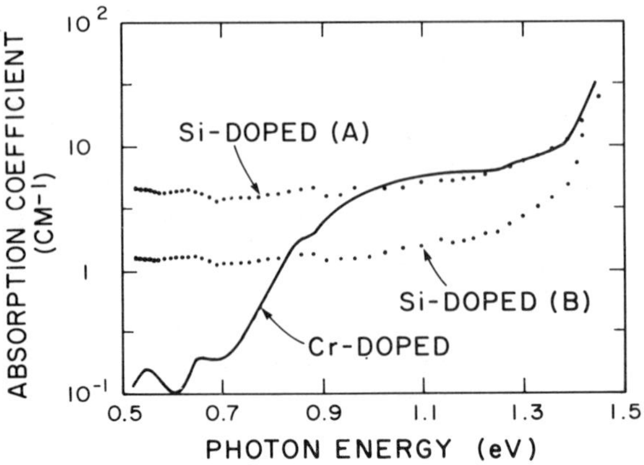

Figure 4 Photothermal absorption spectra of Cr and Si-doped GaAs.

which we tentatively attribute to Fe impurities. The Si-doped sample exhibits a qualitatively different spectrum. The absorption coefficient does not drop at photon energies lower than 0.9 eV as is the case for the Cr-doped material. The spectra of two Si-doped n-GaAs samples are shown in Figure 4. While both are from the same boule, sample A is from near the top of the boule, whereas sample B is from near the centre. The larger thermal gradients and stresses expected near the top end of the boule may account for the increased defect density in sample A.

In Figure 5 we show spectra of SI LEC GaAs and of the same material after being plastically deformed at 400°C. It is clear that the deformation increases the deep-level defect density. This is in agreement with results obtained by other methods such as ESR [7] and conventional infrared absorption [8].

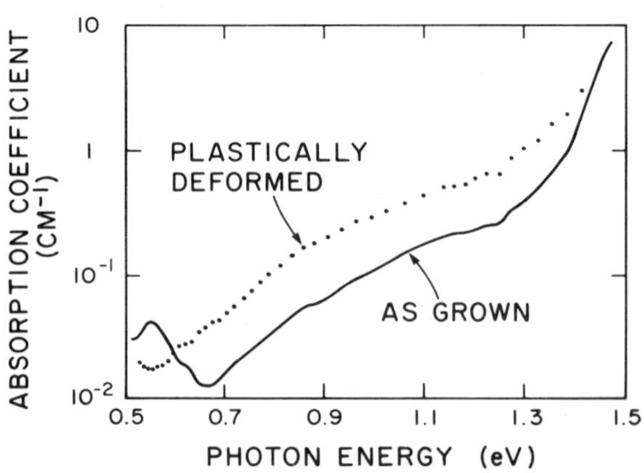

Figure 5 Photothermal absorption spectra of LEC SI as grown and plastically deformed GaAs.

In conclusion, the preliminary results reported here indicate that photothermal spectroscopy and imaging appear to offer a potentially powerful tool for the study and characterization of GaAs and other III–V materials.

ACKNOWLEDGEMENTS

We thank D. Wei and D. Ripple for technical assistance, G. Elliot of Hewlett-Packard for the GaAs wafers and K. Gaugler for the scanned Auger spectra. This work was supported by the Director, Office of Energy Research, Division of Material Science, US Department of Energy, under contract No. DE-AC03-76SF00098.

References

1. Jackson W. B., Amer N. M., Boccara A. C. and Fournier D. (1981). *Appl. Opt.*, **20,** 1333
2. Jackson W. B. and Amer N. M. (1982). *Phys. Rev.*, **B25,** 5559
3. Olmstead M. A., Amer N. M., Kohn S., Fournier D. and Boccara A. C. (1983). *Appl. Phys.*, **A32,** 141
4. Olmstead M. A. and Amer N. M. (1984). *Phys. Rev. Lett.*, **52,** 1148
5. Gant H., Koenders L., Bartels F. and Monch W. (1983). *Appl. Phys. Lett.*, **43,** 1032
6. Martin G. M., Verheijke M. L., Jansen J. A. J. and Poiblaud G. (1979). *J. Appl. Phys.*, **50,** 467
7. Weber E. R., Ennen H., Kaufmann U., Windscheif J., Schneider J. and Wosinski T. (1982). *J. Appl. Phys.*, **53,** 6140
8. Weber E. R. This volume, p. 296

Origin of Impurities at an MOVPE Grown GaAs Epilayer-Substrate Interface

A. M. Huber, G. Morillot, S. D. Hersee, K. Kazmierski, H. Tews and J. L. Pinsard

Thomson-CSF, Domaine de Corbeville, BP 10, 91401 Orsay, France

Abstract

An accumulation of the impurities Mg, Fe, Cr, Mn, Si, S and O was observed at a GaAs (MOVPE grown) epilayer-substrate interface. We present impurity distributions obtained by SIMS (secondary ion mass spectrometry) and capacitance-voltage profiling on a FET (field effect transistor) structure.

Introduction

In recent years, the metalorganic vapour phase epitaxy (MOVPE) technique has been studied for the preparation of GaAs FETs. In spite of improvements in the purity of the organometallic sources the material quality is not always reproducible and variations in threshold voltage are sometimes observed. This problem may be due to the previously reported [1] accumulation of impurities such as Mg, Fe, Cr, Mn, Si, O and S, which were detected by SIMS at the epilayer-substrate interface. The GaAs FET structure that we have studied consists of a 2–3 μm undoped buffer layer followed by an n-type active layer doped at 2×10^{17} cm^{-3}, grown on an undoped SI substrate.

SIMS studies

The accumulation of impurities at the epilayer-substrate interface was believed to be due to one or more of three possible sources:

1. Outdiffusion from the substrate.
2. Incorporation from the ambient gas phase during preheating.
3. Absorption of impurities during chemical surface etching before epitaxy (GaAs etching by 5 : 1 : 1 solution of $H_2SO_4 : H_2O_2 : H_2O$).

It is clear that the phenomenon is partly related to the substrate [2, 3], however, a simple model of diffusion does not explain the observed high concentration of impurities.

The possibility of substrate contamination by exposure to air or heating in the MOVPE reactor was examined as follows: a 1.5 μm thick layer of GaAs was grown and the wafer was removed from the reactor. After 30 minutes in air the wafer was again placed in the reactor and a second epilayer 1 μm thick was grown. Figure 1 shows the 'pile up' of impurities at the epilayer substrate interface. At the epilayer-epilayer interface however, where the growth was stopped, the only concentration change was a slight increase in the level of Si.

The possibility of contamination of the substrate by the pre-growth chemical treatment was tested by growing a thick (5 μm) GaAs layer in the normal way. The pre-growth etch was then used to remove 2–3 μm of this first layer and a second GaAs epilayer was grown. The impurity profile of this sample is shown in Figure 2. One can observe similar accumulations of Mg, Cr, Fe, Si, S and O at the epi-substrate and epi-epi interfaces. We conclude that the majority of impurities result from the chemical etching, even though very pure reagents were used.

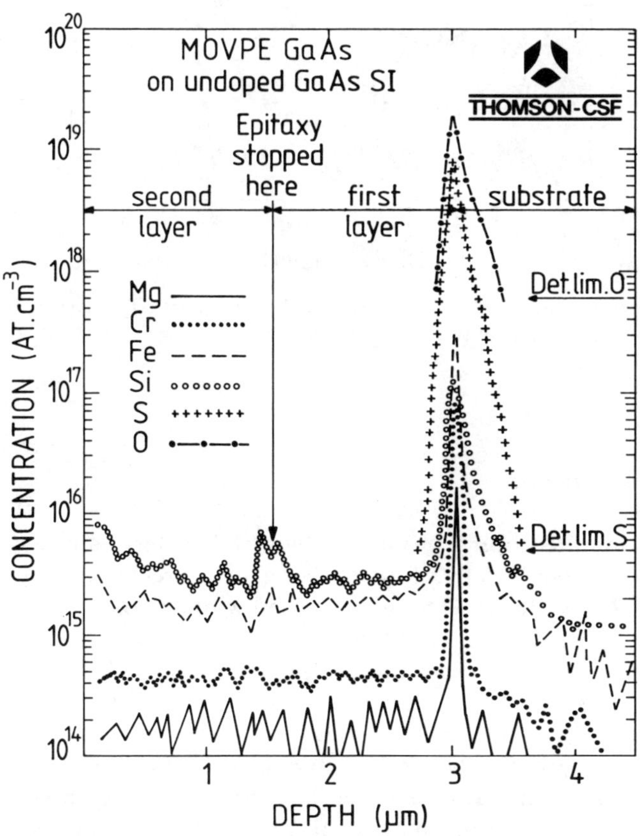

Figure 1 Depth profiles of Mg, Cr, Fe, Si, S and O in an MOVPE layer grown on undoped GaAs SI substrate in two stages. See text for details.

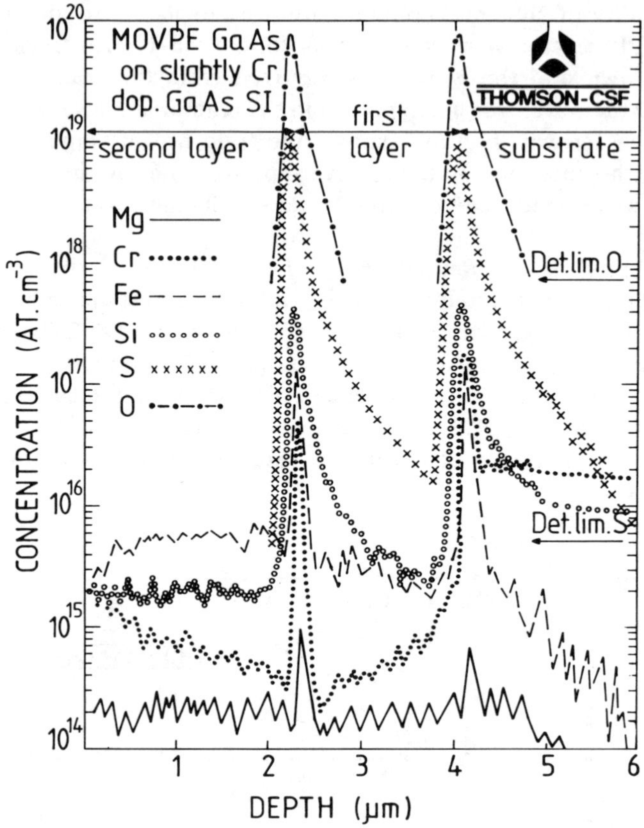

Figure 2 Depth profiles of Mg, Cr, Fe, Si, S and O in an MOVPE layer grown on slightly Cr-doped GaAs SI substrate in two stages different from those in Figure 1. See text for details.

Electrical measurements

Figure 3 shows the doping profiles of two FET wafers as measured by the capacitance voltage C(V) technique using two circular mercury Schottky contacts [4]. A 1 MHz serial circuit was used without correction for parasitic capacitance. The preparation of the wafers and crystal growth was the same for both wafers. In both cases the C(V) profile shows a fall in doping level at the active-layer buffer-layer interface and this is followed by a subsequent increase after approximately 2 μm in wafer 1 and after 5 μm in wafer 2. The thicknesses of the buffer layers as determined by optical microscopy were 3 μm for wafer 1 and 2 μm for wafer 2, respectively. The apparent increase in the electron concentration at a crystal depth larger than the buffer thickness (curve 2) was due to the lateral depletion of the active layer after pinch-off, thus representing a measurement error [5]. The reason for the increase of

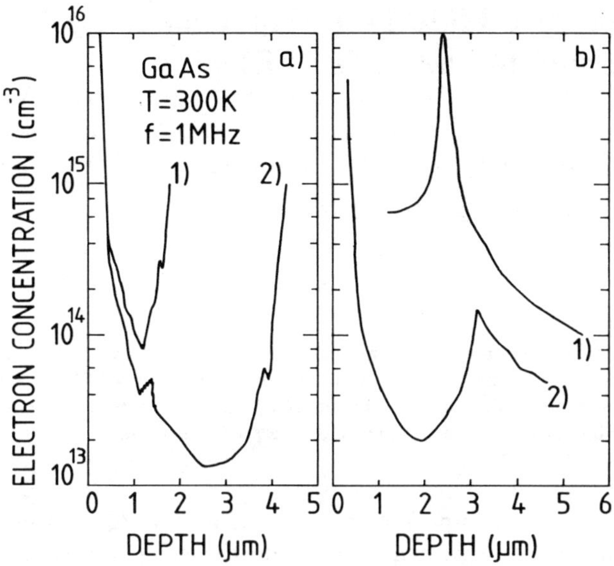

Figure 3 Doping profile of two GaAs wafers with FET structures as obtained by C(V) measurements: (a) before etching; (b) after etching.

curve 1 of Figure 3a, at 2 μm before the metallurgical interface was found when the major part of the active layer was removed by chemical etching. The etching was stopped at the active layer-buffer interface at a doping level of approximately 10^{15}–10^{16} cm^{-3} so as not to deplete the buffer. One observes in Figure 3b a peak of free charges at the buffer-substrate interface which coincides with the accumulation of impurities found by SIMS. Thus these impurities are only partially compensated. Note that the concentration peak reached 10^{16} cm^{-3} in curve 1.

An accumulation of mobile charges at the buffer-substrate interface has severe consequences for the operation of a FET. The negative charges shift the Fermi level upwards towards the conduction band thus reducing the potential barrier at the active layer-buffer interface. The transistor will show a poor pinch-off. Furthermore injection into the buffer at large electric fields will be favoured when these impurities are present.

References

1. Huber A. M., Morillot G., Merenda P., Bonnet M. and Bessonneau G. (1981). *Proceeding 3rd SIMS Budapest 314;* Springer-Verlag
2. Tuck B., Adegboyega G., Jay P. R. and Cardwell M. J. (1979). *Inst. Phys. Conf. Ser.*, **45**, 114
3. Huber A. M., Morillot G., Linh N. T., Favennec P. H., Deve B. and Toulouse B. (1979). *Appl. Phys. Lett.*, **34**, 853
4. Binet M. (1975). *Electron. Lett.*, **11**, 580
5. Heime K. (1972). *Z. Ang. Phys.*, **32**, 374

The Origin of a Hole-Like Capacitance DLTS Peak in Ion-Implanted n-GaAs MESFETs

F. H. Yang, T. K. Plant* and S. J. T. Owen*

Tektronix Inc., Beaverton, Oregon 97077, USA

Abstract

Experimentally observed minority trap-like peaks in capacitance DLTS (deep level transient spectroscopy) measurements of ion-implanted n-GaAs MESFETs are modelled by a trapezoidal doping distribution. The simulated results strongly suggest that the actual doping distribution has a direct influence on the DLTS spectrum.

DLTS has become an important measurement technique in characterizing deep traps in semiconductor materials and devices [1]. Capacitance DLTS is especially useful because the direction of the peak on a $C(t_2)-C(t_1)$ versus temperature curve indicates the type of trap (either majority or minority), where C(t) denotes the diode capacitance at time t after the filling pulse and $t_2 > t_1$. A positive-going peak reveals the existence of a majority trap while a negative-going one shows the presence of a minority trap. For a Schottky diode, however, only the majority traps can be filled by using a forward biasing pulse and, thus, minority traps are not detectable under this electrically filled capacitance DLTS measurement. In this study, capacitance DLTS measurements on an ion-implanted n-GaAs MESFET were investigated under different reverse quiescent biases. The existence of a minority-like negative DLTS peak observed under certain bias conditions is reported. A computer simulation based on the effects of the implant tail shows a similar behaviour for Schottky diodes.

The MESFETs were fabricated by a Si^+ ion implantation process on undoped SI GaAs wafers through an 800 Å SiO_2 protective layer. The implant dose was 5×10^{12} cm^{-2} and the implant energy was 120 keV. Post implantation annealing was done at 800°C for 20 minutes with the SiO_2 layer as a cap. The gate was then recessed 300 Å to avoid surface defects at the gate. The source and drain of the MESFET consisted of alloyed Au/Ge/Ni ohmic contacts on the implanted n^+ region and the gate metallization was deposited Ti/Pd/Au. The MESFET has both gate length and gate width of 100 μm and is called a fat FET because of its large gate area. The gaps between gate and source and between gate and drain are 5 μm. Throughout this measurement,

* Department of Electrical and Computer Engineering, Oregon State University, Corvallis, Oregon 97331, USA.

the source and drain were shorted together and the capacitance of the Schottky diode between the source and the gate was measured.

Figure 1 shows the capacitance DLTS spectrum at $V_R = -0.5$ V. A negative-going peak is visible at $T \cong 170$ K. The negative peak disappears for other quiescent biases. The actual capacitance transient at 167 K as seen from a Tektronix 7603 oscilloscope is shown on the inset of Figure 1. The shape of the transient is consistent with the DLTS results. Similar results were observed on other samples.

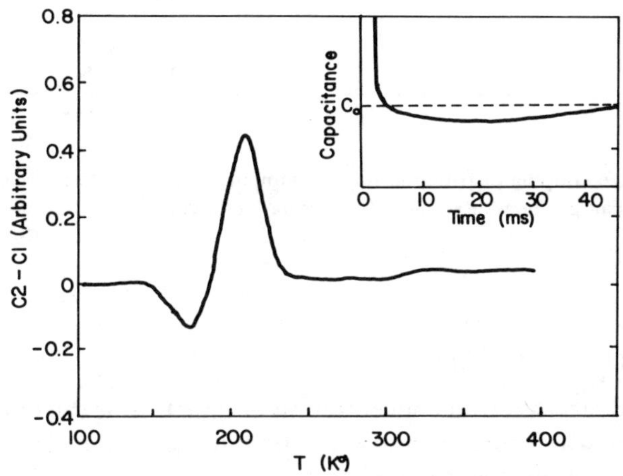

Figure 1 Capacitance DLTS results and actual capacitance transient (see inset) for an ion-implanted n-GaAs MESFET under −0.5 V reverse quiescent bias. The forward biasing pulse is at +0.2 V for ~5 ms. The measurement time delays t_1/t_2 are set at 5 ms/10 ms.

The original DLTS data analysis [1, 2] was based on a constant doping concentration for the semiconductor. For ion-implanted samples, however, the doping profile consists of a near constant level near the surface of the semiconductor followed by a graded tail region deeper into the wafer. Thus, an idealized trapezoidal doping profile is proposed to simulate the actual ion-implanted doping distribution as illustrated in Figure 2. The shallow donor concentration N_s can be described as:

$$N_s = \begin{cases} N_0 & 0 < x < w_0 \\ N_0 - a(x - w_0) & w_0 < x < w_t \end{cases} \tag{1}$$

where N_0 is the constant doping level near the surface, w_0 is the width over which the doping concentration remains constant at N_0, w_t is the depth of the channel and a is the gradient of this graded doping tail. It is assumed further that the trap concentration is related to the substrate only and is not affected by the ion implantation process, thus, staying constant throughout the wafer. Poisson's equation for this system is:

$$\nabla^2 V = -\frac{q}{\varepsilon}(N_s + N_t^+) \tag{2}$$

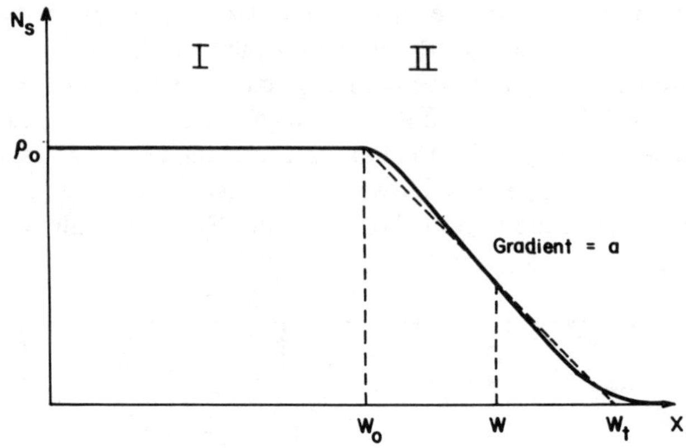

Figure 2 Idealized doping distribution for ion-implanted samples (dashed curve) along with the actual Gaussian profile plotted on a linear scale (solid curve).

where

$$N_t^+ = N_t(1 - \exp(-e_n t)) \tag{3}$$

is the emptied electron trap concentration at time t. N_t is the total concentration of the trap with energy E_t and emission rate e_n. ε is the dielectric constant of GaAs and q is the electronic charge.

The solution of the Poisson's equation is as follows:

$$\Delta V = V_{bi} - V_{ap} = \frac{q}{\varepsilon}\left(-\frac{a}{3}w^3 + \frac{a}{2}w_0 w^2 + \frac{N_0}{2}w^2 - \frac{a}{6}w_0^3\right) + \frac{N_t^+}{2\varepsilon}q(w-\lambda)^2 \tag{4}$$

where w is the depletion width under the applied bias V_{ap}, λ is the distance from the depletion edge w to the point where E_t and E_f, the Fermi level, coincide and V_{bi} is the built-in voltage for the diode. The time dependent capacitance signal for this system is described by:

$$C(t) = \frac{dQ}{dV} = \varepsilon A_0 \times \frac{a(w-w_0) - N_0 - N_t^+ + N_t^+ \dfrac{d\lambda}{dw}}{aw^2 - aw_0 w - N_0 w - N_t^+(w-\lambda)} \tag{5}$$

where A_0 is the junction area. The simple relationship of $C(t) \propto 1/w$ is no longer true under the trapezoidal doping profile unless $N_t^+ = 0$.

A computer simulation based on equations (4) and (5) was performed on a trap with $E_c - E_t = 0.4$ eV and $N_t = 10^{16}$ cm^{-3} with the initial conditions $N_0 = 3 \times 10^{17}$ cm^{-3} and $a = 3 \times 10^{22}$ cm^{-4}. The value of the depletion width is bias dependent. Experimental data on capacitance versus voltage (C–V) measurement shows that w_t and the pinch-off voltage are temperature dependent. This information is incorporated into the simulation program through a semi-empirical temperature dependent w_t. The simulation results for two different biases are shown in Figure 3. A negative-going peak followed by a positive-going one is observed in Figure 3a where the reverse bias has moved the depletion edge into the tail of the ion implanted distribution. In Figure 3b the bias is smaller so the depletion edge is in the more constant doping region, and we see only the single majority-type trap peak. The simulated capacitance transients at 210 K and 192 K are shown in the inset of Figure 3 for the larger applied bias.

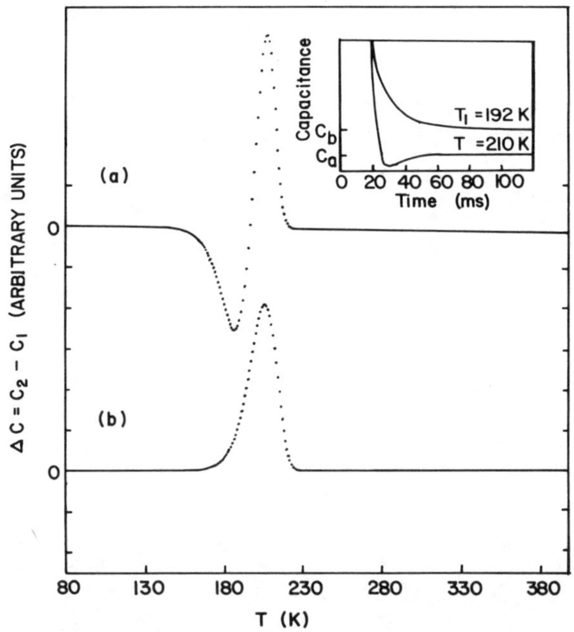

Figure 3 Simulated DLTS results for the single trap level with $E_c - E_t = 0.4$ eV under reverse quiescent biases (a) with the depletion edge into the implant tail and (b) with the depletion edge still in the constant doping region. Initial conditions are $N_0 = 3 \times 10^{17}$ cm^{-3}, $a = 3 \times 10^{22}$ cm^{-4}, $\sigma = 1 \times 10^{-13}$ cm^2, $N_t = 1 \times 10^{16}$ cm^{-3} and $w_t - w_0 = 10^{-5}$ cm.

The negative-going peak on the simulation data occurs at a temperature where w, under the quiescent bias V_R, is very close to w_t. The channel is nearly pinched off at this particular temperature and the trap concentration N_t at the depletion edge can be comparable to the background shallow donor concentration N_s. When this happens, the capacitance transient (governed by equation (5)) behaves like a minority trap in a constant-doped junction (Figure 3).

A minority-like trap peak was observed and modelled for capacitance DLTS measurements on ion-implanted n-GaAs MESFETs. The simulated results strongly suggest that the graded doping tail resulting from the ion-implantation process is responsible for this artifact. It also suggests that the actual semiconductor doping distribution might have a great influence on the DLTS spectrum.

ACKNOWLEDGEMENTS

This work was partially supported by the NSF Division of Material Science Ceramics Program (NSF DMR 7908424). The authors wish to thank R. Koyama and R. Gleason of Tektronix Inc. for the samples and for many helpful discussions.

References

1. Lang D. V. (1974). *J. Appl. Phys.*, **45**, 3023
2. Bleicher M. and Lange E. (1976). *Solid St. Electron.*, **16**, 375

An Improved Technique for Characterizing Dominant Imperfection Distributions in High Resistivity Semiconductor Wafers

Faa-Ching Wang

Santa Rosa Technology Center, Hewlett-Packard Co., 1412 Fountaingrove Parkway, Santa Rosa, California 95401, USA

Abstract

The optically-assisted imperfection profile (OAIP) is an improved nondestructive material characterization technique. The technique measures the relative density distribution of the dominant imperfections in a high resistivity semiconductor wafer at room temperature. The resistance of a small area of the wafer is measured twice under two different illumination conditions. In one case, the area is kept completely dark. In the second case, it is covered by a narrow band optical filter. The band-pass wavelength is set such that it corresponds to the excitation of the dominant imperfection under study. The difference of the two measured resistances is proportional to the density of the dominant imperfections. An example showing how the OAIP is applied to the case of [EL2] distribution in undoped SI LEC (liquid encapsulated Czochralski) GaAs wafers is provided.

Introduction

Characterization techniques capable of detecting dominant imperfection distributions in semiconducting materials are particularly important because these imperfections often determine the device performance. Recently, several nondestructive profiling characterization techniques have been reported [1–13] specifically to study inhomogeneities of SI GaAs materials. The working principles are quite general and could be applied to study other high resistivity semiconductors as well. Those techniques include optically-assisted resistance profile (OARP) [1–3], photoluminescence profile (PLP) [4–7], spectroscopic photoelectrochemical current (SPEC) [8], leakage current profile (I_L) [9–12], cathodoluminescence (CL) [11], and microwave measured conductance change with and without illumination (ΔG) [13]. Most of these techniques, however, are not suitable for direct comparison between a specific material and a specific device parameter, either because they measure the net effect of all intrinsic and extrinsic material properties (e.g. OARP, SPEC, I_L, ΔG) or because of the measuring conditions (e.g. PLP at low measuring temperature). In this paper, we report an improved nondestructive characterization technique which allows a determination of the relative density distribution of the dominant imperfections at room temperature.

It was reported earlier [1] that the resistance nonuniformity of a high resistivity wafer can be determined by using two narrow, highly illuminated areas as conducting paths leading to a small dark area to be measured. This is obtained by use of a mask with two long transparent regions located on both sides of an opaque spot. Because the photoconductivity is several orders of magnitude larger than the dark conductivity, the voltage drop across the two transparent regions is negligible as compared to that across the dark spot area. The measured resistance in this case reflects the net effect of all intrinsic and extrinsic material properties. An OARP of the wafer can be obtained by moving the mask across the wafer.

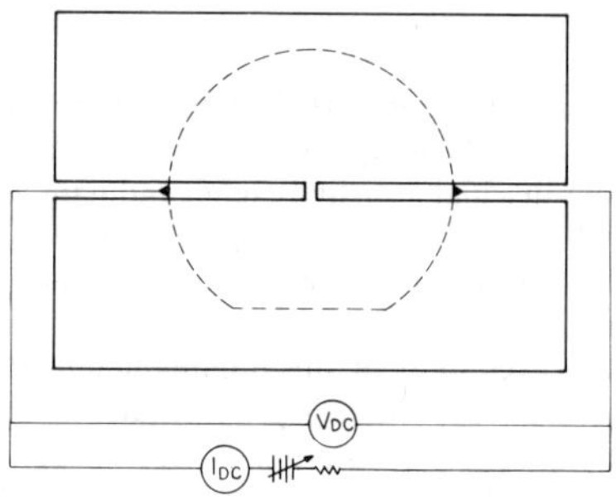

Figure 1 Schematic of the OARP/OAIP set-up.

The schematic diagram of our OARP set up is shown in Figure 1. The dimensions of the square dark spot (2 mm × 2 mm) were chosen to be several times larger than the wafer thickness (~0.4 mm) in order to minimize the fringe effect. In addition, the length of the dark spot was much longer than the diffusion length of the photogenerated carriers in the semiconductors. A ratio of 1000 for the photoconductivity to the dark conductivity in SI GaAs wafers with resistivity larger than 10^7 Ω cm was achieved under illumination from a laboratory microscope lamp. Typical values of a few hundred nA and a few volts were measured. A 5% sensitivity in the resistance determination has been estimated. The OARP has been verified to yield the actual dark resistivity nonuniformity by comparing the OARP results to van der Pauw measurements [2, 3].

Optically-assisted imperfection profile

An improvement of the OARP technique which, in addition to the resistance profile, yields the distribution of the dominant imperfections in the wafer, has been obtained by replacing the dark spot area on the OARP mask with a narrow band optical filter, with wavelength corresponding to the excitation energy of a specific imperfection under illumination. Because of the excess contribution of photocurrent from that specific imperfection, the measured resistance with the optical filter in place is smaller than that with the dark spot for the same material. The difference of the two measurements is proportional to the density of the imperfections for the area under the optical filter. By subtracting the resistance profiles obtained in the two different masking conditions, an OAIP can be obtained.

The incident radiation through the narrow band optical filter excites all imperfections with activation energies smaller than the excitation energy, having significant occupancy and negligible thermal emission at room temperature [14]. However, the main contribution to the photocurrent will come from the 'dominant' imperfection levels. If there is only one dominant deep level to compensate the shallow levels and which therefore determines the high electrical resistivity, the photocurrent will reflect the density of that level [14]. Even for the case of close compensation between two deep levels of similar density in high resistivity semiconductors, with the Fermi level pinned in between [15], a photoconductivity spectral response measurement can help in preselecting two suitable wavelengths for the optical filters for two subsequent OAIP measurements. Therefore, whenever the two photoconductivity peaks can be resolved, the OAIP will still be capable of determining the distribution of the two dominant imperfections.

Experimental results

A typical application of the OAIP technique is the determination of the [EL2] distribution in undoped SI LEC GaAs wafers using a 1074 nm optical filter with band width of 20 nm [16]. Although the origin of the EL2 level is not without ambiguity, it is thought to be the dominant deep level compensating shallow levels in undoped GaAs materials [17]. DLTS (deep level transient spectroscopy) measurements indicate one order of magnitude larger density for the EL2 level than that for any other level (EL3, EL6) in this material. In addition, optical absorption measurements at 300 K show only one primary shoulder at a wavelength of 1 μm in the extrinsic region of the spectrum [16]. The results of OAIP measurements in the $\langle 0\bar{1}1 \rangle$ direction of an undoped SI GaAs wafer are shown in Figure 2. The [EL2] distribution in the same material, as measured by the near infrared absorption technique at both 1.00 and 1.09 μm [16] is also provided in Figure 2 for comparison purposes. The W shape of the profiles obtained with both measuring techniques are consistent with other reports of similar studies of EL2 distributions by near infrared absorption [18]. The dislocation etch pit density (EPD) for the same material is also shown in Figure 2 and a positive correlation is observed. A similar correlation between [EL2] obtained by near infrared absorption and EPD has been previously reported [19]. The measuring areas for OAIP, IR absorption and EPD are 4, 18, and 1.1 mm^2 respectively.

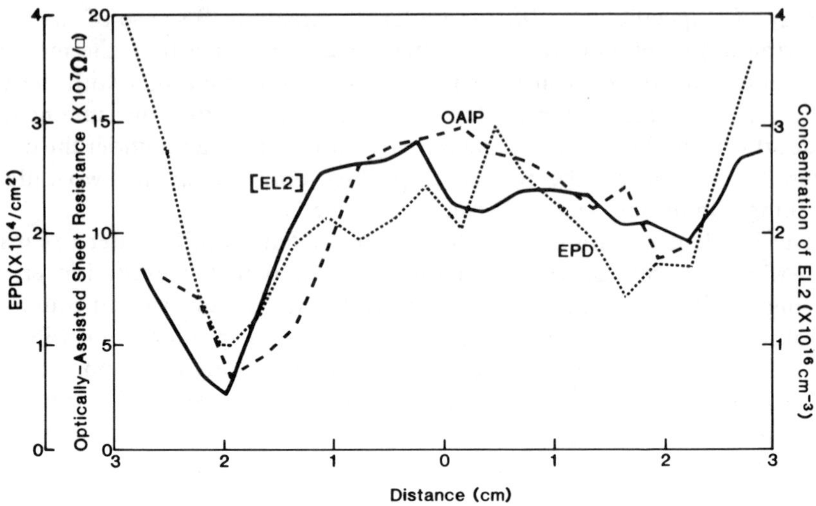

Figure 2 Position dependence of the [EL2] distribution determined by OAIP, using 1.074 μm filter, and by near infrared absorption [16].

Conclusions

An OAIP technique has been developed in order to determine the relative density distribution of the dominant imperfections in high resistivity semiconductor wafers. The technique is nondestructive and can be used at room temperature. As an example, the [EL2] distribution in an undoped SI LEC GaAs wafer has been determined and is shown to be in good agreement with results obtained by other established techniques.

ACKNOWLEDGEMENTS

The help of Katharina Mueller in preparing the wafers is gratefully acknowledged.

References

1. Blunt R. T., Clark S. and Stirland D. J. (1982). *IEEE Trans. El. Dev.*, **ED-29,** 1038
2. Wang F-C. and Li C. (1983). *Seventh Conference on Crystal Growth, Fallen Leaf Lake, California*, Section V-2
3. Wang F-C. and Bujatti M. (1984). *IEEE El. Dev. Lett.*, in press
4. Kitahara K., Nakai K. and Shibatomi S. (1982). *J. Electrochem. Soc.*, **129,** 880
5. Tajima M. (1982). *Jap. J. Appl. Phys.*, **21,** L227
6. Katahara K., Ozeki M. and Shibatoma A. (1983). *Appl. Phys. Lett.*, **42,** 188
7. Kitahara K., Ozeki M. and Shibatoma A. (1983). *Fujitsu Sci. Tech. J.*, **19,** 279

8. Otsubo M. and Murotani T. (1983). *Jap. J. Appl. Phys.*, **22,** L345
9. Matsumoto Y. and Watanabe H. (1982). *Jap. J. Appl. Phys.*, **21,** L515
10. Miyazawa S., Mizutani T. and Yamazaki H. (1982). *Jap. J. Appl. Phys.*, **21,** L542
11. Kamejima T., Shimura F., Matsumoto Y., Watanabe H. and Matsui J. (1982). *Jap. J. Appl. Phys.*, **21,** L721
12. Matsumura T., Emori H., Terashima K. and Fukuda T. (1983). *Jap. J. Appl. Phys.*, **22,** L154
13. Shimizu H., Ohno H. and Hasegawa H. (1982). *Jap. J. Appl. Phys.*, **21,** L786
14. Bube R. H. (1974). *Electronic Properties of Crystalline Solids*, Chapter 12; Academic Press, New York
15. Bube R. H. (1974). *Electronic Properties of Crystalline Solids*, Chapter 9; Academic Press, New York
16. Martin G. M. (1981). *Appl. Phys. Lett.*, **39,** 747
17. Holmes D. E., Chen R. T., Elliott K. R., Kirkpatrick C. G. and Yu P. W. (1982). *IEEE Trans. El. Dev.*, **ED-29,** 1045
18. Brozel M. R., Grant I., Ware R. M. and Stirland D. J. (1983). *Appl. Phys. Lett.*, **42,** 610
19. Holmes D. E., Chen R. T., Elliott K. R. and Kirkpatrick C. G. (1983). *Appl. Phys. Lett.*, **43,** 305

Analysis of the Time Response of Photoconductivity in SI Materials

R. E. Kremer, J. C. Abele* and M. C. Arikan†

Oregon Graduate Center, Beaverton, Oregon 97006, USA

Abstract

Using photocurrent induced transient spectroscopic (PITS) methods, we have studied the deep level states in several SI materials. In contrast to the standard PITS survey technique, which involves monitoring the difference in photoinduced transient currents after fixed delay times, our measurements utilize a computerized signal averaging technique which enables us to measure the entire decay current spectrum at a sequence of fixed temperatures. The transient photocurrent often does not decay with single exponential behaviour; however, by measuring the entire transient, the data acquired can be analysed for non-exponential or multiple exponential decay.

When studying the electrical properties of SI materials such as undoped or Cr-doped GaAs, the effect of deep traps in the band gap far from either band edge must be included. These deep level states essentially determine the minority carrier properties of the material. There are several experimental techniques available to study these deep levels, the best-known being DLTS (deep level transient spectroscopy) which examines transients in the capacitance [1]. For SI materials, however, the capacitive techniques lack sufficient sensitivity and transients in the photocurrent are usually measured (PITS), with the analysis being carried out in a manner similar to that of DLTS [2]. This standard analysis technique involves monitoring the capacitance or current at two times during each transient as the temperature is swept. By taking several temperature sweeps with different sampling times, eventually sufficient data are collected to determine activation energies, concentrations and capture cross-sections for the deep level states involved (see Figure 1). We describe here an alternative method for accumulating and analysing PITS-type data.

Our method, based on computerized signal averaging, not only generates more complete data, but does so with considerably less experimental difficulty. Rather than monitor the photocurrent at only two points during the transient, we record the entire transient. Then, while maintaining the sample at a constant temperature, we average the transient by summing a large number (2^n) of them and storing the result in the memory of the computer. For

* Permanent address: Lewis and Clark College, Portland, Oregon 97219, USA.
† Present address: Tübitak Research Institute, Kocadi, Turkey.

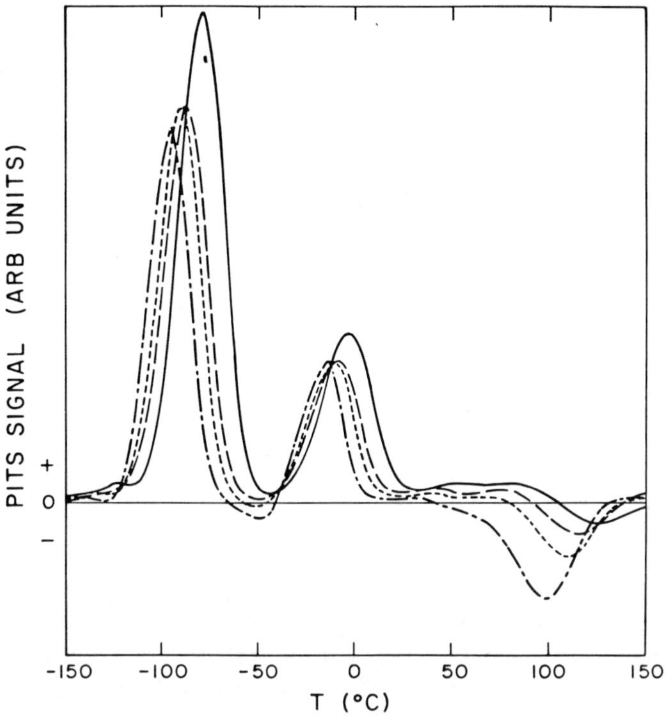

Figure 1 PITS data on a GaAs:Cr sample taken in the standard manner, with times t_1 and $t_2 = 3t_1$ sampled; t_1 progressed from 5 ms (solid curve) to 21 ms in the four traces shown.

relatively stable signals, as few as 32 transients are adequate to produce a noise-free signal, although when the signal is clouded with noise (primarily 60 Hz) as many as 1024 traces must be averaged.

The electronics of our system, shown in Figure 2, are very simple. Signal averaging to eliminate noise also eliminates the need for lock-in detection. We simply record the voltage across a resistor in series with the sample while applying a constant voltage to it. Current transients are provided by chopping extrinsic light (tungsten lamp passed through a silicon filter) at about 1 Hz. The sample is enclosed in a small Joule-Thomson refrigerator which is capable of controlling the temperature to within 0.1 K. After recording enough transients at one temperature to provide a noise-free average, the temperature is raised to the next point. Thus all of the available data (the entire transient) are obtained in a single set of temperature steps.

One of the greatest strengths of this approach is that by analysing the entire transient no *a priori* assumptions need be made concerning the shape of the transient. The standard 2-point approach assumes the photocurrent decay follows a single exponential. Recent work has shown that the transients involved may follow multiple exponentials or even be non-exponential [3] when more than one trap is influencing the signal. Since we have recorded

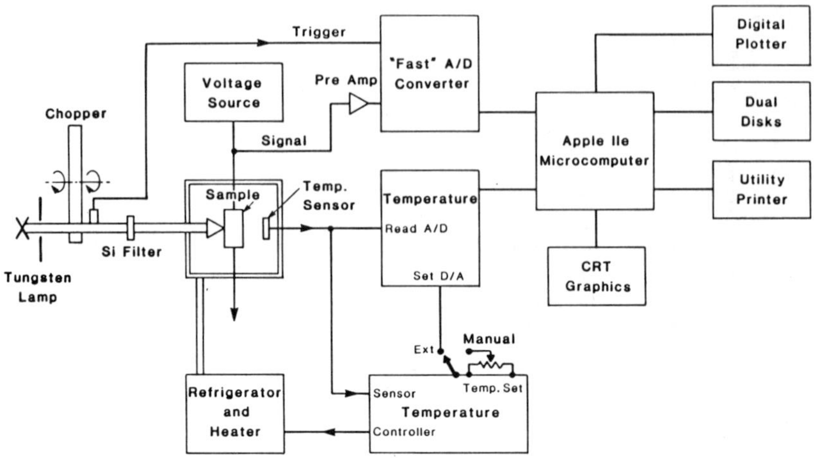

Figure 2 Block diagram of the experimental apparatus.

the entire transient, it is a relatively simple matter to see whether or not the decay follows a single exponential.

Our results show that the transient current does not decay in a simple exponential manner. The decay may be better explained by using several exponentials with different time constants and amplitudes. In Figure 3 we

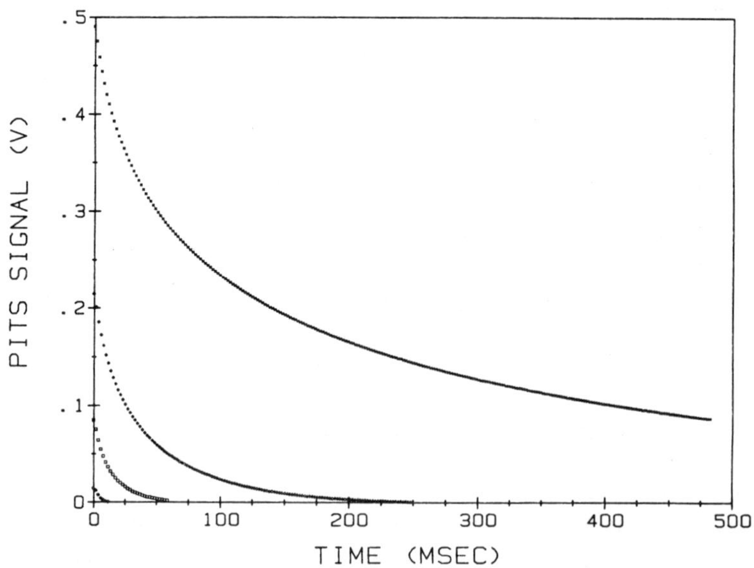

Figure 3 Photocurrent transient of a Cr-doped GaAs sample at −100°C. Top trace—as-recorded transient; second trace—after slowest exponential is subtracted off; third trace—after slowest two exponentials are subtracted; bottom trace—after slowest three exponentials are subtracted. The time constants for these four components are 307, 61, 18 and 4.9 ms respectively.

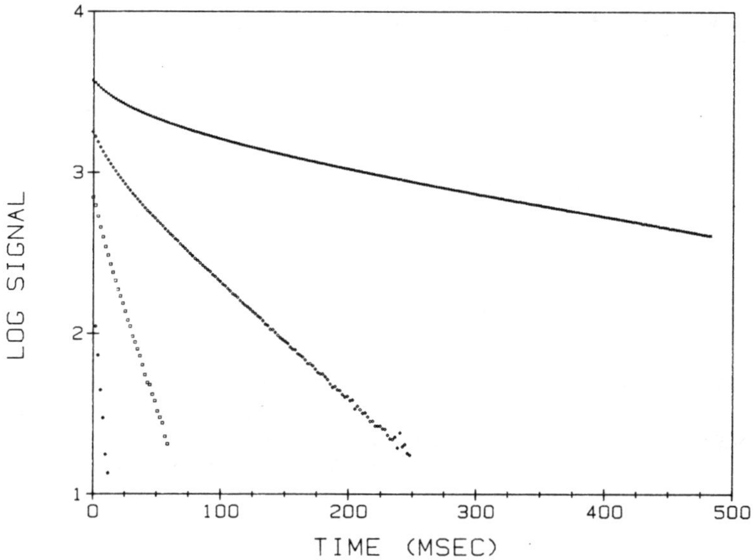

Figure 4 Logs of the traces shown in Figure 3. Top, second, third and bottom traces are the logs of the corresponding traces in Figure 3.

show the components of the transient for a different Cr-doped GaAs sample at −100°C. The top trace is the as-recorded data. Taking the log of these data (Figure 4) reveals a slow exponential dominating the behaviour at times long after the end of the light pulse. By fitting an exponential to the data from 300–500 msec and subtracting its effect from the complete data, the second trace of Figure 3 is obtained. Taking the log of this subtracted data reveals a second exponential which dominates the decay from about 100–250 msec. Repeating the subtraction procedure, we uncover third and fourth exponentials which control the behaviour at short times after the pulse ends. Data taken at other temperatures show similar behaviour.

Using our technique of computerized signal-averaging of photocurrent transients, we have studied the details of deep level states of several SI materials. In addition to the data on Cr-doped GaAs which is reported here, we have studied samples of undoped GaAs and the II–VI semi-insulators, undoped CdTe and $Cd_{0.55}Mn_{0.45}Te$.

ACKNOWLEDGEMENTS

This work was supported by NSF Grant DMR-8305731.

References

1. Lang D. V. (1974). *J. Appl. Phys.*, **45,** 3023
2. Hurtes Ch., Boulou M., Mitonneau A. and Bois D. (1978). *Appl. Phys. Lett.*, **32,** 821
3. White A. M., Day B. and Grant A. J. (1979). *J. Phys. C*, **12,** 4833

Spectroscopy of Vanadium in GaAs

B. Clerjaud, C. Naud, C. Benjeddou*, G. Guillot*, P. Leyral*, B. Deveaud† and B. Lambert†

Laboratoire d'Optique de la Matière Condensée, ERA 133, Université P. et M. Curie, 4 place Jussieu, 75230 Paris Cedex 05, France

Abstract

Electron paramagnetic resonance (EPR), high resolution optical absorption and time decay measurements on LEC (liquid encapsulated Czochralski) and Bridgman grown high resistivity GaAs:V material are reported. All observed results are interpreted in terms of V^{3+} internal transitions. These results are consistent with the fact that the acceptor level of vanadium is very high in the bandgap. The high resistivity behaviour of these materials is discussed.

Introduction

Vanadium-doped SI GaAs is now commercially available. This material has recently been shown to be thermally stable [1] which is a great advantage over the chromium or EL2 compensated material. However the understanding of the role of vanadium in the compensation process is rather poor. The literature on this subject can be divided into two classes: one which considers that vanadium introduces a midgap level [2–4] and the other one which locates the ionized acceptor level about 0.2 eV below the conduction band [5–7]. It is to be noted that the first class of work has been performed on LEC or VPE (vapour phase epitaxial) material while the second one used Bridgman material. In order to improve the understanding of the problem, we have performed a spectroscopic investigation of both Wacker Chemitronic, LEC grown, high resistivity material and weakly n-type Bridgman material, grown at the RTC Company by G. Poiblaud and issued from the tail of the ingot studied in Refs. 6 and 7.

EPR spectroscopy

In the dark both type of samples show the V^{3+} EPR signal identified by Kaufmann *et al.* [3]. Under light illumination the two types of samples have a different behaviour. The Bridgman material is not sensitive to light while in the Wacker material very strong signals appear under infrared illumination

* Laboratoire de Physique de la Matière, LA 358, INSA Lyon, 69621 Villeurbanne Cedex, France.
† CNET (LAB/ICM), 22301 Lannion, France.

in the 1 μm range. These signals occur over the entire magnetic field range available in our experimental set up (0–2.5 T) at X band frequencies. Up to now, we have not been able to interpret these signals. It is worth noting that the appearance of the signals does not correspond to any decrease of the V^{3+} signal which is light insensitive in both materials.

Absorption spectroscopy

For both types of samples the absorption spectrum is similar; its overall shape at 6 K is shown in Figure 1. It is to be noted that we took great care not to stress the samples in our experiments. The main absorption band is identical to the one reported in the literature [2, 6] and to the photoluminescence excitation (PLE) band [7]. The zero phonon line (ZPL) associated with this band marked by an arrow in the figure, had not been previously reported. This ZPL has a linewidth of 3 cm^{-1} and is located at 8131 ± 0.3 cm^{-1}. A sharp line at 10773 ± 0.3 cm^{-1} is also observed in all samples; its intensity is correlated with those of the other lines and its linewidth is 0.5 cm^{-1}.

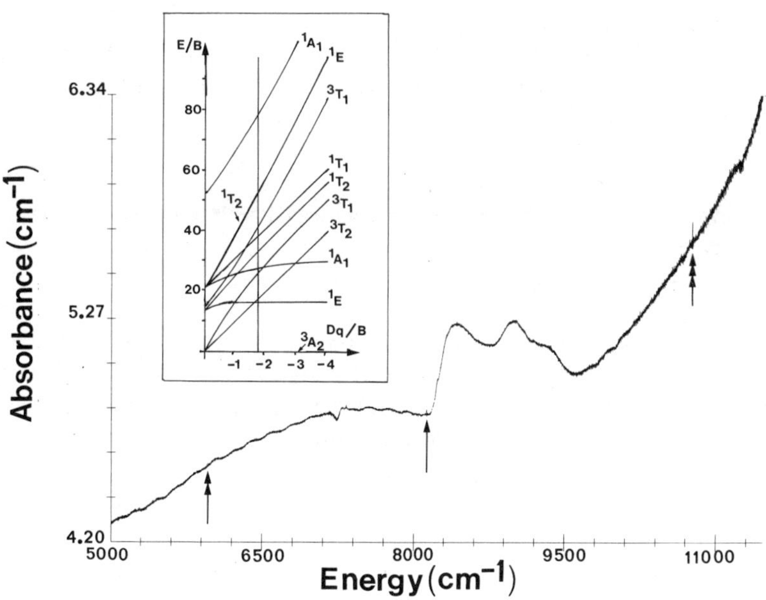

Figure 1 Absorption spectrum of GaAs: V. The V^{3+} Tanabe-Sugano diagram is shown in the inset.

Looking with higher sensitivity in the region of the luminescence ZPL [2, 3, 7, 8] (marked by a double arrow in Figure 1), allows one to observe the spectrum shown in Figure 2. Two features should be noted: the lines in

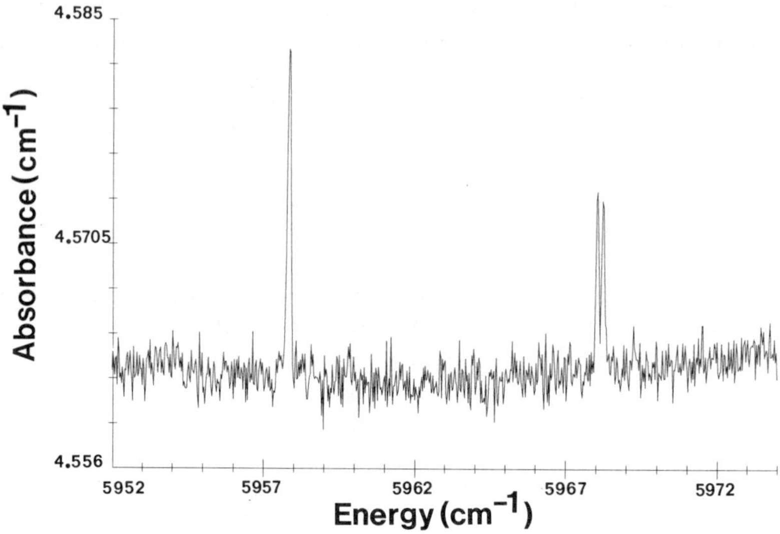

Figure 2 Detailed absorption spectrum in the 5960 cm^{-1} region.

Figure 2 are extremely sharp (0.1 cm^{-1}) and the high energy component consists in fact of two lines, split by 0.2 cm^{-1}. These lines exactly coincide with the ZPL luminescence lines.

Interpretation of the absorption experiments and lifetime measurements

Two features are remarkable:

1. The $5957.85 \pm 0.03\ cm^{-1}$, $5968.05 \pm 0.03\ cm^{-1}$ and $5968.25 \pm 0.03\ cm^{-1}$ transitions are extremely intense in emission and very weak in absorption.
2. The ZPL linewidths are extremely different depending on the transition, the low energy ones being remarkably sharp.

The first feature shows that the radiative transition is forbidden and that non-radiative channels don't exist (otherwise the transition should also be very weak in emission); this suggests a spin-forbidden transition.

The linewidths are indeed inhomogeneous which means that they are governed by the internal stress of the crystals. The second feature means that the response of the various ZPL to the internal stress is extremely different; the low energy lines especially seem to be very weakly sensitive to the stress. Considering the Tanabe-Sugano diagram for V^{3+} ($3d^2$) [9], one can see that the ^1E (^1D) level is almost parallel to the $^3A_2$ (^3F) ground state. As a consequence the $^3A_2$ (^3F)$\leftrightarrow$$^1E$ (^1D) transition energy is very weakly sensitive to symmetrical distortions or strains. On the contrary, the $^3A_2$ (^3F)$\leftrightarrow$$^3T_1$($^3F$)

allowed transition should be quite sensitive to these stresses because the slope of the $^3T_1$ level in the Tanabe-Sugano diagram is rather large. It should be noted that the $^1A_1$ (^1G) level has a weak slope in this diagram and that the $^3A_2$ (^3F)$\leftrightarrow$$^1A_1$ (^1G) transition should have a rather narrow linewidth in our scheme. The Tanabe-Sugano diagram of V^{2+} does not allow such an analysis. We are thus tempted to attribute the 5958 cm^{-1} transition to the spin forbidden $^3A_2$ (^3F)$\leftrightarrow$$^1E$ (^1D) transition, the main band with the ZPL at 8131 cm^{-1} to the $^3A_2$ (^3F)$\leftrightarrow$$^3T_1$ (^3F) transition and the sharp line at 10773 cm^{-1} to the $^3A_2$ (^3F)$\leftrightarrow$$^1A_1$($^1G$) transition of V^{3+} at a gallium site.

The two very sharp lines at 5968.05 cm^{-1} and 5968.25 cm^{-1} can also be interpreted in the frame of our model. In the case of $^1E \otimes \varepsilon$ vibronic coupling, the first vibronic excited state is a $^1A_1 + {}^1A_2$ level; when the vibronic coupling increases, this level approaches the ^1E state [10]; the introduction of higher order vibronic coupling terms ('warping terms') allows the splitting of the $^1A_1$ and $^1A_2$ levels. We can thus interpret the doublet as being due to the first excited vibronic levels $^1A_1$ and $^1A_2$ of ^1E. In principle a ^1E level is non-magnetic, but spin orbit coupling mixes it with other states allowing its splitting by a magnetic field as observed experimentally [3, 11].

In our scheme there should exist a $^3T_2$(^3F) state above ^1E (^1D), but the $^3A_2$ (^3F)$\rightarrow$$^3T_2$($^3F$) transition is dipolar electric forbidden; consequently this transition should be rather weak; moreover, if the vibronic coupling is strong then this transition should consist of a broad band which would be too weak to be observed.

This interpretation of the optical data in terms of V^{3+} is contradictory with many previous interpretations [2, 3, 8], but agrees with the work of Skolnick *et al.* [11] who studied the corresponding luminescence in InP; however these latter authors attributed the luminescence to the $^3T_2$ (^3F)$\rightarrow$$^3A_2$ (^3F) transition.

In order to check our model, we have measured the luminescence lifetime. At 4.2 K this lifetime is 0.15 ms which is quite long but does not definitely establish the spin-forbidden character of the transition.

Discussion of the SI behaviour of some of the GaAs:V samples

From our EPR and optical experiments, it appears that vanadium is in the V^{3+} charge state (i.e. neutral) in both the high resistivity material and the n-type material which proves that vanadium does not act as a midgap acceptor; thus our results are in line with Refs. 5–7.

One possibility for explaining the high resistivity behaviour of some of the GaAs:V material could be that vanadium introduces a deep donor level (V^{4+}/V^{3+}) which could pin the Fermi level in material which has a tendency to be p-type before the introduction of vanadium. There are two arguments against this possibility: (a) the Wacker high resistivity material had been prepared in a quartz vessel [3] and should have an excess of donors (especially silicon); (b) if the Fermi level of the high resistivity material was pinned by a V^{4+}/V^{3+} level then there should be a mixing of V^{3+} and V^{4+} in the samples and it should be possible to change the amount of V^{3+} in the sample by light illumination, which is not the case.

Another possibility is that vanadium does not compensate the samples, but removes the impurity donors, as has been shown to be the case for titanium and zirconium [12]; this effect could depend upon the method of preparation and explain the different behaviour of samples prepared in different ways. This would not give rise to a SI behaviour but rather to a non-uniform poorly conducting material.

Conclusion

Our optical and EPR experiments can be fully explained in terms of the neutral V^{3+} in both SI and n-type. This shows that vanadium does not act as a midgap acceptor in GaAs. The reason why some of the GaAs : V samples have a high resistivity is not yet fully understood.

ACKNOWLEDGEMENTS

Part of this work has been supported by the DRET under contract 81666. We are very grateful to G. Poiblaud (RTC Caen) and U. Kaufmann (IAF Freiburg) for providing the samples and to C. A. Bates, P. J. Dean, F. Gendron, U. Kaufmann, A. Nouailhat, C. Porte, J. Schneider, M. S. Skolnick and A. M. Vasson for very helpful discussions.

References

1. Kütt W., Bimberg D., Maier M., Kräutle H., Köhl F. and Bauser E. (1984). *Appl. Phys. Lett.*, in press
2. Vasil'ev A. V., Ippolitova G. K., Omel'yanovskii E. and Ryskin A. I. (1976). *Fiz. Tekh. Poluprovodn.*, **10,** 571; *Sov. Phys. Semicond.*, **10,** 341
3. Kaufmann U., Ennen H., Schneider J., Wörner R., Weber J. and Köhl F. (1982). *Phys. Rev.*, **B25,** 5598
4. Terao H., Sunakawa H., Ohata K. and Watanabe H. (1982). In *Semi-Insulating III–V Materials, Evian*, p. 54 (S. Makram-Ebeid and B. Tuck, eds.); Shiva Publishing, Nantwich, UK
5. Haisty R. W. and Cronin G. R. (1964). In *Physics of Semiconductors*, p. 1161 (M. Hulin, ed.); Dunod, Paris
6. Mircea Roussel A., Martin G. M. and Lowther J. E. (1980). *Solid-State Commun.*, **36,** 171
7. Litty F., Leyral P., Nouailhat A., Guillot G. and Lannoo M. (1983). *Physica* **117B** and **118B,** 182
8. Ushako V. V. and Gippius A. A. (1980). *Fiz. Tekh. Poluprovodn*, **14,** 336; *Sov. Phys. Semicond.*, **14,** 197
9. Sugano S., Tanabe Y. and Kamimura H. (1970). In *Multiplets of Transition-Metal Ions in Crystals*; Academic Press
10. Ham F. S. (1972). In *Electron Paramagnetic Resonance*, p. 1 (S. Geschwind ed.); Plenum Press
11. Skolnick M. S., Dean P. J., Kane M. J., Uihlein C., Robbins D. J., Hayes W., Cockayne B. and MacEwan W. R. (1983). *J. Phys. C: Solid State Physics*, **16,** L767
12. Stevenson D. A., Ketrush P. I., Chang S. C. and Borshchevsky A. (1980). *Appl. Phys. Lett.*, **37,** 832

Photoluminescence Properties of SI $In_{0.53}Ga_{0.47}As$ Epitaxial Layers Doped with Iron

J. L. Veteran and R. E. Nahory*

University of California, San Diego, EE/CS C-014, La Jolla, California 92093, USA

Abstract

Photoluminescence (PL) measurements have been carried out in LPE InGaAs samples doped with Fe and having free electron concentrations ranging from 8×10^{12} to mid 10^{15} cm^{-3}. Low temperature measurements show that Fe doping does not alter the near-band-edge spectra, while room temperature measurements show a systematic reduction of band-to-band luminescence intensity with decreasing free carrier concentration.

Epitaxial $In_{0.53}Ga_{0.47}As$ lattice-matched to InP is of great interest for a variety of optoelectronic and microwave devices operating at wavelengths near 1.65 μm. Because of recent advances in the growth of this material, high resistivity layers have been obtained using Fe as deep acceptors, with carrier concentrations below 8×10^{12} cm^{-3}. Semi-insulating $In_{0.53}Ga_{0.47}As$ is expected to be useful for device isolation and for high speed devices. Although Fe impurities have been investigated in other III–V compounds, very little has been done with Fe in $In_{0.53}Ga_{0.47}As$. We report on the characterization of this material by PL techniques at both room temperature and low temperature.

The $In_{0.53}Ga_{0.47}As$ samples were grown by LPE (liquid phase epitaxial) in graphite horizontal boat and slider assemblies in two different laboratory systems [1, 2]. Without intentional doping the residual carrier density was usually in the low 10^{15} cm^{-3} range and occasionally somewhat lower. Control of the residual impurities is important if one wishes to obtain lower carrier concentrations by the addition of Fe to the growth solution. This is because only a limited range of Fe concentrations will compensate the residual free carriers without seriously disturbing the layer morphology, an effect related to precipitation of Fe compounds [1].

The PL spectra were measured using a CW argon ion laser at 488 nm, with pump power of approximately 50 mW, focussed to a spot with a 100 μm diameter. The luminescence, after being dispersed by a $\frac{3}{4}$ m spectrometer, was synchronously detected using either a PBS or a cooled Ge photodetector in a lock-in amplifier system. Samples were immersed in liquid helium at 2 K for the low temperature spectra. Free carrier concentrations of these samples were determined by van der Pauw and Miller profile measurements at room temperature.

* Bell Communications Research, Murray Hill, New Jersey 07974, USA.

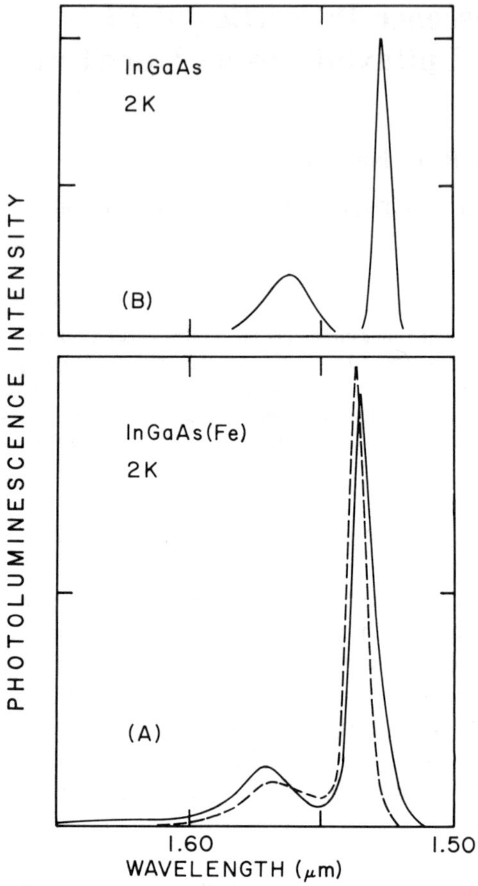

Figure 1 (A) Photoluminescence spectra of two Fe-doped InGaAs samples measured at 2 K. (B) Photoluminescence spectra of undoped InGaAs at 2 K.

In Figure 1A we show typical PL spectra of two Fe-doped $In_{0.53}Ga_{0.47}As$ samples measured at 2 K, which are to be compared to the spectrum of an undoped sample shown in Figure 1B. The spectra show the same structures for each sample. The two peaks observed in each spectrum appear to be the same as those observed by other workers [3–5], and are attributed to transitions involving impurities in the material. The sharper peak at shorter wavelengths has been attributed to transitions involving a donor impurity, while the broader peak at longer wavelengths is characteristic of donor–acceptor pair transitions. Although the identity of these impurity species is not yet known, it is important to note that the same residual impurity transitions seem to occur regardless of the presence of Fe in the material. Small differences in peak wavelengths are attributed to variations in the $In_{0.53}Ga_{0.47}As$ alloy composition, as reported by Penna *et al.* [4].

A typical room temperature PL spectrum is shown in Figure 2 for an iron-doped layer. It consists of one peak which is attributed to band-to-band

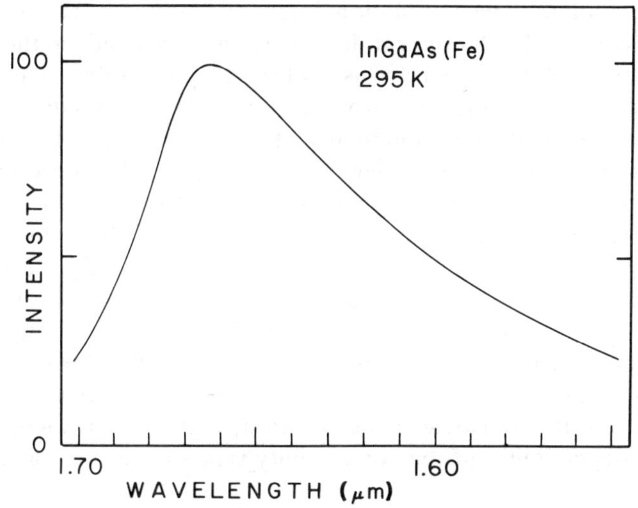

Figure 2 Room temperature photoluminescence spectrum of an Fe-doped InGaAs sample.

transitions. Both the doped and undoped samples used in the present work show the same lineshape and approximately the same spectral width as in Figure 2; however, the intensity varies from sample to sample. Thus, for the purposes of the present analysis we compare the relative peak intensities of these spectra. In Figure 3 we show a plot of these peak intensities as a function of the measured free carrier concentrations, n, for a set of samples with n ranging from 8×10^{12} to 3×10^{15} cm^{-3}. These results show a reduction in intensity of nearly two orders of magnitude over the range studied. The scatter in the data is attributed primarily to the variation which occurs in the residual impurity levels in LPE $In_{0.53}Ga_{0.47}As$. The trend is clear that the near

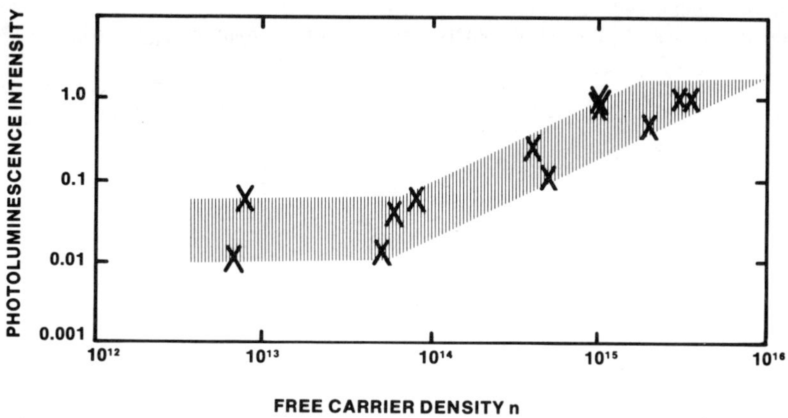

Figure 3 Photoluminescence intensity versus free carrier concentration at room temperature for InGaAs epitaxial layers. The shaded region is drawn to show the trend.

band-edge radiative recombination intensity decreases as the free carrier density decreases. Evidently iron impurities not only reduce the free carrier concentration but they also act as 'killers' of the luminescence. This may occur because these impurities provide very fast recombination which effectively competes with and dominates other recombination processes. The resulting reduction in free carrier lifetime suggests that Fe dopants would have an important role in high speed devices prepared from $In_{0.53}Ga_{0.47}As$.

In conclusion, we have found through the use of PL techniques, that at low temperatures the presence of Fe has not altered the near band-edge spectrum from that of undoped $In_{0.53}Ga_{0.47}As$. From our room temperature data we have observed a systematic relation between intensity and free carrier concentration. Although more work is needed to make this relation precise, these results show a trend which suggests a nondestructive method for the determination of free carrier concentration, with the advantage that it can be used independently of the conductivity type of the substrate.

ACKNOWLEDGEMENTS

The authors are grateful to A. R. Clawson and D. I. Elder of the Naval Ocean Systems Center and to O. G. Lorimor and D. D. Roccasecca of AT&T Bell Laboratories, Murray Hill, for providing the $In_{0.53}Ga_{0.47}As$ layers which they had grown. Part of this work was done at Bell Laboratories Inc., Murray Hill, NJ.

References

1. Clawson A. R., Mullin D. P. and Elder D. I. (1983). *J. Cryst. Growth*, **64**, 90
2. Lorimor O. G. and Roccasecca D. D. Unpublished
3. Chen Y. S. and Kim O. K. (1981). *J. Appl. Phys.*, **52**, 7392
4. Penna A. F. S., Shah J., Chang T. Y., Nahory R. E., Tamargo M., Burroughs M. S. and Cox H. M. Unpublished
5. Bhattacharya P. K., Rao M. V. and Tsai M. J. (1983). *J. Appl. Phys.*, **54**, 5096

A Possible Interpretation of the 0.5 eV Luminescence Band in InP:Fe

B. Deveaud, G. Guillot*, P. Leyral*, C. Benjeddou*, A. Nouailhat*
and B. Lambert

CNET (LAB/ICM), 22301 Lannion, France

Abstract

New experiments have been carried out on the 0.5 eV luminescence in InP: Fe, namely, high resolution luminescence with 1.32 μm YAG laser excitation and time decay measurements leading to a value of 1.5 ms at 4 K. An interpretation in terms of the spin forbidden $^4T_1 \rightarrow {}^6A_1$ internal transition of Fe^{3+} is presented that explains most of the experimental results.

A luminescence band at 0.5 eV appears in SI Fe-doped InP with a ZPL (zero phonon line) at 0.5332 eV and intense phonon replicas at 0.4954 and 0.4901 eV [1, 2]. This emission is obviously related to iron doping as it only appears in Fe-doped materials, whatever their origin (MR, Sumitomo, CNET, RSRE) but up to now no interpretation has been presented for this luminescence. In this paper we shall present results of new experiments on that band (high resolution luminescence and time decay measurements) and give a possible interpretation in terms of an internal transition in an Fe^{3+} ion.

High resolution luminescence experiments were performed using a 1.32 μm YAG:Nd laser excitation source while the detection was standard using a cooled PbS cell. Time decay experiments using a CW argon laser, an acousto-optic deflector and an InSb detector integrating the whole luminescence band, were performed at INSA.

The details of the luminescence spectrum are shown in Figure 1. The 533 meV ZPL shows a barely detectable hot line shifted by 2.1 meV (17 cm^{-1}). The lines at about 495 meV show very complicated structures that would not be expected in the case of usual phonon replica (see V^{3+} [3] or Co^{2+} [4] luminescence bands in InP). The band corresponds to only one centre, however, as excitation at any wavelength or in samples of different origins always gives the same shape. At 4 K, the decay time is found to be 1.5 ms, a very long time which strongly suggests a spin forbidden d-d transition.

Previously published results [1] showed that this band appears in relatively p-type Fe-doped samples and that the onset of PLE (photoluminescence

* Laboratoire de Physique de la Matière (LA 358), Institut National des Sciences Appliquées de Lyon, 20, rue A. Einstein, 69621 Villeubanne, France.

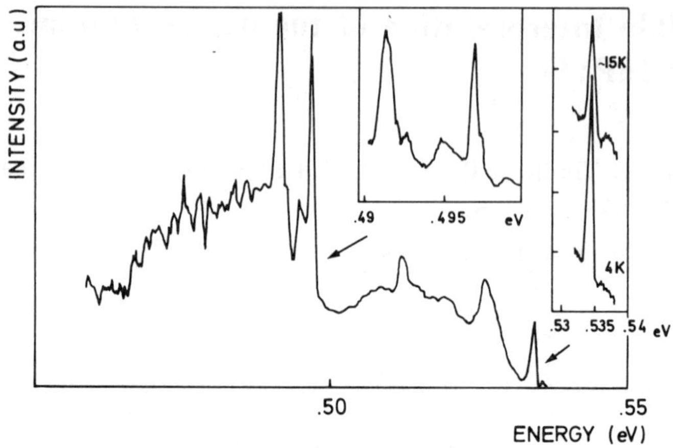

Figure 1 4.2 K luminescence of InP: Fe around 0.5 eV when excited with a 1.32 μm YAG: Nd laser. Details of the structures at 0.530 eV and 0.490 eV are shown in the insets.

excitation) is close to 0.8 eV indicating that luminescence would be excited by the following process:

1st step: $Fe^{3+} + h\nu \rightarrow Fe^{2+} + \text{hole}$

2nd step: $Fe^{2+} + \text{hole} \rightarrow Fe^{3+}$ (excited state)

One set of electron schemes for such a capture is presented in Figure 2 (see [5] for an explanation of these schemes). It can be seen in this figure

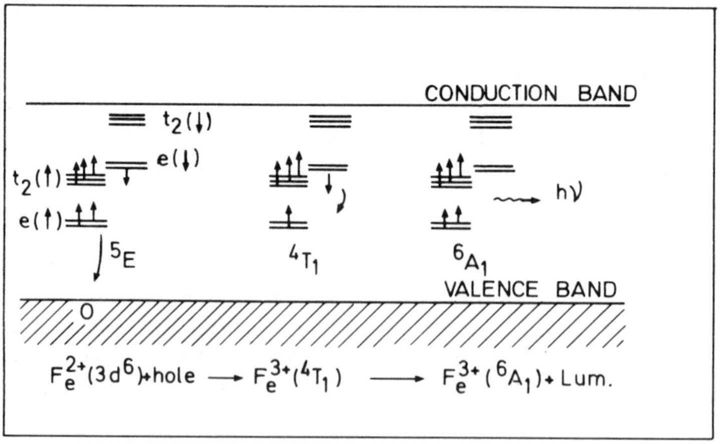

Figure 2 One set of electron schemes giving a prediction of how the capture of a hole at an Fe centre could occur. This hole should be filled by one of the two e(↑) electrons, thus giving rise to an Fe ($^4T_1$) state. The luminescence then corresponds to the spin flip of the e(↑) electron, that is to say, a $^4T_1 \rightarrow {}^6A_1$ transition.

that the most probable process is the recombination of one e(↑) electron in the valence band, leading to the $^4T_1$ excited state of Fe^{3+}. The luminescence transition would then be the spin forbidden $^4T_1 \rightarrow {}^6A_1$ transition. We expect that the most efficient channel is the non-radiative capture of one e(↑) electron which needs only a small amount of energy and leads to an $Fe^{3+}(^4T_1)$ excited state. In the same way electron capture by an Fe^{3+} or a Co^{3+} centre does not give rise to band-level transitions but to Fe^{2+} or Co^{2+} internal transitions [4, 6].

This same transition is not observed in GaAs, as expected, since the distance between the Fe^{2+} level and the valence band is close to 0.55 eV [7] in this material.

The decay time that we have measured, 1.5 msec at 4 K, is very long. Decay times for spin-allowed d-d transitions are usually three orders of magnitude shorter: a few microseconds is found for Fe^{2+} in InP or in GaP [8]. The long lifetime is a strong argument in favour of a spin forbidden transition; however it is to be noted that long decay times have sometimes been reported for spin allowed transitions: 0.86 ms for the $^4T_2(F) \rightarrow {}^4A_2(F)$ transition of Co^{2+} in ZnSe [9]. No spin forbidden transition is expected for Fe^{2+} in this energy region: crystal field splittings would suggest a $^3T_1$ level at about 1 eV above the ^5E ground state, and the intermediate $^5T_2$ state would prevent any direct luminescence. Even if a Fe^{4+} charge state existed in InP, which seems quite improbable, an internal transition to an Fe^{4+} ion ($3d^4$) would be a $^5E \rightarrow {}^5T_2$ allowed transition as exhibited by Cr^{2+} ions [5]. Fe^{3+} is the only charge state of iron that would give rise to a spin forbidden transition compatible with the long lifetime that we have measured.

This centre cannot be an Fe^{2+} complex because then a transition should be observed close to 0.35 eV. It is also difficult to imagine that one could observe an Fe^{3+} complex without also observing the isolated Fe^{3+}; moreover, in such a case, the observation of the corresponding Fe^{2+} complex would be expected.

Crystal field theory predicts for a $3d^5$ ion (Fe^{3+} or Mn^{2+}, for example) a $^6A_1$ ground state and as successive excited states, $^4T_1$, $^4T_2$, ^4E, $^4A_1$. The corresponding transition ($^6A_1 \rightarrow {}^4T_1$) of Mn^{2+} is observed at 1.5 eV in GaP [9] and at higher energies in II–VI compounds. For Mn^{2+}, a long decay time is also observed: 1.5 ms in GaP [10], and the ZPL shows two components separated by about 10 cm^{-1}, the high energy one being a hot line. This is exactly what we observe.

As the Fe^{3+} transition has never been observed in III–V compounds, predictions for the crystal field splitting are very difficult to make. Kaufmann and Schneider [10] assume that energy to be larger than 1 eV. However, the results obtained by Nelkovski *et al.* [11] in II–VI compounds would lead one to predict an energy close to 0.8 eV, not too far from the observed value.

The structure near 499 meV has been interpreted in Ref. 2 as phonon replicas. This can indeed be the case, the structure being very complex due to the occurrence of strong Jahn–Teller coupling in the $^4T_1$ excited state.

As a conclusion, although some details remain unexplained, we have presented strong arguments for the interpretation of the 0.5 eV luminescence band in InP: Fe in terms of internal transition ($^6T_1 \rightarrow {}^6A_1$) in an Fe^{3+} centre. The decay time has been measured (1.5 msec) and found consistent with such an attribution. PLE experiments and also the splitting of the 0.533 eV ZPL are in agreement with our model.

References

1. Leyral P., Bremond G., Nouailhat A. and Guillot G. (1981). *J. Luminescence*, **24/25,** 245
2. Tapster P. R., Skolnick M. S., Humphreys R. G., Dean P. J., Cockayne B. and MacEwan W. R. (1981). *J. Phys. C: Solid State Phys.*, **14,** 5069
3. Skolnick M. S., Dean P. J., Uihlein C., Robbins D. J., Hayes W., Cockayne B. and MacEwan W. R. *J. Phys. C.*, in press
4. Skolnick M. S., Dean P. J., Tapster P. R., Robbins D. J., Cockayne B. and MacEwan W. R. (1981). *J. Luminescence*, **24/25,** 241
5. Deveaud B., Lambert B., Picoli G. and Martinez G. *Phys. Rev.*, in press
6. Bishop S. G., Klein P. B., Henry R. L. and McCombe B. D. (1980). In *Semi-Insulating III–V Materials, Nottingham*, p. 161 (G. J. Rees, ed.); Shiva Publishing, Nantwich, UK
7. Kleverman M., Ohling G., Ledebo L. A. and Grimmeiss H. (1983). *J. Appl. Phys.*, **54,** 814
8. Klein P. B., Furneaux J. E. and Henry R. L. (1983). *Appl. Phys. Lett.*, **42,** 639
9. Robbins D. J., Dean P. J., Glasper J. L. and Bishop S. G. (1980). *Solid-State Commun.*, **36,** 61
10. Kaufmann U. and Schneider J. (1982). *Adv. Electronics and Electron Phys.*, **58,** 81
11. Nelkovski H., Pfutzen Reuter O. and Schritten Lacher W. (1978). *J. Luminescence*, **17,** 419

Polycrystalline Indium Nitride Films

T. L. Tansley and C. P. Foley

School of Mathematics and Physics, Macquarie University, North Ryde, New South Wales, 2117, Australia

Abstract

Indium nitride has previously been available only in impure form with electron concentrations in the range 10^{19}–10^{20} cm^{-3}. Growth of polycrystalline films of n-type material with dopings as low as 10^{16} cm^{-3} and mobilities greater than 3000 $cm^2 V^{-1} s^{-1}$ is here reported for the first time. The critical aspect of the radio-frequency sputtering growth technique is the provision of sufficient atomic nitrogen in the region of the reactive plasma adjacent to the metallic target to sustain a fully nitrided surface layer. Under these conditions, high purity films of the single phase stoichiometric compound are readily prepared.

The semiconducting nitrides of Al, Ga and In are not amenable to growth by standard LPE (liquid phase epitaxial), VPE (vapour phase epitaxial) or, in pure form, organometallic techniques. Radio-frequency sputtering from metal sources in a reactive nitrogen plasma is a plausible alternative, although reports in the literature on related growth methods regularly cite the presence of either nonstoichiometric single phase nitride or mixtures of stoichiometric nitride and pure metallic crystallites. We report, for the first time, preparation of high mobility stochiometric indium nitride in the form of oriented polycrystalline films.

Transport properties for a variety of growth methods reported in the literature are summarized in Table 1 and compared with Hall and optical data for representative samples from the present study. The last three entries in this table compare films grown with metallic indium targets (B), targets in which the surface has been fully converted to the nitride by prolonged exposure to the reactive (atomic) nitrogen species (A) and partially nitrided targets (AB). The combination of low pressure plasma, low growth rate and target nitridation produces films with Hall mobilities one or two orders of magnitude greater, and donor impurity concentrations associated with nitrogen vacancies between two and four orders of magnitude lower than previously reported.

Scanning electron microscopy and X-ray diffractometry show that all films have a preferred orientation, with the c-axis parallel to the direction of growth and randomly oriented a-axes. Films sputtered from metallic targets (type B) also show diffraction peaks from principal vicinal planes while those prepared from fully nitrided targets (type A) are uniquely (00.2) oriented. Studies of the changes in film morphology during target nitridation (type

Table 1 *Comparison of InN prepared by various techniques*

Growth method	$n(cm^{-3})$	$\rho(\Omega\ cm)$	μ_H $(cm^2\ V^{-1}\ s^{-1})$	E_G (eV)	Ref.
r.f.s.	10^{18}	$3–5 \times 10^{-3}$	250	1.7	1
r.e.	10^{20}	3×10^{-3}	20	1.7	2
r.f.s.	$3 \times 10^{18}–10^{19}$	$10^{-3}–10^{-2}$	20–50	2.07	3
d.c.r.s.	$2.5–3.3 \times 10^{20}$	10^{-3}	20	2.05	4
d.c.s.	5×10^{18}	4×10^{-2}	20	1.7	5
r.f.r.s. (B)	7×10^{17}	1.2×10^{-2}	730	2.05	This work
r.f.r.s. (AB)	4×10^{17}	1.3×10^{-2}	1230	2.05	This work
r.f.r.s. (A)	7×10^{16}	2.2×10^{-2}	3980	2.05	This work

n–electron concentration; ρ–resistivity; μ_H–Hall mobility; r.f.s.–radio-frequency sputtering; r.e.–reactive evaporation; d.c.r.s.–d.c. reactive sputtering; d.c.s.–d.c. sputtering; r.f.r.s.–radio-frequency reactive sputtering

AB) show progressive extinction of peaks associated with planes at higher angles from the (00.2) plane (in the order (10.0), (20.0), (10.1), (10.2), (10.3)) until, at complete target nitridation, only the (00.2) peak remains. This we attribute to the change in dominant surface processes during film nucleation described by Thornton [6]. Nucleation during B-type, nitrogen-deficient, growth is determined by a surface mobility controlled by the temperature of the substrate, while type A morphology results from a significant contribution to mobility by chemical heat of formation. Application of the Thornton model to this material will be discussed in detail elsewhere.

Analysis of film surfaces by XPS (X-ray photoemission spectroscopy) shows that nitrogen is either not incorporated in, or is preferentially lost from, the first few monolayers of the otherwise stoichiometric films. However, exposure to air converts the exposed indium monolayer to the oxide In_2O_3, a phenomenon familiar in surface analysis of other III–V compounds.

In related theoretical studies we have calculated the band structure of indium nitride using the Bergstrasser–Cohen technique [7] for hexagonal compound semiconductors with an empirical form factor for nitrogen and a nonempirical indium form factor [8]. Preliminary data are given in Figure 1, while results of a more detailed calculation including density of states and reflectivity data are to be published later.

The dominant feature of the band structure is the principal conduction band minimum at the centre of the hexagonal zone, a feature shared by the nitrides of aluminium [9] and gallium [10] and the hexagonal metal-oxide semiconductors [11], with no satellite minima within several electron volts of this extremum. An almost isotropic effective mass of the order of $0.1m_0$ is indicated, suggesting that a substantial improvement in electron mobility is still possible.

Comparison of the band structure with the only published reflectivity data [12] shows good agreement with the allocated transitions, while values com-

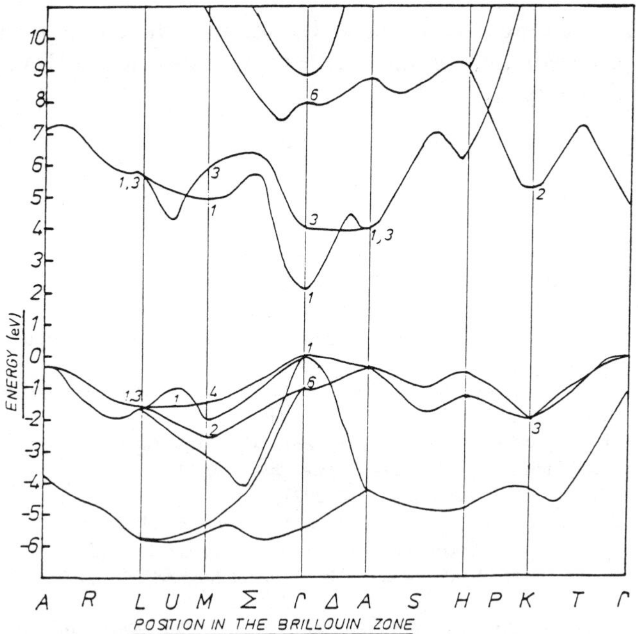

Figure 1 Energy bands of indium nitride.

parable to those occurring in gallium nitride [13] are also evident, as seen in Table 2.

In conclusion, we have developed a technique for the preparation of good quality indium nitride films, a III–V semiconductor about which little is known. Preliminary experimental and theoretical work indicates that the combination of high electron mobility and 2 eV direct bandgap make further investigation worthwhile.

Table 2 *Principal optical transitions*

Transition	1	1a	2	3	4	5	6	7
GaN (Theory) [13]	5.3	6.25	7.05	7.65	7.65	7.65	8.5	9.4
InN (Theory)	4.0	4.1	5.3	7.4	7.3	7.2	7.9	7.9
InN (Experimental) [12]	5.0	5.5	5.8	7.3	7.3	7.3	8.8	8.8

Allocations are: 1–$\Gamma_5-\Gamma_3$; 2–U_4-U_3; 3–M_2-M_1; 4–M_4-M_3; 5–K_3-K_2; 6–$\Gamma_6-\Gamma_1$; 7–$\Gamma_5-\Gamma_6$.

ACKNOWLEDGEMENTS

This work was partially supported by the Australian Research Grants Committee. The authors are presently with the Department of Electrical and

Computer Engineering, Oregon State University, where one of them (TLT) is a visiting professor and the other (CPF) is visiting under a IUVSTA-Welch Foundation Scholarship.

References

1. Hovel H. J. and Cuomo J. J. (1972). *Appl. Phys. Letts.*, **20,** 71
2. Trainor J. W. and Rose K. (1974). *J. Electron. Mat.*, **3,** 821
3. Puchevrier N. and Menoret M. (1976). *Thin Solid Films*, **36,** 141
4. Tyagi V. A., Eustigneev A. M., Krasilio A., Andreeva M. F. and Malatidiou V. Ya. (1977). *Sov. Phys. Semicond.*, **11,** 1257
5. Natarjan B. R., Eltouk A. H. and Greene J. E. (1980). *Thin Solid Films*, **69,** 201
6. Thornton J. A. (1974). *J. Vac. Sci. Technol.*, **11,** 666
7. Bergstrasser T. K. and Cohen M. L. (1967). *Phys. Rev.*, **164,** 1069
8. Cohen M. L. and Heine V. (1970). *Solid State Phys.*, **24,** 208
9. Bloom S. (1971). *J. Phys. Chem. Solids*, **32,** 2027
10. Jones D. and Lettington A. H. (1972). *Solid-State Commun.*, **11,** 701
11. Jarzebski Z. M. (1973). In *Oxide Semiconductors*; Pergamon, Oxford
12. Solobolev V. V., Kroitoru S. G., Sokolov E. B. and Chegrov V. P. (1978). *Sov. Phys. Sol. State*, **20,** 2167
13. Bloom S., Harbeke G., Meier E. and Ortenburger I. B. (1974). *Phys. Stat. Solidi*, **(b)66,** 161

The Effect of Background Fe Concentrations on the Electrical Characteristics of Ion Implanted Si Layers in InP

B. Molnar, P. G. Siebenmann, N. D. Wilsey, H. B. Dietrich and J. D. Oberstar*

Naval Research Laboratory, Washington DC 20375, USA

Abstract

The influence of the SI bulk InP substrate on the electrical character of implanted and annealed layers has been studied by transport, SIMS (secondary ion mass spectrometry) and EPR (electron paramagnetic resonance) techniques. It has been found that the electrical properties of Si implanted layers correlate closely with the substrate Fe content. That is, high Fe content leads to low mobilities, large compensation ratios and low implant activation. This is taken as evidence that the quality of implanted layers is primarily determined by the unintentional background impurities and the Fe concentration in the InP substrates. The deep acceptor Fe is, of course, needed to control the SI properties of the material by compensating the residual unintentional donors.

Semi-insulating substrates are necessary for ion implantation processing of devices such as FETs (field effect transistors). Iron acts as a deep acceptor in InP [1] and is generally used to compensate n-type background impurities and render the material semi-insulating [2]. It has been shown that the highest quality layers in GaAs are obtained when either high purity LPE (liquid phase epitaxial) or high purity bulk material is used in place of Cr-doped material as substrates for implantation [3]. Since Fe plays the same role in InP as Cr in GaAs, it is reasonable to assume that high purity InP is superior to InP: Fe substrates for ion implantation. The purpose of the present investigation is to establish the extent to which InP bulk substrate properties affect the quality of implanted and annealed layers. The study includes correlation of electrical properties of layers produced by Si implantation into bulk LEC (liquid encapsulated Czochralski) grown substrates with Fe doping levels determined by SIMS and EPR measurements.

Silicon ions were implanted into (100) polished wafers held at room temperature and oriented to minimize channeling. Shallow (5000 Å) or extended ($\sim 10^4$ Å) layers with Si concentrations in the range of 1.8×10^{16}–1.8×10^{17} were examined. Thermal activation was performed in a capless annealing system [4] at 750°C for 15 minutes. The properties of the layers were determined using Hall measurements employing the van der Pauw geometry and differential Hall measurements were made with chemical stripping to obtain carrier concentration and mobility profiles.

* US Army EDT Laboratory, Fort Monmouth, New Jersey 07703, USA.

The ^{56}Fe concentration in the bulk was determined using a Cameca IMS-3f ion microscope. Calibration was obtained from a standard prepared by Fe implantations. The absolute values of SIMS from repeated measurements were estimated to be accurate within a factor of 1.5. Concentrations of Fe^{3+} were determined from EPR measurements [5] with an uncertainty of 50%. In order to assess the SI ingots, at least four wafers from each ingot were randomly selected and implanted with multiple energy Si atoms. The implantation schedule, layer thickness and resulting average Si concentration is given in Table 1. The silicon concentrations listed in the table were obtained from LSS (Linhard, Scharff and Schiøtt) calculations [6]. The layer thicknesses were chosen to avoid the effects of surface and interface depletion for free carrier density determinations [7].

Table 1 *Implantation schedule*

600 keV dose (cm^{-2})	240 keV dose (cm^{-2})	100 keV dose (cm^{-2})	[Si] (cm^{-3})	Layer thickness (Å)
0.0	6.45×10^{12}	2.16×10^{12}	1.8×10^{17}	4500
4.0×10^{12}	1.8×10^{11}	6.0×10^{11}	5.0×10^{16}	9000
2.0×10^{12}	9.0×10^{11}	3.0×10^{11}	2.5×10^{16}	9000
8.0×10^{11}	6.5×10^{11}	2.2×10^{11}	1.8×10^{16}	9000

The electrical properties of the implanted layers were found to be substrate dependent. Table 2 summarizes the average and standard deviation of parameters determined for some representative ingots. Ingot 1 with implants giving 1.8×10^{17} cm^{-3} Si concentration yielded layers with 80–90% activation and Hall mobility equal to similarly doped n-type bulk InP. From this, it is clear that direct implantation into high quality SI InP can produce state-of-the-art FET channel layers. However, even in this material the substrate effect becomes significant as the Si concentration level decreases below 5×10^{16} cm^{-3}.

The compensation ratio is also a very important parameter for evaluating n-type InP. We have compared the experimental mobilities with theoretical calculations [8] and derived compensation ratios shown in Table 3. We have assumed that the dominant compensation is related to the concentration of ionized deep acceptors $\bar{N}_{AA}$. Their corresponding values, together with the concentration of Fe as determined by SIMS and by EPR measurements are shown in Table 3. The EPR measured Fe concentrations are in good agreement with the deep acceptor concentrations, while those derived by SIMS are somewhat uncertain in absolute values. In either case, a close correlation between the deep acceptors and the Fe concentrations has been established.

As shown in Tables 2 and 3, the electrical properties of lightly-doped, n-type layers produced by implantation of Si into SI InP are strong functions of the properties of the starting substrate material. We have established that

Table 2 *Average and standard deviation of parameters for representative ingots*

Ingot	[Si] (cm^{-3})	Effective sheet resistivity ($\Omega/\square$)	Effective Hall Mobility ($cm^2\ V^{-1}\ s^{-1}$)	
			77 K	300 K
1	1.8×10^{17}	328 ± 12^{b}	3600	2900 ± 90^{b}
2	1.8×10^{17}	345 ± 70	3000	2660 ± 100
3	1.8×10^{17}	475 ± 110	2700	2500 ± 110
4	1.8×10^{17}	577 ± 156	2300	2320 ± 120
5	1.8×10^{17}	1130 ± 500	750	900 ± 350
	1.8×10^{17}	$(255)^{a}$		$(2960)^{a}$
1	5.0×10^{16}	400		3070
2	5.0×10^{16}	1060		1800
3	5.0×10^{16}	890		2300
4	5.0×10^{16}	2600		1700
1	2.5×10^{16}	2000		2300
2	2.5×10^{16}	7000		1200
1	1.8×10^{16}	10^{7}		

[a] calculated values; [b] standard deviation.

Table 3 *Compensation ratios*

Ingot	[Si] (cm^{-3})	Compensation ratio 300 K	$\bar{N}_{AA}$ (cm^{-3}) 300 K	$[Fe]^{a}$ (cm^{-3})	$[Fe^{3+}]^{b}$ (cm^{-3})
1	1.8×10^{17}	0.1	1.8×10^{16}	4.9×10^{16}	2.8×10^{16}
3	1.8×10^{17}	0.3	5.4×10^{16}	7.1×10^{16}	3.7×10^{16}
4	1.8×10^{17}	0.4	7.2×10^{16}	1.5×10^{17}	6.3×10^{16}
1	5.0×10^{16}	0.2	1.0×10^{16}	4.9×10^{16}	2.8×10^{16}
3	5.0×10^{16}	0.75	3.8×10^{16}	7.1×10^{16}	3.7×10^{16}
4	5.0×10^{16}	0.8	4.0×10^{16}	1.5×10^{17}	6.3×10^{16}
1	2.5×10^{16}	0.85	2.2×10^{16}	4.9×10^{16}	2.8×10^{16}
3	2.5×10^{16}	0.98	2.5×10^{16}	7.1×10^{16}	3.7×10^{16}

[a]SIMS; [b]EPR.

the sheet resistivity and Hall mobility characterizing such layers correlate with the Fe concentration of the starting material. We have consistently seen that the highest quality layers are obtained in the material with the lowest Fe concentration. Not only are the activations and mobilities higher in the substrates with lower Fe concentration, but the electrical parameters are also more reproducible. This situation is analogous to that of ion implantation into Cr-doped GaAs substrates where it has long been known that the electrical properties of the implanted layers are adversely affected by the background Cr concentration. It has been tacitly assumed that the Fe in InP is equally detrimental to the activation of implanted layers and this work demonstrates this in a quantitative manner. For device purposes, it would be preferable to have substrate material with no deep traps. However, since the highest purity bulk InP has an n-type background impurity concentration of greater than 10^{15} cm^{-3}, SI material can only be produced by the introduction of traps near the midgap. It is clear that further advances in the InP implantation technology will require a reduction in the concentration of background shallow donors and acceptors. Such a reduction will allow a concomitant decrease in the level of Fe doping required for the production of SI material.

We conclude that, in order to improve the implanted channel mobility, the concentration of the unintentional shallow donors and acceptors have to be further reduced and, consequently, a lower concentration of Fe will be needed for maintaining the SI property.

ACKNOWLEDGEMENTS

This work has been supported in part by the US Air Force Office of Scientific Research.

References

1. Mizuno O. and Watanabe H. (1975). *Electron. Lett.*, **11,** 118
2. Johnson E. J., Kafalas J. A. and Davies R. W. (1983). *J. Appl. Phys.*, **54,** 204
3. Stolte C. A. (1980). In *Semi-Insulating III–V Materials, Nottingham,* p. 93 (G. J. Rees, ed.); Shiva Publishing, Nantwich, UK
4. Molnar B. (1980). *Appl. Phys. Lett.*, **36,** 927
5. Kaufmann U. and Kennedy T. A. (1981). *J. Electron. Mat.*, **10,** 347
6. Gibbons J. F., Johnson W. S. and Mylroie S. W. (1975). *Projected Range Statistics*; Dowden, Hutchinson and Ross Inc., Stroudsburg
7. Chandra A., Wood E. C., Woodard D. W. and Eastman L. F. (1979). *Solid St. Electron.*, **22,** 645
8. Walukiewicz W., Lagowski J., Jastrzebski L., Rave P., Lichtensteiger M., Gatos H. C. and Gatos H. D. (1980). *J. Appl. Phys.*, **51,** 2659